Web Based Enterprise Energy and Building Automation Systems

Web Based Enterprise Energy and Building Automation Systems

Editors
Barney L. Capehart, Ph.D., CEM
and
Lynne C. Capehart

Associate Editors
Paul Allen
David Green

Routledge
Taylor & Francis Group
LONDON AND NEW YORK

Published 2020 by River Publishers
River Publishers
Alsbjergvej 10, 9260 Gistrup, Denmark
www.riverpublishers.com

Distributed exclusively by Routledge
4 Park Square, Milton Park, Abingdon, Oxon OX14 4RN
605 Third Avenue, New York, NY 10017, USA

Library of Congress Cataloging-in-Publication Data

Web based enterprise energy and building automation systems / editors,
Barney L. Capehart and Lynne C. Capehart ; associate editors, Paul Allen,
David Green.
     p. cm.
Includes bibliographical references and index.
ISBN 978-0-8493-8235-2 (print) -- ISBN 978-8-7702-2252-5 (electronic)
   1. Buildings--Mechanical equipment--Automatic control. 2.
Buildings--Energy conservation--Automation. I. Capehart, B. L. (Barney L.)
II. Capehart, Lynne C.
 TH6012.W43 2007
 658.2--dc22

                            2006041266

*Web based enterprise energy and building automation systems/editors, Barney L.
Capehart and Lynne C. Capehart ; associate editors, Paul Allen, David Green*
First published by Fairmont Press in 2007.

Routledge is an imprint of the Taylor & Francis Group, an informa business

0-88173-536-1 (The Fairmont Press, Inc.)
978-0-8493-8235-2 (print)
978-8-7702-2252-5 (online)
978-1-0031-5123-4 (ebook master)

While every effort is made to provide dependable information, the publisher,
authors, and editors cannot be held responsible for any errors or omissions.

Table of Contents

Foreword

Michael Ivanovich

In the fields of science, technology, and voodoo that comprise the inner workings of the built environment, none are more dynamic and ephemeral as controls and building automation systems. Aside from the physical manifestations of network physical layers—the cables and connectors (and now airwaves) through which data and instructions flow—it's the gray cells of the human brain, translated into logic and instructions which are translated into bits and bytes traversing the physical layers, that make things work. Or not.

All is dependent on the knowledge and expertise of the engineers, owners, contractors, installers, commission providers, and operators along the food chain and lifecycle of procurement, design, installation, commissioning, and operators & maintenance. If everybody is good at what they do and the owners get systems right for them and their buildings and staff, and all the proper steps are taken to ensure a good design is properly installed and activated, and that staffs are prepared to take over a working system, then it's likely all will be well. However, any weak link in the chain can doom the lifecycle to high costs and poor performance—and I'm not just talking about the controls or the building automation system—it's the entire building and everyone and everything in it that suffers. Comfort and indoor-air-quality complaints; premature equipment wear-and-tear; high energy bills; poor security; disastrous plant or data center outages; frequent high-cost maintenance actions… the list goes on.

Controls and building automation systems are a form of information technology. They are the computer-controlled, networked motors that turn on or off chillers, boilers, fans, cooling towers, and emergency generators; that adjust dampers, valves, and variable-speed drives; that sense alarm conditions and activate the alarms; that know who should be in the building and when or that a fire is starting in the storage room. Controls are where action starts and stops in building systems, and they are the record of those actions and the quality of those actions. They are the workstation monitor—they eye of the building looking into the eyes of the operators.

And because controls and building automation systems are information technology, they are the leaves of the tree that is the built environment. They undergo the most change in the shortest time; and sometimes are blown away by the winds of IT revolution.

This book—the third volume of knowledge compiled and edited by the Dr. Barney Capehart, is a field guide to the science, technology and voodoo integrating information technology with controls and building-automation systems. It comes during a season of wildfire change in the buildings industry—as wireless controls are seeing their dawn; as mergers and acquisitions of controls firms are running rampant to no one's surprise in hindsight; as our energy security and national security are becoming a unified goal; as the world is getting flattened by global cooperation and competition; and as the Arctic and Antarctic ice shelves are melting from what seems to be the effects of human-induced global climate change.

There are many reasons to read this book and to translate its knowledge into your practice. Your place in the food chain; the place of your work or systems in the lifecycle of controls and building-automation systems will determine what, where, and when this book and the previous two volumes will have impact. What's important and what's cool is that it's in your hands. You are showing leadership and initiative to have found and picked up this book—enough that it's likely you'll do something with it. Godspeed.

Foreword

Ken Sinclair
Editor/Owner AutomatedBuildings.com

The Internet, IT convergence, wireless networks, globalization and the whole M2M thing is causing a Real Renaissance that is Reshaping the Building Automation Industry.

A renaissance is defined as the revival of learning and culture "a cultural rebirth" resurgence; revitalization.

A renaissance is certainly what is happening in the world of Building Automation Connectivity. We must completely rethink how we do everything while immersing ourselves in a new IP/IT culture. As we move from a physical world to a virtual world with our products and services it is a true renaissance that will change us and our industry forever. Of course the renaissance has already begun for the masses in their transition from a physical world to an anything anywhere virtual world powered by powerful web services delivered by the pervasive internet. Our task is to provide all our future building services, products, existing infrastructure and services virtual connections so that they seamlessly mesh with the new world renaissance.

Large IT companies such as Cisco are coming into our space. In their Cisco Connected Real Estate Whitepaper they talk of harnessing the power of Internet to turn traditional building construction and management paradigms on their heads. In so doing CCRE will bring huge financial and operational advantages not only to the construction, real estate and property services industries but also to downstream constituencies – such as hotel operators, multiplexed retail outlets, and corporate tenants – in sectors as diverse as leisure, healthcare, education and retail finance.

This will bring with it IT integrators, consultants and for the first time for years, will allow top level executives to get a new way to look at their facilities. The IT people are coming with the next wave of convergence: BAS along with Voice, Video & Data.

Our Building Automation Industry is now talking IP, walking IP and providing products and services based on Internet Protocol. The BAS market is at a point called the "Decision Zone". Wireless technology has been shown to be a cost effective solution for building control systems, enabling wireless mobility for building personnel, and bridging several networks for interoperability.

Evolving IP technologies such as Power over Ethernet (PoE), Voice over internet VOIP, Video over internet are all just part of the new power we have inherited as part of IP and IT thinking.

The first ever industry Connected Real Estate Roundtable in Palm Springs occurred as a well planned prelude to BuilConn's *ConnectivityWeek. BuilConn* provided the only logical venue for these extremely necessary roundtable discussions that crossed over and cross pollinated information of the many industries presently interacting in our converged environment. I was extremely pleased to be invited to the table. The outcome was that we all have to go forward and think about what having our data as part of the data utility really means, and to set in place a guideline on how that can be done with best practices. The IT solution with one building backbone was identified as the only way to future proof our projects, because rapid web service developments will radically reshape our future. Areas of core competence must be created to allow all to connect to the fourth utility—the data backbone. We must build strong IT foundations.

There are several global issues such as Energy, Green Buildings, GridWise and open communication standards that are fueling the renaissance. With all the developing countries around the world needing more fuel, and the increasing concern for a green planet, the energy subject is high on the agenda. Forces such as GridWise, an entirely new way to think about how we generate, distribute and use energy, are creating killer applications for our new web-based abilities. The ASHRAE BACnet™ standards committee announced that the protocol has been amended to include a Grid-Wise Object. The use of technology to resolve energy problems is here and building automation is right in the middle of it all, holding many of the cards.

The Open Building Information Exchange (oBIX) Web services specification combined with IT protocol, BACnet, LonMark, and other leading open standards will become the substance of the weave of the renaissance. Function and form will meld together.

IT integration is shaking the foundations of the Traditional Control Theory and is a large part of the renaissance. The increased amount of available measured variables and web-based information for control relationships is creating many new opportunities.

Relational Control is a control technology developed for digital network controls. It is focused on replacing PID (Proportional + Integral + Derivative) Control when the control platform has suitable networking features. Relational control techniques can be employed to operate building energy systems much more efficiently and more effectively than has ever been possible with PID based controls. Control of 18 buildings on the Hong Kong waterfront in an integration of architectural lighting, laser effects, pyrotechnics and accompanied by a sound simulcast provides a very graphic demonstration of what city wide control integration might look like.

A blended city is defined as commercial real estate that brings together a physical and digital environment to enable human capital to contribute to the global economy. In these communities intelligent buildings are helping the tenants in them to live and work better.

Companies are providing online independent data analysis and portfolio-wide consolidated reporting to the facilities and energy departments of building owners. These reports analyze and improve energy efficiency and reduce energy, operations, maintenance and regulatory costs, mining data for value.

The results are a wealth of ongoing unbiased recommendations and management information that enable firms to reduce energy and operational costs, identify potential equipment problems in order to avoid downtime, and enjoy the benefits of a continuously commissioned facility. This process of data acquisition, analysis, and reporting is called Infometrics, and is a key part of Enterprise Energy and Facility Management. This latest book provides an overview of many of these IT, connectivity, integration and Enterprise Energy and Facility Management issues, and helps us see where we are headed along the path to Web Based Enterprise Resource Management to operate our facilities in much better ways. This book is an important resource for helping us all proceed into the future of the Intelligent Building.

Foreword

Anto Budiardjo
President & CEO, Clasma Events Inc.

Energy is without question the single most important enabler *and* driver for the modern world. Energy is an enabler because without it there is no life in this digital age we now live. Energy is a driver because it is a resource in short supply causing many of the challenges we face today from politics and wars to the impact on economics on a personal and global level.

The continued evolution of energy management in the in-built environment through control systems is the core purpose of this book. As in all almost all walks of life today, building control systems are becoming enabled and in a way also being driven by the progress of information technologies, specifically the evolution of the so-called Pervasive Internet (the Internet that reaches all things beyond human-centric computers).

For while the fundamentals of energy management are about the management, distribution and use of energy in whatever form (electricity, gas, oil, etc.), the control of energy is about information and this is where Information Technology comes in. IT has an intriguing way of usurping everything it touches; commerce is now ecommerce through IT, inventory management is now Supply Chain with the help of information technology, sales management is now CRM (Customer Relationship Management) powered by IT.

In many walks of life, IT now provides key technology for the control and management of the subject matter. CRM, for example, is about the management of relationships as opposed to the relationship itself. This often obscures the subject through the immense influence and visibility that the IT is bringing.

In energy this is no different. In what many are now calling EEMS (Enterprise Energy Management), IT is making the discipline of energy management a mix of science and fine art that can tap into the rich and powerful tools available to all energy professionals eager to better their profession. With IT, and the as yet untapped power of the Pervasive Internet, energy professionals can have at their disposal a level, broadness and depth of information and tools never before available.

For truly open minded and forward-looking stakeholders, this is a boon that will be rewarding beyond imagination, but there is a catch. IT, or more importantly the possibilities of using IT for energy management, requires an understanding of the technology both present and future. This is because IT itself is not standing still. To rely upon technology that is currently in use will cheat energy professionals from the true potential of what IT can contribute to this important topic.

Energy professionals are also encouraged to truly understand the importance and value of information about energy as much as the energy itself; this is often a difficult issue to grasp and maybe an example will help. While Supply Chain is about the accurate delivery of materials from factory through distribution and retail (as in the well publicized examples of Wal-Mart), it is the information about where things are and the ability to control where things should be that is of value to organizations such as Wal-Mart. The information is as valuable as the subject.

I am pleased with the opportunity to provide this introduction to this book, and would encourage readers to take every advantage of the positive contribution that IT is making and is going to make to the discipline of energy management.

Energy is far too important of a subject not to throw in the full weight of IT.

Section I

Introduction to Web Based Enterprise Energy and Building Automation Systems

Chapter 1

Introduction to Web Based Enterprise Energy and Building Automation Systems

Barney L. Capehart, Professor Emeritus, University of Florida

THIS IS THE THIRD—and most likely the last—book in the series on information technology for energy managers and web based energy information and control systems. This book concentrates on web based enterprise energy and building automation systems, and serves as a capstone volume in this series. The thrust here is that the highest level functions of a building and facility automation system are provided by a web based EIS/ECS system that provides energy management, maintenance management, overall facility operational management, and ties in with the enterprise resource management system for the entire facility or the group of facilities being managed. If there were ever to be a fourth volume in this series, it would follow the logical progression of the first three volumes, and would probably be titled Web Based Enterprise Resource Management Systems. This is where we are headed with our use of IT, TCP/IP, XML and web based systems to help us operate our facilities better; where better relates to higher energy efficiency and lower operating costs, improved occupant satisfaction, and higher productivity of our facilities through better indoor environmental quality, and a more direct tie-in to the business functions of our facilities.

Improved enterprise energy management and improved overall enterprise resource management begin with data collection. All three volumes of this series have emphasized the need for metering, monitoring, and measuring devices to obtain data that are then sent to a centralized data base for processing and storage. TCP/IP and XML are critical data transmission protocols and structures to obtain and transmit the huge amount of data that are necessary to provide the information needed to operate our facilities better. Data are the starting place. But it is only when we can turn data into information to help us make decisions which will result in our facilities and our enterprises operating better, that we are making real progress toward accomplishing our goals.

In these three volumes we have progressed through the basic ideas of information technology, TCP/IP, and web based EIS and ECS systems, to the top of the pyramid in dealing with enterprise energy management in particular, and leading up to the implementation of enterprise resource management. The real news at this point in time is not that IT, the internet and web based EIS and ECS are the new wave of technology for buildings and systems, but that the application of these systems to improving the overall operation of the enterprise is the even newer and more powerful wave of change! The clear shift to the use of web based EIS and ECS systems for intelligent buildings, smart buildings and building and facility automation systems, including focusing on the enterprise level of control and management is the really new news. This is the future that all of our authors are talking about. When players like Cisco and Hewlett Packard start entering the market for applying enterprise resource management to our hotels, office buildings, schools, hospitals and manufacturing and industrial facilities, it should be clear to anyone that there has been a sea change—or paradigm shift—in our business.

Visualizing this relationship between IT, the internet, our facilities, and the enterprise is greatly assisted by the BuilConn Pyramid, shown below and provided by one of our authors, Anto Budiardjo who is with Clasma Events. From our side, starting with the building systems and facilities, we can clearly see how our job has changed so dramatically over the last several years. Just the facility's information and control tasks are broad enough and complex enough to occupy our full attention—as it has been in the past. But now we also will have to focus on integrating the basic building systems together; and finally to integrate into the business side of our facilities.

With the help of 55 authors and 40 chapters, this book sets out to provide real-world assistance to the energy managers and facility managers who are now hav-

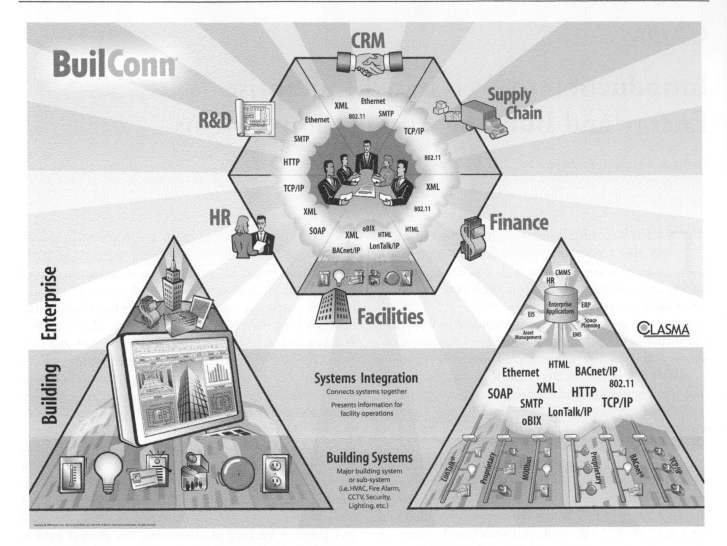

ing to deal with these greatly expanded and significantly more complicated tasks. The majority of chapters in this book are either case studies or applications related to actual facilities. Authors of these chapters are facing these new problems themselves, and solving them—or at least parts of them—every day. Huge strides have been made from the time the first book in this series came out. And even more great progress since the second book came out.

All of us associated with this book in particular—and those associated with one or both of the first two books—are pleased and proud to have helped so many people learn about this new technology and new applications of this technology. We all feel that our initial goal of this project has been met, and that was:

To help prepare energy and facility managers to understand some of the basic principles on IT, so that they can successfully:

- purchase or develop
- install
- operate

- improve, and
- capture the facility operational cost savings and improvements from web based energy information and control systems, including BAS systems and enterprise energy management systems.

The one over-arching theme that is common to each of the chapters in this third book is a critical need for detailed operational data from our facilities. Various authors then go on to tell us what they see as a specific application of the use of that data to help operate facilities better. Other authors address the problems of operating the individual building systems more efficiently and more cost-effectively. HVAC systems are the main ones, but lighting and heating systems are also important.

Starting with the introductory section of this book, we are fortunate to have the views and visions of some of the most well-known technologists in the building automation and intelligent buildings field. Anto Budiardjo, Paul Ehrlich, Bill Gnerre, Kevin Fuller, Michael Ivanovich, and Ken Sinclair all provide us with exciting and challenging ideas and opportunities in the areas of enterprise energy and facility management systems. These chapters give a broad overview of these areas, as well as giving us an overall assessment of this technology.

Nine of the chapters provide detailed case studies for the development and use of enterprise energy and facility management systems. The range of these case studies includes, among others, hospitals, office buildings, schools, entertainment parks, and manufacturing plants. Another nine chapters discuss more specific applications of enterprise energy and facility management systems. For the first time we have a chapter completely devoted to the aspect of training operators to work with the new digital monitoring and control systems.

For new facilities, or for retrofit projects, several authors discuss the use of facility data from web based EIS/ECS systems to help commission these facilities or projects. Automated commissioning has the potential to dramatically reduce the cost of commissioning, making it even more cost-effective process. Next, Paul Allen, one of our associate editors, gives us a comprehensive behind-the-scenes look at the enterprise energy management systems at Walt Disney World in Orlando, Florida, a large entertainment park, and discusses the web based EIS/ECS system that is the heart of the park's enterprise energy management system.

Three chapters report on case study results for demand response applications using web based EIS/ECS systems. Another six chapters explain important tools needed for providing the information needed for making facility operational decisions—including weather normalization of utility bills, and a detailed energy and cost savings calculation system. Finally, the last two chapters are visions of the future from two of the major technologists in the area of building automation systems—Ken Sinclair, who owns and operates the automatedbuildings.com web site, and Dave Clute from Cisco. Ken has been providing a wonderful service to all of us by hosting the automatedbuildings.com web site which gives us almost real-time information on new developments in the building automation systems area. This web site is absolutely "must reading" for anyone trying to keep up on technology developments in this area.

The last chapter is, in many ways, the most eye-opening, startling, and "Wow!" chapter in the book. Dave Clute from Cisco Systems tells us about the Cisco Connected Real Estate System (CCRE), which is one of the newest and most all-encompassing IT and web-based enterprise resource management systems to come into existence. The CCRE system not only incorporates an enterprise energy management system (EEM), but it also incorporates the aggregation of all IT functions—computers, voice, data, fire, security and the tie-in to all of the other enterprise resource management functions—such as CRM, ERP, and so forth. The scope of this Cisco CCRE system is so broad, and so far beyond any of the other web-based building automation systems that we have seen, that it truly represents a sea change, or paradigm shift, for those of us who have been working in this area for the last few years. We're fortunate to have this chapter from Dave Clute and Cisco to give us an early look at the radically different level of involvement of a large IT company in this area.

EVOLUTION OF THIS PROJECT

My original objective was to just produce the one book on information technology for energy managers, and to finish that book by showing that many facilities were already using this technology to greatly improve the energy management function of their building or facility. That book was intended to provide the technological discussion of principles of information technology for energy managers who were not yet comfortable with this new area. Both the use of IT in general, and the use of web based energy information and control systems was quite foreign to most energy and facility managers at that time. The book was intended to help ease these newer potential users of this technology into the more advanced DDC and BAS systems. As input from editors and authors came in, it became apparent that this was a seminal project, and that with this first book we were only touching the basic levels of need for information and education on building and facility management using web based systems.

Thus the idea and need for the second book emerged early on during the production and collection of chapters for the first book. The second book would further expand the knowledge and comfort level of the energy and facility managers with more complex ideas from IT and the internet. But, the major focus of this second book would be on the practical applications of this new technology, and would serve to show both interested and skeptical readers that this technology was available, was working, and could produce some outstanding improvements in facility energy and operational management.

One of my objectives for the second book was to present the case for maintenance management to be considered as an integral part of energy management. Good energy management and good maintenance management go hand in hand. Energy management cannot be totally successful without significant attention to maintenance management. Once a highly energy efficient piece of equipment or system has been installed, all of the benefits can be lost if the equipment or system fails because it is not properly maintained. Several of the detailed case studies in the second book showed how commercially available energy information systems received large quantities of data to be analyzed for examining the operation of building and facility systems, and performing automated diagnostics to determine equipment problems.

The second book essentially shifted in purpose from primarily an expanded technical volume to an almost complete application volume; hence the title Web Based Energy Information and Control Systems: Case Studies and Applications. A few chapters were devoted to more detailed and technical topics in information technology, but most of the chapters were examples of existing web based systems performing a wide variety of facility energy and operational management tasks. The case studies and applications discussed in this second volume were detailed enough and powerful enough to convince most readers that this was the technology of the future; that it is here now; and that it is cost effectively helping both small and large facilities operate more efficiently and with lower energy costs. However, even now, the scale of adoption of this technology is still so small that most facilities are failing to take advantage of the cost savings and operational improvements that implementing these web based EIS and ECS systems can provide.

As the second book was mid-way toward completion it was clear that the direction this project was taking was further away from the programs, software and IT principles that initially drove it, and were originally considered to be the crux of the books. Already in this second book, the terms building management systems, facility energy management systems, enterprise energy management, and even enterprise resource management systems began to appear in many chapters. The next change in direction was underway. This change would now result in the third book being not only about more sophisticated ways to control energy in single buildings and facilities, but to start talking openly about controlling maintenance, safety, comfort and productivity of single buildings and facilities up to large complexes of buildings and facilities.

This third book does have more IT explanations and principles in it, and it does have more case studies of web based EIS and ECS systems controlling the use of energy in single and multiple buildings and facilities.

This book has emerged in a similar manner from the same evolutionary process where the second book editors and authors have stressed the need to go further into overall enterprise energy management and resource management systems. Thus, the real contribution of this third book is in its movement beyond the pure technical aspects of the IT and web based technology and to talk in detail about the application of the technology to operate both single and multiple buildings and facilities—as enterprises—in a much better manner.

THE BUSINESS CASE FOR ENTERPRISE ENERGY AND FACILITY MANAGEMENT

There is another huge change in emphasis coming in to this field of application of IT, the internet and web based information and control systems—it is the business tie-in! Some of our authors in this third volume are already making the "business case" for enterprise energy and resource management systems. This emphasis is helping to point us further away from the hardware and software technology and more toward the application of the technology to the business functions of our facilities—or enterprises. Certainly energy, but also all other purchased utilities such as gas, oil, water, sewer, steam, hot and chilled water, compressed air as well as self produced utilities. But this is only a start to the list of what we will be using IT, the internet and web based systems to manage. We will also be managing the comfort, safety and productivity of the occupants of our facilities. And all of these will be related to the business success side of our companies and organizations.

The "business case" for enterprise energy management and enterprise resource management is not just about energy and utility savings and cost control, but also about the increased productivity of our workplaces obtained as a result of better controlled lighting, HVAC temperature and indoor air quality, and general Indoor Environmental Quality. The "business case" comes from identifying the energy and utility savings as well as the increased profits from higher productivity of our occupants because of a greater level of comfort and satisfaction with their indoor environmental quality. This increased cash flow from our buildings and facilities is then evaluated in its ability to increase the asset value of our enterprise—our buildings and facilities.

One of the things we need to understand and appreciate is encapsulated in a very short and simple statement "We are all in business." Operating a facility is a business even though we might not have recognized it as such in the past. The facilities industry has focused its technology advancements on hardware efficiency and control technology, but still has not seriously embraced information technology (IT). Facilities management today is not just about how the HVAC system or other individual building systems perform, but how the business performs. Today the demands on our facilities continually increase in many ways, and we have many different jobs to do. Our old job that's always been there is to help reduce operating costs in our facility. This includes energy costs—which in spite of increasing on their own we are asked to reduce the actual energy costs at our own facility. And we also have some other parts of our old job that deal with controlling the costs of other facilities resources, such as water and sewer costs. We are also being asked to continually reduce the cost of maintenance, and the cost of maintenance training. These two maintenance cost goals are not really compatible since we can't do cost-effective maintenance without well-trained people.

In addition to all these areas that we have always had to work with and perform, we have a new expanded group of requirements dealing with improved indoor environmental quality (IEQ). IEQ not only contains IAQ (indoor air quality) that we have had to deal with in the past, but now IEQ includes improved over all comfort, as well as improved air quality, and often includes improved productivity. Many of these jobs are new for us over the last few years, but will become a much more significant part of our overall job as facility manager in the future. We need help doing these new jobs, as well as help doing our own jobs which we are now expected to perform with fewer people and people typically with fewer skills.

Help is here in the form of the use of IT, TCP/IP, XML and web based systems to help us operate our facilities better. Where better relates to higher energy efficiency and lower energy and other operating costs, improved satisfaction and productivity of our facilities through better indoor environmental quality, and a more direct tie-in to the business functions of our facilities. The technology is here, it is available, and most facilities need to begin implementation of this new and powerful technology. In the past many facilities have had a very hard time justifying the capital expenditure needed to put in a comprehensive metering system and metering information system.

A useful accounting principle can be applied to our problem, and this might be just what is needed to attract the attention of our fiscal management leaders, and to convince them that making a capital investment in an EIS/ECS or EEM system that provides a stream of annual operational savings is a good investment for them.

That principle and the magic number that goes with it is something that the accountants and business people call the capitalization factor. For our application, it is a multiplier on the annual cash flow from the benefits of our EIS/ECS, EEMs or our ERMs. The capitalization factor is a number typically from 5 to 10 that comes from taking the present value of these annual cash flows over the lifetimes of our various improvement projects or activities. As an example, let us assume our company has a capitalization factor of 7 for one of our projects involving installation of a web based EEM system. We have calculated that the new EIS/ECS system will save our company $100,000 a year in energy and operating costs.

Thus, the new web based EEM system which saves $100,000 a year has increased the asset value of our company by:

$$
\begin{aligned}
\text{Asset Value Increase} &= \text{Capitalization Factor x} \\
&\quad \text{Annual savings} \\
&= (7) \times (\$100,000) \\
&= \underline{\$700,000}
\end{aligned}
$$

An energy and operational cost savings project that increases the asset value of our building, facility or enterprise by almost three quarters of a million dollars is going to get somebody's attention from the highest level of our owners or property management operators! This is now a direct business tie-in for our efforts in implementing and running the EEM system to help make better decisions about operating our facility.

THE NEED FOR DATA AND FOR DATA DRIVEN FACILITY OPERATING DECISIONS

The use of information technology, mainly in the form of TCP/IP, XML, and the web, has revolutionized the facility and building energy control systems industry; and a web-based information and control system can significantly enhance your facility operations. Many authors of chapters in this third volume have been making a very credible case that implementing enterprise energy management and facility management systems can result in significant improvements and cost reductions associated with operating our facilities. The basis for this case is that using detailed and comprehensive data about the utility use in our facility, and about all aspects of the facility's operation allows the facility operators to make decisions that are data based decisions that result in the facility operating better.

We need to have a lot of facility operational data. As one of our authors states, we need "all the data all the time." Our need for huge amounts of operational data about our facilities operation comes from the formal or informal use of the Deming Continuous Improvement Cycle (CI), or the Motorola Six Sigma method; and one of their fundamental principles—make only data driven decisions.

Most of us realize the need to make data driven decisions even if we've never heard of Deming or the Continuous Improvement cycle, or Six Sigma. Deming and others have formulated this process for us, and have helped convince us that this is what we should be doing in our organizations. A fundamental part of our job as an energy manager or a facility manager is to manage. And we need data in order to make data driven decisions. Management by facts—using data—is a management concept for preventing management by opinion. The analysis of relevant data allows informed decisions to be made and significantly reduces the risk of decisions being made only on someone's opinion, and not based on existing facts. Facts are unknown until they are established through the collection of measurement data that show verifiable results. A person without data is just another person with an opinion. However, someone with legitimate experience may have some good data in their head.

Measurements are required. We have all heard the statement that "You can't manage what you don't measure." A more positive version of this statement is that "If you measure it, you can manage it." One of the professional statistics associations has a wonderful quote that I like, and it is "In God we trust. All others bring data." Having all the data all the time is the starting place for good data driven decision-making. Data driven decision-making provides a structure—a set of guidelines for knowing what decisions to make. Effective use of IT, web-based EIS/ECS systems and EEM systems allows the collection of large amounts of data, and results in the capacity to share data and information leading to a deeper understanding of facility operational decisions for success.

Data can be extremely powerful. Data allow us to assess our current and future needs of our facility and its operation. Data give us the basis to decide what to change in our facility operation. Data tell us if our goals are actually being met. Collecting all the data all the time allows us to engage in continuous facility improvement programs. Data help us identify the root causes of problems in our facility. And data allow us to promote accountability for meeting the goals and objectives of our facility's operation.

There are a number of principles of data driven facility decisions. Principle number one is to collect and examine data first. The second principle is to emphasize what is important, by using the Pareto 80/20 rule. The third principle is to set goals that can be measured—use benchmarks and key performance indicators. The fourth principle is to focus on what is effective. And the fifth principle is to align the goals of the facility with facility operational strategies.

To have a successful facility improvement process, we need to set measurable goals and targets toward key performance indicators and benchmarks. Each facility needs to set its own key performance indicators, and needs to identify what benchmarks are appropriate to use for assessing the performance of the facility. For example is Btu per square foot per year the appropriate benchmark to use, or is dollars per square foot per year the benchmark to use. Data must be collected by electronic methods. There is no other way to get the massive amount of data collected and processed in order to use the data to make data driven decisions. After data are collected and processed to produce actionable information, they must then be delivered to appropriate decision-makers who can take action based on that information. The last step in the facility improvement process is to determine the results of key performance indicators and benchmarks, and to evaluate the opportunities for improvements.

We have now moved from discussing the need for data, to the need to provide information from which data driven decisions can be made. There is a huge difference between raw data and information from our facility information and control systems about the status of operation and of our facility support systems. We often hear facility operations staff make a statement "I don't have enough data to operate my facility as well as I would like to." It is possible that these facility operators may not have enough data, or enough of the right kind of data. But it is much more likely that the facility operators have too much data. What they don't have enough of is INFORMATION.

What is the difference between information and data? Information comes from processed data. Information is what helps us make decisions to operate our facility better. Getting the data is the starting place for getting the information. Information is really needed to allow us to make operating decisions that let us run our facilities more effectively, more efficiently and more cost effectively. Too much raw data overwhelms us. It paralyzes us. Almost every facility has a massive amount of data coming from thermostats, sensors for pressures, flows, levels and status of equipment. It is simply too much data for any one—or more—operating personnel to deal with.

What is needed in our facilities is a highly capable, highly connected system to take these massive amounts of data, process it and change it into a few selective pieces of information. Compare the old operator workstation of the past which was a screen full of numbers, to a modern GUI which shows a visual schematic of the system of interest. And under the schematic is a message area that says "Change the filter in AH3B," or "Valve V5 in chilled water line CW2 is stuck open."

With today's information technology providing high-speed and high-capacity computers and communication networks, we have the ability to take thousands of data points as inputs, and process and store them in huge relational databases in our EIS/ECS and EEM systems. Using TCP/IP over our local area network and over the internet lets us easily collect the data from these hundreds or thousands of data points, and our EIS/ECS system can process this data to create information that can help us operate our facilities much better.

With the information produced by these web-based EIS/ECS and EEM systems, we now have the hardware and the software technology to help us accomplish energy management, maintenance management and overall facility management. Now, the facility management staff has enough information to operate the facility efficiently and cost effectively.

The technology for overall facility management is here today. It is the wave of the future. The move to use web-based systems and modern information technology is a wave of even greater magnitude than the DDC wave of 20 years ago. Previously, those who did not get on the wave of DDC got rolled over by that wave. It is now time for facility energy managers to get on the wave of IT and web-based systems in order to capture the benefits of reduced operational costs and improved facility productivity. This is now not only enterprise energy management; it is enterprise resource management.

EPACT 2005, METERING, AND EIS/ECS AND EEM SYSTEMS

In many ways we are still well ahead of our time with the material in these three books, and even in the first book on basic ideas of information technology and web based energy information systems. The recent Energy Policy Act of 2005 contained a provision that has now required all federal facilities to have a meter on each building by 2012. What kind of meter will be installed in these buildings? Who will collect the data? Will the meters be read manually? What will be done with all this data that will now be available from the one meter

on each building? How will the meter data be processed and delivered to the facility managers? And finally, how will the data be used to make better decisions about the operation of the buildings at the facility?

These are all basic questions that have been addressed in the first and second volumes of this three book series. Getting the meters installed is certainly a huge starting point. But the next question is how will that data be collected? Hopefully not by someone traveling around the facility looking at individual meters, and writing down the meter reading once a month. The only reasonable answer that can be given is to use the technology of IT and the web—including using TCP/IP as the major data transmission protocol, and transmitting the data over the facility's local area network. Then the data can be stored in a centralized data warehouse and processed to provide at least bar and line charts in Excel to provide information to the facility energy manager.

Just this level of use of IT and TCP/IP would be a major step forward for federal facilities that have thousands of buildings with no meters, and no meter data collection and processing system. But as even volume one in this series showed, there is a much greater potential to use the data collected to provide information to operate buildings and facilities better. Volume two presented many case studies of facilities using these web based EIS and ECS systems, and showed the kinds of operational benefits that could be achieved with them. This third volume expanded this list of case studies showing the cost savings potential of these systems. Commercial vendors have developed these web based EIS/ECS and EEM systems and there is a large offering of products at various levels of sophistication and cost.

It is surprising to many of us that the current requirement for federal buildings is to just have one meter! At this point in our understanding of energy and facility management, the expectation had been that some level of submetering would have been required for larger buildings containing equipment such as chillers, large lighting loads, data centers, boilers, etc. It seems like some size of load should have been specified for the additional requirement of submetering. That way small buildings could have had one meter only, but larger buildings should also have to have submeters. There is certainly a need for a master meter on each building, but there is also a need for much more detailed submetering data on larger and more complex buildings.

Another striking limitation of the federal requirement for a meter on each building, is that it only requires one ELECTRICAL meter! This is probably reasonable for many buildings, but also very unreasonable for many other buildings that have boilers, gas heaters, gas and oil water heaters, and other large, fuel supplied thermal loads. It would have been easy to specify a minimum level of Btu/h of equipment loads or use that would have triggered the requirement for a fuel use and thermal submeter. This is the basis for such a decision in most of our larger buildings in the commercial, industrial, educational and institutional sectors.

Fortunately, many of the energy managers at federal facilities have already recognized the value in putting in multifunction, smart, and networked meters that will collect a wide range of data automatically, and send it to a computer system for processing and storage. Even using a basic EIS/ECS or EEM system can have tremendous savings potential, as well as facility operating improvements from utilizing the data to make data driven operational decisions.

SOME THOUGHTS ON EIS/ECS AND EEM SYSTEM IMPLEMENTATIONS—PAUL ALLEN, WALT DISNEY WORLD

As energy managers our primary mission is to reduce energy consumption and costs for our clients. My experience has shown me that a sustainable energy management program requires three essential items to be successful. The first and most important is a *commitment* from management and employees to save energy. This sounds like a simple thing but it turns out to be the most difficult to sustain. Regardless, it is absolutely the most important step toward the goal of reducing energy cost in any organization.

The next two items are technology-based and have benefited from the web-based standards and innovations that are discussed in detail throughout this book. The first system is a building automation system (BAS). The BAS is a *control* system that turns equipment on/off and adjusts HVAC setpoints to maintain building comfort conditions. The next system is an energy information system (EIS) that is used to *measure* energy reduction efforts by transforming mountains of utility data into easy-to-understand information.

Owners today are looking for BAS/EIS systems that are reliable, low-cost, easily expandable, and easy-to-use. Incorporating web-based technology is just one desirable feature. The web provides the means to share information easier, quicker, and cheaper than ever before. However, the reality is that most owners have an assortment of legacy BAS systems that they already have an enormous investment in. Jumping to the latest and greatest technology is an expensive proposition for most.

So how do you bridge from the old systems to the new systems? The answer is to create a strategic plan and to transform that plan into good BAS designs for new/existing projects. By establishing a strategic plan that shows how new buildings and renovations are integrated with their existing BAS/EIS, owners should take the bull by the horns and set their own destiny. Owners need to work with their design engineers to establish plans and specifications that set their strategic plan in motion. Instead of putting a comment on the drawings that says "Connect to existing BAS," the engineer should show more of the "how it's done" instead of passing this responsibility to the BAS controls contractor. The goal is to have the new building BAS integrate seamlessly with the owner's existing EMS and is both cost-effective and takes full advantage of web-based technologies.

Since web-based EIS grew up with the web, most owners have the ability to pick the solution that is best for them. Owners have two options—(1) do-it-yourself or (2) the pay-for-the-service. Each has their benefits and it really boils down how much will each approach cost. Remember EIS rule number 1—an EIS will not save anything by itself. Instead it is the actions that are taken based on the EIS information that result in energy savings. That's why it's important to start sharing the utility data with your energy partners. The more eyes on it the better to look for energy waste. A web-based EIS is the best solution.

Stay focused on the mission—reduce energy consumption and costs. Take control over your BAS. Develop a strategic plan to integrate new technology into your buildings and work with the design engineers to transform this vision into plans and specifications. Transform the mountains of utility data into a web-based EIS that measures your energy progress and share it with management and staff using reports/graphs that can be easily understood. When all three parts of a sustainable energy management program are working together, (1) a commitment from management and staff, (2) controlling equipment with your BAS and (3) measuring energy progress with your EIS, it's magic.

Acknowledgements

Major thanks go to each of the other 54 authors who took many hours of their personal time to write these chapters in the book. Their reason and reward for that effort has been to help explain and promote this new wave of web based EIS/ECS systems for building automation and enterprise energy and facility management. This unselfish willingness to help educate energy and facility managers about the tremendous benefits of this new wave of technology is a pleasure to see and appreciate. And for most of us, this is the fun part of our professional lives.

Thanks also go to Paul Allen and Dave Green who served dual roles of authors and associate editors. Their help was crucial to the success of this third book, as well is that of the entire project. Special thanks go to Lynne Capehart who has served as associate editor for the first book, and coeditor for the second and third books. Her editing skills have greatly improved the reading and understanding of the material in all three of these books.

Thanks also go to Al Thumann, the executive director of the Association of Energy Engineers who encouraged me to produce the first book—and then the second and third books. The folks at Fairmont press were also very easy to work with, and helped ensure quality books that got published on schedule. Linda Hutchings at Fairmont Press, and the layout specialist, Joy Garland, made sure of this.

And finally my thanks to all of you readers who have the interest in learning about the huge potential for improving the operation of your facility using these web based EIS/ECS and EEM systems. I hope these books encourage you to use the systems and technology to operate your facility better. My own reward will come in seeing this technology implemented widely and rapidly. Good luck to us all!

Barney Capehart

Chapter 2

Building Automation, Beyond the Simple Web Server

Anto Budiardjo
President & CEO, Clasma Events Inc.
Email: antob@clasma.com—Phone: (972) 865-2231
Dallas, TX—May 2006

THE QUEST FOR TECHNOLOGY to enable intelligent buildings has been around for almost two decades. This promise has come under many disguises, including integration of building systems, a quest for open systems, and the use of IT network communication standards for the devices necessary to make buildings more intelligent.

While what people call intelligent buildings may vary, visionaries generally agree that technology *can* automate the functions of a building, which in turn can improve the comfort and security of buildings. This is in addition to a new more controversial view that intelligent buildings can better serve their owners regardless of what the purpose of the building is: Comfortable and safe buildings make for better occupants (workers) and is thus good for business.

Over the past two decades, many technology initiatives have come and gone to create networking and data standards in a hope that a standard can be widely adopted by all those participating in building systems. The surviving initiatives today are basically LONWORKS™, BACnet and EIB/Konnex, as well as those closer to industrial applications such as ModBus and OPC. While these are achieving some degree of success, the scope of their adoption has been limited to certain functions in buildings. None of these standards can architecturally provide the answer to all of the requirements necessary for the dream of intelligent buildings.

While these standards have been battling it out in the building systems space, the internet has blossomed into one of the most disruptive changes not only to core technology but also to information distribution, commerce and almost every facet of life in the 21st century. Internet technologies, specifically TCP/IP, Web Browsers and more importantly XML and Web Services have become (or 'is becoming' in some industries) the only credible network and data architecture for human, systems and device-centric connectivity.

While most understand TCP/IP and Web browsers, the true impact of XML and Web Services lies silently below the surface. XML is a mechanism to describe information in a manner that is flexible and easily communicated across TCP/IP networks. Web Services can best be described as a mechanism for applications (both system- and device-centric) to communicate with each other across the internet and intranets. The combination creates a powerful network-centric architecture that is being adopted by almost all of the world's industry sectors including banking, media, commerce, science, engineering and retail to name a few. Chances are you have used XML and Web Services if you have booked a flight or hotel room on sites such as Expedia or Priceline, or if you do online banking. It is without doubt very pervasive, but its true role is really hidden from view.

FLEXIBILITY

The key to the architectural elegance of XML and Web Services is its inherent flexibility based on a very solid and simple set of definitions. This flexibility is especially interesting because it allows specific industries to create a set of data and behavior standards based on the needs of the applications for that industry yet conform to the basic standards that apply to all industries.

In the building automation area, a number of XML and Web Services standard initiatives are taking root. oBIX (part of OASIS) is tackling the problem much more from the enterprise and IT perspective, adopting many of the Web Services standards being developed within OASIS and other IT centric groups. BACnet Web Services, while adopting IT standards, is taking a much more buildings centric view in line with the objectives and mission of the BACnet standard started over 15 years ago within ASHRAE. A number of initiatives are taking place in the security arena as well as other aspects of the total building automation picture. Coordination between these groups is there but not comprehensive.

The question most often asked within the scope of Building Automation is how XML and Web Services will impact products, systems and applications for buildings. Before diving into that, it is worth noting that in this chapter, the term "Building Automation" is used to refer to any and all systems that provide some form of automation for buildings, including HVAC, security, lighting, energy, vertical transport, etc.

CHALLENGES

Today there are two major district challenges in Building Automation as far as enabling the vision of an intelligent building. First is how the different elements of a building can cooperate together in an integrated manner, and second, how the building can be connected in real-time to the enterprise systems that now run all organizations.

While proponents of network standards mentioned above may still wish that their technology becomes *the* technology to link systems together, even a cursory analysis of the development and spread of internet technologies clearly shows that the only technology that will achieve that is a combination of TCP/IP and a browser-based user interface as well as XML and Web Services.

Many in the building automation industry will quickly criticize the above assertion, stating numerous challenges, including cost of CAT5 (or similar) wiring, the existing availability of a large array of building automation products and solutions, and the fact that many existing systems will remain in buildings for years, if not decades, due to the typical life cycle of building automation systems.

It is true that today the cost of cabling and management of an ethernet-based device is significantly more than the cost of running a [cheap] single, twisted-pair typically used in building automation to network devices that are designed to be self-managed. But this is changing rapidly as ethernet and internet network infrastructure and management products become commonly available and the skill base required to install and maintain them becomes widespread. The prospect of PoE (Power over Ethernet) will dramatically change the balance of this equation because eliminating the need to run power to building automation devices will bring

significant benefits overcoming most, if not all, the negatives of ethernet. IPv6 will also further drive IP-based architectures for smart devices in buildings.

GATEWAYS ARE IN

The large number of existing and available non-IP devices as well as the large number of installed legacy non-IP systems will dictate the need for gateways and other protocol translation devices. The only logical architecture of such interfaces is that data and connectivity on the ethernet and IP side is structured using XML and Web Services.

Anyway you look at it, in the next few years of Building Automation, Web Services will play a very important role both in native IP solutions and adopting the IP infrastructure as an integrating backbone of building systems.

As previously mentioned, the second most important requirement for tomorrow's buildings is their inclusion into the enterprise systems of today's corporate and institutional organizations. Buildings need to be plugged into their enterprise systems so that the enormous amount of data that exists in buildings—information that has until this point been inaccessible to the enterprise—can be liberated and made available in real-time to information-hungry enterprise systems, keen to squeeze the last drop of efficiencies out of their assets.

XML and Web Services is the only way this can be done.

One of the challenges facing stakeholders is figuring out the value of having buildings as a real-time citizen of the corporate information network. Just consider for a moment that the cost of facilities is the second largest line item for most organizations today, second only to people. The value of controlling facilities at the enterprise level is two fold. First, it will enable organizations to cut operations costs be it energy, staff costs, maintenance or many other areas brought about by this level of super integration. Second, and most important, there is significant potential for correlating the effect of facilities to the performance and effectiveness of people the occupants (the highest cost line item).

Whether through comfort, safety or bringing about a more rewarding occupant experience, building systems, when tied to and driven by business objectives can make the behavior and performance of staff and customers more in-line with the business objectives of the building owner. This can only come about by enterprise-level integration in a similar way that supply chain, human resource and customer relationship management are now tied closely together by enterprise systems, something that is best done using XML and Web Services.

WEB SERVICES SCENARIOS

Consider a retail scenario where the success of a retail unit (measured by POS data) can be correlated to inside temperature, lighting, time of day and the weather outside. Even the most fundamental expert system could analyze patterns, which could feed the staff schedule so that the appropriate staffing level is maintained based on weather, date, time-of-day and other data. Such predictive behavior could also be tied into the store's CRM (customer relationship management) system to automatically discount certain items in real-time, send Email invitations to local customers for "daily specials," and set the temperature, music and lighting to the most suitable setting for selling *those* products. This scenario requires a very high level, enterprise-centric integration, using tools and systems from many disciplines—all of which use XML and Web Services.

In airports systems exist to manage the arrival and departures of flights. If this is connected to the environmental, lighting and security systems, the airport could automatically turn systems into standby mode when the system knows that occupancy in that area is going to be very light, and into full mode minutes before the arrival of an aircraft. A pleasant environment at the arrival gate is very customer friendly to travelers; airports can use this to improve how travelers feel about their experience at that airport without having to worry about that unnecessary expense burden incurred when the areas of the airport remain unused. Creative airports could also link this information to their billing system so that the airline in question is billed for out-of-hour usage.

Occupant experience is yet another area where enterprise level integration can score benefits. Imagine a hotel chain that connects their loyalty program and back-end systems to their in-room systems so that the room temperature, music type, default news channel, and wake-up preference is automatically set as soon as the guest registers. Hotel chains would be hard pressed to achieve this with the plethora of building systems in their hundreds and thousands of hotel properties, but given XML and Web Services, this becomes an achievable scenario and one of immense benefit to hotel chains vying to lock in their customers.

While the adoption of XML and Web Services in Building Automation can bring benefits directly to the discipline of building automation systems, the real benefit is how buildings can be integrated to other systems that can be impacted by the facility's environment.

So we need to return to the effect of Web Services on existing standards, specifically LonWorks, BACnet, KNX, ModBus and OPC. The short answer from the perspective of the author is that in the near term, XML and Web Services will supplement these standards and not in any way replace them. In the longer term, as new technologies such as PoE, IPv6, PLC (Power Line Carrier) and wireless mesh become standard, some of these existing standards will likely diminish, and over the next decade or two some will cease to exist.

INDUSTRY DYNAMICS

The building automation industry is today undergoing significant changes. Consolidation is the name of the game as major multi-national corporations acquire technology, expertise and global distribution with one goal: provide *all* of the systems required for an integrated building system. Whereas five or ten years ago there were HVAC companies, security companies, CCTV companies, lighting companies, etc., today the major players are providing all of these products from a single source. In a short period of time there will be a handful of major vendors such as Honeywell, Siemens, JCI as well as maybe GE, Schneider Electric, Tyco, UTC and others that would provide complete solutions for the intelligent building.

But since many of the skills necessary to provide intelligent building solutions are going to be IT based, don't leave out IT-centric systems integrators and consulting services companies in this picture, companies such as EDS, Accenture, IBM Consulting, and Cisco as well as newer, more nimble integrators will vie for a piece of the intelligent building pie. And many would argue that these players are in the best position to take the lead from the enterprise and IT perspectives.

Indeed XML and Web Services is the combination of technology that links the two major vendor groups above, and it is probably this, more than the fundamental technology, that will make XML and Web Services a very disruptive change to buildings as we know them today.

Look at XML and Web Services as an undercurrent that will sweep the building industry far away from the construction mentality of days past.

BUILDING AUTOMATION TECHNOLOGIES

Provided below is an outline of the three phases of networked building automation systems.

	Before Internet Technologies	With Internet Technologies Without XML & Web Services	With XML & Web Services
Typical User Interface	• "Thick client" applications running on PC's. • Typically OS locked (Windows & DOS before) • Non-standard graphics engines • Non-standard operation for alarms, schedules, trending, etc.	• Web browser based interfaces common • Web server applications on devices and PC based servers on site • Hyper links between pages residing on different sites	• Rich dynamic web browser based interfaces • ASP and hosted servers, connected to sites via Web Services • Rich information from multiple sites and non-buildings sources (e.g. weather data).
Integration	• Single vendor integration of systems • Gateway based hardware or software solutions • OPC and DDE interfaces	• TCP/IP based gateways and protocol translators • Web pages do not typically display information from multiple sites	• ASP based web pages not limited to a single site of information • Inter-system communications by XML and Web Services
Security	• Ad-hoc use of security implemented mostly in non-standard ways • "Security through obscurity" of stand-alone system unconnected to corporate networks	• Leverage SSL and other packet based security schemas • Web server password based security common military, eCommerce, etc.	• Leverage security technologies developed for Web Services • Technologies secure enough for applications such as banking,
Database	• Typically proprietary relational databases, or PC centric DB's such as dBase	• Web server centric Access type databases • Distributed enterprise level databases	• Rich use of SQL and XML and Web Services interfaces
Proprietary Systems	• Dominant from field level • Dominant at management, configuration and operation stations	• Remain significantly dominant at the field level despite open systems • Operator stations still use proprietary systems	• Converted to XML & Web Services gateways to extend life of functional systems • Can be beneficial for application logic
Remote Operation	• Via difficult to deploy dial-up modems. • Slow and costly to maintain.	• Via TCP/IP based gateways and device • Serial servers (RS-232 tunnel infrastructures	• XML and Web Services signaling and eventing • Use of wired & wireless
Enterprise Integration	• Custom interfaces • Use of DDE via gateway PC's • Very rare to be real valuable applications for data	• Special software to convert proprietary protocols tunneled through TCP/IP • HTML parsing and mining • More centric toward business objectives • Building becomes an enterprise "sub-system"	• Use of XML and Web Services is standard • Used by enterprise integration professionals
Open System Standards	• LonWorks and BACnet protocols provided promise of open systems • PC based technologies (OLE, DDE, etc.) • Dial-up standards object definitions	• LonWorks and BACnet provides standards for certain functions in the architecture. • TCP/IP standards (HTML, HTTP, SSL, etc.).	• LonWorks devices, LonMark • BACnet system centric standard functions • XML schemas based on existing standards • Layering of standards from IT standard bodies
Infrastructure Products	• RS-232 based products • Dial-up modems that were constantly changing	• Use of TCP/IP connectivity products such as routers and hubs firewalls, software, etc.	• Chosen from plethora of available IT centric routers, hubs, servers,
Tools	• Proprietary and "home grown" tools • Use of standard office applications (Excel)	• Use of Web server tools (FrontPage) • Use of infrastructure tools, sniffers, etc.	• Use of standard XML and Web Services tools • Development platforms such as .NET
Skill base	• Segmented to vendor centric groups • Difficult to transfer to different systems • Typically buildings or controls centric base	• Increasing availability of web browser and TCP/IP skill base available • Broad training available	• Much more IT centric skills required • Very broad education (courses, books, etc.) • Less concerned with buildings or controls

Chapter 3

What Is an Intelligent Building?

Paul Ehrlich
Building Intelligence Group

ABSTRACT

I N THIS CHAPTER we look at the definition of an intelligent building and what is necessary in the design process to ensure that intelligent attributes are incorporated. We also look at what is involved in the crucial phases of construction and operation of an intelligent building. As an industry, we know how to deliver intelligent buildings. Many of the products and solutions are already readily available and understood within our industry. Others will be available within the next few years. The challenge that we face is to move forward together to start delivering the truly intelligent building.

INTRODUCTION TO INTELLIGENT BUILDING DESIGN

Over the last 20 years, there has been a lot of discussion and debate about the concept of an "intelligent building." Work has gone on in many forums to define and quantify what the term really means. The end result of all of these efforts is that an intelligent building is not just one thing. My definition of intelligent buildings is as follows:

"Use of technology and process to create a building that is safer and more productive for its occupants and more operationally efficient for its owners."

The results from implementing these technologies and processes are buildings that cost less to operate and are worth more to their occupants. For projects that are owner occupied, such as Corporate, Government and Institutions, the benefits of an intelligent building provide an immediate ROI in terms of higher employee productivity and reduced operating expenses. For commercial developments, these projects are expected to result in above market rents, improved retention, higher occu-pancy rates and lower operating expenses. All around this is a win-win situation!

So what are the technologies and processes that are required to create such a project? There is a long list, starting with design and going through long term operations, retrofit and eventual decommissioning. See the sidebar article for a brief summary of the attributes.

Let's start by looking at the design process for intelligent buildings. The decision to make a project "intelligent" needs to come early in the design process. Making the decision to create a new or retrofit an existing project to make it intelligent is similar to what goes into creating a LEED certified project. There needs to be a commitment from the owner and their design team to invest in a project with superior performance and value. Once this occurs, the design process can continue as usual. But it is important to keep the focus on creating a superior project and avoid the temptation to "value engineer" out the intelligent components.

Project Scope and Purpose

One of the first attributes in an intelligent design is to carefully evaluate the current and future use of the project. This starts by clearly identifying the purpose and needs of the targeted building occupants. This process will vary depending on whether it will be an owner occupied or a commercial development. For an owner occupied building, surveys and focus groups can be held with the building occupants, analyzing and prioritizing their needs to select proper project features. For a commercial development, the project target market needs to be identified and attributes designed to suit. For example, an office building might target technology companies that would benefit from an urban environment, high speed network access and 24/7 availability.

It is important to realize, however, that few projects are used as originally envisioned. A good intelligent design should incorporate flexibility to allow for easy change. Examples of this type of design characteristic include CLA (communications, life safety, automation) structured cabling design and open space with movable or demountable partitions. An intelligent building needs to be designed to meet the needs of initial occupants and be flexible to meet the needs of future occupants.

Concept and Budget

When setting initial project budgets, intelligent attributes must be included. Creating an intelligent building does require an investment in advanced technology, processes, and solutions. An up-front investment is required to realize a significant return later on. It is unrealistic to expect to make a project intelligent unless there is early buy in on investment. Again, these decisions need to happen prior to the start of design work. One of the challenges is to educate owners need to be educated on the benefits of an intelligent building design. Waiting until the MEP is brought on to the design team may be too late. This makes education of both owners and architects as to the benefits of intelligent solutions critical for success.

Site Selection and Integration

An intelligent design begins by looking at the site as it integrates with the community. Is this a location that is a new "green-field" location or a reuse of an existing "brown-field" site? Can the project be sited for maximum solar efficiency? How will it fit in with community land and space planning? Does it integrate with existing (or planned) public transportation? Site integration and impact are critical for environmental impact, and strongly impacts how the building occupants interact with the

building. At a macro scale, community integration is determined by community space planning and zoning regulations. An intelligent building should go beyond that with consideration as to how this fits in with the communities needs, transportation, and amenities. The combination of the two makes the building more marketable with a lower impact on the environment.

Environmental Design

An intelligent building starts with an environmentally friendly design. Creating a project that is environmentally friendly and energy efficient ties in closely with many of the intelligent attributes. Intelligent buildings are designed for long term sustainability and minimal environmental impact through the selection of recycled and recyclable materials, construction, maintenance and operations procedures. Providing the ability to integrate building controls, optimize operations and enterprise level management results in a significant enhancement in energy efficiency, lowering both cost and energy usage compared to non-intelligent projects.

Intelligent buildings are intended to be the preferred environment for occupants. This requires focused attention to environmental factors that affect occupants perception, comfort and productivity. An intelligent design finds the balance providing a superior indoor environment and minimizing energy usage and operating labor. This is where the technology becomes valuable. Using integration and automation we are able to implement solutions that both provide a superior environment and minimize energy. Examples include:

Feature	Benefit
Dimmable fluorescents lighting integrated with sun blind control	Optimal lighting level and quality can be determined by the occupants.
Lighting control with motion sensors integrated with security	Only provide lighting as needed. Reduces energy use and increases security.
Natural and displacement ventilation	More efficient and effective distribution of ventilation.
Use of economizers for free cooling	Energy efficiency.
Individual temperature and lighting control	Improved comfort is shown to improve productivity. Addresses the number 1 concern of tenants as found in BOMA surveys.
Radiant heating and cooling	Improved comfort, reduced energy use.
Optimized control algorithms	Reduce energy use with little or no impact on comfort.
Combined heat and power plants	Improved energy efficiency and sustainability.
After hours control of lights and HVAC integrated with security	Improved security while reducing energy use.
Monitoring of IAQ and contaminants	Improved comfort, safety and productivity.

USGBC/LEED

The US Green Building Council—Leadership in Energy and Environmental Design (USGBC-LEED) program provides an excellent mechanism to promote, measure and quantify environmental and energy efficiency in both new and existing projects. There is a very strong synergy between an intelligent building design and a LEED certified design. Intelligent buildings demand reduced energy usage through optimization, system integration and enterprise applications. LEED certification requires energy efficiency, monitoring, validation and control of all building systems. The goals and benefits of LEED and intelligent building design go together arm and arm. An intelligent building program should start with LEED certification and work to improve the building beyond that.

Building Modeling

An intelligent design needs to start with a complete model. This modeling begins early on with CAD designs that evolve into project renderings. Using new standards such as AEC-XML and GB-XML, this information can readily be shared with HVAC and other system models. Modeling of an intelligent building will be used not just in design, but will continue through into construction and operation. In the past, building modeling has been widely used as a design tool and often for construction as well. In an intelligent building we would expect that this model will be used by new sophisticated tools that will actually be able to use the original modeling information to make decisions about optimization and continuous re-commissioning of critical building systems. Ideally, the model will follow through the life span of the building, be updating as necessary and serving as a digital document of the building.

Building Circulation and Networking

Buildings exist to enable collaboration, allowing occupants to be productive, efficient and creative. Intelligent buildings provide for improved occupant circulation, interaction and collaboration. From a design perspective this means attention to how the occupants will circulate through the building. How will they enter the space? How will they move efficiently vertically and horizontally through the space? Can we incorporate digital signage to improve navigation and circulation?

Collaboration can also be improved through the use of design elements to encourage networking in both formal and informal spaces. Formal collaboration spaces are conference rooms, break rooms, classrooms and seminar rooms. Informal collaboration spaces include niches and seating spaces in corridors, coffee shops, outdoor seating areas, and other places where building occupants can get together for brief planned or unplanned interactions.

Project Negotiation and Value Engineering

Once an intelligent building is designed, the first challenge is to make sure that the team remains committed to keeping it intelligent through construction. There are a lot of discussions about the best way to contract for an intelligent building. The traditional model is to hire an architect who, in turn, builds a team with a mechanical, electrical and plumbing (MEP) consultant and they, in turn, create project documents. General contractors are then asked to bid on the construction documents and assemble a project price using their team of sub-contractors. Unfortunately during this process the desire to make a building intelligent is often lost in the confusion of keeping the project within the allocated budget.

Often we try to bring projects within budget through a process called "value engineering," implying that we are enhancing the value of a project. In reality this process often removes areas of high value in a desperate attempt to keep the project under budget. It is important that the design and construction team agrees early on as to the importance and priority of the intelligent attributes and works to keep them in scope, even when project budgets become a challenge.

There are a few alternative construction models that may work better for intelligent buildings. The simplest is a design build process where the owner contracts with a design build contractor to provide a project with the desired features for a set price. Another alternative is to break out the intelligent building portion of the project. This can be as nominal as breaking it into a separate specification section (Division 17 has often been used and Division 25 is designated by the new CSI-2004). In many cases, the intelligent building portions may also be split out in a separate RFP or on the bid forms for a supplier decision directly by the owner. The most radical approach, which is being used on projects outside of North America, is to hire a firm to design the buildings technologies and they, in turn, hire an architect and the rest of the project team. These approaches have the same desired result: to focus on creating a better building and not be overcome by the challenges of budget and schedule.

Construction Processes

On an intelligent project the design document files should readily extend into the construction process. The goal is to keep as much of the construction process as "paperless" as possible. Keeping updated electronic documentation is valuable, not just because it reduces cost during the construction process, but also because it forms the basis for continuous documentation of the project. In reality the mission of many buildings is constantly in flux, resulting in the construction process is never being totally done. Having accurate documentation of how the building is constructed and modified provides the ability to bring this information into operations.

From an environmental or green perspective we want to use construction processes that are sustainable. This means looking to minimize construction waste, utilize environmentally friendly materials, develop on brown field sites and recycle materials whenever possible.

Facility Operations

At the heart of an intelligent building is the benefits and changes that will occur in operations. The reason that we look to implement intelligent attributes is to improve the efficiency and effectiveness of the operations staff. The operation of most commercial buildings today is a challenging and often frustrating task. Operations staff are bombarded with phone calls, meetings and other tasks that leave little time for planning and strategic operations. In addition many facility managers have received little, if any, formal training on how to successfully complete their complex and demanding jobs. As a result, the operation of a facility has become a craft, one that is often self taught, rarely well documented, and often, not repeatable. A facility with a strong operating engineer may run efficiently and have satisfied occupants, yet the building across the street with a less experienced staff may run poorly and be uncomfortable. Add in that these buildings are often poorly commissioned and have been constantly modified, and the result is many buildings that do not operate properly. Unfortunately for the business managers, owners, and investors of these buildings, there are no checks, balances, or controls in place today to let them know if their building is well operated or not.

One of the primary goals in an intelligent building is to provide the technology, tools, and processes to improve the operation of the facility. In many cases, these tools will be provided for the existing on-site operations staff. In other cases on site operations staff will be replaced or augmented with services from a centralized operations center or service.

Building Management Tools

An intelligent building will typically have thousands of pieces of data available. The goal, however, is not to present all of this information to the facility operator. Rather, tools will be used to evaluate, and prioritize, presenting only the required information to the operator. Examples of these tools include:

- **Complete Integration**—System integration of all critical building systems including HVAC, Electrical, Fire Alarm, Security, Video Monitoring, and Digital Signage. All of these systems will be integrated on the building network and will share this infrastructure with other applications including data, voice, and video. This information will need to be secured and will travel on both the private and public networks. Open standards are at the heart of enabling the intelligent building as solutions including BACnet, LonTalk, oBIX, Modbus and other protocols used to enable integration.

- **Tenant Portals**—One critical element of an intelligent building is providing a method for the building occupants (tenants, employees, associates, students, patients, etc.) to interact with the building and building management. In the past, this has been done with phone calls, face to face meetings, and faxes. Today, it is most effectively done with an internal website called a portal. Tenant portals provide information about the facility, contact information, directories, energy efficiency, emergency preparedness and a central place to enter issues. Information from the portal can then be used to drive maintenance and operation requests. Since the portal is a two way communications channel, it can also be used to collect critical feedback on occupant satisfaction and comfort levels.

- **System Dashboards**—Like the dash of your car, a system dashboard provides a summary of critical building alarms, energy information and key maintenance items at a glance. The dashboard is responsible to summarize all of the critical building information and present it in the proper format for different members of the facility management team. For example, the operating engineer requires detailed information about specific mechanical systems, while the property manager needs a summary of energy and operating expenses over the last 4 weeks. The difference between a system dashboard and a typical user interface for an integrated building automation system is in focus. Dashboards are more focused on sorting and filtering data to provide the information needed to perform specific roles. Building automation systems tend to be much more generalized and designed for the operating engineer.

- **Next Generation Maintenance Management Programs**—Maintenance management systems typically track work orders and schedule repairs and preventative maintenance. In the future, these systems will be closely integrated with the building systems allowing for critical data evaluation from equipment, determining if it is operating properly and what maintenance is required. These systems will also be used to deliver requests from building occupants by integrating with the portal. Operating personnel, both in house and contractors, will automatically be dispatched using wireless communications to their cell phones and PDAs. This allows for more proactive operations and for increased efficiency.

- **Enterprise Energy Management**—Most energy management systems today are focused only on the operation of the building. This includes functions such as demand limiting, scheduling, and system optimization. Intelligent buildings take this one step further by incorporating real time utility rate information and making energy management decisions not just for a single building but for groups of buildings. By managing energy in concert with the utility, there is the ability not to only reduce energy usage but more importantly, to dramatically decrease energy expenditure.

Operating Staff Efficiency

Today most buildings have several operations groups. One group is charged with property management, a second with security, another with building maintenance, while staff or contractors are used for custodial, HVAC service, and fire protection. There can be significant improvements with the use of technology to allow the operating staff to do more with fewer resources. In some cases, this will mean providing tools to be more efficient. In other cases, it is a matter of centralizing certain services and providing them for groups of buildings. Here are a few examples:

- **Mobile Operations**—The operating personnel within an intelligent building will be equipped with wireless devices (cell phones, PDAs, laptops) that will allow them to readily access all building systems as well as receive and process work orders and tenant requests. This will allow the staff to be in the building responding to issues, dealing with maintenance tasks, and evaluating security issues all while "on line." The result is less time spent looking at systems, ordering parts, finding draw-

ings and building documentation and more time getting tasks accomplished.

- **Centralized Operations Center**—Major retailers, large school districts, and health care centers have worked for years to centralize their operations. Providing a centralized operations center (often called a building operations center or BOC) provides the ability to consolidate functions across facilities. The BOC typically provides a central location with a call center and hosting for enterprise applications. In addition to the tools specialized staff members with in depth expertise on systems, energy, security and other systems work at this center. Using a centralized facility stretches the ability of these experts to impact all of the facilities. Since communications with the BOC uses Internet technology, the center can be virtual and be located anywhere in the world. For example one center might monitor operations during the day and a second one located on the other side of the world might provide similar functions at night and on weekends.

- **Virtual Concierge**—It is even possible to centralize "high touch" functions such as visitor management. Services are available that utilize video and voice over Internet communications to allow for real time interaction with visitors, allowing them to be greeted, present identification and receive building badges all from a centralized, often remote, location!

Summary of Attributes of an Intelligent Building

Intelligent buildings mean many things depending on your perspective and role. The following list is one summary of these attributes.

Process:
- Design
  - Flexibility—designed to change
  - Energy efficient design (LEED)
  - Complete building modeling
  - Focus on building circulation and Feng Shui and common spaces for networking
  - Integration with transportation and surrounding community
- Construction
  - Sustainable construction practices
  - Electronic project documentation
  - Modeling extended into construction
- Operations
  - Integration of all systems

— Remote operations and optimization
— Tenant portals
— After hours operation
— Maintenance management and dispatch
— Energy information and management systems
— Real time energy response
— Continuous comfort monitoring and feedback

Technology:
- General
  — Tenant amenities
      Concierge
      Shopping
      Restaurants
      Lodging
      Parking
      Restrooms
  — Optimized vertical transport
  — Personal comfort control
      Temperature
      Humidity
      IAQ
      Lighting
      Acoustic
- Networking/Telecom
  — Common network infrastructure
  — Structured-maintainable cabling
  — Wifi
  — VOIP
  — Digital signage
- Security/Life Safety
  — Digital video monitoring
  — Access control and monitoring
  — Automatic fire suppression
  — Fire detection and alarm
  — Egress support (lighting, signage, smoke control, etc.)
  — Contaminant monitoring and containment
  — Proximate security/guard services
- Mechanical
  — Energy efficient equipment
  — Thermal storage
  — Combined heat and power
  — Controls optimization
      Extensive sensing

      Energy efficiency
      Indoor air quality
      Comfort monitoring
      Internet enabled controls
      Enterprise integration
      Water and gas metering/sub-metering
- Electrical
  — Energy efficient lighting
  — Lighting control
  — Distributed generation
  — Dual power feeds/emergency power
  — Power quality monitoring
  — Sub-metering/billing

CONCLUSION

In the start of this chapter, we defined an intelligent building as:

"Use of technology and process to create a building that is safer and more productive for its occupants and more operationally efficient for its owners."

In simpler terms the goal is to provide a better building. The result is a facility that uses less energy, has dramatically lower operating expenses and provides an improved indoor environment and better responsiveness for the occupants. All of this provides a significant return on investment. As an industry, we know how to deliver on intelligent buildings. Many of the products and solutions are already readily available and understood within our industry. Others will be available within the next few years. The challenge that we face is to move forward together to start delivering on the truly intelligent building.

The goal of having an intelligent building only starts with early planning in the design stage. In many ways this mirrors the design and fulfillment of many green or LEED projects today, but uses technology to provide for a superior space. There are enormous benefits to be gained by creating intelligent buildings. We need to continue to work as an industry to quantify these benefits, educate owners and consultants and to deliver a superior product to the market.

Chapter 4

How Can a Building Be Intelligent If It Has Nothing to Say?

The Need for Business Systems in Today's Buildings

Bill Gnerre, Greg Cmar, and Kevin Fuller
Interval Data Systems, Inc.

S O, YOU HAVE AN INTELLIGENT BUILDING. Congratulations. Web-enabled. Operational systems connected on a shared IP backbone. All the coolest technologies—Web services, SOAP/XML, etc. You can sit on your couch, connect wirelessly to your control systems and... accomplish nothing that you couldn't before.

What? Everything is connected—it has to be better.

Your building might be "intelligent," but all too often it has nothing to say.

IT'S NOT ABOUT TECHNOLOGY STANDARDS

In all the discussion about building-IT convergence, it seems that everyone forgot about the "I" and focused on the "T." After all, get a bunch of technologists together and the conversation is bound be become about technology. Only problem is, the buyers (building owners, facilities executives, CEOs, CFOs) don't really care. They care about business value—operational and financial information, productivity, verifiable savings, accountability—not technology standards.

The IT world has repeatedly demonstrated, through both its successes and failures, that technology without a clear business purpose is a complete waste. The buildings world needs to heed that lesson, not repeat the mistakes.

Let's take the focus off IP, except as an enabling technology and implementation standard. IP itself is not the answer, at least not to any question a CEO ever asked (John Chambers excluded, of course).

VALUE IS IN THE INFORMATION

The business value of Building-IT convergence comes from information. More specifically, from being able to extract actionable information from operational data. No data, no information—no information, no business value. All the technology infrastructure in the world won't change that. IP connectivity without data is like a superhighway without on ramps.

If you're thinking to yourself, "I can get any data I want out of my building automation system—what's the big deal?" then start considering these questions:

- Can you collect *all* the data, simultaneously, from every point in your building systems, or from just a few dozen (or perhaps a few hundred) points?

- Could you view the operational data for the past year, for any piece of equipment in a building or the physical plant, if you needed it right now?

- Are the data from all operational systems (BASs, meters, utility data, fire and safety, etc.) available in one place, time synchronized for easy comparisons?

- Are the operational data integrated with other business systems, such as space planning systems and CMMS?

- Does everyone who could benefit from the information—inside and outside facilities—have access, and is it organized to meet each user's individual needs?

- Are the IT applications (designed for the business of running a facility) in place to achieve the productivity gains, cost savings, and other business benefits possible?

If you answer "no" to most of these, you're not alone. After talking to well over 300 facilities people in the past two years, less than two percent are doing anything to address the need for data. But those two percent are reaping the rewards.

MASSIVE PRODUCTIVITY GAINS

It is often said that if you can raise the productivity of the entire workforce by just one percent that the benefits far outweigh energy/operational costs to make that happen. While conceptually interesting, these arguments typically have enough holes in them to vent a boiler room. Not what we're talking about.

Instead let's look at making dozens of people across facilities and maintenance organizations more productive by 70, 80, even 90%. Take a senior engineer for example...

A Web-based interface to the control system means the engineer can operate from anywhere. What's the value? Well, it means they can override a setpoint from their living room while watching "24" and wondering why Chloe has schematics to every facility in existence, but they still can't get "as built" drawings for their latest building. Convenient? Yes. Did it change what they could do, or their productivity? Only a little—there is some value to not having to return to the control room for everything.

What did that same engineer do all day? They spent four hours trying to collect data from various sources to do some analysis. They took spotty data from control systems, data loggers, and threw in some estimates, combining six spreadsheets so that the time-stamps matched and they could finally do the analysis. Then they did 20 minutes of actual engineering. This, unfortunately, is the norm.

Stop Wasting Time

Whether it's internal staff or a contracted engineering firm, engineers spend 4 – 12 minutes collecting data for every minute of actual engineering. It is such an accepted way of life that organizations don't even realize how much time is wasted. If the building had something to say, it reverses that ratio, improving our engineer's productivity by as much as 90%.

That's just the tip of the iceberg. There are dozens of commonly performed tasks, ranging from simple equipment information requests and performance measures to complex financial analysis and energy audits, where the productivity gains can be multiple orders of magnitude in scope when all the data are available.

NEW VALUE FROM OLD DATA

A permanent record of facilities operations is an asset, just as the physical structures are. Its value is not just in having history, but in how it can be used. When all the data are available, the building has a lot to tell about past, present, and future operations.

First, a simple case. For diagnostic purposes, the data values at the time of equipment malfunction, or after the hot/cold call came in, is of minimal value. The historical data leading to the problem is where the information lives to identify and fix the root cause. The old data deliver new value by way of solving today's operational problems.

But there is more that you can do. One great thing about collecting building data into an IT application is that you can do things with it without interfering with ongoing operations. You're not limited to just mining the data, you can add to them. You can build calculations on top of the raw data. Instead of building models based on engineering assumptions and design specs, run those same equations against actual operational data. Want to change the model? Go ahead and run it again. Compare the two results. Manage cost, consumption, comfort. Normalize for weather or inflation. The beauty of having a complete operational record in a data warehouse is that you're not limited to analysis or modeling going forward, but you can also apply them to the past. With the data, buildings have an endless supply of information to tell you.

ENSURING YOUR BUILDINGS
HAVE SOMETHING TO SAY

Unfortunately, it is still hard and/or expensive to get data out of the underlying building systems and accessible through IT applications. There are proprietary systems still shipping today. Concepts like "proprietary BACnet" exist. Remember, the data belong to the building owner, not the systems vendor.

Even open systems don't necessarily allow an IT application to collect all building system data. Architectures were developed for control, not information access (a reasonable decision given that control is the system's primary function), which sometimes results in the case where you can collect data from any point, but not from all points. Don't forget, most existing buildings aren't equipped with the latest open technologies; they have systems that are a decade old.

While overall this situation is slowly getting better, there are vendors doing the open systems Moonwalk—taking steps with the illusion of going forward while actually moving backwards—making data harder to collect. To cover up their shortcomings, some manufacturers will question why you need the data or disparage the cost of collecting and storing data (today's cost of storage is insignificant). Don't be fooled.

ANY VERSUS ALL

We hear it frequently—especially with newer, open control systems—you can access any point in the system or trend any point in the system. Actually, that's been mostly true for a while. Most DDC systems can view or trend any point, even those systems we now refer to as legacy. That works OK for control systems, but falls totally flat when it comes to building information systems.

Compare "access to any data" with possessing all the data, all the time, for every point. The historical record of how systems operate, how they interact, how they respond to various external conditions such as weather and occupancy, creates the foundation for a facilities business system.

Think about other business systems. What would a sales information system be like without all the sales data? Could you run a retail business with only today's sales data? How about historical data, but from only two percent of the stores? These fit the "access to any point" or "trend any point" information model, but it's absurd to imagine running a $50 million retail business this way. So why run a $50 million facilities operation with "any" data?

In contrast, look at a sales system with all the data. You can trend and analyze sales in any dimension, look at correlations with weather, see how different stores perform compared to each other or to industry benchmarks, identify the impact of exemplary (or poor) performance. In short, you can make informed business decisions and ensure that stores are run the way you intended. Now, change "sales" to "operations" and "stores" to "buildings," and the same is true for running a facilities business.

SEPARATE REPRESENTATIONS, SEPARATE DATA

Now that we've established the business need for *all* the data, does that mean you should turn on trending for every point in your building systems? No. Odds are that will bring the control function to its knees. Instead, what is needed is an information system that extracts data from the control system into a separate data warehouse—an IT representation of the building. By doing so you take the burden for data collection and data management off the control system, which isn't designed for it in the first place, and create the basis for a facilities business system.

This approach is necessary even for the few control and metering systems that are capable of logging all the data into their own database. Metering systems are used to trending a lot of meter points, but can you afford to meter everything—every fan, pump, cooling coil, etc.—not a chance. But it's not just a cost issue. There are several other advantages to separately collecting data for business applications:

- Many control systems only keep data for a few days or a few months. This way you can keep the data forever, creating a historical record of how your facilities operated.

- All prior data are available, no matter when you need it or what you need it for.

- You can combine and correlate data from many different building systems, or even other business operational systems, for analysis.

- The data warehouse architecture is not tied to the needs of the building systems, enabling a different set of business applications.

- An unlimited amount of modeling and analysis is possible using actual past operational data to improve future operations, designs, and financial performance.

- The data become accessible by a wide variety of users across the enterprise and external contractors, instead of limited to control system users, increasing the knowledge base of the entire facilities organization.

USING IT TO ADD BUSINESS VALUE

The advantages of a separate building representation, collected into its own data warehouse are examples of how an information technology (IT) system adds business value. They are built around the business needs of their users, supporting the way you run your business, but also facilitating the transformation from service delivery to operating as a multi-million dollar business.

Don't assume the word "business" only refers to finances. Managing a facilities operations business requires tackling many issues—service-level agreements (explicit or implied) for comfort, health and safety, utilities services, building maintenance, and several hundred employees to manage. There are new construction projects, renovations, and qualitative issues such as accountability and credibility. And of course, there is the financial side of managing utility costs, budget cycles, capital requests, etc.

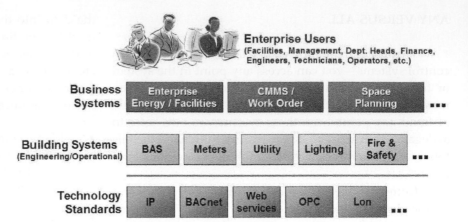

Figure 4-1. Business systems exist at a level above the engineering- and operations-oriented building systems. They serve different uses and functions, and require separate technology architectures to deliver their intended business value.

IT systems can completely change how you approach every one of these business challenges. The data collection discussed above is just the first step—a critical step, but only the first of many. The IT system must organize data to make it easy for end-users to extract actionable information. The organization must be flexible to enable many different views, as diverse users have unique needs. These organizational requirements inform the data representation, which conversely allows or limits the flexibility of information organization and presentation.

In the end, the business value from an IT-based building system comes from the ability to make better decisions, increase productivity, improve customer service, and other business. The IT representation needs to translate these high-level benefits into detailed implementation decisions. For example, how data trees are constructed, what reports exist (and what data feeds them), the level of interactivity for ad hoc analysis, the ability to layer calculations on top of raw data, are all defined by varying users and uses of the information system.

The information to transform your facilities operation into a business is not possible with today's control systems, metering systems, or other engineering-level applications. This is why you need to collect all the operational data separately—to build a business-level system (see Figure 1) capable of delivering entirely new levels of value.

REAL-WORLD EXAMPLES

Enough with the theory. Let's look at some typical examples where the world changes if you have an information system in place.

Hot/Cold Calls

You get a hot or cold call. What happens? In most cases the technician that responds will check the current conditions, adjust a thermostat or override a setpoint, and that's about it. If the room has habitual problems, perhaps there are a few trend logs running, but what data is available?

An information system provides operational details for every room (or zone) in the building. The technician not only sees current conditions, but also knows how long the space has been uncomfortable, how well the air handler is running, how the terminal box is operating, cooling and reheat valve behavior, etc. In most cases, it only takes five minutes to determine the real cause of the comfort problem so that the proper correction is made the first time.

Taken a step further, facilities has the information to show its customers what happened. Historical data showing the space becoming uncomfortable, the extent of which is measured in a comfort index (calculated from the captured data) that non-engineers can easily understand. The information is there to show when the call was logged, the corrective action taken, and exactly how long it took to become comfortable again.

The IT system helps the business of meeting the comfort obligation, providing fast and accurate customer service, and communicating with the customer in a way that they understand.

Business Metrics

Facilities leaders and executives have entirely different information needs. They need to understand how the business is running—how current operations stack up against last month, last year, industry standards, and organizational goals. Utility bills are a terrible way to manage the business. Too few data points that are far removed from the time they account for.

A big advantage of separating the collected operational data from the control systems is the additional processing that IT systems can perform. Take a supply fan, for example. Most control systems can tell you if the fan is on or off, and the percent of full-load amps. That is all you need to calculate the kWh for the fan. The manufacturer's specifications will supply horsepower, motor efficiency, and any other necessary parameters. Add utility rate information and you can report the cost in dollars/hour at any point in time.

Similar capabilities exist to measure the energy consumption of pumps, heating and cooling units, exhaust fans, or any other mechanical systems. These are the building blocks that most facilities executives only dream about. With them you can produce reports that show the metrics of your choice for the building, broken down by air handler and each piece of equipment, or broken down by floor, zone, or room. Metrics that normalize across multiple buildings, such as MBtu/SqFt become simple. Accounting for weather is similarly easy. The information is there to see what's running well and what isn't, what's improving and what's not.

The ability to set and meet concrete goals, measure successes and document issues, and prioritize work are just some of the real-world advantages a true information system provides that control systems cannot.

Growing Staff Knowledge

Some facilities organizations perceive adding an information system as yet another thing to do. Who has time to look at the information? The short answer—almost everyone.

A major goal of IT systems is to deliver the right information to the right users, and do it fast. Instead of spending four hours in spreadsheet hell, you can spend 40 seconds pulling together the information you need. You can, that is, if the underlying data is complete and the system was designed to meet users' requirements. You're not likely to train 100+ users to use the control system. Even if that were easy, you wouldn't want to. IT systems are designed to handle hundreds, even thousands, of different users' information needs.

When more people start using information to understand how the buildings and systems they work with daily actually function, good things happen. Issues are caught before they become problems. Communications improve. Learning happens. Staff can see the impact of their work. The total knowledge base of facilities team rises. Growing the staff's knowledge is good business as tomorrow's challenges will be greater than today's.

CONCLUSIONS

How can a building be intelligent if it has nothing to say? It can't.

How does this happen, that buildings thought to be intelligent sit there quietly, saying nothing? By focusing solely on the technology. By forgetting that it's the information that provides business value. By assuming that the data is always available.

Intelligent buildings must talk. The business value is only achieved when they share what they know, communicating between building systems and with their owners. They do this through data. Without the data you limit the building's intelligence and you limit what you can accomplish. Technology infrastructure doesn't change that.

Architects say that form follows function. The IT equivalent is that technology follows business function. Otherwise, you get technology in search of an actual problem to solve. This has happened many times in the past and unfortunately, is doomed to repeat itself. In the end, it's always the same result—no business value equals eventual failure.

Building control systems exist at the engineering or operational level. They are not designed as business systems, nor should they try to be as their primary function is of critical importance. The control system representation of a building is different from an IT representation at both the technical and end-user level.

IT systems, when done properly, have changed the way businesses function. Every other function in a company/institution has changed over the past 25 years due to the availability of information—finance, administration, sales, customer service, marketing, manufacturing, R&D, distribution, you name it.

Only facilities operations has yet to take this step. It's a step that requires a business mentality and a new approach in both technology and management.

References

Gnerre, B.; Fuller, K. "How Can a Building be Intelligent if it has Nothing to Say?" AutomatedBuildings.com; Sinclair, K., publisher; May 2006

Gnerre, B.; Fuller, K. "Stop Trying to Solve Business Problems with a Control System" AutomatedBuildings.com; Sinclair, K., publisher; June 2006

Chapter 5

Ten Technology Takeaways for 2006

Ken Sinclair, Owner
AutomatedBuildings.COM

INTRODUCTION

I HAVE JUST RETURNED from the largest ever AHR Expo in Chicago which was a kaleidoscope of change. The energy and feeling of the transition into a new era was everywhere. Industry synergy abounded and the feeling of all the pieces fitting together was never greater. I wish to share with you 10 observations that I formulated while at this event. The exposition is a true melting pot for our industry which is converging in several different ways and several different directions simultaneously. If you were unable to attend, put this on your "must be there list" for 2007 in Dallas and 2008 in New York City.

2006 AHR EXPO CHICAGO SETS NEW RECORD
AS BIGGEST EVENT IN 76-YEAR HISTORY

MY 10 TECHNOLOGY TAKEAWAYS
IN NO PARTICULAR ORDER

#1: BACNET INTERNATIONAL—A FRIENDLY POWERFUL GLOBAL COMMUNITY—WAS BORN JUST BEFORE THE SHOW

All the hard work of achieving a consensus standard has come to a brilliant finish. The energy, no the synergy in the BACnet International booth was amazing, the sense of industry cooperation with the demonstration of a myriad of connected manufacturers products. The notion that was obvious was the fact as each of these manufacturers innovates with new BACnet products and services they add to the strength, the reach plus the cost effectiveness of their international community. Another feather in BACnet International cap was the support by the major control companies of their inter-

national standard. This is discussed in **The New Face of BACnet** below.
EMAIL INTERVIEW—Ben H. Dorsey III & Ken Sinclair
Ben H. Dorsey III, VP of Communications KMC Controls, Inc., and newly elected Marketing Committee Chair for BACnet International; http://www.bacnetinternational. org/

THE NEW FACE OF BACNET

We will be a resource to the greater body of BACnet users, product providers, service providers, and building professionals.

Sinclair: I saw a new look for the BACnet Community at the recent AHR Expo. Can you explain what organizational changes have occurred?

Dorsey: Just before the opening of the show, members of two former organizations voted to consolidate and rename their collective assembly. BACnet International was born. The two organizations had been known as the BACnet Manufacturers Association (BMA) and the BACnet Interest Group of North America (BIG-NA). The former boards of these organizations have now been combined into a single and smaller board that will reflect the diversity of membership and the agility and responsiveness that will permeate the new organization.

Sinclair: What brought about the change?

Dorsey: Some time ago we began to see the limitations of our two organizations, especially in regards to reaching a greater audience of those interested in the implementation of BACnet. The BMA, as its name suggests, was limited to those providing products or services to the industry. And the BIG-NA simply lacked the resources to effectively communicate to the entire community of users. So, we began discussions many months ago to redefine our joint interests and goals. At the BACnet Conference and Expo, held last October, we put the matter before members of both groups as a proposed plan and sought further input. Obviously, there were necessary legal steps and some agreements that had to be reached. Then, on January 22, we voted to formalize the decision.

Sinclair: What do you hope to achieve as BACnet International?

Dorsey: We will be a resource to the greater body of BACnet users, product providers, service providers, and building professionals. The new organization will begin to take on the diversity of all interested parties. Education and training will become a major focus. And, because our membership is now open to a greater audience, we expect to grow as well.

Sinclair: Who is now eligible for membership?

Dorsey: In actuality, a scalable and flexible membership structure attracts all interested parties. There are different levels of membership for manufacturers and service providers, for example. We also anticipate that some building owners, facility managers, and consulting engineers will see the benefits of membership and join our ranks. We've even created a membership for BACnet sites, facilities whose owners wish to receive recognition for the integration or interoperability showcased therein.

Sinclair: You mentioned education. Can you explain further?

Dorsey: The educational resource we will provide will vary by the intended audience. At the recent AHR Expo, for example, many wandered into the BACnet International booth simply wanting to learn more about what BACnet is and what it could do for them. Then, there are those seeking product information, or news about the work of the ASHRAE Standing Standard Process Committee (SSPC-135) which maintains the standard. We'll provide the necessary resources for all such audiences.

Sinclair: I'm glad you mentioned the BACnet International booth at the AHR show. I was extremely impressed by the cooperation I sensed there. How was it achieved?

Dorsey: I'm glad you noticed. Our booth included kiosks for 12 different member companies of the new organization. Quite frankly, we would have had more had the space not limited us. In any event, as you intimated,

many of us who were there are direct competitors. Yet, we have become united by a common purpose: to advance the implementation of BACnet. To the extent we do so, all of us will benefit. By the same token, I have been impressed by the level of cooperation I've experienced on the marketing committee of the former BMA over the last two years in which I've been involved.

Sinclair: Will there be any changes to the product certifications previously administered by the BTL (BACnet Testing Laboratories)?

Dorsey: Yes and no. The BTL will continue to represent the mark of quality in BACnet products. And, the BTL will continue to provide rigorous testing for all BACnet device profiles in terms of compliance to the standard and interoperability. What is changing is that the BTL will be governed by BACnet International. In fact, a new test facility partnership is now being established and should be operational in the second quarter of this year.

Sinclair: Any early indications of how the new BACnet organization is being received?

Dorsey: Well, if response at the AHR Expo was any indicator, we'll have bright and busy future. The BACnet International booth was constantly busy. Attendees had lots of questions—some intended for specific vendors and others more general in nature. We secured contact information for about 150 individuals and handed out dozens of applications for membership. What seems clear to me is that the interest in BACnet continues to escalate and that all parties have been seeking a definitive resource for BACnet information. BACnet International is now that resource.

Editor's note:

A new web site is in development for BACnet International (www.bacnetinternational.org). In the meantime, those interested in further information can call or email the organization (312.540.1200; info@bacnetinternational.org).

#2:   WIRELESS WILL WIN

Lots of real \ products at this year's Expo. The picture on the right shows several vendors' equipment having a wireless chat in a casual gathering on the table at the Kiyon booth. AHR "Innovation Award" Winner—Kiyon led the wireless way

EMAIL INTERVIEW—John Edler & Ken Sinclair
John Edler, VP, Kiyon Inc.
To view product go to Kiyon Booth 3829 in Chicago
Wireless BACnet Router, winner AHR Expo 2006 Innovation Award.
John Edler is VP for Building Automation at Kiyon.

Three factors that differentiate Kiyon are standards compatibility, ease of use, and state-of-the-art wireless network technology.

Sinclair: You must be pleased to win the AHR "Innovation Award." Tell me about the product that won.

Edler: It's a great honor to win the Innovation Award and be recognized by peers in the industry as having done something special. The AHR show is always a great event, attracting the best companies in the industry and customers from all over the world. With many new products introduced each year, we feel this award not only recognizes our product, but also the strong demand for wireless in the building automation industry. Our KAN 254B Wireless BACnet Router is being well received in the market. It is the first BACnet wireless mesh network solution in the industry and converts all standard field controllers or supervisory controllers using BACnet MSTP, BACnet IP, or Ethernet IP to a wireless mesh network. We have tested our products with all the major vendors either individually or at the BACnet sponsored Plugfest. Since the KAN 254B is broadband, it can handle many MSTP and IP devices simultaneously on the same network.

Sinclair: How is Kiyon's wireless product different than others in the industry?

Edler: Three factors that differentiate Kiyon are stan-

dards compatibility, ease of use, and state-of-the-art wireless network technology.

First, Kiyon utilizes standards compatibility. From watching the market place, we observed that proprietary products limit customers on how they can apply them. By using 802.11 a, b and g, we are taking advantage of the same broadband technology applied by companies such as Linksys and D-link. This allows us to be compatible with laptops, PDA's and a host of other mobile devices being used to enhance worker productivity. By staying compatible with off-the-shelf hardware, we are able to provide lower costs product platforms and a wide range of choices for the customer.

Second, we have focused on ease of installation and operation. You don't have to be an RF engineer to install and operate our wireless network. You simply connect your BACnet devices and power. Everything else is handled for you. The KAN 254B's automatically find each other and dynamically determine the best path for routing communications, adjusting in real-time for interference and roaming. Kiyon's self managing network features fast-failover alternate path routing to maintain high connection reliability and fills-in coverage shadows common to access point type wireless networks. It also uses distributed intelligence in that each node is aware of all other nodes in its area so that they manage traffic without need of centralized switches. We also provide network monitoring software that shows the current status of the entire wireless network. This makes network setup and on-going network operations easy for a typical building installer and operator.

Third is the state-of the-art technology we put into our wireless mesh network. BACnet MSTP has timing constraints that would normally prevent it from being used with a wireless networks. We have developed some unique ways of handling the communications that make it compatible with BACnet specifications. Also, as data is communicated over any wireless network, the bandwidth normally goes down with each hop. We have developed methods to maintain the highest bandwidth in the industry, even after several hops to provide users greater flexibility in extending network range without wires.

Sinclair: What markets are you targeting?

Edler: We are focused on in-building broadband applications. Three specific markets we are focusing on are Building Automation, RFID, and Wireless Broadband. These markets all deal with harsh RF indoor environments that greatly benefit from our high reliability, low cost per node and high user density applications.

Sinclair: What Building Applications can the KAN 254B be applied?

Edler: KAN 254B can be applied to Building Automation such as VAVs, Field Controllers, and Supervisory controllers. It can also be used for security systems, video cameras, lighting systems, fire, and internet applications. Our customers have applied it in many types of buildings and even to connect buildings together when running wires were prohibitive. Wireless is ideal for retrofit applications. Although it was designed for traditional building automation applications, our creative customers continue to come up with uses we never thought of. For example, our wireless network provides the ability for mobile troubleshooting. By using laptops or PDA's, a user can connect anywhere in the building to monitor, troubleshoot, or upgrade software. This provides a lot of flexibility to building operators.

More information about this state-of-the-art wireless mesh network can be found on: www.kiyon.com.

There are many other wireless products like Tour Andover's wireless BACnet extender (more to follow on that) and of course the announcement of Siemens *First Wireless Building Automation System*.

Siemens Introduces First Wireless Building Automation System

BUFFALO GROVE, IL—Siemens Building Technologies, Inc., announced the introduction of the industry's first wireless building automation system. Based on proven mesh topology for reliability, APOGEE® Wireless field level networks are ready to provide facility builders, owners and managers increased occupant comfort, greater flexibility, and optimized, efficient control of heating, ventilation and air conditioning (HVAC) systems.

"Our customers continue to rely on us to provide innovative building automation technologies that deliver true value to their operations," said Rick LeBlanc, Sr. vice president, and head of Siemens Building Automation Division. "While the potential of wireless should be obvious, the technology only has value to our customers if proven and ready for the rigors of real-world application."

Ready for challenging building environments, Siemens APOGEE Wireless ensures network integrity by creating multiple redundant paths of communication. A mesh topology field level network is inherently self-healing, so users won't have to worry about losing communication with control devices across the building automation system.

"Simply put," said Jay Hendrix, manager, Wireless Solutions, "the network can't be compromised because the signal is automatically able to circumvent obstructions and find its target."

It's widely acknowledged that wireless building automation can have a far-reaching impact on facility performance, cost efficiency and return on investment. Operational synergies begin at implementation because wireless environments easily adapt to changing business needs or new facility requirements. Also, removing wiring and its related hassles accelerates installations and simplifies retrofits and system extensions.

A true "future-proof" technology, wireless-based systems offer building owners and facility managers more choice and fewer constraints, including: Simpler more flexible system design; Faster, less disruptive installations and retrofits; and Smoother, less costly migrations staged to accommodate budgets and schedules. With APOGEE Wireless, owners and builders can increase and protect the value of their facility investment, improve the overall marketability of their properties and take advantage of opportunities brought by future technologies.

As a leading provider of building controls, fire safety and security system solutions, Siemens Building Technologies, Inc., makes buildings comfortable, safe, productive and less costly to operate. The company focuses on improving the performance of its customers' buildings, so that its customers can focus on improving their business performance. With U.S. headquarters in Buffalo Grove, IL, Siemens Building Technologies employs 7,500 people and provides a full range of services and solutions from more than 100 locations coast-to-coast. Worldwide, the company has 29,000 employees and operates in more than 42 countries. Visit www.sbt.siemens.com for more information.

In the Building Intelligence Tour all day educational seminar Wednesday, (for more insight read; "The Time is Now for Intelligent Buildings—Learn why at AHR Expo" also see Building Intelligence Tour at AHR Expo 2006 Chicago) the discussion was hot about how wireless would dominate the last 300 feet adding thousands of points of intelligence to our systems. It was interesting to see that the structure cabling folks support wireless as it increases the need to have a structured back bone to accept all this new data that will be wirelessly collected.

The Building Intelligence Tour was a great success with over 100 folks showing up for this full day event. Kudos to both Paul Ehrlich, P.E. "The Time is Now for Intelligent Buildings—Learn why at AHR Expo" also see Building Intelligence Tour at AHR Expo 2006 Chicago).

A special thanks to all who volunteered their time and resources to bring this tour of intelligence to Chicago.

This one-day educational seminar gave attendees a chance to network and learn directly from industry experts. The seminar included in-depth talks, panels, and workshops with great examples of projects from around the world, details on technologies, current research, value propositions and economics. Leading suppliers who sponsored the event had a mini trade show in the coffee area. We all left with a solid understanding of the benefits of being involved with Intelligent Buildings. More information on the seminar including a complete agenda can be found at www.buildingintelligencetour.com

After Paul set the scene and defined what is a intelligent building and why they are cost effective, the day started with technology visionary, Jim Young as the keynote speaker. As the President of RealComm, the leading event for real estate and technology, Jim has traveled around the world researching next generation and intelligent buildings. BOMA International views RealComm as their source for information on technology. Jim shared his findings about the need for intelligent buildings, warned us of how far we are falling behind and challenged us if we did not believe him to take his Realcomm 2nd Annual Asia NextGen Best Practices Tour May 5-16, 2006. He talked about how other countries use the cell phone as a credit card and many other examples of how we have become digital lagging.

I copied this picture from his presentation as the one to show where Asia is in comparison to North America.

This was just one of many pictures that drove home the point that we live in the land of dumb buildings. We talk about intelligent buildings but do not build them. We must work harder to correct this or we will fall even further behind.

Next was a Case Study: The Future of Integrated Buildings: From Myth to Math

Deploying a whole-building IP infrastructure, the client model showed how the integration of building automation, structured cabling, Voice over IP Telephony, Digital Signage and fully Networked IT applications produces a more financially attractive building. The model showed how Net Operating Income, Asset Valuation and Shareholder Values all improved, while concurrently yielding impressive Rates of Return and becoming more competitive at Retaining and Attracting new tenants.

Presented by: Doug Smith, Richards Zeta and
Tom Shircliff, Intelligent Buildings

This was a great presentation but the deeper message for me was that the true power was in the team.

Next I was moderator for the **Intelligent Building Technology Panel.**

Intelligent buildings become possible due to enabling technologies. The panel featured experts highlighting several of these technologies. Panelists were: BACnet's Jim Lee; LonMark's Jeremy Roberts, LonMark International; oBIX's Tim Huneycutt, GridLogix; Internet Technologies' Dave Clute, Cisco Systems; Structured Cabling's Jed Barker, Panduit; Wireless Mesh Networking's Rob Conant, Dust Networks.

The session went very well with lots of information exchanged and the interaction from the floor was great.

See what Bob Metcalfe chairman and interim CEO for Ember Corporation has to say in our interview. Ember and STMicroelectronics partnership boosts the

ZigBee market:
EMAIL INTERVIEW—Robert M. Metcalfe & Ken Sinclair
Robert M. Metcalfe, Chairman, Ember Corporation
Bob Metcalfe is chairman and interim CEO for Ember Corporation as well as a high-tech venture capitalist at Polaris Venture Partners in Waltham, Massachusetts. He serves on the boards of Polaris-backed start-ups including Narad, Ember, Paratek, SiCortex, and Mintera. In 1973, he invented Ethernet at the Xerox Palo Alto Research Center. In 1979, he founded 3Com Corporation and took it public in 1984. During the 1990s, he wrote a weekly Internet column in InfoWorld for over 500,000 information technologists. Metcalfe graduated from MIT, got his PhD from Harvard, taught at Stanford and Cambridge, and was elected in 1997 to the National Academy of Engineering. He received the National Medal of Technology in 2005 and is a Life Trustee of MIT.

Ember and STMicroelectronics partnership boosts the ZigBee market.

Our mission is to network embedded micro-controllers with standards-based radio semiconductors and protocol software.

Sinclair: What is the recent partnership between Ember and STMicro all about?

Metcalfe: Ember and STMicroelectronics announced what may be the first reciprocal second source and co-development partnership for ZigBee/802.15.4, the embedded micro-controller networking standard. STMicro, one of the world's largest semiconductor companies, will work with Ember to jointly develop complete semiconductor solutions for the fast-growing ZigBee wireless networking market.

Sinclair: What is ZigBee, particularly in its relevancy to building automation?

Metcalfe: ZigBee is a wireless, standards-based radio technology that addresses the unique needs of remote monitoring, control and sensor network applications. It will play an increasing role in enabling embedded networks for building automation. ZigBee is not just low-power radio standards (802.15.4), but also wireless mesh radio protocol stack standards.

Ember is short for "embedded radio." Our mission is to network embedded micro-controllers with standards-based radio semiconductors and protocol software. What Ethernets did for PC networking is what we plan to do for embedded networking of HVAC, lighting, refrigeration and access control, to name a few building automation applications.

Sinclair: Where does ZigBee and embedding networking fit into existing networking technologies?

Metcalfe: Embedded networking is what venture capitalists call a huge "space." The top dozen micro-controller suppliers finely divide 72% of 2004's $12 billion market. And now the embedded networking space is beginning to develop some structure. Let's call the structure "CSI," for (c)ontrol, (s)ensor, and (i)dentity.

Above CSI embedded networking, WiFi for PCs and Bluetooth for cell phones require megabits per second and batteries recharged every day.

Just below WiFi and Bluetooth is CONTROL networking for embedded micro-controllers. Control is where Ember mainly plays, with its ZigBee/802.15.4-based radio semiconductors and mesh protocol software. Building control network nodes typically require lower speeds, for example ZigBee's 250Kbps, among many battery-powered micro-controller network nodes, with batteries lasting out beyond five years.

Below ZigBee/802.15.4 are SENSOR and IDENTITY networks (RFID). Roughly speaking, sensor network nodes are less capable than control nodes, often waking up intermittently to report changes. Identity nodes are less capable than sensor nodes, often having no micro-controller and even no batteries—getting their power over the air.

Of course it's quite early in embedded networking, and this CSI structure for the diverse embedded networking space may not emerge cleanly. For example, it is possible WiFi and Bluetooth personal networking technologies will move down into embedded CSI applications. Or, sensor and identity technologies will move up, giving Ember and ZigBee/802.15.4 some competition in control networking. Or vice versa. Ember ZigBee-based networks are already being deployed not only in building automation applications, but also sensor applications, and even some higher-end identity applications.

Invading these huge spaces is on the mind of semiconductor suppliers and their customers everywhere, including especially now STMicro.

Sinclair: Why is this STMicro different from other partnerships?

Metcalfe: STMicro is not Ember's first embedded networking partner, nor will it be the last, but ST is certainly Ember's deepest partnership so far, by far. Our non-exclusive partnership is quite different from Texas Instrument's swallowing up of Chipcon last month. Ember's strategy is not to compete with the suppliers of the world's 10 billion micro-controllers. We do not think that's what the users of micro-controllers need. We think micro-controller users need us to partner with the many and varied existing suppliers to get them ZigBee/802.15.4-enabled, or as we say in Boston, to get them Ember-enabled.

So Ember remains independent of STMicro and intent on working with all micro-controller vendors, including TI. We intend our partnership not only to benefit our respective customers, but also—through "coopetition"—to accelerate development of the whole ZigBee market. Our products will be reciprocal second sources. The ZigBee market is ready for and now has its first semiconductor and software second sources.

Sinclair: What do manufacturers and integrators of building automation systems stand to gain from this partnership?

Metcalfe: Ember's EM260 ZigBee co-processor will soon be available from ST mostly for their OEM customers and from Ember for everyone else's. For example, Siemens, Hitachi and Andover Controls are all developing control systems based on our technology. Our central offer to these customers is to put the EM260 next to existing micro-controllers and connect them up with the standard serial interface. We are offering an open-source programming interface to the Ember ZigBee Serial Protocol (EZSP). Code running on application micro-controllers designed for building controls can use EZSP to talk to their EM260s and thereby become ZigBee-enabled.

Watch Wireless Win in 2006!

#3: HONEYWELL KEEPS TRIDIUM'S NIAGARA PATH FROM FIELD TO ENTERPRISE OPEN

A concern by the complete industry of what would happen to the open path from field to enterprise provided by Tridium was answered in last month's interview "The Recent Acquisition of Tridium by Honeywell" by John Petze, President & CEO and Marc Petock, Director of Marketing Tridium.
EMAIL INTERVIEW—John Petze, Marc Petock & Ken Sinclair
John Petze, President & CEO and Marc Petock, Director of Marketing Tridium

The recent acquisition of Tridium by Honeywell Our customers should see a renewed commitment and investment along with the same great technology and support.

Sinclair: Why did Honeywell acquire Tridium?

Petze/Petock: Honeywell made the investment because it believes in our technology, people and business potential. The company believes in the vision and technology that has been guiding Tridium for the past 10 years. With Honeywell, we are in a good position to accelerate our vision and expand our growth. Furthermore, there are significant opportunities to implement the Niagara Framework and our technology across several lines of businesses within Honeywell's operating units.

Sinclair: What do you see as the immediate benefit of this relationship?

Petze/Petock: Tridium will become a stronger company. We see many exciting opportunities that will enable us to accelerate the development efforts that have been underway on our product roadmap, such as lighting, security and the rapidly growing machine-to-machine market. We will be able to enhance and expand our products to better serve all of our partners and customers. Honeywell's global position will clearly help us to accelerate our growth internationally. We believe the company will flourish as Honeywell is fully committed to Tridium's vision of the Niagara Framework as a pervasive, open software platform for building automation and device-to-enterprise applications.

Sinclair: How will Tridium align itself within Honeywell?

Petze/Petock: Tridium will operate as a separate business entity. Our leadership and employee base will stay the same. The visionary team that conceived of, developed and delivered our exciting technology will remain focused on enhancing and pursuing broader opportunities and applications. Our strategies and development paths will continue as we have planned.

Sinclair: Will there be any changes to Tridium's brands and products?

Petze/Petock: No. Our multiple channels along with the Niagara and Vykon brands will remain and will continue to grow with their respective customer bases.

Sinclair: Will your customers see anything different?

Petze/Petock: Our customers should not see any change at all. We are committed to maintaining and growing our relationships with every single customer and potential customer. If fact, our customers should see a renewed commitment and investment along with the same great technology and support.

Tridium is the inventor of the Niagara Framework®, a software framework that integrates diverse systems and devices—regardless of manufacturer, or communication protocol—into a unified platform that can be easily managed and controlled in real time over the Internet using a standard web browser. The general the feeling by most was that this open path will not become a closed corporate standard and that the large Niagara community will strive to keep it open. Growing the Niagara open community is of benefit for both Honeywell's Comfort Point Control and Enterprise Integrator product, both of which provide a valuable bridge to bring their legacy systems into the new era. It should be a good marriage.

EMAIL INTERVIEW—Mike Taylor & Ken Sinclair
Mike Taylor, Vice President of Marketing, Honeywell Building SolutionsMike Taylor is the Vice President of Marketing for Honeywell Building Solutions. In this role, he manages strategy, marketing, product management and training for service offerings to the commercial buildings market. Mike has been with Honeywell for more than 25 years, and has held a variety of sales, marketing and general management positions. Mike has a bachelor's degree in industrial distribution from Texas A&M.

Honeywell Acquires Tridium

Honeywell invested because we believe in Tridium's technology and its people.

Sinclair: Why did Honeywell acquire Tridium?

Taylor: Tridium has developed a proven and highly successful software platform that cuts through the barriers of connecting and operating multiple systems and devices by providing easy, real-time communication between devices in homes, commercial buildings and industrial applications. We don't know of anyone else offering such cutting-edge communications and data-gathering capabilities.

Honeywell invested because we believe in Tridium's technology and its people. We can help expand the company's vision and business through the implementa-tion of their technology within several of our business lines, more aggressive investments, and rapid access to new industries, vertical markets and large, global customers to its core business.

Sinclair: How will this acquisition affect your business?

Taylor: Their technology has broad applicability across the Honeywell Automation and Control Solutions (ACS) businesses. For example, Honeywell Building Solutions (HBS) is using the Tridium Niagara^AX Framework in Honeywell ComfortPoint—our new BACnet-compliant, Web-enabled control system.

Sinclair: Tell me more about ComfortPoint. What applications is it targeted at?

Taylor: ComfortPoint is a best-in-class HVAC control and energy-management system consisting of a family of controllers and a simplified interface for monitoring and managing mechanical and electrical equipment. The system is designed to optimize occupant comfort and reduce operational costs for a single building, a campus or a multi-location operation. ComfortPoint is native BACnet and has been listed by the BACnet Testing Laboratory (BTL).

Operation of the system is simple. That's because it is accessed via a single interface—regardless of whether the user is an administrator who wants to change the occupied schedule for a particular floor or a control engineer who wants to change the sequence of operation for a VAV system. Menu choices are customizable based on user type, authorization level and the like.

The interface also is the same whether the user is on a desktop PC in the building or on the Web half way around the world. The Niagara framework makes this possible by transforming data from the BACnet devices into uniform software components and enabling the devices to communicate with the Web-based interface.

This offering is really targeted at contractors and consulting engineers who are looking for an easy-to-use, cost-effective BACnet control platform for HVAC applications.

Sinclair: Does this mean you're moving away from LON as an open protocol?

Taylor: Definitely not. ComfortPoint just adds to Honeywell's portfolio of open solutions. We provide a LON-based offering using our XL10 line of controllers or a BACnet solution with ComfortPoint. Both are excellent protocols and we're giving out customers the choice to pick the best solution for their application.

Sinclair: What's the difference between ComfortPoint and EBI?

Taylor: ComfortPoint was designed—and offers the most cost savings—for standalone HVAC applications. The Honeywell Enterprise Buildings Integrator (EBI) is an integration platform for bringing together HVAC, fire alarm and life safety, security, digital video, asset tracking, and other building system into a single, powerful user interface. And the two systems are built to work together. Contractors and owners can tie ComfortPoint HVAC controllers into an EBI system—giving them the option of standalone HVAC control or an integrated building management system.

Sinclair: Is ComfortPoint compatible with other manufacturers control systems?

Taylor: Yes. Any building management system that uses BACnet or LON can be expanded with ComfortPoint. This includes control platforms from JCI and Siemens.

Sinclair: What trends do you see having a significant impact in 2006?

Taylor: The first baby boomers are turning 60. And the industry will begin to see a decline in skilled laborers as a result. So facility managers will have fewer people to do the work in an area with increasing responsibility. As a result, technology will play a larger role in keeping facilities running smoothly and efficiently.

Plus, there are numerous benefits to augmenting labor-based services with technology beyond filling a void in the workforce—e.g., lower cost of operations achieved through technology cost-savings and access to real-time, actionable information.

That's the thought behind Honeywell Service Online (HSO), a new suite of online tools, information and system support. Among other features, HSO includes an automated software update service for timely delivery of control system updates and patches. Using built-in intelligence, the service assesses the components and requirements of individual facilities to deliver tailored system updates—Microsoft hotfixes, policy enforcers, etc. This keeps systems current and secure with minimal effort.

It's also why we've invested millions in our Global Service Response Center (GSRC).The GSRC is a centralized dispatching center for service calls. It is one of the most established and technically-advanced monitoring and control facilities in the world. Staffers at the center and our service technicians in the field are using the latest tools to help customers save time and manage their facilities more efficiently.

Technicians, for example, carry handheld computers that provide job-specific information and access to complete service records. After a call is complete, they use the handheld to enter details on the job and the data is uploaded to a database at the GSRC. That information is available to customers in real time through an HSO application called ServicePortal.

We'll see more Web applications, more remote connectivity and more of an emphasis on leveraging the latest technology advancements in the coming years.

#4: SOFT CONTROL OF HARD EQUIPMENT IS THE FUTURE OF NEW CONSTRUCTION

The announcement earlier this year of the Johnson Controls to Acquire York merger reminded us of the Carrier—Automated Logic merger. Carrier Corporation: To Acquire Automated Logic Corporation, Technology Leader in Commercial Controls. Why are hard equipment manufacturers and control companies merging? Two major reasons:

First the new major construction business is not in North America but in India, China, Dubai and other off shore locations. Projects will be looking for complete solutions to HVAC comfort needs that include soft intelligent control of hard equipment, i.e. chillers, air handling systems, terminal units etc.

Reason number two is that hard equipment becomes more powerful and cost effective with soft controls. Look at today's auto industry: it is becoming that the car that has the most soft toys and best integrates with your ipod wins.

Although 2005 was a year of mergers I feel 2006 will be no different. Trane has bolstered their knowledge of soft control but Siemens and Honeywell potentially could be left without access to hard equipment. Schneider Electric may still be in a buying mood so look for more mergers of hard and soft companies.

Open standards and connection to them will play a major roll in the development of the independent hard equipment manufacturers as they work to supply products to fill holes in the product lines for the major equipment suppliers.

#5: RETROFIT WILL DOMINATE THE NORTH AMERICAN MARKET

Most North American buildings have been built, but are in dire need of retrofit to be able to cope with the new era. It is likely that the independent integration contractors will excel in this high touch engineering and labor market. Movements like Gridwisean entirely new way to think about how we generate, distribute and use energy. Using http://www.gridwise.com/advanced communications and up-to-date information technology, GridWise will improve coordination between supply and demand, and enable a smarter, more efficient, secure and reliable electric power system. Green Buildings will fuel the growth in this market. The actual definition of what is our market will increase market size and reach. In this article Defining The Intelligent Building Control Market I provides insight into understanding the true volumetric size of our new industry as we converge in a new era.

Understanding the true volumetric size of our new industry gets much deeper as we converge. Will biosensors that are tied to the security databases be part of our intelligent building control market? I certainly hope so. Who will do the necessary network knitting of the open protocol devices into a coherent strategy for the buildings? Do any of us really understand the width, the depth, and the resulting volume of our potential market in this complex convergence equation?

Defining New Markets

The single most asked question I am asked as editor of AutomatedBuildings.com is, "How big is the intelligent building control market?" Although it is a very complex question, I always defer it by asking this second question: "What do you think is included in this market?" I am often surprised about what the asker thinks is the market. Although the equation was blurred in the past, it is even blurrier now as our industry becomes almost inseparable from the broad brush of the vague term that is information technology.

What do you think is included in our intelligent building control market for the future? Labor and component costs are shrinking with the advent of wireless and open protocols, which are turning controls into commodities. Although this is bad news for the control traditionalist, it brings a whole new marketplace of system integration services to our doorstep.

We are in for a radical change as our industry moves from being wire pullers to system and Web service providers. If traditionalists do not embrace these changes, they will lose their market share. Those that "get it" will experience phenomenal growth. In my "Building Automation" column from the January 2006 issue of Engineered Systems titled, "Virtual Value Vision," I cited several examples of the radical changes our industry will encounter as we morph into these new markets.

Let's start by discussing just what might be part of the new markets. An obvious part that comes to my mind is digital signage: the vision providing the virtual interactive interface of the future. Narrowcasting over this media will bring alive our interfaces with traditional building services. Who will be the supplier of this valuable service? Our industry has the wisdom of what clients and building interfaces require, but there are others that have command of the new technology. Obviously some cross-pollination will be necessary, but if we perceive that this is not part of our market, it clearly will not be.

What else might be a bit grey to our control traditionalist? What about the interfaces that will build on voice over internet protocol (VOIP)? Who will determine how we use the inherited capabilities that come with the convergence to IT ways? This is just the tip of the IT iceberg, as this service connects to cell phones and hand-held data devices, all which transmit our services and information to millions.

Are you confused yet? Understanding the true volumetric size of our new industry gets much deeper as we converge. Will biosensors that are tied to the security databases be part of our intelligent building control market? I certainly hope so. Who will do the necessary network knitting of the open protocol devices into a coherent strategy for the building? That has to be part of our market, correct?

Identifying New Players

Will the virtual architect who uses IT systems and the correct underlying protocols to bring virtual value to the enterprise be part of our market? Will the affected community? Yes. What about virtual value mentors who will show us the available virtual systems and information so we can understand how to best use these new ways to provide our own optimized and personalized best virtual value? Surely they will be part of our industry.

As intelligent buildings evolve to intelligent communities and these communities become intelligent cities, intelligent electrical grids, and even intelligent countries, will our folks be part of this market? In my new world they are. Not all of these people are in our industry yet, but they will come and clearly redefine what our industry and our market is. They will come because the intelligent building control market is something everyone wants to be part of; it just needs to be redefined.

The Web will continue to lead the way with strong examples and models to follow for adding virtual value to everything in our industry. Most of the methods of presenting information now exist on the Web: on-demand video, sound, radio, news feeds, narrowcasting, PowerPoints, PDF, Flash presentations, etc.

Help me, and the industry in general, redefine what the intelligent building control market is.

#6: **LONMARK LEADS IN RE-EDUCATING OUR INDUSTRY BUT WE ALL MUST INVEST IN EDUCATION TO MOVE TO THE NEW ERA**

LONMARK membership soared in 2005 with growth of nearly 30% over the previous year. This increase is due to the successful 50+ city seminar program hosted by the LONMARK organization last year, attracting over 4,000 people in North America, Europe and Asia; and the addition of five countries to the LONMARK Affiliation Program. LONMARK affiliated organizations are active in Denmark, France, Italy, Switzerland and the United Kingdom. Several LONMARK affiliates are forming around the globe in Australia, China, Germany, Korea, Russia, and Singapore.

Chicago, IL—January 23, 2006—LONMARK International announced today that it is developing a new program to test and certify industry professionals of LONMARK® Open Systems. The new program will target individuals that install, integrate and maintain systems utilizing LONMARK certified products.

Chicago, IL—January 23, 2006—LONMARK

International and its members created an impressive demonstration of open, interoperable products at AHR Expo 2006. The demonstration contained 36 products from 21 different vendors, linked together with eight different Human Machine Interfaces (HMIs), showing the dynamic nature of open, interoperable solutions.

In addition LonMark offered several free education sessions at AHRExpo.

Kudos' to the Lon folks this is a lead we all must follow.

#7: **CISCO AND OTHER IT GIANTS ARE STUDYING THE PIECE PLAYERS OF OUR INDUSTRY TO HELP US FORM THE BIG PICTURE**

"Many are asking why Cisco is interested in building automation. The quick answer is a $25b worldwide opportunity. What's behind that is Cisco's desire to partner with the industry and understand how it can participate and contribute to this emerging IP centric market."

As our industry turns IT we will need guidance in this new world of Internet Technologies, We somewhat comprehend the Core Technologies– Switching & Routing– Software & Services but are babes in the wood when it come to Advanced Technologies– Storage Networking– Wireless– Network Security– IP Telephony– The Networked Home– Optical. The IT masters will lead us as well as all other converging industries into this new world. Below is a slide from the Building Intelligence Tour from Dave Clute, Cisco Systems, Inc. Advisory Services—Customer Solution Manager CCRE—Cisco Connected Real Estate

It shows several interrelating services that are not yet on our radar screen.

For more on Cisco thinking be sure to read this month's interview with Anto called "AHR and the Road to Palm Springs."

EMAIL INTERVIEW Anto Budiardjo & Ken Sinclair
Anto Budiardjo is President & CEO of Clasma Events Inc., the organizer of BuilConn Americas in Palm Springs, May 16-18, 2006 (www.builconn.com) and the co-located M2M Expo & Conference (www.m2mexpo.com).

Please send comments and questions to antob@clasma.com.

———————

AHR and the Road to Palm Springs

It was a perfect start of 2006, a year that will be a

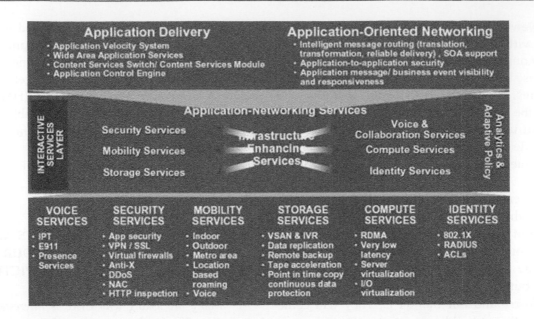

critical year for convergence. Cisco's participation will make BuilConn Palm Springs a must-attend event for vendors, integrators, consultants and owners.

Sinclair: What was your impression of AHR in Chicago this year?

Budiardjo: The pre-show anticipation was high, but the reality was even more interesting, specially regarding the road to convergence. I met up with and spoke to around a hundred people, and it was a very accepted notion that convergence, Ethernet, IP are things that are happening now.

Sinclair: Many were asking why Cisco is interested in building automation.

Budiardjo: Yes, that was a dominant question being asked; why is Cisco sponsoring BuilConn in such a major way, what are their motives? The quick answer is a $25b worldwide opportunity. What's behind that is Cisco's desire to partner with the industry and understand how it can participate and contribute to this emerging IP centric market. Cisco feels that BuilConn is an ideal way that they can enter this market.

Sinclair: Does the Roundtable at BuilConn seem critical?

Budiardjo: Yes the by-invitation-only Roundtable on the pre-day of BuilConn is important. It will gather key stakeholders around the convergence subject, to help understand the current issues faced by owners, integrators and vendors, as well as to discuss where the conver-

gence market is likely to evolve and grow.

Sinclair: How does the industry at large hear about the outcome of the Roundtable?

Budiardjo: The reason for doing it on the pre-conference day of BuilConn is that members of the roundtable can participate in panels during the formal part of BuilConn. This is where the outcome will be disseminated and further discussed with the broader audience in Palm Springs.

Sinclair: Will organizations such as BACnet and Lon-Mark be at the Roundtable?

Budiardjo: They have been invited, as are the major system vendors and a number of Integrators and building owners. The objective is to get a broad group of key and influential people that understand the convergence space.

Sinclair: Turning back to Chicago, what was the highlight for you?

Budiardjo: Clearly the time for BACnet has arrived; I was very impressed with BACnet on a number of levels. The adoption of BACnet was strong by almost all vendors on the floor. Secondly, their message on their strength at the system level was spot on, the demonstrations at their packed booths was strong about system level interoperability. Also, after many years of being an internally driven standards group, the BACnet Community is now focused on external business development and marketing activities that will, if they continue as they did in Chicago, make them very quickly a dominant force. Lastly, though not very visible on the floor, they are taking IP and Web Services very seriously.

Sinclair: How about LonMark?

Budiardjo: The LonWorks story was also strong. Though to an extent this is a mature subject, so it was not as much of a buzz at the show, most people know what LonWorks is, where it is strong and where to use it. The LonMark organization continues from strength to strength in signing up affiliate organizations around the world. There is also significant behind the scenes work regarding integrator certification within LonMark. This promises to be something significant also.

Sinclair: How about wireless in Chicago?

Budiardjo: Most vendors had wireless products, solutions or were communicating wireless plans. Some of these were very strong; such as Siemens, Kiyon, JCI, ZigBee and others. Wireless is clearly the way to go, just not quite mainstream yet. There are still many standard issues to resolve in this space, but I saw Gridlogix demonstrate an approach to fix this at a system level; a very impressive concept.

Sinclair: How does this lead to Palm Springs?

Budiardjo: It was a perfect start of 2006, a year that will be a critical year for convergence. Cisco's participation will make BuilConn Palm Springs a must-attend event for vendors, integrators, consultants and owners. We have invited both the LonMark and BACnet communities to be in Palm Springs to help fit these pieces together so we can start growing this business segment and deliver owners what they need.

Sinclair: I heard you are organizing a GridWise Expo.

Budiardjo: Yes Ken, the GridWise Expo is now a third element for Palm Springs, so we now have BuilConn, M2M and GridWise. All co-located in Palm Springs May 16-18. If you go to one, you get all three.

Sinclair: How does this differ from the GridWise Constitutional Convention?

Budiardjo: The convention in Philadelphia was necessary for the GridWise community to establish a set of guiding principles, which we now know as the GridWise Constitution. The Expo is a next phase; it is education,

business development and evangelizing the GridWise vision as well as technology and products surrounding this very important subject.

Sinclair: How about your plans for the rest of 2006?

Budiardjo: As you know, we will go back to Amsterdam 3-5 October for BuilConn and M2M. We are looking at planning other events in key markets of the world, your readers should keep tuned in to AutomatedBuildings. com as well as subscribe to the BuilConn Update by visiting www.builconn.com/bcupdate.

Sinclair: 2006 is turning out to be an exciting year!

Budiardjo: Yes it is, or as Cisco puts it, it is now the "Decision Point" for IP convergence. For many of your readers, it should also be a decision point about their involvement.

In a second interview with Cisco
EMAIL INTERVIEW—David W. Clute, Mike Lavazza & Ken Sinclair
David W. Clute Customer Solution Manager, Advisory Services Mike Lavazza Manager of Operations and Engineering Cisco Systems

———

Next Wave of Convergence: BAS along with Voice, Video & Data

If the design of the converged network occurs early enough in the building design and construction life-cycle, there are actually capital expense savings.

Sinclair: If we assume that Building Automation and Physical Security systems (such as surveillance and access control) are moving towards convergence with voice, video and data, how are companies like Cisco Systems handling this internally?

Clute/Lavazza: We have had segments of our Building Automation System (BAS) over our Corporate IT Network since 1999. We collaborated with our IT Infrastructure and LAN Operations groups to detail the requirements such as the equipment we would be connecting to the Corporate IT Network, bandwidth impact, security and static IP address requirements. The IT Infrastructure and LAN Operations groups are responsible for approving device connectivity and static IP address assignment.

Sinclair: How does the CIO/IT organization address the concerns of Facility Managers and Building Engineers that they may lose control of their systems to the IT group, as BAS moves to the IT backbone?

Clute/Lavazza: Communications and collaboration between IT Infrastructure and Building Automation System engineers will evolve and grow. The dependencies will drive these relationships to ensure standards and guidelines are followed and new requirements are communicated and understood. Open dialogue between the groups with regard to intent and future direction will ensure that both groups' objectives and initiatives are maintained.

Sinclair: Realistically, how long will it take for the industry to produce native IP devices (IP-enabled endpoint devices such as VAV controllers, light fixtures, etc.) that will allow full migration to a converged IT network?

Clute/Lavazza: This could take a few years. The amount of information that is collected or passed to these devices is small and the bandwidth requirements are minimum. IP connectivity to these devices is like taking a drink from a fire hose, but as technology advances and price points drop, these device will eventually be IP enabled.

Sinclair: What are the tangible cost benefits associated with BAS convergence over the IT network, in other words, what are the "bottom-line" reasons that might influence an owner of a building to invest in moving their BAS systems to an IP-backbone?

Clute/Lavazza: The cost of implementing a converged BAS system over an IP network does not necessarily mean greater capital investment. If the design of the converged network occurs early enough in the building design and construction life-cycle, there are actually capital expense savings. If the design of the converged network occurs late in the design cycle, then yes, there are additional costs associated with changes to the structured cabling system and building control systems. The tangible cost savings that can be attributed to converged BAS over IP, if adopted early enough in the building life cycle, include the following:

- Potentially lower cabling costs for the HVAC control systems by using wireless thermostats and PoE (Power Over Ethernet) control circuits.

- Lower energy costs associated with improved optimization, scheduling and coordination

- Improved comfort due to individual control of lights and HVAC systems through web-based control devices on their PCs, hand-held PDAs or IP Telephone control consoles—leading to greater employee productivity

- Plug and Play integrated Technologies—The key to adapting building automation subsystems and a BMS network to tenants is plug and play integrated technologies. The ideal system offers cost effective turnkey solutions for building owners. This translates into reduced costs for modifying a space as users' requirements change. Turnkey solutions are defined as workplace and building environmental changes required for accommodating each new user (or existing user with revised operating needs). This allows for a "box or briefcase" move rather than a retrofit or renovation with each new user.

- Reduction in technological obsolescence by facilitating the easy and economical introduction of new technologies, or replacement of dated technologies.

#8: QUARTER BAKED THINKING COUPLED WITH TRADITIONAL MARKETING PRACTICES IS ERODING OUR MARKETS.

Many of our control and automation companies seem pre-occupied with making a profit in the next quarter and are not focusing on the fact that we are in the middle of an era change. If they do not extend their planning window beyond the next quarter they will lose their market share and possibly their complete companies to new IT companies and/or off shore folk who have fully baked plans in place. Looking to our existing market for help is of little use; they have only a slim idea of what is the potential available. We must lead our markets into the new era.

In my Virtual Value Visions for Building Automation in 2006 I wrote; "I love this quote from Henry Ford, the automotive pioneer: "If I had asked people what they wanted, they would have said faster horses." I believe there are those in the traditional building automation industry who are designing faster horses, rather than trying to get their minds around the fact we are moving into a new era."

It was ironical that while I was in Chicago at the Exposition, the Ford Motor Company announced massive cuts back. I am sure that their marketing plan was a careful survey of their existing Ford owners who when asked "What do you want for the future?" responded with "a better Ford." Ford believing this line of bull gave them what they said they wanted and built them a "Taurus" to be exact. It was a disaster and the rest is history. What did the market really want? Something that looked like a Lexis and performed like a Toyota Prius. Why did they not ask for it? The conventional market did not know that competitors would provide better value cars and the gas prices would double. I wonder if Henry would have made the same mistake? I think not.

Industry! We are moving into the information age, a new era of integrated everything, not just information but everything which will lead us to a new way of doing everything. We cannot move into this new era selling dusted off dinosaurs for control systems because we think we can squeak one more quarter out of them. The new IT stuff coming down the pipe is slick and works well. Off shore markets are leapfrogging us with fully baked plans that provide connection to the new era. We must concentrate on providing a complete plan and lose that quarter baked thinking to provide new era connections to all of the pieces.

#9: CABA PROVIDES THE CATALYST AND GLUE TO HELP BIND OUR INDUSTRY.

The following headlines from the CABA web site show how they do this.

December 7, 2005 CABA TO SUPPORT BUILDING INTELLIGENCE TOUR SEMINAR AT AHR EXPO

January 9, 2006 CABA COMMISSIONS CONSORTIUM TO BUILD INTELLIGENT BUILDING RANKING TOOL

December 6, 2005 CABA TO CREATE TECHNOLOGY ROADMAP FOR INTELLIGENT BUILDINGS UPDATE

December 9, 2005 IPSOS REID TO DEVELOP CABA CONNECTED HOME ROADMAP

CABA has also started a new web sight *www.buildingintelligencequotient.com.*

The BIQ™ site will be an affordable and easy to use online Intelligent Building ranking tool. The Web site and the subsequent Certification Process will increase market penetrability of Intelligent Building technology with building owners, operators, managers and designers by demonstrating value and providing guidance.

To learn more about BIQ™ download this document http://www.caba.org/biq/BIQC-AHRHandout.pdf

Watch and help this site grow.

The CABA Intelligent & Integrated Buildings Council works to strengthen the large building automation industry through innovative technology-driven research projects. The Council was established in 2001 by CABA to specifically review opportunities, strategize, take action and monitor initiatives that relate to integrated systems and automation in the large building sector.

Our projects promote the next generation of intelligent building technologies, plus a whole-building approach that optimizes building performance and savings.

The Building Intelligence Quotient Steering Committee provides oversight over the development of an intelligent building ranking tool.

The Life Cycle Costs Steering Committee provides guidance over the development of an online life cycle costing tool.

Check out this valuable organization and see how you can become a member and get involved in their many valuable projects http://www.caba.org

#10: AHR EXPO'S INTERNATIONAL EXPOSITIONS CO. PROVIDES THE ANNUAL MELTING POT FOR OUR CONVERGING COMMUNITY

AHR Expo is always a field for the cross pollination of ideas but this years it exceeded our expectations. With at least 57,673 people attended AHR Expo, according to preliminary results, every section of our industry was well represented. This was our 7th year of AutomatedBuildings.com providing free education sessions for the industry. We would not be able to do this without the support of International Expositions who produces and manages this show. The show has become the place where our industry can introduce new products, innovate, grow and change. This year's event was a record breaker for all of that. It was great to see the major control companies back on the floor in Chicago. I am sure lots of new and old players will now be planning to come to the Dallas 2007 show. Please join me in inviting our convergence partners to be part of our AHR Expo. There is only one way to grasp the size and the dynamics plus the energy that is our industry—come to AHR Expo 2007 Dallas, TX.

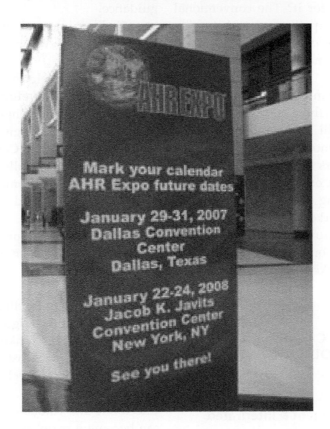

Chapter 6

The Business Value of Enterprise Energy Management at DFW Airport

Rusty T. Hodapp, P.E., CEM, CEP, LEED™ AP
Vice President, Energy & Transportation Management
Dallas/Fort Worth International Airport Board

ABSTRACT

THE DALLAS/FORT WORTH INTERNATIONAL AIRPORT has a long track record of success in conventional energy management. For 20 years, this technically oriented program existed principally as an initiative of the airport's maintenance department and flourished in a stable environment characterized by plentiful resources and little competitive pressure. Although successful in producing technical accomplishments and cost reductions, the program never achieved broad corporate impact.

In the mid-1990s, under the leadership of a new CEO, DFW adopted a business-oriented posture focusing on service quality and competitiveness. Although slow to adapt to the changing internal and external environments, by 1999 the maintenance department succeeded in reinventing itself by radically changing its business model and adapting its structure and processes to the new competitive landscape. New department leadership leveraged existing core competencies to recreate the energy management program with an enterprise orientation. They were subsequently able to demonstrate to executive management how the new model supported strategic business objectives and directly contributed to DFW's competitive advantage. Enterprise energy management was represented as a core business function that supported internal objectives (business growth, customer satisfaction, asset renewal) and addressed external factors (electric industry deregulation, environmental issues) by virtue of its positive impact on cost effectiveness, asset productivity and performance, resource utilization, and regional public policy.

Having established credibility and demonstrated the business value of enterprise energy management, the department received unparalleled support from the DFW executive team and board of directors. Corporate policies were enacted to mandate energy efficiency, commissioning, clean fuel vehicles, and energy efficient building codes. New business strategies were developed including energy master planning, evaluation of large-scale onsite power generation, adoption of sustainable practices in investment evaluation, design, construction, operation, and procurement, and development of an integrated/interoperable technology approach to enterprise energy and asset management. Substantial financial and human resources were committed to support program objectives, and the maintenance department was renamed to Energy & Asset Management to signal its new stature and enterprise orientation. These outcomes reflected DFW Airport's renewed, top-down commitment to enterprise energy management as a source of competitive advantage, and their persistence over time has confirmed the validity of the basic value proposition.

BACKGROUND

The DFW International Airport, which first opened to traffic a few minutes past midnight on January 13, 1974, is jointly owned by the cities of Dallas and Fort Worth and governed by the DFW airport board. Today, DFW is the world's third busiest airport serving over 59 million passengers a year. The airport maintenance department manages the airport board's multi-billion dollar facility and infrastructure asset portfolio and provides a variety of services including energy management, thermal energy production and distribution, potable water and sewer system operation, transit system operation, facility management, fleet management, and infrastructure repair and renewal. The original airport maintenance department organizational structure was purely functional in design, and its business processes were dominated by a task orientation.

An energy management program was initiated at DFW soon after the airport opened. This program was managed as an airport maintenance department function, and it achieved many notable technological successes. The program's objectives consisted solely of sound operating and maintenance practices, retrofits of existing systems, and incorporation of energy efficient technology in new construction. Significant economic benefits and reductions in energy consumption were produced; however, energy management remained a department-level initiative with modest recognition of its value at the corporate level.

Little change occurred within the department over the next 20 years as competitive pressures were virtually non-existent and resources relatively plentiful. The energy management program remained focused on applications of technology and achieved relatively little visibility at the corporate level.

MOTIVATION AND OPPORTUNITY TO CHANGE

The early 1990's saw the entire commercial aviation industry experience severe financial losses resulting in strong pressure on airports to reduce operating costs. In 1993, a new chief executive officer assumed leadership of the airport board. He established a vision of "running the airport like a business." The next several years produced corporate-level reorganizations creating various business development oriented units, an aggressive program of diversifying airport revenues, and a continuation of the cost containment focus.

The airport maintenance department was slow to respond to these changes in the internal and external environment. Consequently, the department came to be viewed by senior management as resource intensive, inflexible, dominated by an internal perspective (e.g., maintenance as an end in itself rather than a means to achieve a corporate goal), and out of alignment with the evolving corporate culture and business objectives. The tenure of a new Vice President selected in 1994 produced limited improvement in airport maintenance, and the author; a 12-year member of the department's management team, replaced him in 1997. The CEO personally communicated to the author the need to change the department and bring it into alignment with the new corporate model.

Under this new leadership and with a mandate to change, a comprehensive performance improvement program was designed and implemented, resulting in a near total revision of the department's structure and business processes.

Reinvention

The result was a comprehensive program designed to reinvent the airport maintenance department at its most fundamental levels including:

*   Role and direction
*   Strategies and processes
*   Structure and image
*   Culture

A major objective was to establish a clear direction and shared set of values. A comprehensive situation analysis was performed and from it goals, strategies, and objectives developed and implemented. One major outcome involved changing the department's core business model to one employing a total asset management approach. The total asset management model incorporates the full lifecycle of an asset from acquisition through decommissioning and thus requires an enterprise orientation. From this broader perspective, the department began to evaluate its core business functions and competencies.

Assessing the Existing Energy Management Program

From his background in energy engineering, the author was strongly committed to energy management as a core business function and a core competency of the department. Viewed from a total asset management (i.e., lifecycle) perspective, he considered energy management to be a potential source of strategic value to a capital asset- intense enterprise such as a large commercial airport.

In evaluating the department's energy management program a basic SWOT analysis was conducted. It concluded that major strengths consisted of expertise in technical energy management, best-in-class district energy system operation, and willingness to change in order to improve. Major weaknesses included the basic lack of alignment with corporate organization and strategic business objectives and the resulting lack of internal credibility.

The mandate to change established by the CEO introduced a major opportunity to recreate the existing energy management program as an enterprise business function. A second was identified in the impending deregulation of the electric utility industry in Texas and its foreseeable impact on electricity consumers. Ironically, deregulation also created a certain amount of threat as business entities evolving or materializing to operate in the future deregulated markets began approaching key decision makers with various alternatives. The result was a series of mixed messages relative to the viability of energy management as a core airport business function.

SELLING ENTERPRISE ENERGY MANAGEMENT

Upon concluding that energy management, if applied with an enterprise orientation, would contribute fundamentally to the airport's core business success, it also became apparent that selling this concept could provide a critical means of demonstrating the maintenance department's value added contributions and corporate alignment.

The basic strategy developed to sell enterprise energy management involved leveraging a currently successful business operation against a new opportunity. This strategy would establish credibility internally and then explicitly link energy management outcomes to key business objectives. In this case, the maintenance department's district energy (DE) system operation (thermal energy production & distribution business process) was leveraged against the opportunity presented by deregulation of Texas' electric markets. The airport DE system's full cost of service had been benchmarked at best in class levels for years demonstrating efficient operation and capable management. Numerous business entities positioning themselves for the post-deregulation market place were approaching the airport about selling or outsourcing the DE system. The physical facilities and operating/financial records were opened to all business entities desiring to make a proposal to purchase or contract for O&M of the DE system. The low number and limited nature of responses demonstrated forcefully the success of the airport's DE system operation in the competitive marketplace. Consequently, the department now had a platform for initiating a dialog with the CEO regarding energy management and airport business objectives.

Gaining an Audience

The opportunity to sell the business value of enterprise energy management to the CEO came during a presentation on the results of the DE system acquisition and/or outsourcing proposal process. The publicity associated with impending deregulation in Texas and the associated business offers being conveyed directly or indirectly to the CEO stimulated his direct personal interest in the internal analysis. His interest in evaluat-

ing the changes being implemented in the maintenance department also predisposed him to participate actively in the dialog.

As the department Vice President, the author along with the utility business unit manager delivered the presentation and key messages to the CEO and the Executive Vice President of Development.

The results of the process established several critical assurances to the executive team:

1. The department management team's willingness to evaluate and embrace change.
2. Their comprehension of corporate business objectives and ability to manage in alignment with them.
3. Energy management as a core competency.

The circumstances also provided an opportunity to extend the discussion and present potential energy management outcomes in the specific context of the airport's $2.5 billion expansion program, then in its initial programming stages. Energy management strategies supporting four vital strategic objectives were proposed:

1. Airport development (expansion and redevelopment of the DE system)
2. Infrastructure renewal (renew 30 year old assets)
3. Electric utility deregulation (position the airport to operate cost effectively in competitive energy markets)
4. Air emission reductions (reduce point source emissions to comply with regulatory mandates)

Two principal challenges had to be overcome in order to convince the CEO of the business value of maintaining ownership and management of the DE system as well as significantly increasing the capital invested in that particular enterprise. Countering the differing viewpoints of other influencers, primarily outside firms, and demonstrating that risks (real and perceived) associated with the proposed changes were manageable proved to be essential in selling the results of the analysis and the proposed changes.

Key Messages

The airport strategic plan developed in 1999 identified two key elements of success for DFW: the capacity to grow by developing its facilities and infrastructure; and a low operating cost structure. To communicate to the CEO how enterprise energy management would contribute to key business objectives, program outcomes were linked to these basic success factors.

An inherent factor in a large commercial airport's ability to grow is the need to attain necessary environmental approvals. Virtually all major commercial airports are located in urban areas with moderate to severe air quality issues. Thus, the emission reductions that would be created by decreasing energy consumption, including those originating on the airport and those resulting from regional power generation, constituted a key message linking energy management explicitly to enterprise business objectives. Similarly, reducing the demand on the airport's energy production and delivery infrastructure would result in improved asset utilization, thus enabling additional development from the existing fixed asset base and deferring capital expansion.

The contribution of an effective energy management program to a low overall operating cost structure was fairly easy to demonstrate. The existing program's track record of demonstrated success in reducing cost and consumption and the additional benefits that would result from broader application formed another key message.

Finally, a number of other strategic business objectives were eventually shown to benefit from enterprise energy management outcomes, including:

Objective—Customer Friendly Facilities
Benefit—Improved asset performance and occupant comfort/satisfaction

Objective —Industry Leading Environmental Programs and Practices
Benefit — Emission reductions, reduced natural resource use, energy efficient building code, purchasing policies

Objective —Revenue Growth
Benefit —Tenant energy supply chain management, expanded thermal energy services

Objective —Total Asset Management
Benefit —Commissioning, lifecycle cost analysis

Objective —Superior Management
Benefit —Industry leading energy and environmental programs

In some cases, the success metrics proposed to evaluate the effectiveness of enterprise energy management were qualitative in nature and described as enhancements along the airport's value chain (i.e., industry leadership, reduced resource use). In others, explicit quantitative measures were offered (i.e., percent reduc-tions in lifecycle cost due to commissioning, energy use reductions resulting from efficient code and purchasing policy, emissions reduced, etc.).

Outcomes

In general, the presentation of the results of the DE system acquisition/outsourcing proposal process produced four explicit outcomes:

1. CEO and Executive agreement with the proposal to continue internal O&M of the airport's DE system.
2. Approval of the proposed alternative reconfiguration of the DE system expansion programmed in support of the airport capital development program to also address renewal of the aging energy infrastructure, flexibility required to operate effectively in deregulated energy markets and reduce regional air emissions. This resulted in an increased investment in the project of approximately $88 million.
3. Direction to pursue detailed engineering and economic analysis of incorporating combined heat and power to supply 100% of the airport's total electric power needs.
4. The opportunity to provide a full briefing to the airport board of directors on energy management accomplishments and proposals.

More important, an ongoing dialog with the CEO and Executive team on energy issues was initiated, as was their appreciation of, and commitment to, the business value of energy management. This commitment was initially signaled to the organization through the approval of the DE system expansion and reconfiguration recommended. In approving a four-fold increase in the capital investment originally programmed and conferring control over the project's design intent and operating business plan to the maintenance department, a powerful and unmistakable message of support was delivered.

The CEO's continued commitment was further communicated in both formal and informal ways. Formal corporate statements of policy were adopted establishing principles in support of energy efficiency, commissioning of all new airport construction, industry collaboration to expand the availability of efficient technologies, and support for legislative action creating incentives to stimulate deployment of clean and efficient technologies. Revisions to a number of existing business processes and adoption of new ones were also authorized, including:

- Reorganization to create an Energy & Facility Services business unit.
- Participation in a retail electric competition pilot program.
- Strategy for energy procurement and management of this function with internal energy staff.
- Adoption of the International Energy Conservation Code
- Development of a strategic energy management plan.

In addition, it became easier to secure approval for resources and participation in initiatives to elevate the visibility and influence of the airport's energy management accomplishments. Examples include:

- Addition of energy engineer and energy analyst positions to augment the energy manager's staff.
- A full time staff position to function as the airport's commissioning authority.
- Annual investments in energy efficiency projects identified in the airport's 10-year capital program.
- Application for, and acceptance of, numerous grants for energy audits and demonstration or acquisition of clean and efficient technologies.
- Participation in federal, state, and NGO initiatives to study airport energy use, sustainable practices, etc.
- Memberships in high profile organizations and initiatives including Energy Star, Rebuild America, U.S. Green Building Council, Texas Energy Partnership, etc.

Informal means included continued visibility of energy management objectives, accomplishments, and plans at the CEO and board of director levels through regular briefings.

The airport's reconstituted energy management program has since been recognized by industry and governmental organizations, the trade press, and with regional, State, and international awards. In a telling measure of the CEO's continuing commitment, in 2003, he directed the maintenance department be renamed to Energy & Asset Management to more accurately reflect its enterprise orientation.

DE SYSTEM EXPANSION PROJECT

The approved reconfiguration of the DE system expansion project addressed each of the four strategic objectives noted previously by installing new chilled water, thermal energy storage, heating, preconditioned air (PCA) and controls systems.

- Chilled Water—(6) 5,500 ton chillers and 90,000 ton-hour stratified chilled water storage
- Steam—(4) 40 MMBH and (1) 100 MBH medium pressure boilers with ultra-low NO_x (9ppm) burners
- PCA—(5) 1,350 ton chillers and (6) 1,130 ton heat exchangers for precooling or heating
- Cooling Towers—35% capacity increase using existing structure with optimized fill, water flow & distribution, and increased airflow
- Controls—new industrial distributed controls

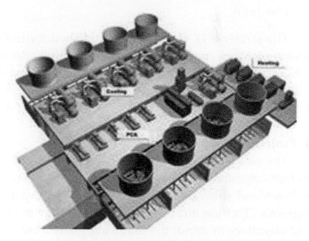

Figure 6-1. DFW District Energy Plant Layout

Thermal energy service was provided with the new system to existing loads in 2004 and substantial completion of the project coincided with the opening of the new international passenger terminal in July 2005.

DE System Controls and Automation

To coordinate, monitor, control and troubleshoot the complex thermal systems a robust and sophisticated automation system was required. An industrial grade distributed control system (DCS) was selected to provide the multi-level redundancy necessary for a business critical operation.

Information Technology and Airport Expansion

As noted previously, DFW was engaged at this time in a multi-billion dollar expansion program which was the principal driver for the DE system expansion project. The program centered on a new, 2.1 million square foot international passenger terminal and a new automated people mover system. Both project elements would involve significant information technology (IT) components.

The scale and complexity of IT applications coupled with the advent of entirely new business processes (for DFW) dictated a strategic approach to technology planning for the program. The program's technology (or technology enabled) goals included:

- Common use equipment
- High levels of customer service and amenities
- Ability to effectively manage assets and resources
- Integration with campus IT infrastructure
- Integration with enterprise business applications
- Enhanced information availability and access
- Enhanced situational awareness
- Operation efficiency
- Cost effectiveness
- Flexibility

The strategic IT vision thus established for the expansion program created another opportunity to demonstrate alignment of enterprise energy management and to leverage project specific IT applications to achieve strategic business objectives.

DE Project Automation Goals

The new DCS replaced an existing 18 year old system that consisted entirely of vendor specific, proprietary hardware and software. In the context of the expansion program's IT vision and with enterprise energy management objectives in mind, the following goals were established for the DE project's automation system.

- Replace the existing obsolete DCS
- Maintain use of industrial grade automation system for the DE production and distribution processes
- Specify a system based on open architecture
- Substantially improve the overall automation of DE plant equipment and unit operations
- Leverage the centralized monitoring and operator interface facilities and capabilities
- Interoperability with campus building management and process control systems
- Integrate with enterprise business applications to achieve a high degree of information sharing
- Provide DE customers with access to billing and usage information

While certain of these goals were specific and limited to the DE project, (DCS system replacement, automation enhancement), the integration and access goals were developed with a view towards improving energy management capabilities at the enterprise level.

The nature of the expansion program itself—huge scale involving hundreds of design firms, contractors and dozens of project elements, coupled with the public procurement environment that DFW is obligated to operate in prevented a single technology or technology company solution to all IT applications, or in some cases (building automation systems for example) to similar applications in different construction elements. With that being the case, the energy management team saw the DE project's DCS as means to achieve integration of disparate automation systems (new and existing) as well as certain enterprise business applications such as asset management, service requests and dispatch, procurement, performance measurement and reporting, and emergency operations management.

With this direction the project team developed functional block diagrams showing the DE plant DCS as the platform for integration of building management systems, other process control systems and high value enterprise business applications.

Similarly, network architecture diagrams established connectivity criteria for the project which included use of DCS communication bus features within the DE plant, corporate network for external connectivity with building management and other systems/applications and internet access for DE customers.

Finally, with no building automation system integration standard (e.g., BACNet or LONWorks) having been established for the expansion program as a whole (nor was a de facto one established through existing airport systems), the DCS specifications were developed around open, interoperable standards-based technologies to facilitate connections across the enterprise.

DCS Solutions

To implement the project automation goals the specifications provided for a single bid for the DCS product and integration services. The selected vendor was required to provide the hardware and integration services as one turn-key solution.

The system/service procurement was by bid under the in-place construction contract and resulted in the selection of Emerson's Delta V automation system for the hardware solution with integration services being provided by a major player in the integration services market.

DCS Based Enterprise Integration

The Delta V system makes extensive use of commercial off-the-shelf (COTS) technologies which facilitates easy connections across the enterprise and reduces the dependency on proprietary hardware and software.

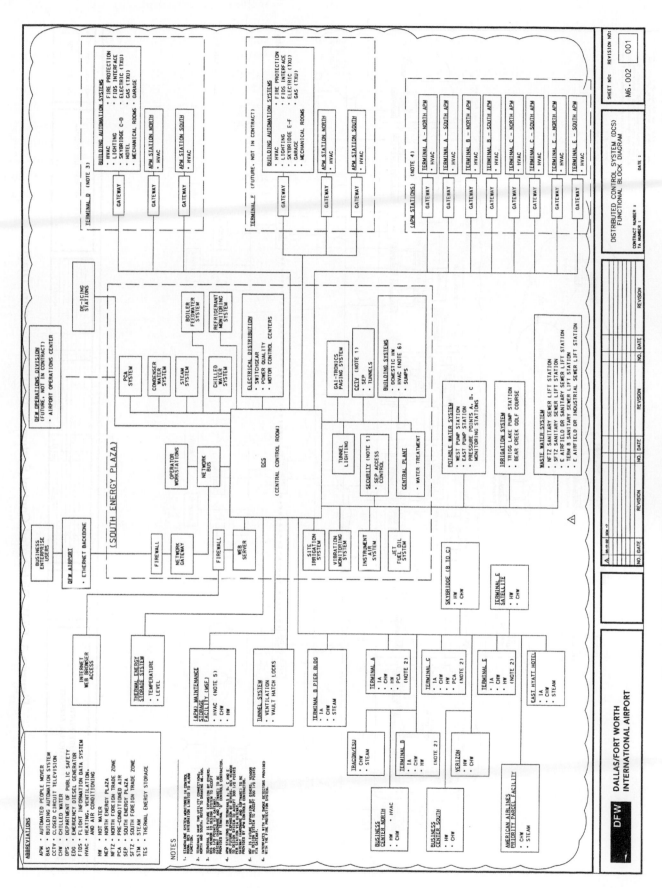

Figure 6-2. DFW Distributed Control System Functional Block Diagram

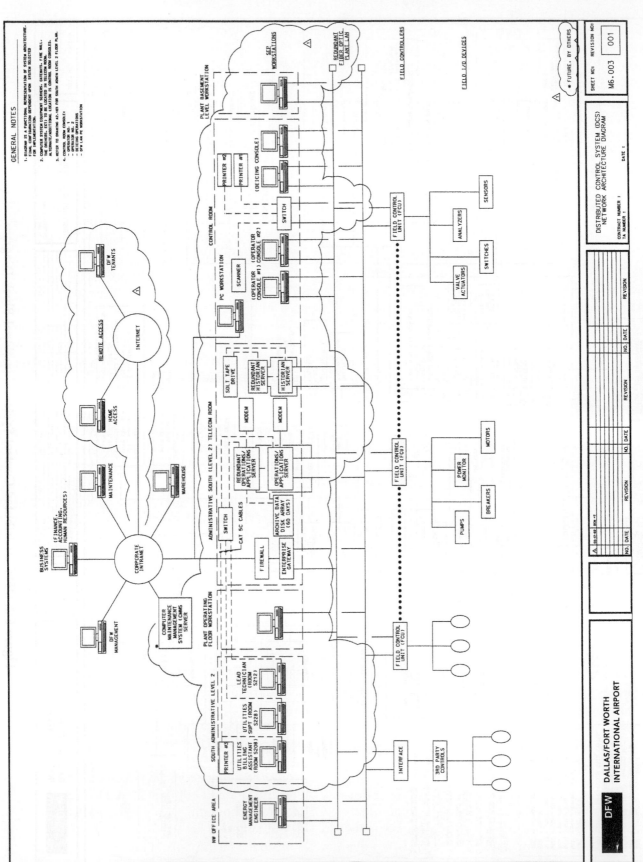

Figure 6-3. DFW Distributed Control System Network Architecture Diagram

Embedded Foundation fieldbus and other digital busses provided relatively simple plug-and-play integration solutions within the DE plant and other process automation applications, particularly where programmable logic controllers (PLCs) were used. Open, interoperable standards including Extensible Markup Language (XML) and Object Linking and Embedding for Process Control (OPC) provided easy connections with Microsoft Office applications and enterprise business systems, and accessibility over the internet.

Numerous PLCs were utilized for process automation throughout the DE plant and other utility processes on the airport including water distribution, wastewater collection, and collection, storage and treatment of spent aircraft deicing fluid. These systems were easily integrated with the Delta V using open bus standards (Modbus or Data Highway).

Integration of building automation systems with the Delta V DCS was more challenging. Building automation systems had been installed as part of the construction of ten new facilities totaling over 2.7 million square feet. In conjunction with building automation systems in existing airport facilities, products from a wide variety of vendors were represented. Given the diversity of products installed, neither BAC-Net nor LONWorks would provide a single integration solution. However all systems targeted for integration were OPC compliant. Consequently, OPC provided the interoperability solution necessary to integrate both new and existing building automation systems via the Delta V DCS.

A redundant Delta V network provides connectivity within the DE plant while an OPC network services BAS applications. Remote systems utilize the DFW LAN for access to the OPC network. Remote users have access via the internet.

OPC and XML provided the primary means of integrating the DCS (and other integrated automation systems) with enterprise and local user business applications.

Enterprise Asset Management

The integrated system developed through the DE project delivers significant value through its information sharing capabilities in addition to the efficiencies of automating energy and utility process operations. Experienced DE plant operators now have process control and operating information from systems across the airport and enterprise—thermal energy, water, wastewater, irrigation, HVAC, weather data, status of maintenance actions—at their fingertips and all through a common user interface. Summary reports of production, distribution, performance, exceptions, history, etc. are available

for management and engineering—and may be shared via the network. The airport's asset management system (DataStream 7i) interfaces directly with systems and facilities through the integrated DCS allowing for automated tracking of critical asset information, generation of work requests, emergency response, etc. In addition, the asset management system itself accesses information from throughout the enterprise and interfaces with numerous other enterprise business applications supporting a wide array of users and critical business processes.

DE service invoices are automated and electronically transferred to accounts receivable and customers have web access to their real time and historical consumption and cost data. The strategies, technologies and techniques employed have effectively created an enterprise asset management system that improves efficiency through automation and interoperability and delivers value by connecting islands of information.

DE Project Results

The DE project has delivered impressive results through a combination of sound design principles, efficient equipment technology and a sophisticated automation strategy, all driven by enterprise energy management objectives. Nitrogen oxides (NO_x) emissions from combustion operations were reduced by 86% exceeded the regulatory mandate (70%). While adding 2.9 million square feet of service area, the DE plant energy consumption per square foot served has decreased by 47% ($5 million annually at current energy prices). The thermal energy storage system and associated operating strategy have demonstrated the ability to shift over 15MW of electric load off-peak. The combination of off-peak commodity pricing and reduced transmission and distribution charges has reduced annual operating costs by as much as $750,000.

Perhaps as important as the directly quantifiable

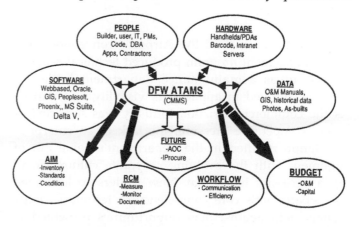

Figure 6-4. DFW Automated Asset Management System (ATAMS)

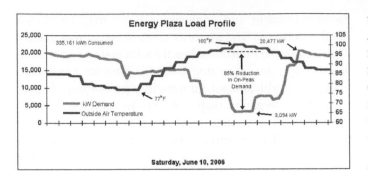

Figure 6-5. DFW District Energy Plant Load Profile with TES

economic returns accruing from the project is the creation of an enterprise enabled platform through which DFW will achieve multiple benefits in managing its energy, physical and information assets, resources and environmental footprint. The open standards-based approach enables a centralized and remote approach to monitoring, maintenance, control and management of the operating environment. The use of COTS and interoperable technologies provides a pathway for continuous integration of existing and new facilities, systems and business applications at the enterprise level. Capital and operational expenditures will be reduced by limiting the need for multiple proprietary networks and enhancing the ability to manage large amounts of data while ensuring persistence of energy management measures. The productivity of assets and other resources will be increased through more flexible work process and scalable tools that enable improved collaboration and connectivity. The connections created between uses and users of formerly disparate islands of information will greatly enhance situational awareness and facilitate effective management of key business processes.

CONCLUSION

For 20 years, the DFW Airport's conventional energy management program produced technical achievements but as a maintenance department initiative, remained limited in application, visibility, and impact. As the commercial aviation industry was subjected to severe competitive pressure, the lack of an enterprise orientation resulted in the department's being out of alignment with the evolving corporate culture, creating a necessity to change. New leadership leveraged the department's core competency in technical energy management against new opportunities presented in the pending deregulation of Texas' electric market and the airport CEO's desire for performance improvement

to recreate the program with an enterprise orientation. An initial audience with the CEO evolved into an ongoing dialog regarding the strategic implications of energy issues for the airport. Ultimately, the business value of energy management applied on an enterprise basis was successfully demonstrated and a top-down commitment to it as an important contributor to the airport's competitiveness realized.

The application of open/interoperable standards based information and commercial off-the-shelf technologies provided a cost-effective and efficient means of developing an enterprise energy and asset management system. The technologies and strategies deployed produced significant efficiency gains, improved productivity and operational flexibility and have provided a pathway for continuous development of integrated enterprise solutions for management of key business processes.

The process of reinventing DFW's energy management program and gaining the CEO's commitment to it may offer a few lessons for consideration by other similarly situated organizations:

- Critically evaluate (situation assessment) existing energy management programs and practices.
- Seek alignment of energy management program and practices with corporate strategies and business objectives.
- Identify energy management contributions to the corporate value chain.
- Link energy management outcomes to key business objectives (key messages).
- Demonstrate (or establish) and then leverage credibility (personal and/or organizational) for access and/or to reinforce the key messages.
- Know your audience (CEO's perspective, critical issues, success metrics).
- Understand associated risks (real and perceived) and show they are manageable.
- Understand the value of information as a core asset of the enterprise—and how that value is enhanced when its uses are connected.
- Consider the value potential inherent in the convergence of IT, energy management systems, building automation systems and the web to manage effectively across the enterprise.

Acknowledgement
An earlier version of the material in this chapter appeared in the article, "The Business Value of Enterprise Energy Management at DFW Airport," by Rusty T. Hodapp, *Strategic Planning for Energy and the Environment*, Summer 2005.

Section II

Web Based Enterprise Energy and Facility Management System Case Studies

Chapter 7

Innovating the Business of Facilities Operations
A Framework for the Next Major Advancement in Facilities Operations

Bill Gnerre, Gregory Cmar, and Kevin Fuller
Interval Data Systems, Inc.

PREFACE

FACILITIES LEADERS are under budget pressures that manifest themselves as a need to do more, limit headcount, go faster, and reduce energy costs. There exists an urgent need to improve operations to meet these constantly increasing demands. Sound familiar? Then this chapter is for you.

There are two main business facts that form the core of this chapter.

First, businesses need to change, and make no mistake, operating a facility is a business. Incremental improvements happen continuously, but every once in a while there needs to be a quantum leap forward—something more revolutionary than evolutionary. Often, technology is an agent of change (PCs, Internet, DDC, CAD), but it is only a catalyst. For major progress, facilities leaders must realize the need to do something very different, embrace the idea of innovation, and be willing to make fundamental changes to how the group operates.

Second, data are immensely valuable. This is true for every business function within every industry. Facilities operations is no different, just a little behind the curve. The facilities industry has focused its technology advancements on hardware efficiency and control technology, but still has not seriously embraced *information* technology (focus on information). The few that have operate in a different world than the rest.

We have proven the value of using data and information technology to advance the business of facilities operations. The examples and situations in this chapter are real, taken directly from actual implementations at facilities organizations. The use of data is a foundation that affects much more than just operational efficiency. We are more than happy to show you the proof, just ask.

The staff of Interval Data Systems

FOREWORD

The following is not an endorsement of Interval Data Systems or its products, but of the ideas presented within this chapter and the importance of data, information, and knowledge to the business of facilities operations.

Today's facilities are more highly instrumented and have more sophisticated control systems than ever before. So much so, that we are generating more data than our people can assimilate. In addition, our occupants are changing the way they do business in shorter and shorter cycles as they strive to be on the leading edge of their business sector. It becomes increasingly important to truly understand how our buildings are interacting with and meeting the needs of the occupants.

This means that we must find ways to "expose" the data our buildings generate and find ways to present that information to building staff who can then diagnose operational issues before they become building problems, before the occupants even know anything is wrong. Our goal should not be to throw more people at the problem, but to find ways to use the power of technology to deliver better information to our existing staff to allow them to "virtually travel" through their building space and to quickly identify malfunctions. We must not be satisfied with concepts and practices that set up service desks and phone lines to receive complaints and then tell occupants that we will "be there when you need us." We must do our work closer and closer to "real time."

The good news is that we are starting to see software products that do not control things, but that gather data with the idea that we should be doing something with that data. The technology piece may well be the easier part of this transition. But, it is a transition we must make if we are to create "knowledge workers" of our building staff and in the process engage many more people in solving problems where they occur—in the buildings.

David J. Miller
Director, Facilities Planning & Management Operations
Iowa State University

INTRODUCTION

Demands on facilities operations continually increase—improved comfort and air quality, lower costs (despite rising energy prices), new construction, green initiatives—but progress is not keeping pace with the demands of business management. A few innovative facilities directors, unwilling to accept the status quo, have developed a strategy to fundamentally reinvent how their organizations operate. It is not just about how the HVAC systems perform, but how the business performs.

These pioneers are leading their teams out of reactive mode, becoming proactive organizations. They understand that they are running a services business and know that to advance it they must perform and progress on many fronts:

• Control HVAC systems operations to deliver space comfort
• Reduce the annual utility bill
• Improve the productivity of staff, from senior engineers to area mechanics
• Teach technicians to fix problems instead of treating symptoms
• Increase departmental credibility
• Hold contractors and facilities staff accountable

Budget pressures have made it necessary for facilities leaders to examine where they spend their limited dollars. Spending on individual productivity tools should be re-examined to see if there are more effective, strategic ways to spend those same dollars and advance the business in multiple areas.

There is a common denominator among the facilities innovators—they have embraced an advanced use of operational data throughout their organizations. This is not just data provided by a couple hundred data loggers or trend logs, but a complete historical record of all the operational data. This is a big difference, but periodically it takes this scale of change—in philosophy and approach—to transform operations rather than adding one more incremental improvement.

This chapter is about quantum improvements in departmental productivity, operational efficiency, de-sign engineering, lower construction costs, preventative maintenance, accountability, and credibility—and how to go about achieving it.

THE CRITICAL ASSET REQUIRED FOR CHANGE

What asset are we referring to? In a word—**data**.

It is extremely rare to find a facility in any environment—academic, medical, commercial, residential—with anything approaching a complete record of operational data. Many organizations operate under an erroneous perception that they have the data they need, because they have multiple systems that each collect some data and generate reports. In fact, building automation systems (BAS) typically offer woefully deficient tools, masked by an illusion of adequacy. Metering systems and utility companies are no better. Combining data from multiple sources, well, that is almost unheard of.

Consider this, almost every facility out there has only one or two percent of the operational data easily accessible, and the majority of facilities directors don't even realize it. Most decisions are based on engineering assumptions and educated guesswork. Imagine a finance department operating on even half the financial data and guessing at the rest (they'd be in jail), or a customer service group with only a fraction of the actual customer and order data. Even auto technicians have data readily available; the car has data collection built in. To service it, the mechanic plugs in the computer and reviews the data—before ever picking up a wrench.

In contrast, the facilities world accepts its lack of data. This is partially due to the lack of understanding of its value, and partially because until recently, there really were no viable methods to have all the data all the time.

That is the critical asset required for change—all the data, all the time. Every single point from the BAS(s), meters (automated & manually read), utilities (interval consumption data & purchased utility bills), weather, maintenance and space planning systems, collected at consistent intervals (15-minute intervals work well), in a centralized data warehouse designed for efficient, flexible access to the information.

There is substantially more gained by applying data and information systems to operations than by adding more people. Information is an asset that can leverage the existing knowledge of the whole operations staff. An information-rich system makes an order of magnitude difference. Information enables facilities teams to make more and better decisions in less time, make the business go faster, and operate differently.

Facilities leaders and facilities staff, who do not yet understand the value of data, and the information that is derived from them, often offer up various excuses why facilities should not collect the operational data and build business-oriented information systems on top of them. Table 7-1 lists eight commonly heard excuses or myths about taking advantage of information.

To reiterate, in order to achieve major advancements and revolutionary changes in facilities operations, one *must* have the operational data—period.

OPERATIONS AND MAINTENANCE

Operations and maintenance (O&M) represents a large portion of the daily work within facilities. Keep the occupants comfortable, maintain indoor air quality, and do so as cost-effectively as possible. Dozens of engineers, technicians, area mechanics and others, combine to perform necessary monitoring, diagnostic and maintenance tasks. Most operate in a very reactive mode (and have always done so), responding to customer service calls and addressing problems. This can change. A proactive facilities team is possible, and information is the basis to make it happen.

Table 7-1. Operational Data Myths

Myth 1: We already have all the data.
After talking to over 300 sites, we found only two that actually do have all the data. Others who perceive that they do are fooled by limited trending or report features that rarely can access more than 1% of all the operational data points.

Myth 2: BAS trend logs are sufficient.
Where to start... see Myth 1. Trend logs degrade control system performance, severely limiting the number of trends possible. They don't combine data from other systems. They only keep data for a week or two. Hence you have a subset of the operational data, and most are hard to work with.

Myth 3: Many of the BAS points are not important.
So why pay the BAS vendor to include them? Usually these "unimportant" points are for control. It is not sufficient to see what has happened, but you also need to isolate why, and can only do so by capturing control changes in addition to monitoring points.

Myth 4: There is too much information to look at.
The point isn't for one person to read a 2,000 page listing of data points, but to make all data available when needed, and provide methods to monitor data quickly and easily, converting the raw data into information in the process. By involving dozens of people from across the facilities organization, it's easy to cover all critical operations, and you build the knowledge base of the staff in the process.

Myth 5: We don't have time to look at the data.
The reality is that you already spend the time. Engineers typically spend 5X-10X more time gathering data than doing engineering work. The availability of information improves productivity. People will look if the time spent returns enough value and saves labor hours. Just look at the examples throughout this chapter.

Myth 6: Existing systems already use a database, so we're covered.
There is a huge difference between having a good database manager, such as SQL Server, and a complete, functional data warehouse with all operational data made accessible.

Myth 7: You can't trend everything without hurting control performance.
While true for BAS trend logs, collecting interval data and providing trend data for analysis is a separate issue. Software and communications standards, such as OPC, BACnet, LonWorks, and Web services are up to the task.

Myth 8: Metering solves all data needs.
Metering does provide useful data, but it is expensive, slow to implement, and provides no operational insight explaining why consumption levels are where they are. Comprehensive BAS data collection provides far more actionable information, and can alleviate or postpone the need for meters with better return on investment and cash flow.

There is a perception by some facilities directors that they don't have the time to look at data—too much work, not enough hours, not enough staff. Wrong. Looking at data, which has been organized and presented as actionable information, is the only way to get ahead. Pick your cliché—work smarter not harder, do more with less, plan your work and work your plan—they all apply. Information is the way to leverage existing staff, get out of reactive mode, and provide superior service.

Comfort is Not a Luxury—
Overpaying for It Is

Most facilities have standards for comfort that define acceptable ranges for temperature and humidity. Some, such as hospitals, have far stricter air quality requirements. Improving comfort is typically more important than reducing energy costs. People's comfort complaints trump cost-cutting practices.

On a day-to-day basis, comfort is what influences the perception of facilities by the rest of the organization. Image and credibility rise when everyone is comfortable, and drops with the need for frequent hot and cold calls. Of course, while the facilities customers (occupants) are most influenced by comfort, upper management scrutinizes the costs. Information is the key to serving both masters.

Here is an example: An air handler serves adjacent spaces, one- and two-person offices, and printing/copy rooms where the equipment generates significant heat. Despite efforts by the technicians, there were often cold calls from the offices and hot calls from the printing rooms. Examining this air handler's data uncovered a discharge air temperature trying to maintain 50°F to provide cooler air to the printing rooms, which caused reheat to occur at the VAV boxes supplying the offices. The real problem wasn't temperature or air flow, which is what technicians had adjusted to respond to hot/cold calls, it was a balancing problem. The engineering assumptions for load were wrong.

Using information allowed the actual root problem (engineering design) to be identified and fixed, resulting in improved comfort for both spaces and lower costs by requiring less cooling for the air handler and less reheat by the VAVs. Plus, the data continuously verify the fix, enabling the facilities director to hold the engineering firm accountable to get it right this time.

As for the analysis effort to figure this out—to solve the comfort problem, which had been ongoing (and wasting money) for a year, took about 15 minutes looking at the data.

Cost-based Optimization—
The Only Kind That Matters

Few facilities departments actually know the run rate (i.e. $/hour) to operate their mechanical, lighting, and other systems. Due to the sorry state of information tools commonly in use today, most organizations pay 15-20% more to their utility companies for HVAC than necessary.

Operators take actions every day that can affect cost and comfort, but don't have the information to make informed decisions. A well-intentioned change to a chiller may impact cooling towers, secondary pumping, or air handler operations. The only way to know is to be able to see the data—see how operations and costs are affected system-wide. The goal is not to make equipment run better, but to make the system run well.

Cost-based optimization is a fantasy without complete operational data. Missing information skews the results and can lead to poor control decisions. However, data provide the information to know, for example:

- which chiller combinations operate at the lowest cost per ton-hour at on-peak versus off-peak hours,

- when to use gas versus electric chillers,

- the best control strategies for cooling tower fans, or

- the impact of addressing a hot call by increasing air flow versus decreasing air temperature (or if the real problem is something else, like a broken reheat valve or a design problem as discussed above).

The basis for cost-based optimization is conceptually simple, but complicated to implement. First, use the operational data available from the BAS(s) and meters to calculate energy consumption for electricity, gas, and other utilities. Then, model the utility rate structures in all their complexity (multiple rates, time-of-day pricing, and other determinants as appropriate) and calculate operational costs. With this model, facilities directors can know what the cost per hour is to run the facility, each system, sub-system, and piece of equipment, recalculated every 15 minutes. Operational changes can be tracked to verify not only the impact where the adjustment was made, but of ripple effect changes anywhere in the system. Knowing costs every 15 minutes is a different world than receiving a utility bill at month's end—with immediate information you can take action to affect the problem, thirty days later is too late.

INTERACTIVE COMMISSIONING— VERIFIABLE RESULTS AT A FRACTION OF THE COST

Commissioning of all forms (new construction, retro- or re-commissioning, Continuous Commissioning®) suffers from the same basic three issues: 1) the labor cost to do it is very high; 2) building owners have yet to see enough proof of the value and ROI to risk that cost, and; 3) there is even less proof that gains realized through commissioning will persist. Interactive commissioning uses data to combat all three of these concerns.

One person can interactively commission more equipment in a week through the data than a typical commissioning team can during the entire project. Commissioning agents often sample equipment, for example, commissioning only 10% of the VAV boxes because the labor cost to do them all is prohibitively high. Facilities directors should not tolerate this. Find an agent willing to use the data and do a thorough job. Hold the commissioning agent (and the contractor if there is new construction) accountable to make the equipment meet operational requirements, not merely pass some functional tests. Pay the contractor after systems pass operational tests. This is not affordable with labor-based process, but you can with an information-based approach.

This example shows the impact of information on retro-commissioning 115 air handlers. An air handler optimization had resulted in a $1,200/year savings (cooling, heating, and fan savings). The estimated savings across 115 AHUs was $500-$1,500 annually, with an average of $750, per air handler. That is $86,250 currently going to utility companies that is wasted.

Using data, one person evaluated 115 AHUs in less than three days. The evaluation documented whether each AHU was meeting comfort requirements and operating as cost-effectively as possible, and listed recommendations for improvements where needed. While most directors suspect problems exist, evidence allows them to take corrective action—to create a plan to systematically address each issue and stop throwing money at the utility companies. The data provide the evidence.

Compare those results to not using the data. A commissioning firm (and/or internal staff) working to produce the same analysis would spend 50+ person-days crawling around the building with data loggers and in the end have less information and no good way to verify results. That labor cost is prohibitively high and the commissioning never actually happens.

The only way to attack this kind of problem is with information. Verification that fixes had the intended results is fast and based on hard engineering facts, not estimates. Ongoing monitoring of the data ensures persistence. At the same time, optimizing the air handlers likely reduces hot/cold calls and improves customer satisfaction.

Facilities directors who are happy with the status quo can continue to send the utility companies an extra $7,000 a month. Those looking to reinvent their operations can do so through information, and keep that "bonus" money in their own budget or return it to the bottom line.

Preventative Maintenance Today Stops Disasters Tomorrow

There's no doubt that preventative maintenance is beneficial. From a financial point of view, various institutions have calculated that spending a dollar on preventative maintenance eliminates the need to spend four dollars later on deferred maintenance. Then there are the less tangible benefits—avoiding disasters such as the heat failing on the first cold day of winter or earning a reputation as a facility that constantly breaks down. Continual monitoring of operational data will better inform and prioritize preventative efforts so that fewer issues are ever notices by occupants and tomorrow's disasters never materialize.

Add 20% to Equipment Life

Even while meeting comfort specifications, it is not uncommon to find equipment trashing, cycling, and hunting once one looks at the data. Well maintained equipment that runs smoothly will last as much as 20% longer—no rocket science there. But without information, it is almost impossible to know if equipment is being beaten into an early grave. With data, you not only can see if problems exist, but you can easily diagnose, fix, and verify the results.

MANAGEMENT AND CREDIBILITY

The facilities director's role is in large part about leadership. It is not limited to engineering leadership, but includes managing a business environment that makes constant demands:

- Can you eliminate some headcount?
- Here's some additional space to manage.
- When are you going to upgrade that legacy equipment?
- Meet these new environmental requirements.
- Can't you go faster?

All of this goes on while energy costs rise and the systems managing energy get more complex. Spending limited available dollars wisely is an important skill for facility leaders. Money spent ineffectively compounds the problem, but spent intelligently allows one to navigate their way to smoother sailing.

Leadership is more than budget allocation. It includes the management and effective utilization of staff, and shaping the group's image viewed across the organization. The credibility of the facilities group rests with the director, whether through direct interaction with executives or management practices that dictate how the team serves its customers. Tapping into the operational data asset is a powerful tool to lead the group to major improvements, increased respect, and credibility on all fronts.

Lead Staff Out of "Reaction Mode"

Operations staff spends too much time working on the wrong things. This is what happens when you live in "reaction mode." Area mechanics spend hours walking through their building(s) gathering what is actually a small amount of information. Engineers spend much more time collecting the data that they need for analysis than doing actual engineering. Everyone chases today's problems and treats the reported symptoms. But, there is a path out of reaction mode, and it is in the data.

For example, with a well-structured view of building data, area mechanics can perform a "walk through" in about eight minutes of looking at the information. They can see what happened overnight, know how each air handler and terminal box is operating, know if the lights came on at the proper time, etc. Data provide the information to properly prioritize their work and free up time for preventative maintenance tasks (which are also well-defined and prioritized using the data).

In a world with data, area mechanics, engineers, technicians, and operators all spend more time implementing proactive tasks, and significantly less time gathering the data to make the decisions.

Win Back Control of the Operations Budget

CFOs like numbers—it's the world they know. Facilities directors have lived in a world of estimates. No wonder it is often hard to deal with the folks that hold the purse strings. Time for a change.

With a complete record of operational data, with irrefutable facts, facilities directors can change their relationship with the CFO and be in control of the operations finances instead of being controlled by them. Think how differently the next annual budget cycle or capital improvement request would go if armed with hard data that:

- Quantifies the waste to be corrected by a redesign
- Verifies that a retrofit will have the intended result
- Proves that a chronic comfort issue needs funding to address
- Measures operational savings achieved by the facilities team in the past year
- Demonstrates an unprecedented level of customer service

Directors can use information to consistently measure and improve operations in ways that were impossible before. When armed with such facts, they can not only demand accountability from subordinates, but also accept accountability for operations' performance when dealing with upper management. The director that does this will change the rules of dealing with the CFO, win many more budget battles, and see their credibility soar with executive management and across the organization.

Build Departmental Credibility and a Culture of Accountability

The chapter has repeatedly made the point that operational data provide a means for accountability, and that improvements made will lead to credibility. However, do not leave this to chance. Accountability must become part of the culture and credibility nurtured once earned. Data can help.

The advancements discussed above in comfort, customer service, operating costs, proactive maintenance, staff utilization, and fiscal management offer major gains in credibility. Some are more visible to customers, some to management, all are important. With this kind of success, there is nothing wrong with some self-promotion—especially when the promotion provides a valuable service and even further adds to the group's credibility.

Energy consumers, department heads, building executives, and senior management all have vested interests in certain operational data (whether they express it or not). Publishing information suitable for each of these constituents is another path to increasing facilities credibility. It shows that the facilities group knows how systems are running and has the facts to back them up. It also sends a message to upper management that the operations group is proactive and can be responsive to additional information requests in the future.

Beyond the communications aspects of publishing

pertinent information, engaging management and users is likely to affect their behavior. With increased credibility, facilities leaders can completely transform relationships and build a culture of accountability within and beyond the operations group, for example:

- Reassign partial responsibility for energy consumption to the users
- Change the ground rules for dealing with finance and purchasing, and justify budgets more easily with the CFO
- Promote optimization gains achieved per building and within the central plant, recognizing both behavioral changes by occupants and operational improvements by facilities staff
- Directly account for, and show the value of, each staff member
- Verify the effectiveness of design engineers and commissioning agents
- Educate customers about the consequences and costs of their decisions and actions
- Develop the support and admiration of the trustees

All of this is only possible in a world of data that provides irrefutable facts. Otherwise, it is a land of estimates and conjecture that will soon be unacceptable as top management demands more.

ENGINEERING (RE)DESIGN AND CONSTRUCTION

Estimates are that half of all redesign projects do not address the problems originally targeted. That suggests that either half of the design engineers are incompetent (unlikely), or that they work with too little factual operating data to precisely know the problem. Like technicians, they often address the symptoms.

Engineers design for worst case conditions. They err on the side of over design. It is easy to see how this happens, after all, an over designed system might be wasteful, but an under designed one won't perform and that will come back to haunt them. Plus, there is very little accountability for poor design. The bottom line to the facilities director is that the chances of paying for an over designed system are pretty high, which will insure higher construction and operating costs. All for the lack of good information.

Facilities directors should require design engineers to use operational data and prove that redesigns are effective at solving existing problems and are not over

designed. The same is true for new construction, where over designing and over building mechanical systems is commonplace. With operational data, designers have the ability to truly assess existing designs to see how they work in the field. This feedback, which does not exist today, will influence the use of systems and equipment in new designs.

Data are the key to design and construction accountability. If you have operational data to feed the design process and later validate the results, design engineers and mechanical contractors have the information to do the job correctly, and can and should be held accountable for their results.

COST ALLOCATION AND UTILITY BILLING

Operational (demand side) and utility data (supply side), combined with space planning information, can enable an organization do perform detailed energy cost allocation by space, department, or cost center. The data must be present to determine space consumption, and then rates can be applied to generate accurate costs for each space. This approach more accurately assigns costs to actual energy consumers as opposed to a simple square footage allocation.

One advantage to accurate allocations is that it holds high energy consumers accountable for their consumption instead of subsidizing them by the low energy consumers, the typical approach. This approach places the responsibility for energy consumption squarely at the feet of the users. However, it requires a utility bill with operational data that show the user where issues may exist and what actions to take to reduce costs in the future. The tie between operational and billing data also equips the facilities team to assist departments/cost centers with questions about their allocation and further explore strategies to reduce consumption—another chance for the facilities group to earn a reputation for outstanding service.

When asking for departmental accountability, timeliness of information is important. A system that only supplies data in the monthly utility bill puts everyone at a disadvantage. While the facilities group may have no problems correlating billing data to operations for any specific period of time, it will be nearly impossible for department heads to correlate operations to actions from weeks ago. Publishing month-to-date (updated every 15 minutes) billing estimates based on operational data, makes the information available to all concerned. Now, facilities customers can review current information, associate actions with consumption, and change behavior.

UTILITY PURCHASING

Purchasing energy is a complex task that requires a thorough understanding of demand. Minimizing energy costs requires insight into how to manipulate the timing and height of peaks in demand. Also, because energy pricing structures change regularly, effective purchase negotiations require a sound knowledge of the energy market's price drivers to "lock in" the best price.

Using operational data can provide knowledge and insights into the institution's consumption patterns, the drivers of these patterns, and the impact of the weather. Additionally, the information can help participation in demand response programs, many of which have financial incentives. Utility buyers can negotiate far better energy rates and avoid expensive ratchet charges when armed with the data.

Another area where data can assist in driving down utility costs is sewer charges. Cooling towers evaporate significant volumes of water. Operational data can, quite accurately, calculate this water loss, which never enters the sewer system. When water and sewer are not separately metered, some organizations have successfully negotiated reduced sewer charges using data.

Finally, many organizations deploy expensive metering systems that parallel utility meters for the purpose of verifying the utility bills. Instead, operational data can summarize energy usage, identify losses, and provide supporting documentation to reconcile discrepancies with the utility—all at a fraction of the time and cost.

CONCLUSIONS

HVAC and electrical systems account for 20-25% of the value of building assets (based on replacement costs). Yet it is rare to find a facilities department with a complete historical record of operational data that show how those systems perform. Ironically, this is exactly the path to solving many of the largest problems facilities departments face today.

The business climate for facilities is the most difficult it has ever been. More demands, more space to manage, less staff, rising costs. Facilities directors find themselves trapped in a fog of issues without the ability to see a clear path forward. Information enables them to map a course and drive the business forward.

The investment in information is one of the few ways a facilities director can advance the business on multiple fronts. Most dollars spent for improvements go to individual productivity tools where the value and return on investment is one-dimensional. With the asset of data, not only does it provide value directly, but it also enables better informed decisions and prioritization for other expenditures.

Data, and the actionable information they provide, are the tools for a facilities director to demand accountability from internal staff, external contractors, customers, and themselves. When operational performance, customer service, and information distribution make leaps forward, the credibility of the facilities operations group ascends to new levels as well.

Chapter 8

An IT Approach to Optimization and Diagnosing Complex System Interactions

Bill Gnerre, CEO
Greg Cmar, CTO
Kevin Fuller, EVP
Interval Data Systems, Inc.

ABSTRACT

MOST FACILITIES PERSONNEL do not know what is really happening with their building operations. They can't because they see only a tiny fraction of the operational data that would give them real insights. Today, optimization happens at a component or sub-system level with little or no information on how the parts work together. Diagnosing complex problems is often impossible because you cannot see how all the affected systems interact. Actual hourly operating cost data, and any ability to see how control changes, weather, occupancy, etc. influence costs is simply beyond most organizations.

This case study of a hospital in the southeastern U.S. illustrates how an IT approach, which makes all operational information available, quickly uncovers the interdependencies of the chilled water plant and building systems. The diagnosis explores VAV box, AHU, secondary pump, and chiller operations. Interval data provide an in-depth look at the unintended side effects of control engineering, operator decisions, and resultant energy waste caused by addressing the wrong issue—and the verified savings from addressing the correct issues.

Critical to identifying and solving challenges such as these is the availability of continuous, historical operational data from both the plant and buildings. This chapter describes how data are used to show cause-and-effect relationships between (and within) building and plant operations, and unnoticed system instability. It also demonstrates how continuously available data reduces the time demands for operations and maintenance, making such analysis both feasible and cost-effective.

INTRODUCTION

A recent news story told how UPS is saving millions of dollars by reducing gas consumption of its delivery vehicles. They use data to analyze dispatch plans and optimize delivery routes and times, saving UPS almost 14 million gallons of fuel annually. It's the data that make the improvements possible, and enable them to measure and verify results (such as reducing mileage by 100 million miles) as well. Mike Eskew, UPS's chairman and CEO commented, "At UPS, we're never satisfied with the status quo."

What does UPS's gas consumption have to do with your HVAC system? A lot. It's all about using the data to improve operations.

1. Both UPS and your facilities operations group are service businesses. Whether formally structured that way or not, facilities operations is a services business, and the more it is run like a business, the better it will perform.

2. Both strive to provide a high level of service (package delivery or building comfort) that consumes a substantial amount of energy to deliver.

3. Both have customers that take the service somewhat for granted, expect nearly perfect reliability, and give little thought to the costs incurred to make it happen.

4. Both employ technicians/mechanics to keep their equipment (delivery trucks or HVAC systems) running, and benefit if the personnel can work more efficiently and proactively leading to labor savings and extended equipment life.

5.   Both can optimize performance, using data and information technology (IT) to calculate detailed operational costs. But, in all probability, UPS does a much better job of this than your facility.

Yes, there are some differences. UPS uses IT to leverage their operational data to great benefit. Most facilities operations groups don't. Without the data, you're guessing, and optimization is not possible on any meaningful scale. Without the data, you're accepting the status quo.

BACKGROUND

Before we get started, here's some background information to help give context to the diagnostic information presented later in the chapter.

The Facility

The hospital in this case study is a 2,000,000 square foot facility. The hospital campus includes a half dozen buildings with a total of 115 air handlers served by the physical plant. The plant, partially shown in Figure 8-1, has a total of 5,700 tons of chilling capacity, a bi-directional bypass, and a secondary loop with four 75hp pumps that service seven zones. The chiller plant also includes three cooling towers and a 400 ton absorber with its own cooling tower.

In Figure 8-2 you can see the standard configuration for an air handling unit. A VFD fan provides the air flow and two cooling coils chill the air. Dampers regulate return air, exhaust, and outside air. This AHU has 24 VAV boxes, each with a reheat coil.

Data Collection

A building automation system (BAS) will contain thousands of monitoring and control points (over

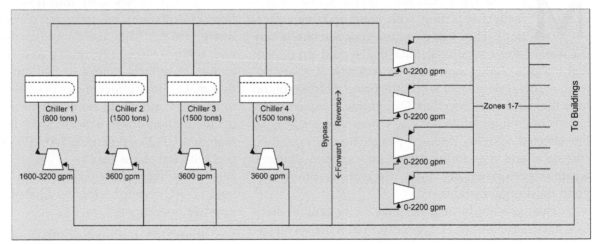

Figure 8-1. Chilled Water Plant—Chillers with Primary and Secondary loop.

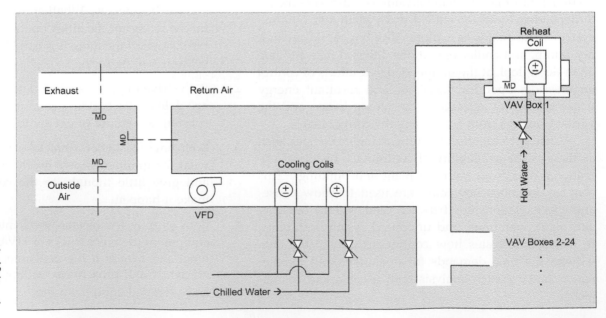

Figure 8-2. Air Handling Unit and VAV boxes.

Table 8-1. Sample of Interval Data Points from Hospital Systems

System	Partial Points List
Chillers & Absorber	tons output; CHW flow, setpoint, supply temp, return temp, ΔT*; CW flow, supply temp, return temp, ΔT*; % full load amps; efficiency*; kW/ton*; production balance*; pump brake horsepower*
Cooling Towers	CW supply temp, setpoint, low limit, high limit; CW return temp; CW flow, % full load amps; fan speed, load, kW*, kW-hour; fan motor speed, volts, amps, estimated make-up water*; estimated make-up water BTU*
Secondary Pumps	CHW supply, return, & differential pressure; zone differential pressures & setpoint; average valve % open; secondary CHW supply temp, return temp, flow; bypass flow & direction; kW*;
AHUs	return air humidity, humidity setpoint, temp; discharge air temp, setpoint; cooling valve positions; outside air CFM, CFM setpoint, damper position; supply fan static pressure, setpoint; average VAV heating valve position
VAVs	zone temp, heating limit, cooling limit; discharge air temp; heating valve position; CFM min, max, heating, calc & actual; damper position
Weather	outside air temp, humidity, wetbulb

*calculated point

100,000 for larger campuses). You need them all. The hospital is collecting data from about 18,000 points (see Table 8-1 for a partial listing). Ideally, you want data not just from the BAS (although it's a better place to start than meters), but from every data source:

- Building automation systems (all of them if you have more than one)
- Advanced meter systems
- Manually read meters
- Utility interval data
- Utility billing data (including all rate information)
- Weather (from local airport as BAS weather sensors are often faulty)

The hospital collects each point at standard 15-minute time intervals to enable synchronized views, for example, showing chilled water supply temperature, AHU discharge air temperature, and outside wetbulb at the same time. Saving the data indefinitely (for the life of the equipment, 20 years or so) allows the review of past operations and provides a historical record of all the equipment in the facility.

THE IMPACT OF A SINGLE PLANT CONTROL DECISION

There's a saying in this business that chiller plants are run by legend. Unfortunately there is a lot of truth to it. Instead of a disciplined engineering process, based on complete operational data, facilities rely on estimates based on a tiny data sample and educated guesswork.

Nowhere else are decisions with such a large financial impact based on so little information. Even small decisions get more thorough analysis. Imagine buying a computer only knowing the processor speed and hard drive size—it would be a crap shoot. You wouldn't make a $2,000 PC decision not knowing if the screen was large enough to be readable, if there was enough memory to handle your workload or just enough to boot the operating system, if it came with an operating system at all, or even if it were a laptop or desktop. But facilities groups make decisions with 100 times the financial consequences of buying the wrong PC every day as part of standard operating procedure.

Facilities business decisions don't have to be left to random chance. You can put the engineering back into your operations and make informed decisions based on a complete set of facts. You do it with data and an IT approach that turns the raw system data into actionable information.

Inside the Chilled Water Distribution Loop

During an investigation of the chiller plant, it was found that the secondary chilled water flow exceeds the primary flow quite frequently. When this occurs, the bypass flow reverses and the secondary pumps, of course, work harder creating the additional flow. This one situation has an impact across a wide assortment of air handling units—this chapter focuses on just one.

Experience has shown that every time you collect data and thoroughly examine them, you will find scores of problems. It has also shown that building HVAC systems are complex, interactive, and self-compensating, requiring a systemic view to really identify root causes of problems. Once you have a starting point, diagnostics is a stream-of-consciousness process that requires immediate interactive access to data or else the effort gets derailed. You'll see this in practice in the sections that follow.

Note: the starting point, in this case chilled water flows in the distribution loops, is not particularly important. It's just where this diagnosis started based on what was noticed first. The data exploration through a business-oriented IT system is the important part, and would have led to the same conclusions if the process had begun inside the air handler or with the VAV boxes.

Figure 8-3 illustrates the flow problems that are commonplace. You can see the primary chilled water flow drop suddenly at the same time the secondary flow

jumps. At the same time, the bypass flow increases and reverses direction.

There is a cost associated with this behavior—that of the secondary pumps that are now working overtime. The pumps are maxing out as can be seen by the total pump kW. This isn't as costly as it might be, as this is all happening during off-peak hours when electric rates are lower, but the consumption is notably up.

What is causing the demand for greater flow? The air handlers start to tell the story.

Cooling Valves and Discharge Air

Looking at data for one of the hospital's rooftop units during the same time period shows more details. The cooling valve opens to 100% instead of the 45-60% range where one sees it at other times (Figure 8-5). This demands more chilled water, making the secondary

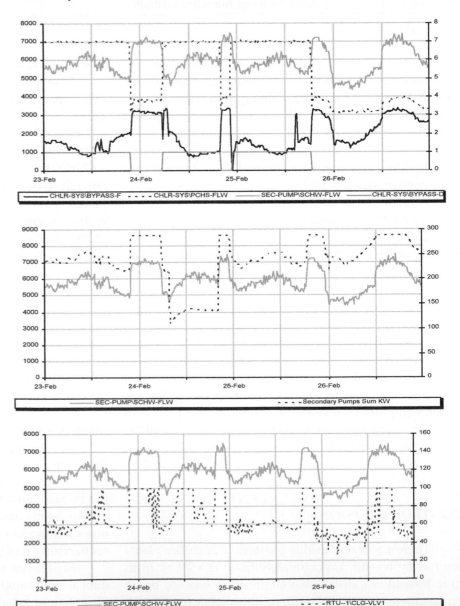

Figure 8-3. Coincident Flows in System—the Primary CHW Flow (dotted line) drops as a chiller is shut down, and at the same time, both the Secondary CHW Flow (gray line) and Bypass Flow (thin black line) jump. You can also see the Bypass direction reverse (1 is forward, 0 is reverse).

Figure 8-4. Pump Electrical Consumption Maxes Out—each time the Secondary CHW Flow (gray line) jumps to 7000 gpm, the total kW (dotted line) of the pumps reaches their maximum.

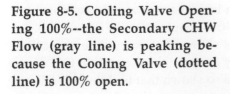

Figure 8-5. Cooling Valve Opening 100%--the Secondary CHW Flow (gray line) is peaking because the Cooling Valve (dotted line) is 100% open.

pumps work harder to deliver the increased flow. The correlation between this single valve and total secondary system water flow indicates that the issue is systemic.

The cooling valves need a reason to open up, and that reason is the discharge air temperature (Figure 8-6). It fails to meet its 55°F setpoint at the times in question, and calls for more cooling, causing the cooling valves to open to meet that request.

All of this activity is happening at night or on the weekend. The outside temperature and the building load are low most of these times. So why can't discharge air meet setpoint?

Chilled Water Temperature

In order to determine what was causing the discharge air to exceed setpoint the chilled water temperature was examined. Figure 8-7 shows the discharge air

temperature and setpoint and the secondary chilled water supply temperature. At the times in question, the secondary system is delivering 47-50°F water instead of around 42.5°F, where it stays most of the weekday hours.

Chiller Operations

February in Florida does require cooling, but the load is not high, especially nights and weekends. During those off hours, the hospital ran only one chiller, as it was assumed it would save money. Figure 8-8 shows that every time the system cuts back to one chiller, all the other issues start to arise.

Now look at the supply temperatures in Figure 8-9. The chiller 4 and primary loop supply temperatures hold tight, as they should, except in our problem areas. At those times the primary loop loses about 1.5°F. This

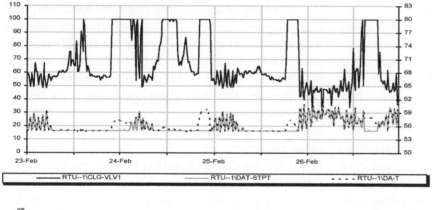

Figure 8-6. AHU Discharge Air Can't Make Setpoint—the Cooling Valve (thin black line) is at 100% open because the Discharge Air Temperature (dotted line) does not maintain Setpoint (gray line).

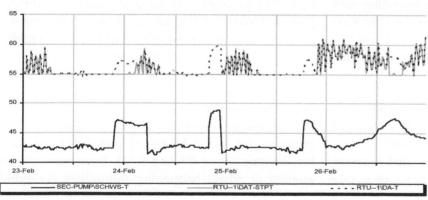

Figure 8-7. Secondary CHW Supply Temp Rises—the Discharge Air Temperature (dotted line) does not maintain Setpoint (gray line) when the Secondary CHW Supply Temperature jumps.

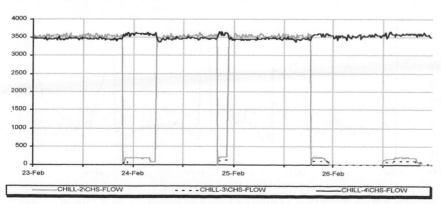

Figure 8-8. Chiller Flows—Chiller 4 (black line) stays on at a fairly constant 3,500 gpm flow, while Chiller 2 (gray line) is turned off during the times in question. Note the small flow amounts for Chiller 2 and Chiller 3 (dotted line) as there is some drag flow through those chillers.

is due to the drag flow through chillers 2 and 3 (Figure 8-8), where 200gpm of warm return water is mixing with the properly chilled output from chiller 4.

The additional 2°F jump between primary and secondary supply temperatures is the result of the reverse bypass flow. When the secondary loop flow is greater than the primary loop, the bypass reverses direction (Figure 8-2) mixing return water with supply water, increasing the temperature.

That circles us all the way back to Figure 8-3 where you can see, of course, that going from running two chillers to one cuts the primary flow in half since chillers 2, 3, and 4 are all fixed-capacity 3,600gpm units (Figure 8-1). When the chiller shutdown occurs, even though the space load is low, there is a chain reaction due to the interdependencies of the chiller plant, chilled water loop, and air handlers.

Did the Strategy Really Save Money?

Throughout this little ride, the VAV boxes never noticed anything going on, and operate steadily (Figure 8-12). One could argue that saving $34-$37/hour (the cost of running one of these chillers during off-peak hours) is worth a little instability as long as space comfort isn't affected. But the truth is that the savings are only a fraction of that.

Figure 8-10 shows the hourly cost of operations for various plant components and the total plant cost. While you do cut the cost of one chiller, and you get an extra $2/hour in reduced cooling tower operating cost, the remaining chiller is working harder, costing $20/hour more to run than when running as one of two (A). Then when the secondary pumps start working harder to compensate, they increase another $3/hour (a 25% jump). In total, there is a small operational savings of about $10/hour (B), but on a ton-hour basis, costs actually increased 7% (C). Factor in the instability introduced and the added wear on equipment—is it worth it?

This is great. The analysis has shown what occurred, how the plant and building systems interact, how to take the instability out of the operations, and what the cost implications are of these choices. There's only one problem... none of this actually fixes anything.

"AS DESIGNED" VERSUS "AS NEEDED"

What do we mean that it didn't fix anything? This is a classic case of treating the symptoms instead of the real root cause. The one-chiller-or-two decision affects behavior as far out as the air handler. But the real problem with the discharge air not maintaining setpoint isn't too little chilled water; it's that the AHU's discharge air setpoint is too low in those circumstances. The issue isn't originating at the plant, but at the other end of the system... at the VAV boxes.

Space Comfort and Air Flow

Looking at this system from the space end, at first glance everything seems fine—temperatures hold within defined comfort ranges (Figure 8-11) and air flows have steady CFM readings (Figure 8-12). Most engineers/technicians (including some at the hospital) would report that the VAV boxes are operating right on design spec—no problems—since the VAV is supplying the space with proper temperature and air flow. Gathering more data (there are 288 points associated with the VAV boxes for this one air handler) without probable cause is usually too labor intensive to look further.

"No problems" is an understandable conclusion without looking at the data in detail. Unfortunately, it's also wrong. A properly built IT system has the information to allow a closer analysis of the VAV information, leading to a better picture of how things should operate.

Too Much Reheat

As can be seen by looking at the discharge air temperatures for each VAV box (Figure 8-13) and the VAV box heating valve (Figure 8-14), there is a substantial amount of reheat happening throughout the building.

Figure 8-9. Chilled Water Supply Temps—there is an increase in chilled water temperature as it circulates through the system. The Chiller 4 Supply Temperature (gray line) is the lowest, with the Primary CHW Supply Temperature (dotted line) above it, and finally the Secondary CHW Supply Temperature has the highest temperature.

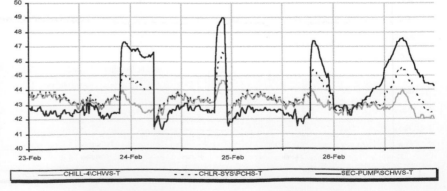

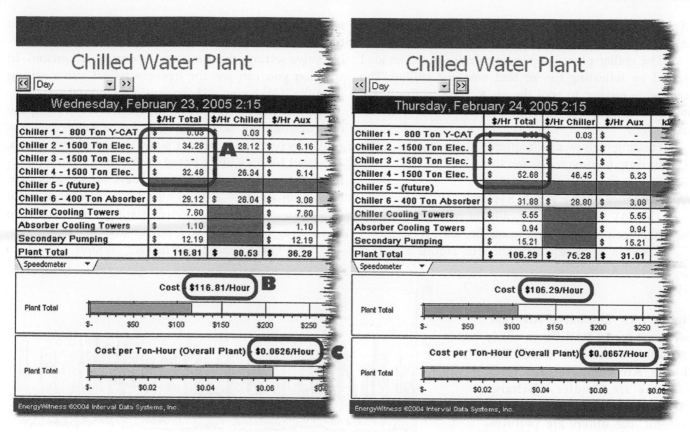

Figure 8-10. Hourly Cost of Chiller Plant Operations

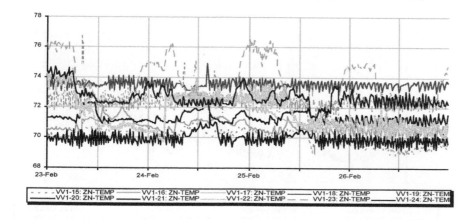

Figure 8-11. Space Temperatures at 10 VAV Boxes—the Zone Temperatures are all fairly constant and within a comfortable range of 69-74°F. The one temperature that jumps up to 76°F does so in response to occupants adjusting the thermostat.

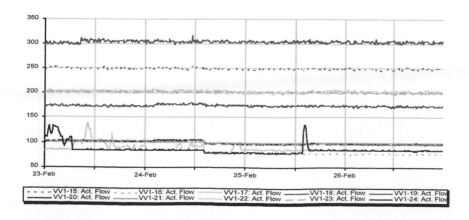

Figure 8-12. Air Flows (CFM) for 10 VAV Boxes—only two of the VAV boxes ever operated above their minimum CFM levels.

Of the ten VAV boxes shown in Figure 8-14, only two are not supplying a significant amount of reheat.

The chiller plant is operating under the false load created by reheating the air that was just chilled. The hospital is paying to cool the air to 55°F in the AHU, then paying again to warm that air to 70-100+°F in the VAV boxes.

Heating valves are frequently opening to 100%, even during daytime hours for some VAV boxes.

VAV Influence on the Whole System

The VAV behavior, specifically the frequent reheat, is the real culprit in the system instability shown earlier.

There is a lot of information in Figure 8-15 that serves to pull the story together. From the AHU there is the discharge air temperature and setpoint, and the cooling valve setting (there are two, operating in unison). In addition you can see the average reheat valve setting from the VAV boxes and the outside temperature.

Even though there is reheat occurring during the day, the outdoor temperature is low enough that the AHU discharge air setpoint stays at its 55°F minimum. The need for 55°F air is based on meeting summer daytime cooling demands, but these data are for February. Even with air flow at minimum CFM, discharge air at 55°F causes reheating to occur, creating a feedback loop demanding more cooling.

One chiller could not deliver enough cooling to

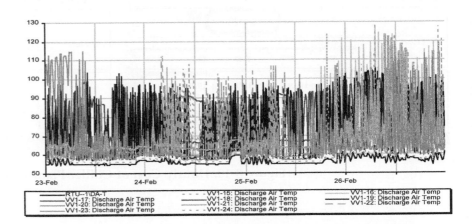

Figure 8-13. Discharge Air Temperatures for 10 VAV Boxes—only a couple of the VAV boxes Discharge Air Temperatures track near the air handler Discharge Air Temperature (black line across bottom), the others are performing significant reheat, but tend to spike rather than provide steady temperatures.

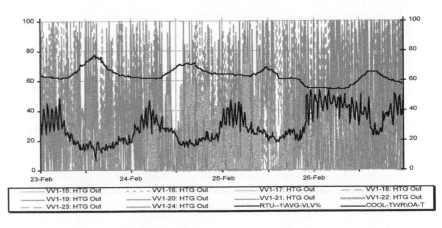

Figure 8-14. Reheat Valve Position on 10 VAV Boxes—many VAV box Reheat Valves are opening to 100% and cycle widely between 0-100%. The chart also shows Outside Air Temperature (upper black line) and the Average Heating Valve Position (lower thick black line).

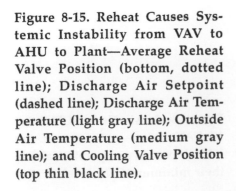

Figure 8-15. Reheat Causes Systemic Instability from VAV to AHU to Plant—Average Reheat Valve Position (bottom, dotted line); Discharge Air Setpoint (dashed line); Discharge Air Temperature (light gray line); Outside Air Temperature (medium gray line); and Cooling Valve Position (top thin black line).

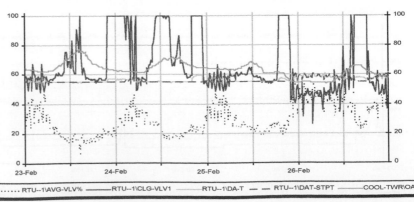

keep the AHU discharge air at setpoint, causing the increased secondary loop flow and bypass reversal shown earlier. Eventually, the average reheat valve positions impact the discharge air setpoint, moving it up 3-5°F, although oscillating by about 3°F due to the cycling reheat valve positions. The discharge air temperature reaches setpoint again either when a second chiller comes online or when reheat pushes the setpoint high enough so that the cooling valves don't need to be 100% open and driving the demand on the chilled water loop.

Implementing the Solution and Measuring the Savings

The AHU and VAV boxes needed rebalancing based on actual needs for operating the space properly (as opposed to original design specs). Balancing is an iterative process—make changes, re-examine the data, make more changes as necessary, etc.

The hospital changed the control program on the AHU's discharge air setpoint to stop using the average reheat setting. They employed a simpler approach based on outside air temperature and a higher baseline setpoint. We monitored the affected spaces and also adjusted air flow settings for several of the VAV boxes.

It's worth noting that this whole scenario and strategy is based on winter conditions. The summer will bring a new set of circumstances and needs, which the data will make clear when it happens.

The results of the optimization didn't change space comfort (which was fine and never an issue). It did, however, have a significant impact on the operating costs, as seen in Table 8-2.

All of the adjustments made were within the buildings, to the air handler unit and the VAV boxes. Yet as you can see from the chart above, the majority of the savings are realized in the central plant, with approxi-

mately equal portions coming from cooling and heating savings. While much of the optimization work happens in the buildings, most of the dollar savings occurs in the plant.

Repeat the Process 115 More Times

That optimization did a lot to improve operations and lower costs for the air handler. In the big scheme of things, it is only a start at reducing the hospital load enough so that next winter they could indeed run with just one chiller. Another 115 air handlers need optimization as well.

The implementation of the changes to the rest of the AHUs is still underway as this chapter is being written. However, the hospital's facilities director and staff have a plan in place to prioritize and systematically address each one. The director suspected that issues existed for some time, but prior to this analysis didn't have the evidence and facts to put an action plan into place.

Using the data we were able to evaluate every AHU, determine if space comfort requirements were being met, analyze the operations (whether meeting comfort levels or not), and make recommendations where needed. The plan exists because the data exist.

Over one third of the 115 air handlers had issues. A partial list includes:

- AHU discharge air temperature is 50°F but cooling valve is closed.
- AHU discharge air temperature setpoint is 50°F. Cooling valve at 100% all the time
- Discharge air temperature resets to 49°F based on humidity, but humidity rises because space is overcooled.
- Unit provides 51°F discharge air temperature, but not making return air temp setpoint.

Table 8-2. Cost Savings Calculation

Temp Range	Differential DA Temp	Hours at Temp	Delta Enthalpy	Avg Tons Saved	Cooling Savings	Heating Savings	Fan Savings	Total Savings
45-50°F	4.64°F	314	1.13	967	$70	$80	$17	$172
50-55°F	4.74°F	371	1.15	1,170	$90	$97	$20	$208
55-60°F	3.55°F	617	0.86	1,457	$113	$121	$33	$267
60-65°F	2.54°F	854	0.62	1,442	$112	$119	$46	$277
65-70°F	2.29°F	1,005	0.56	1,531	$118	$127	$54	$300
Totals		3,160		6,567	$ 508	$ 544	$171	$1,223

Total system operating cost: $6,612   Percent reduction: 18.5%

- Unit is not making discharge air temperature. Outside air pre-cooling has capacity.
- Unit is not making pre-cooling temperature, but is making discharge air temp. Pre-cooling valve is 100% open.
- Unit is not making return air temperature, set at 72°F, despite cooling valve @100%.
- Cooling valve is locked at 20% open and space is below setpoint.
- Zone temperature is 68°F but cooling valve is closed.

Oh, did I forget to mention, one diagnostician did the entire analysis and report in less than three (count 'em, 3) days?

THE ANSWERS ARE IN THE DATA

The exploration of the hospital's operations shows two main points:

- First, it showed in detail how the system operates; how changes have a systemic impact since the overall system is self-compensating; and how "as designed" is not "as needed" regarding how the buildings should operate.

- Second, the exercise of looking at interval data, in detail, was the key to unlocking real insights about this facility in an extremely efficient manner.

It doesn't really matter what the questions are, the answers are somewhere in the data. This case—identifying where problems existed, tracking them throughout the facility, measuring the cost impact, finding the true root cause, determining a plan to fix the problem, and measuring the results—took only three hours of engineering/diagnostician time. Another four to five hours went into communicating with the hospital over the course of a week. And, it was done without ever stepping foot in the hospital—completely diagnosed from 1,000 miles away using the data.

The optimization discussion in this chapter is only the tip of the iceberg in terms of what you can accomplish with an IT approach to optimization and diagnostics. Effective IT systems are driven by data. Data collection is not a waste of time, but does represent a lot of time wasted. Let me explain. Collecting data and having them at your fingertips is invaluable. However, in most facilities today, data collection consumes a huge amount of time by senior staff, a situation that is either not realized or accepted as "part of the job." Conservatively speaking, an engineer is likely to spend 5-8 times as long collecting data as doing real engineering analysis. Think of the value your organization would realize if the data collection to analysis ratio were reversed.

Today's HVAC systems are complex and the components are highly interdependent. The self-compensating control logic often masks problems so they go unnoticed or show up as symptoms in an entirely different part of the facility. A complete view of the data from the plant all the way to the terminal boxes, which can track and show the interdependencies, is the only way to effectively manage and diagnose these systems. The artificial walls that organizations create separating the plant from the buildings, or operations from utilities, are worse than counterproductive, they impede the facilities from operating at peak efficiency. Business-oriented IT systems that sit above the engineering/control systems break down those walls of inefficiency.

Operational and cost data are the cornerstones to getting a complete understanding of how your facilities operate. It provides you with the engineering facts to establish a systematic plan to make real progress. And once you look at the data, you will want to establish a new plan—any that you had before will be tossed once you actually see what is going on. Data create a different world for operations that completely changes what you can do and how long it takes.

Or, as Yoda might say, "The answers you seek, in the data they are."

References
Cmar, G.; Gnerre, W. "Defining the Next Generation Enterprise Energy Management System" In *Web-Based Energy Information and Control Systems: Case Studies and Applications*; Capehart, B., Capehart, L., Eds.; Fairmont Press: Atlanta, GA, 2005; Chapter 32.
Cmar, G.; Gnerre, W.; Rubin, L. "Diagnosing Complex System Interactions (from 1000 miles away)," Proceedings from the 13th National Conference on Building Commissioning. New York, 2005.
Cmar, G.; Gnerre, W. "Optimizing Building HVAC through Data and Costs," Proceedings from the 2005 World Energy Engineering Congress. Austin, TX, 2005.
Mills, E., Friedman, H., Powell, T., Bourassa, N., Claridge, D., Haasl, T., Piette, M., 2004, "The Cost-Effectiveness of Commercial-Buildings Commissioning: A Meta-Analysis of Energy and Non-Energy Impacts in Existing Buildings and New Construction in the United States."

Chapter 9

How to Automate Strategies
That Make Companies Energy Savvy

Dirk E. Mahling,
Richard Paradis,
William O'Connor, and
Selly Kruthoffer. WebGen Systems, Inc.

ABSTRACT

THIS CHAPTER PRESENTS an intelligent energy management system that provides benefits to individual buildings and portfolios of buildings, thus realizing enterprise-wide energy management. The specific system introduced is the IUE (intelligent use of energy) system. A case study is reported that shows how the IUE system responded in 2002 to a 2 MW demand response call in a 10 MW portfolio of 78 buildings. The system responds automatically without human intervention, delivering the required demand reduction at the least possible decrease in comfort. The chapter further shows continuous peak load avoidance and permanent load reduction across the portfolio. The various functional categories of the system are reviewed, as is the applicability to a large number of buildings management systems.

INTRODUCTION: WHY ENERGY MANAGEMENT IS CRUCIAL FOR TODAY'S ENTERPRISE

Energy is becoming a scarce and increasingly expensive commodity. Many enterprises today are feeling the sting of sharply rising energy prices and reduced availability. At the same time, every area of the enterprise—at some point in time—became the focus for optimization over the last decades. A quick glance at the business press shows these efforts: business process re-engineering, total quality management, just-in-time warehouse management, supply chain optimization, consolidated sourcing, or knowledge management. Cost and inefficiencies have been squeezed out of area after

area. Yet with all these efforts, energy is still one of the largest "out of control" line items for many companies' operations budgets. Increasingly this line item is gaining the attention of top-level management, who are looking for solutions to reign in the energy expenses.

In some geographical areas, supply of energy can barely keep up with demand. While private households certainly feel the consequences of rising energy costs on their pocketbooks, enterprises may actually be vitally challenged by the lack of energy available. Production may be curtailed; expansion or growth may be impossible in the region.

In addition to the questions of pure availability of energy, volatility becomes a crucial factor that enterprises, as large consumers of energy, have to manage. Many utilities have introduced tariffs that have elements of real-time pricing molded into them. Sometimes this is a fairly simple line item, sometimes it is opaque as in demand charges that are computed on the grids'—not the customers'—coincident peak. Understanding these issues in an enterprise context and acting accordingly is becoming increasingly difficult, yet imperative for enterprises.

Absent a well-developed set of tools and middlemen, similar to the one that has evolved in the stock market, volatility is hard to cope with. If we liken a continually changing real-time price of energy to a stock price, we can gain a quick insight into the inadequacy of the current tools and infrastructures around energy management. Imagine a stock market that still has rapidly changing stock prices but no "management structure," such as mutual funds, financial advisors, prospectus, news shows, internet brokers, etc. around them. Who of us would be willing to manage our own 401k by following the bare-bones, minute to minute moves of stocks in our portfolio without any help? Yet this is what we ask of energy managers when we introduce real-time pricing, or elements thereof.

As if the problems of scarcity and volatility were not enough, remnants of the regulated world survive into the current hybrid of regulation and deregulation. Effectively, any participant in the energy market has to play according to two very different sets of rules (regulated rules vs. deregulated rules) at the same time. Yet these two sets live side by side in the same geographical region, governing the flow of the very same electrons over the same wires and transmission lines.

Faced with these problems, enterprises have three choices. They can either ignore the problem and continue with "business as usual," hoping that the problem will somehow "go away" or that they can get lucky and squeeze by. This is not a very attractive option for any diligent businessperson.

The second choice is to start a large, hands-on initiative to gather data, survey, compile, and analyze energy costs across all facilities that the enterprise owns in various utility territories. Equipped with the results from this physical and financial analysis, action plans to respond strategically and tactically must be drawn up, implemented, continually enacted, supervised, and tweaked. This is a daunting order for any enterprise, which more often than not disintegrates into "paralysis by analysis."

The third choice is to leverage modern communication and decision-making technologies to automate the management of energy as much as possible. Analogous to the example we gave earlier of the stock-market, tools to monitor energy consumption and energy prices enter the scene at this stage, as well as the automation of energy management techniques known to energy managers for many years. Other elements that allow enterprises to thrive in this new world of energy are services provided by outsourced technologies and service bureaus. This chapter shows in detail that such an approach is not only possible, but has been practiced by global leaders in the industry for many years.

In the predecessor to the current volume [1], we introduced the Intelligent Use of Energy (IUE) system. We showed how IUE leverages the Internet to communicate bi-directionally with building management systems and how it automates energy management strategies, such as "static pressure reset" or "pre-cooling" for single buildings.

In this volume we show how the IUE system goes beyond energy management for a single building. Thus the role of the building management system (BMS) becomes very clear: it runs devices according to schedules or control loops in a single building ignorant of energy cost, weather forecasts, or tariff structures. The energy management system adds energy awareness to a BMS and works in tandem with it for a single building.

In this chapter, we first show that an enterprise energy management system has the advantage of bringing many different buildings with different building management systems, reporting structures, utilities, metrics, etc. under one umbrella and thus makes them comparable. We define an enterprise energy management system in its most encompassing meaning:

• an automated, internet-based system, which collects data from meters

• adds select data from many BMSs from diverse manufacturers into a normalized real-time database

• allows end-users via the internet to trend and graph these data

• provides analysis tools to interpret the data

• has a financial engine to translate physical data into financial data, not just for one building, but enterprise-wide

• provides remote control via the internet into the connected BMSs

• provides advice for continuous commissioning and one-time projects

• provides automated, enterprise-wide deployment of best energy management practices that cannot be done in a single building but require the degrees of freedom given by a portfolio of buildings.

We also show that an additional set of strategies for energy management becomes available when the IUE system has a set of buildings or portfolio of buildings under management. Additional degrees of freedom become available, while the complexity of different tariffs and rules complicates the picture. All these elements bring about the concerted "bounding" of the energy cost line-item in the operations budget.

We further show how demand-response (DR) programs set forth by utilities or ISOs (independent system operators) can become a strategic element of the companies energy management mix. We show through a case study, that the automated IUE system can handle complex DR calls faster and more reliably than any manual alternative. The case study is drawn from a global 500, national retail bank, which as a global leader in energy management and conservation, implemented these practices as early as 2002.

Thirdly, we show how enterprise energy management (EEM) optimizes energy usage under business-as-usual conditions, in particular via permanent load reduction and peak load avoidance. Based on the energy event and the continuous operation, we list the features and functions of a modern EEM system. Finally we give an outlook on current developments and trends in the arenas of technology, politics, and economics that will drive the ongoing development of advanced EEM features. The last section summarizes the results and capabilities of this specific EEM from the perspective of economic benefits to the owner.

AN ENTERPRISE-WIDE DEMAND RESPONSE EVENT: SHEDDING MEGA-WATTS, NOT COMFORT

WebGen's IUE system emerged as early as 2001 as an intelligent, automated system that leverages the Internet for energy management at an enterprise level. This section reports one of the earliest experiences with the IUE system in responding to massive demand response (DR) events in a time of acute energy sensitivity.

In 2001 and 2002 WebGen installed the IUE system [3] in 78 California branch banks and administration centers of a nationwide, American retail bank. The IUE system was connected to a total load of more than 10 MW at these 78 buildings. The total square footage was in excess of four million square feet. The building management systems ranged from Siemens and Johnson to Enercon to simple DDC units. The locations ranged from San Diego in Southern California to the San Francisco Bay area.

The IUE system connects via the Internet to the lo-cal building management system (BMS), which allows it to upload data from the BMS and to send optimized setpoint values to the local BMS. (The IUE system also provides trending, reporting, analysis, advising, financial, and many other functions, which go beyond the scope of this current chapter but are detailed in [3]). Figure 9-1 shows the overall architecture of the IUE communications chain.

Figure 9-2 shows the IUE system as an umbrella, connected to many branch banks via different connections (directly to BMS, wireless, or DDC overrides). The Universal Connect section will go into more detail for connectivity and specific BMSs.

The IUE system has numerous energy management strategies for single buildings and for portfolios of buildings, i.e., the enterprise. All energy management strategies are implemented as expert system and do not require manual intervention. The general idea for each strategy is to save energy, while keeping the loss in comfort un-noticeable to the majority of tenants. Examples of these strategies are well known in the energy management community. They include simple strategies, such as "supply air temperature reset," moderately complex ones, such as "Round-robin Load Rotation," and very sophisticated ones, such as "Enterprise-wide precooling." These strategies use psycho-physical research results such as the one, that 80% of the population does not notice a 1.6°F change in room temperature over a 1-hour time span.

Equipped with this system, the bank confidently participated in the 2002 demand response program of the California Energy Commission (CEC), which was created as a response to the energy shortages in 2000 and 2001 in the State of California.

On June 27, 2002, at 10:45 AM PDT, the CEC called for a 2 MW reduction from 11:00 AM to 3:00 PM on that

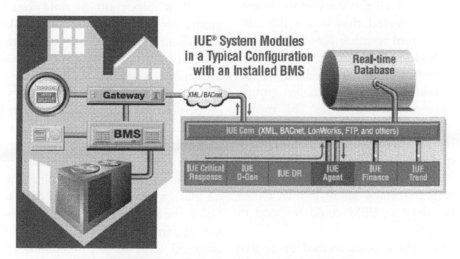

Figure 9-1. IUE Communications Chain

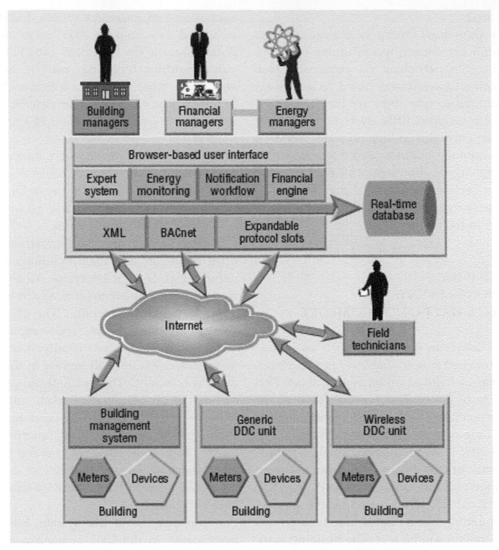

Figure 9-2. IUE Connected to many Branch Banks

very day. With only 15 minutes warning, the IUE system went from "business as usual" mode into "demand response mode."

Table 9-1 and Figure 9-3 tell the story of the event. Obviously, the agents discovered that under the current load and with the current weather forecast and the predicted occupancy, shedding load on a few large air handlers would give the fastest demand reduction. The system automatically created three load rotation groups of air-handlers that have similar properties. This means that on the fly, the system picked large air handlers from different buildings that have a similar degradation rates in their associated spaces. This can only be done in a very large building with many air handlers or in a portfolio of buildings. With the many small branch banks connected in this case, only an EEM could respond to the demand reduction call.

In other, later cases, the system created up to five load rotation groups. The number of groups depends on the amount of demand reduction called for, the number of devices available, the current and forecasted weather conditions, etc.

It is important to note that these load rotation groups are not preconfigured. The IUE system keeps data on each air-handler and the associated building zone. The data for each air-handler and its associated zone are continuously updated as new data are acquired by the IUE system. In this fashion, IUE knows about the changing properties and operating characteristics of air handlers, chillers, heat-pumps, etc and the effects these have on their zones. This allows IUE to create load-rotation groups that contain similar devices, thus getting the most from each device before terminating the demand reduction on that group and switching over to the next group. By 11:20h the IUE system had accomplished the 2MW reduction in demand across the full portfolio of buildings. This occurred without any telephone calls to local building managers, enactment of predetermined "switching off or

Table 9-1. Sequence of Events on the D/R event Day

Time Stamp	Action
10:45	DR signal received
10:46	IUE shifts from Business as Usual to Demand Response-mode
11:00	Load-Rotation agent deploys "Speed Reduction" strategy on largest air-handlers
11:10	1 MW reduction across the 78 buildings
11:15	"Load Rotation" agent releases first group of air handlers, assembles second, similar, group of air-handlers in different buildings; deploys
11:20	2 MW reduction across the 78 buildings; process keeps repeating across 3 groups
12:00	"Supply Air Temperature" agent raises SAT at select buildings to relieve Load-Rotation and Speed-Reset agents
12:15	"Load Rotation" and " SAT" agents shift deployment groups to create relieve
12:30	Combinations of "Load Rotation"
12:45-15:00	Actions repeat
15:00	End of event; system switches back to "BaU" (business as usual)

resetting" of equipment. It was done merely by the best set of actions determined at that moment by the agents in the system, which then automatically enacted those actions via the connected BMSs and DDCs.

By about 12 noon, the Load-Rotation (L/R) Agent started to run into its natural limitation, since the reduction in fan speed started to allow the buildings to heat up more than the psycho-physical JND (just-noticeable difference) allowed. Thus, L/R throttled its activities back and the SAT (supply air temperature) agent kicked in to aid the L/R agent. SAT pushes the load reduction further back into the stack by putting less demand on the chiller. The 1 pm short spike followed by the deep valley at 2 pm shows the workings of these two agents (see Figure 9-3).

Apart from L/R and SAT other agents aided in the demand reduction. Their activities were mainly a function of the equipment and the BMS at the local site. Speed-reset and Static-duct-pressure-reset are just a few of the strategies that flanked the major efforts delineated in Table 9-1. Other strategies, such as "pre-cooling," were not participants in this event, since the "heads-up" time of 15 minutes was not sufficient for this strategy to work. We will return to precooling in the next section.

The final result, which was verified by an external energy auditor, was a 2.1 MW reduction for the 4-hour time span, i.e., an 8.4 MWh reduction. During that period of time there was no noticeable increase in hot/cold calls. This case study shows that an EEM, such as IUE, can reliably deliver demand reduction without severely

impacting the comfort of tenants. It also shows that EEM is in a different league than a simple BMS, which only focus on one building.

CONTINUOUS ENERGY MANAGEMENT ACROSS A PORTFOLIO: PERMANENT LOAD REDUCTION, PEAK LOAD AVOIDANCE, AND THE BENEFIT OF BUSINESS AS USUAL

A BMS usually controls a single building. One or more meters provide the utility bills for this building. This creates the relationship between energy management measures and the effect they have on energy cost.

Managing energy in a portfolio makes things easier and harder at the same time. Easier, because more air-handlers, heat-pumps, lights, exhaust-fans, and other such devices can be addressed, thus providing more flexibility and degrees of freedom. Harder, because the complexity of multiple bills and interaction between devices grows exponentially while billing is no longer restricted to a single building.

This section shows how the IUE system described in the previous section work under everyday conditions or what is called "Business as Usual" (BaU) conditions. While the previous sections showed how the IUE system handles a one-time event, in this case the CEC's 2MW call, the current section concentrates on the measures the IUE system takes everyday to manage energy in a

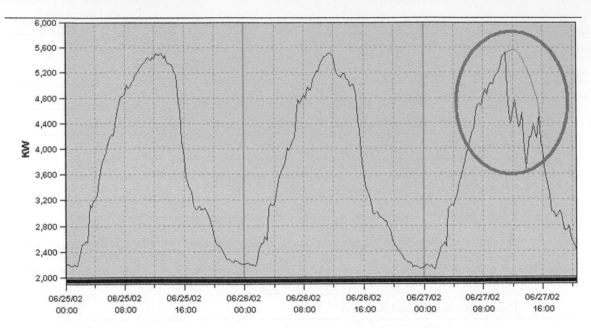

Figure 9-3. Demand for half the portfolio on event day and two previous days.

portfolio. In a way, the demand reduction event merely uses the basic services and strategies that go to work everyday in the portfolio.

Automatic Peak Load Avoidance (PLA)

One of the most important BaU cases is the avoidance of peaks in any building. Usually utility bills have demand components (measured in kW) and consumption components (measured in kWh)—both are paid in dollars. For some tariffs, the demand component can make up to 50% of the utility bill (before taxes and other such charges). By limiting the singularity of the peak on the meter for that building, a major saving can be accomplished. While consumption savings must be accomplished every minute, every day, demand savings can be accomplished by limiting that one time interval where the meter would peak. In that way it is almost the inverse of consumption savings. While consumption savings accrue little by little, demand savings require little action. But that action must happen at the right time thus the demand curve must constantly be monitored.

The IUE enterprise energy management system is connected to all meters on all buildings in the portfolio, thus spanning the current energy consumption of the enterprise. Stored in its database are energy baselines. Therefore, the system knows what the peak kW number for the same month in the previous year (or the baseline year) was. IUE can also calculate the energy cost for any given month across the portfolio.

Since IUE uses neural networks to predict energy demand in a building, it is not only constantly vigilant, but knows up to 24 hours ahead what will happen to a building in the conditions that will most likely present themselves the next day. Figure 9-4 shows the demand curve and the neural-network-based demand forecast for a building.

This allows IUE to leverage such strategies as precooling, which need ample lead-time. In the case of precooling, more kWh are used than normally would, yet the effect of a precooled building on a peak demand day is offset by the savings in demand (kW). Predicting energy use across a portfolio thus allows the IUE EEM to select short term and long-term strategies that best fit the tariffs and thus optimize demand across the whole enterprise.

Another way to leverage peak load avoidance in a portfolio is the aggregation of individual meters into one aggregate meter. Two dynamics contribute to the benefit of this measure. First, is the fact that it is highly unlikely that any two meters will peak in the same time period. Nevertheless, when utility bills are paid, the peak of each meter is used for the bill calculation.

To make this point very clear, assume meter A peaked on the 23rd at 1:45 with 340 kW. Let meter B peak on the 25th at 2:30 with 280 kW. If both meters are on a similar tariff, the combined peak is 620 kW. Yet, when the 15-minute kW readings of both meters are summed up for each 15-minute reading a different picture emerges. Table 1 shows that the combined readings peak on the 24th at 2:00h with only 580 kW. This *coincidental enterprise peak* is lower than the artificial *non-coincidental peak* of 620 kW. A full 40 kW is created by the non-coincidental addition. Having an enterprise system that allows the creation of aggregate meters in this way allows the customer to be billed on the aggregate meter

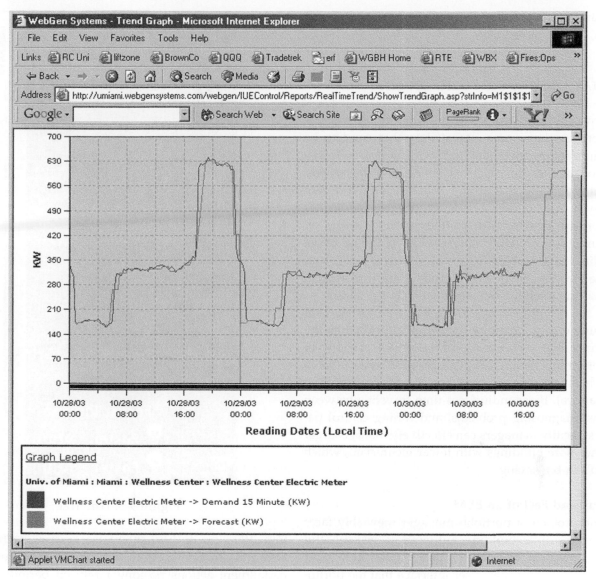

Figure 9-4. Neural Network predicts demand curve

is an immediate benefit of an EEM. Adding PLA on top of this aggregate meter multiplies the benefits.

Table 9-2.
Coincident vs. Non-Coincident Peaks on two meters.

Date/Time Stamp	kW Meter A	kW Meter B	kW Meter A+B
23rd 1:45	**340**	220	560
24th 2:00	310	270	**580**
25th 2:30	270	**280**	550

Permanent Load Reduction (PLR)

Leveraging IUE for permanent load reduction means reducing the cumulative number of kWh used in the portfolio of buildings. As explained above under PLA, permanent load reduction works very differently from peak load avoidance. In this case the neural networks assist the energy saving agent strategies by predicting the effect agent action will have upon spaces and equipment. In this way, each agent strategy can bring the best temporary setpoints for supply air temperature, static duct pressure, etc. to bear.

Employing this PLR idea for a portfolio becomes most clear when many buildings in a campus-style setting are on a chiller loop. By merely optimizing energy consumption in each single building, only a local maximum can be achieved, since the chiller plant is not taken into consideration. By using an enterprise-wide strategy that takes the aggregate energy consumption (on a virtual aggregate meter) into consideration, the local maximum issue can be avoided. In this respect the local maximum resembles the non-coincident peak issue for PLA, except in reverse.

WHY HAVE YOUR PORTFOLIO ON-LINE

Having a portfolio of facilities, of any size, means that there are large amounts of data that need to be turned into information. Having a portfolio on-line gives access to much of the data that are needed for controlling and managing cost at each site. Turning that data into easy to read information is a necessary function of the enterprise system. With limited capital to spend on maintaining/improving comfort while reducing costs, the on-line data provide a way of selecting where and how the dollars can be spent, and, more importantly, track that the intended outcomes are being achieved.

In the day-to-day operation, alarms and hot/cold calls can be more readily reviewed, minimizing the disturbance that such events can cause. Dispatching of the appropriate repair team can also be more reliably done when site data are available for review prior to on-site arrival. By spending less time traveling between locations, diagnosing problems, and tuning control parameters, facility managers can effectively manage larger areas and more buildings with fewer technicians, which can result in big savings.

The Look and Feel of an EEM

One problem a portfolio manager inevitably faces is having multiple building automation systems and learning how deal with them. Each BMS manufacturer will have its' own graphic user interface that the portfolio manager has to learn to extract data to be analyzed in order to make effective energy operation decisions. The EEM provides a standard portal into each facility regardless of BMS type, giving the user the same look and feel to all sites. Having a standard look and feel will improve the users efficiency, making more time available for using the information to make economic prudent decisions. The IUE System is interfaced over the Internet using the explorer browser so no new skill or tool is required to access each site. The IUE Navigation is based on a tree structure with items grouped in logical hierarchy based on their relationships to each other: Customer > Portfolio > Building > Meter > Device. The tree structure can be organized to fit the company's need, be it by region, manager, building type, etc.

Demand Response

Demand response programs can bring tremendous savings, but they also represent a risk to the organiza-

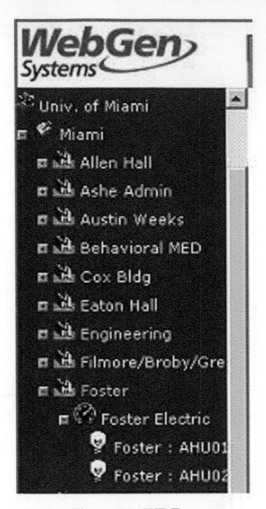

Figure 9-5. IUE Tree

tion. With an on-line enterprise system, such as IUE Demand Response, that risk approaches zero since curtailment actions no longer rely on periodic manual intervention but precise data optimized by the web based system, as seen in the previous CEC example. Being able to offer a portfolio or groups of buildings from a portfolio decreases the impact at any one site while maximizing the overall contribution to the DR program to the benefit of the portfolio owner, ISO, and others. The ability to compare disparate BMS data from any meter, building portfolio, or campus along with real-time environmental and economic data (e.g., energy price feeds) gives this tool its power to deliver on these types of curtailment contracts while maintaining acceptable comfort conditions at each site. In addition to providing the facilities management team confidence to participate in these programs, IUE Demand Response automatically manages the demand response calls from the utility or the Independent System Operator (ISO), so each contract is fulfilled in the most accurate and time-efficient manner. At the same time, IUE Demand Response maintains the highest comfort across all buildings by devising the most optimal curtailment regimen based on the most current data.

Peak Load Response

Electrical energy is billed on highest demand (kW) and by energy usage or consumption (kWh). By using weather forecasts, occupancy schedules, etc., and feeding these data into the neural networks a highly accurate forecast for demand can be predicted. Steps can then be taken to ensure that a peak is not set by utilizing various 'agent' actions such as, load rotation, static pressure reset, supply air temperature reset, fan speed control, etc. This same type of action can be triggered by a real-time price feeds from the ISO for facilities on real-time, day-ahead or critical peak pricing where the cost for a select period of time will exceed the norm significantly. A portfolio manager will not have to watch each individual building, only be sure that the correct tariff is assigned and that the peak load agents have been selected for the site.

Critical Peak Pricing (CPP)

This same strategy described above for peak load response can be utilized for facilities that are subject to the critical peak pricing tariffs. While these tariffs cost less for most hours during the year, for the ISO or Utility defined critical periods, energy price can be 10 times higher than the 'typical' rate. The few hours a year, usually less than 120 hrs/yr, at this high price should not be enough to offset having the remainder of the year at below normal tariff rates thereby reducing overall costs. Set up to receive the warning from the ISO or Utility, the EEM can trigger a peak load response to all facilities subject to the critical peak pricing tariff thereby limiting the exposure to the high cost associated with the CPP time period. The real time data can be viewed to ensure that the appropriate actions have been carried out, thereby providing peace of mind to economic decision makers during these events.

Alarming

Standard reports can be created that allows you to review a building or several buildings key performance or comfort indicators. As such, a quick glance can show which building(s) are operating out of range and therefore might lead to a hot/cold call or other comfort

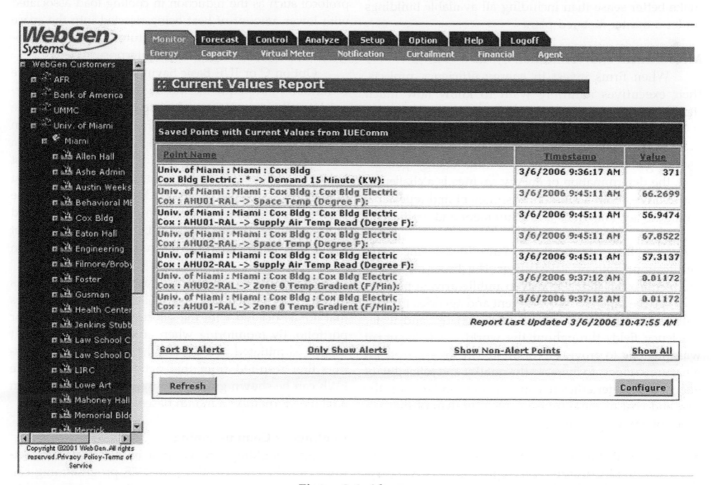

Figure 9-6. Alarms

complaint. The overview allows for a quick high-level review of all your buildings, which then helps determine the order of a more focused evaluation. All performed remotely in bathrobe and coffee cup in hand!

Analysis

Among the various analyses that can be performed with an enterprise system, one of the more interesting financial analyses involves the virtual meter. Virtual meters allow the aggregation of any number of physical meters. This means that a collection of buildings can be grouped under one virtual meter, and users can create any number of virtual meters to meet various needs. For example, virtual meters may be created on the basis of geographic area, or building use (labs vs. classrooms) or by building managers. Aggregating several meters and assigning a tariff to it can perform a cost comparison of a master meter set-up versus individually billed buildings. Having the metered data allows one to do what-if scenarios to find the optimum aggregation of buildings that yields the least utility-billed cost. Sometimes due to individual load profiles, eliminating some buildings may make better sense than including all available buildings under a master metered plan.

Measurement and Verification

When firms invest in energy efficiency projects, their executives naturally want to know how much they have saved and how long their savings will last. If the installation had been made to generate energy, then measurements would be trivial—install a meter on the generation equipment. Unlike energy generation, the determination of energy savings is a challenge, and requires both accurate measurement and repeatable methodology, known as a measurement and verification protocol.

The "International Performance Measurement and Verification Protocol" (IPMVP or The Protocol)[4] was the result of an industry driven public-private partnership between the U.S. Government and agencies in other nations, and business leaders in the energy and water efficiency fields throughout the world. That document was intended to provide a common way to carry out and measure savings, to increase the quality and reliability of energy and water efficiency investments and reduces the risk and cost of these investments. The two of the four types of M&V approaches are:

- Option B: End-Use Retrofits—Measured Capacity, Measured Consumption approach. This is the approach utilized in IUE Summary.

- Option C: Whole-Facility or Main Meter Measurement approach. This is the approach utilized in the IUE Basic Savings Report.

The Option C protocol is generally the option of choice when the bulk of the savings are coming from the implementation of energy control strategies while Option B, End-Use Measurements is the protocol of choice when system component efficiencies are being improved. Since all electrical energy decisions and activities that occur in a building are ultimately reflected in the utility bill, an installations who's intent is to constantly change or otherwise modify how these decisions are made would appear to be an ideal candidate for the Option C protocol.

Option B protocol is more readily suited, but certainly not limited to, changing to high efficiency motors or more energy efficient lamps/ballasts in existing lighting fixtures. A direct measurement before and after the retrofit to establish the reduction in energy usage, kW per fixture/kW per horsepower, and monitoring of its runtime will produce a fairly accurate savings verification report. Some 'stack effect' will be lost from the Option B protocol such as the reduction in cooling load associated with lower amount of heat being rejected into the space by the reduced wattage lighting fixture. Both of these options are available on-line in the enterprise system.

Option C or IUE Basic Savings report involves the continuous, automatic monitoring and recording of meter data allowing IUE to calculate real-time bills, create savings reports in relation to an energy consumption baseline, and to trace cost spikes. All these actions can be taken at different levels of aggregation. On the lowest level you can analyze each meter; at the next level you can analyze a whole building. One level higher is the portfolio level, a logical collection of buildings arranged in whatever fashion you prefer. Ultimately, you can get a financial picture from the top or portfolio level (given permission).

Option B or IUE Summary accumulates the results of action performed by all agent activity for the devices that are associated with the selected meter, building, or portfolio. By monitoring when an energy saving agent activity is initiated at a device, the EEM calculates savings that resulted from that activity. The cumulative kWh can be shown graphically by day or month and in a tabular format on a month bases.

Continuous Commissioning

Since buildings are dynamic in nature, optimal settings can change over time, or BMS programming can be changed, as occupancy or space function changes, so control points no longer are optimal. Perhaps a previously unused area becomes occupied. Perhaps a new building goes up across the street, shielding a building from afternoon sun. Ideally, a facility manager could determine a new "optimal" setting. Unfortunately, it is not cost-effective for an energy manager to tune a building every day, or even once a quarter. That is where an advanced enterprise system comes in handy. A computer does not get tired of observing set points, fan speeds, space temperatures, etc. This means that as conditions change, an enterprise system adapts. An advanced enterprise system can include many hundreds of rules and standard solutions based on collective experience of the best energy and facility managers in the world. The end result—rising energy costs—is held in check without sacrificing tenant comfort.

Real-time Critical Management

The challenges of mission-critical infrastructure management drive the need for real-time information. Effective data center operations requires synthesizing the vast amounts of available information into its most

critical components for intelligent decision-making, planning, and forecasting. The mission-critical site facility manager needs an IT tool that can provide connectivity and data from the wide variety of disparate electrical and mechanical systems. This information must be available in real-time to display, analyze, and present—quickly and easily. And the tool must store the data for forensic and historical trending purposes. The EEM allows this to occur not only for each site manager, but also for the regional or corporate level view of facilities. The WebGen EEM System ties together all the disparate systems into a single system, providing trend and historical data in a downloadable format.

Best Practice Transfer

With energy usage information available from all sites, benchmarking can be performed to see how facilities compare against each other as well as against industry standards, like IFMA (International Facility Management Association) or Energy Star. This type of comparison, when normalized for weather and facility type, can quickly identify the star facilities from the dogs. A

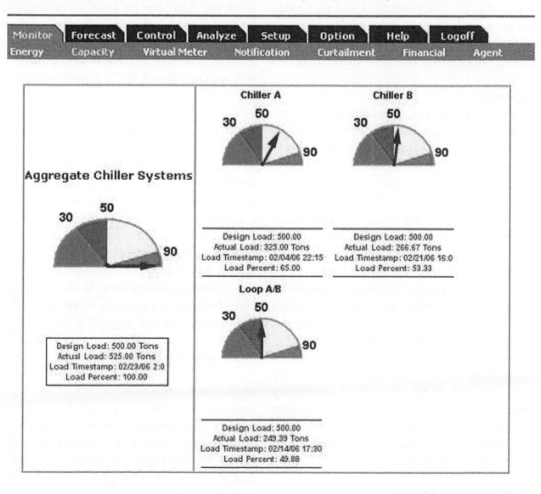

Figure 3: Analyze capacity.

program can then be established that will increase all facilities to a minimum energy performance standard. The cost savings from achieving said standard, can also be determined from these data, giving management the information they need for forecasting a cost effective budget for the required improvements. The operations utilized at the best buildings, become the standard or scope-of-work for the rest of the portfolio. This will streamline the process from conception to implementation. The EEM can then be used post-project installation, to track the actual performance versus the projected, as described earlier, thus validating the expenditure.

Energy Procurement

In states where deregulation of electric utilities has occurred, the EEM can give the portfolio manager a better barging position when purchasing electrical power. Information on a facility's load shape can generally garner a better deal; information at a portfolio level, even more so. Having the ability to automate peak load avoidance across the portfolio reduces the risk the third party provider has to assume, thereby impacting the cost that they will charge. The portfolio's energy profile could also be run against the published price feeds to determine the economic risk/benefit of switching from a more typical utility tariff to a real-time-price tariff.

Predictive Maintenance

Many of today's BMSs can be coupled to a computerized maintenance management systems (CMMS). A simplified description of this interface has the CMMS produce a work order to perform maintenance on a piece of equipment based on run-time, vibration analysis or other predetermined parameters. The CMMS would also have device specific information, so the technician being dispatched has the right parts and tools to perform the task assigned. As part of the EEM, an enhanced version of CMMS can be conducted over a wider geographic area. Tracking the maintenance issues across several buildings, when a breakdown occurs, how often equipment fails, etc. a pattern recognition analysis can be performed to supplement the building level BMS/CMMS. This predictive maintenance will result in less equipment down time, fewer hot/cold calls caused by failed equipment and reduce the amount of emergency repairs required. The more maintenance that can be performed on a scheduled bases that also only provides the amount of maintenance required for optimum performance is the least cost approach.

Unified Remote Control and Reset

A very obvious advantage of an EEM is the ability to centralize energy programs and operations centers. The EEM allows direct addressing of every BMS that is connected to it. On a lower level this means, that central policy decisions concerning energy can be rolled out, monitored, and even enforced centrally. Raising the space air temperature during the summer months by 1 degree Fahrenheit, or turning off all common space lighting after 10:00 pm are examples of such central policy decisions. In the past a vast army of consultants and engineers had to be mobilized to drive these directives into the local buildings and enact the changes in the respective buildings management systems. With an EEM, the set-point overrides for space temperature or lighting schedules can be set and enacted centrally. This saves time and money. It also ensures an almost 100% compliance with the directive. Of course local overrides should be allowed to give flexibility and responsiveness to local personnel and local exceptions. The number and type of overrides by local personnel can again be seen and reported by the EEM. That way a nice system of checks and balances is set up that satisfies enterprise energy goals, while addressing local needs. On a higher level, the EEM can be used to centrally re-program the local BMS, rather than merely override certain set-points for certain hours. This feature is somewhat harder to achieve, since the representation of time and date in the various constituent BMSs varies wildly.

UNIVERSAL CONNECT

In order to establish a true enterprise energy management system, it must be able to communicate to all the buildings in the portfolio. Most buildings within a portfolio will have a Building Management System. Building Management systems (BMS) have come a long way in the last decade. One of the most important advances technology has brought to the BMS world is in the communications area. In the past, your BMS was a closed system. Point data came in via I/O, and stayed there. Printouts and exporting data to files typically was the extent of using the amount of data stored and available within the BMS.

The reason these data stayed put, was in part, due to the fact the almost all BMS manufacturers viewed their communications protocol as a high priority commodity. The mindset was that by being proprietary the customer was forever locked into their system. It would be far too expensive to replace the system should the customer become unhappy with the system itself or the manufacturer.

The beginning stages on integration started as BMS

manufacturers saw the need to communicate directly with large pieces if equipment. As chillers and large Air Handling Units became available with built in computerized controls, it obviously would be much more cost effective to be able to communicate directly with the equipment rather than install redundant sensors so the information could be read on the BMS. Proprietary modules with drivers such as MODBUS allowed this type of communication. Now your BMS could access all the data the built-in sensors could provide.

Integration really took off with ASHRAE developing their BACnet protocol [5] and the Echelon Corp [6] developing their LON protocol. The intent of these protocols was to allow manufacturers to build product that could connect to any BMS which uses the open protocol while at the same time, allowing customers the flexibility to change both manufacturers and contractors as desired without needing to replace systems.

With the industry moving toward open protocols, the natural progression was to use open protocols to allow the BMS to communicate to the outside world. This is where we stand today. Many BMS systems can be integrated so their data can be transferred electronically in two ways. Many third party gateway devices are available, which can be thought of as translators. They communicate to the BMS via their proprietary protocol, and connect to the Internet using standard communications protocols such as HTTP and XML. Many BMS manufacturers, even though they may use a proprietary protocol throughout their system, have gateways available to allow third parties to connect to their systems via protocols like BACnet and OPC. Details on general connectivity were addressed in an earlier paper [3].

Now that technology has advanced to the point where BMS systems have reliable integration tools available, it opens the door to an entire new realm of application services that can be applied to the data on these systems—data that until now, were not easily accessible. Not only can application services now be provided for individual buildings, they can be applied across entire portfolios of buildings, regardless of the buildings individual BMS system manufacturer. By having one common server hosting the application and connecting to multiple buildings via gateways and standard protocol

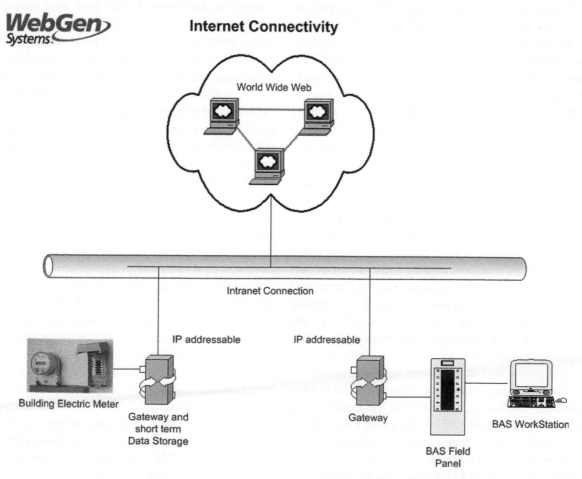

Figure 9-8. Typical BMS and electric meter Internet connectivity

interfaces, a building owner can gather data, compare building criteria across his portfolio and manipulate his entire portfolio of buildings to be as productive and efficient as possible.

By applying similar solutions to a building's electric utility meters, that data can also be captured by the hosting server.

Once these data are captured across a portfolio on one common platform, enterprise wide comparisons to other buildings within the portfolio as well as detailed energy analysis on individual buildings can be done. With the knowledge contained in these data readily available and easily accessible, the depth of analysis and modeling now possible raises the target of efficiency to a whole new level.

For the Enterprise project discussed earlier in this chapter, multiple variations of the above were implemented to accomplish the connectivity required for the 78-building project.

Many different methods of Internet connectivity needed to be installed for the many types of buildings in this project. For the larger buildings, which had well-established BMS systems, upgrades and gateways could connect directly to the existing system. For example, for JCI Metasys systems, a JCI Network Integration Engine (NIE) provided a direct BACnet interface to the system. Siemens Apogee systems were fitted with BACnet server workstations, while Siemens System 600 systems made use of Tridium JACE gateways with System 600 drivers. One of these approaches is available for nearly all BMS systems today and for this project were applied to Alerton, Automated Logic, Honeywell, Trane, Invensys and more.

In the smaller buildings, typically branch offices less then 50,000 sq ft, full BMS systems did not exist, so a wireless solution was used. Packaged roof top unit thermostats were replaced with networkable thermostats. In lieu of running network wiring throughout these older occupied buildings, a self-organizing wireless mesh network technology was used. Chips installed in thermostat bases, allowed wireless communications to a gateway located in the remote mechanical space. This allowed the IUE optimization to perform set-point changes and lock out compressors to shed load remotely, while tracking space temperatures and maintaining tenant comfort.

For Electric meter Internet connections, there were again different solutions applied. With the existence of, or installation by the utility of, a kWh pulse block (KYZ) contact, a meter storage device could be added to read the data, store them, and transmit them over the Internet when polled. The Tridium JACE gateway, MuNet's WebGate or Echelon's ILON were all used for this project to gather, store, and transmit the meter data, real time, over the internet to WebGen.

By adding simple programming logic to the BMS system to allow for remote changes, usually by the buildings existing preferred vendors, you have now upgraded the BMS system architecture to one that can be read from and written to by an enterprise energy management system which can apply services via new technologies to optimize buildings in a way you never could before.

WHAT THE FUTURE HOLDS
FOR ENTERPRISE ENERGY MANAGEMENT

The last three sections have shown the current features and benefits of today's best EEM systems. Looking into the future, a number of clear additional developments can be discerned, each one multiplying the value of an EEM system.

One of these developments in the energy markets is the return of real-time pricing. The 4CP program in Texas and the coincidental-peak tariff in Georgia are harbingers of this trend. In the past, real-time pricing was not the success it's originators had hoped for. One of the many reasons is the lack of tools and infrastructure to deal with a rapidly changing price signal on the side of the energy consumer. In the introduction to this chapter we drew the analogy to the stock market with its many, rapidly changing price signals; we hinted at the tools that must be given to energy consumers, large and small, to successfully leverage a real-time price signal. The EEM described in this chapter is one of those tools. Since the EEM can monitor the real-time price signal in many different utility districts, the EEM can chart the best course of action for the buildings in that district. Having more than one building connected, the EEM could initiate a "voluntary, internal demand response" event when it observes grid demand peaking. This action would take advantage of the definition of the coincidental grid-peak pricing in Georgia and Texas. With real-time price feeds, the EEM can load-rotate among many buildings to adjust to minute changes in the price, going well beyond the coarse set of actions delineated today.

A second, major area of progress for EEM is communications technology. In particular power line communications (PLC) and wireless mesh networks are of interest in this regard. Cheaper, more reliable, and more widely available communications lines drive the value proposition of EEM ever deeper into the enterprise.

Power line communications (PLC) is a technology that leverages existing power lines in a building (or even long-distance transmission lines) to relay data signals. Looking at the building first, PLC turns every electrical outlet in the building into an Internet drop. The entity that is to be controlled and measured (electricity) becomes the medium that relays the necessary information. The traditional separation of energy delivery (through electrical wires) and informational clearing (via meter reading and utility bills) is now co-joined. Every room and every device becomes directly IP-addressable without the need for an intermediary BMS. Or a secondary, redundant sensor network for air quality, to give just one example, could be built this way, augmenting the BMS. The BMS data and the data from the PLC based secondary sensor network both feed into the EEMs, which in turn can make crucial decisions based on many different, independent information stream. In return, PLC can be used to access devices, which are traditionally stand-alone devices, working independently of the BMS. Examples here are VAV-boxes that are controlled by a local thermostat loop. Cheap PLC can make these VAV-boxes IP-addressable and add them to the EEM logic.

Wireless communications, in particular wireless mesh networks, open a different avenue for extending the value proposition of EEM. Wireless solutions have already existed in the BMS and facility management space. All these solutions were of a point-to-point nature, we a wireless hub was installed, that communicated with satellites directly. The communications pattern was a hub-and-spoke type configuration. The problem with such networks is its vulnerability along the spokes. To overcome this problem and to facilitate installation of such networks, wireless mesh networks were invented. These types of networks consist of many wireless nodes, where each node can communicate with each other node. Such networks are self-configuring, driving down installation costs. They also provide many paths from one node to any other node, thus providing redundancy of communication paths, which yields higher reliability. Such networks are applicable in environments were right of ways do not exist (e.g. a meter in a parking lot across the street), or where building idiosyncrasies disable other solutions (e.g., pulling communications wire in asbestos-ridden buildings). Controlling individual ceiling lights or outside signage are just a few of the energy management applications made possible in the future by wireless mesh networks.

With the federal energy policy act of 2005, intelligent meters have been mandated for many buildings. The idea of bringing energy consumption and information flows together has already been outlined above and intelligent meters are but one embodiment of it. Intelligent meters, in addition to being Internet readable, add the ability actuate control signals. Simple actions, such as switching air-handlers on or off via a relay that is part of an intelligent meter extend the capability of the EEM in the future, as intelligent meters become more common.

Measuring the impact of energy consumption and energy savings are key objectives of the Kyoto protocol and the national energy star program. An EEM is perfectly suited to aid in this endeavor, since it has many of the necessary information streams already pointed to it. Energy consumption and energy savings are directly related to their equivalents in NO_X, SO_X, and CO_X. The EEM can become the prime tool to tally up such emission credits or to start automatically trading them.

Continuing in the direction of green technologies and energy, intelligent EEMs in the future can help to make the best decision in the selection of energy mix. Based on current weather conditions and predicted usage patterns, the EEM may decide to leverage PV units and feed the energy directly into the building, rather than storing the PV electricity in UPS systems for later usage. This may help with PLA at the moment in time. Similar decisions across the portfolio can be made for the addition of power from BUGs (back up generator) and from hydrogen-cells (once they become commercially viable).

A current research trend in information systems will certainly find its way into the EEM arena: self-healing systems or systems that can withstand the loss of critical components. As energy and energy management becomes a crucial element in fortifying our infrastructure against terrorist attacks, the ability to detect such an attack and quickly react to it becomes crucial. EEMs can pool the intelligence from air quality sensors in many different buildings and even from public spaces. Having an instantaneous map of an attack allows the intelligence in an EEM to decide between such actions as stopping the ventilation for an affected buildings zone or running all fans in reverse to dump the toxins out of the building (unless the exhaust is located over the local evacuation assembly point).

A second aspect of creating robustness in our energy infrastructures is that of self-healing systems. In the wireless mesh networks paragraph we already encountered an example of a network that can withstand the loss of some of its nodes and re-route communications along a different path. An EEM system, by definition, consists of numerous sub-systems, i.e. the building management systems, which are sophisticated stand-alone systems in their own rights. Further research and development in the area of EEM will show us how to leverage advanced communication structures and local intelligence to create self-healing or at least sufficiently coping systems.

One last glimpse into the future is the consolidation of various modalities of energy and other resources. Already today, many EEMs can handle the output from gas, water, steam, and electrical meters, thus providing a multi-modal energy database. Moving this trend a few steps further points to the integration of EEM with ERP (enterprise resource planning) systems. Energy reading from meters can now be matched to budgets and line items from the general ledger. Energy, one of the largest "out of control" line items in the operational expense section, becomes a controllable entity, just as warehousing or value chain have before it. This development closes the arc that spanned this chapter, starting with the premise of EEM in the introduction and ending with the integration of EEM and ERP in this chapter.

SUMMARY

In this chapter we presented the current state and the future of enterprise energy management. We did this from a number of different perspectives, including technology, process, operations, and policy.

We presented the "state of the art" in EEM (enterprise energy management) via the IUE (intelligent use of energy system). We showed how the IUE system connects to a multitude of BMSs (building management systems) from different manufacturers. With IUE connected to different BMSs and DDCs, we presented in a case study, how the IUE system automatically, without human intervention, handles a large scale demand response call involving 78 buildings, delivering over 2MW for 4 hours from a 10 MW portfolio. Details on the sequence of actions was given, that showed that these

actions are not pre-established, but are computed at the time of demand to ensure that the contracted demand can be delivered, while keeping any decreases in comfort to a minimum. For more detail on the workings of the strategies, we referred the reader to the companion chapter in the previous volume of this series [3].

From this case study we generalized to the "business as usual" scenario, which mainly comprises "peak load avoidance" (PLA) and "permanent load reduction" (PLR). These two value propositions are currently at work in all EEM installations of the IUE system. A comprehensive list of EEM features and functions was given, showing the various conceptual problems that EEM systems have solved beyond the connectivity issue. Some of these issues were the conceptual representation of different BMSs in the EEM, the associated user interface, or the handling of different time and tariff zones.

In total, the value of an automated EEM should have become clear, for both cases, the demand response call case and the business as usual case. While many of the benefits and features introduced in this chapter may seem advanced, they have actually been around since the start of the 21st century and delivered provable value to many enterprise customers.

Bibliography

[1]   Capehart, B.L. (editor): *Information Technology for Energy Managers*; The Fairmont Press, Lilburn, Georgia & Marcel Dekker, New York; pp, 2004.
[2]   Capehart, B.L. and Capehart, L.C. (editors): *Web Based Energy Information and Control Systems*; The Fairmont Press, Lilburn, Georgia; 2005.
[3]   Mahling, D., Noyes, M., O'Connor, W. and Paradis, R.: *Intelligent Use of Energy at Work*; in Capehart 2005, pp. 173 to 188; 2005.
[4]   *International Performance Measurement & Verification Protocol, Concepts and Options for Determining Energy and Water Savings* Volume 1, Revised March 2002, DOE/GO-102002-1554, www.ipmvp.org
[5]   www.BACnet.org —Accessed March 30, 2006
[6]   www.echelon.com — Accessed March 30, 2006

Chapter 10

Bringing Building Automation Systems Under Control

James Lee, CEO, Cimetrics Inc.

EXECUTIVE SUMMARY

ENTERPRISE-WIDE THINKING has been applied to many aspects of business: supply-chain management, manufacturing, inventory control, quality control, human resources, etc. Now it's time to apply this model to enterprise-wide facilities management. Instead of thinking of building automation as merely a series of controls, it is time to think of it as part of the overall IT infrastructure.

Significant structural problems exist today in the operating management of institutional real estate. These issues result in overspending on building maintenance and energy, undercutting asset profitability. Most real estate owners do not realize the economic impact of these inherent structural problems.

In general, energy costs for buildings are one of the largest variable components of the cost structure and are becoming a larger component as energy prices increase. Over 80% of buildings are never commissioned and the potential energy efficiencies never fully understood. Building management is sometimes assessed based on energy utilization, however tenant comfort is often achieved without a real understanding of energy cost impact.

Additionally, the management and operation of most buildings (even newer and highly sophisticated "smart buildings") is performed by facility personnel or is outsourced completely. While these personnel may be knowledgeable about operating basic mechanical equipment and making simple repairs, they often lack the training and the capacity to actively diagnose the performance of a building and to understand its implications for profitability. This lack of understanding of performance also results in reactive rather than proactive repair and maintenance operations, ultimately a more costly approach for the building owner.

Building owners are beginning to understand the inherent inefficiencies that exist in the operations of their real estate assets. They are proactively looking for ways to improve their return profile. Cimetrics provides independent data analysis and portfolio-wide consolidated reporting to the facilities and energy departments of building owners. These reports analyze and improve energy efficiency and reduce energy, operations, maintenance and regulatory costs. Cimetrics links into a facility's mechanical equipment (Heating, Ventilation, Air Conditioning, Lighting, etc.) through the building's automation system and utility meters, acquiring an ongoing flow of operational data. In the past, these raw data have been discarded by facility departments due to their volume and complexity. Now, however, Cimetrics collects this information portfolio–wide, and transmits it to a centralized database where Cimetrics' engineers use a set of proprietary algorithms to analyze and mine it for value.

The results are a wealth of ongoing unbiased recommendations and management information that enable firms to reduce energy and operational costs, identify potential equipment problems in order to avoid downtime, and enjoy the benefits of a continuously commissioned facility. This process of data acquisition, analysis, and reporting is called Infometrics. The Infometrics service provides a means to understand and manage building operations and their implications for cost structure and economic returns. By requiring little capital up front, Infometrics can create an immediate and lasting reduction in the operating cost structure of a building.

Infometrics has numerous potential benefits to building owners, including the following:

- Reduced energy consumption and energy cost
- Prioritization of equipment maintenance
- Reduced downtime caused by mechanical equipment failure
- Improved facility operations
- Ongoing commissioning of mechanical systems and control systems
- Reduced risk of indoor air quality problems
- Identification of profitable mechanical retrofit opportunities

- Improved occupant comfort
- Knowledge of facility energy consumption patterns and trends

Building controls companies, equipment and systems manufacturers, energy providers, utilities and design engineers will face increasing pressure to improve performance and reduce costs. These pressures drive the development, adoption and use of Infometrics.

CIMETRICS INC. AND INFOMETRICS

Cimetrics Inc. provides high-value energy and facilities management services to owners and occupants of commercial, institutional and industrial buildings. Through its Infometrics suite of products and services, the company collects real-time data from a customer's building automation systems (i.e., HVAC, lighting, fire control, etc.), integrates information from multiple facilities, applies proprietary algorithms, and generates regular and highly detailed reports for the building owner's facilities and energy departments. These reports identify opportunities to reduce energy, maintenance, operational and regulatory costs; uncover potential equipment problems; point to profitable retrofit projects; improve occupant comfort and enhance facility operations and uptime.

Cimetrics links into a facility's mechanical equipment through the building's automation system and utility meters, acquiring an ongoing flow of operational data. In the past, these raw data have been discarded by building operations staff due to their volume and complexity. Now, however, Cimetrics collects this information portfolio–wide (across multiple buildings) and transmits it to a centralized database where Cimetrics' engineers use a set of proprietary algorithms to analyze and mine it for value. The results are a wealth of ongoing unbiased recommendations and management information that enables firms to reduce energy and operational costs, identify potential equipment problems in order to avoid downtime, and enjoy the benefits of a continuously commissioned facility. This process of data acquisition, analysis, and reporting is called Infometrics.

The ability to create "smart buildings" is taking shape rapidly due to the proliferation of new technologies and the internet revolution, and Cimetrics has developed the technology and services to make the "smart building" concept a reality for the property owner/manager. To reduce its clients' energy costs and improve productivity, Cimetrics implements its Infometrics solutions by: (i) connecting to clients' building automation systems; (ii) analyzing the data produced (through proprietary algorithms and other software technology); and (iii) assessing system performance in order to better manage utility and facilities costs. Infometrics allows building owners to integrate building automation systems and energy equipment at the building level with information systems at the corporate level.

Infometrics provides commercial, institutional and industrial building owners the ability to improve substantially the operating control, costs and efficiency of their buildings through greater communication and efficiency of building systems. Cimetrics was instrumental in developing and implementing the Building Automation and Control network ("BACnet®"), the dominant open standard (ISO 16464-5) in building automation communications world-wide. Cimetrics is the world leader in the development of BACnet® communication software, network analyzers and routers which, along with their analysis and recommendations, enable the Infometrics solution.

Infometrics offers the only complete remote monitoring solution in the marketplace, leveraging Cimetrics' depth of knowledge in system connectivity, proprietary analysis algorithms, engineering and high-touch consulting. A dedicated analyst is assigned to each client, providing unparalleled access and responsive service on a range of issues from reviewing periodic reports to maintaining communications and problem solving.

Cimetrics' professional team of engineers, project managers and analysts offers expertise in energy management and building operations, and in all aspects of optimizing facilities for energy and operational efficiency, maximizing clients' potential for significant savings. Cimetrics provides services ranging from energy cost savings analysis to long-term monitoring, analysis and reporting of building data. Cimetrics provides a complete, unbiased solution for a facility's needs by working solely for the building owner.

THE INFOMETRICS PROCESS

Collecting Data from Building Systems

The Infometrics system links into a facility's mechanical equipment through the building's automation system and utility meters, acquiring an ongoing flow of operational data. In the past, these raw data have been discarded by facility departments due to their volume and complexity.

Now, however, Cimetrics collects this information

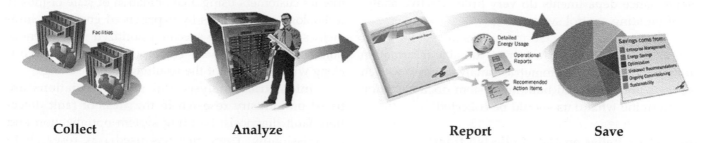

Collect   **Analyze**   **Report**   **Save**

facility-wide from multiple disparate systems through the BACnet® protocol, and transmits it securely via the internet to their analysts. This scalable data processing technology is capable of collecting and analyzing information from thousands of buildings worldwide over long periods of time.

Data Analysis by Software and Engineers

After connecting to a building automation system, data relevant to the analysis are transmitted over a secure internet connection to the Infometrics data center. Cimetrics energy, electrical and mechanical engineers use proven algorithms and software to analyze building efficiency. Analysis algorithms use static data (equipment specs, system topology) and dynamic data (weather information and operational data collected on each piece of HVAC equipment as well as the entire building's mechanical systems).

Infometrics algorithms have been designed based on standard industry techniques and academic research. The analysis team has 100+ years of experience in energy engineering, controls, communications, and software development. Infometrics provides an independent measurement of building systems' efficiencies. The depth of the team's expertise enables Cimetrics to mine value from data for building owners. The Infometrics analytical approach, examining both static and dynamic data, focuses exclusively on adding value to customers' portfolios.

Complete Report Delivery

Periodically, Cimetrics analysts create a report on facility performance based on data which have been collected and processed by the Infometrics system. This report includes management information on energy consumption and mechanical system performance as well as specific prioritized recommendations. Target energy consumption and operational characteristics are identified and variances from predicted results are analyzed, problems identified and appropriate measures for remediation recommended to the owner.

The results are a wealth of ongoing unbiased recommendations and management information that

enables firms to reduce energy and operational costs, identify potential equipment problems in order to avoid downtime, and enjoy the benefits of a continuously commissioned facility. Infometrics engineers work with facility staff and owners' contractors to deliver maximum value to the building owner.

Implementation of Recommendations and Savings

Infometrics' prioritized recommendations uncover hidden maintenance issues, providing a road map to immediate savings. Building staff can now act effectively to create value by optimizing equipment performance, reducing costly downtime and improving comfort.

Building Data

The value that Infometrics delivers is primarily derived from data collected from sensors and actuators that are connected to building control systems and building mechanical equipment. When combined with system and equipment set points, the data can tell a great deal about how well the building control system is performing and where the problems are, including issues that can't be detected through simple equipment observation.

There are vast amounts of data available in a large building—if you were to read just one sensor every 15 minutes, you would have 35,000 data samples per year from that one sensor alone. Some buildings have thousands of sensors and actuators. Special tools and expertise are needed to collect, manage and analyze all of this information. It is not surprising that most facility maintenance departments do very little effective analysis of building control system data.

Cimetrics has invested many person-years of effort into understanding how to extract valuable information from building data. Infometrics analysts review each facility's points list and building system documentation to determine what data should be collected.

Infometrics Relies on Data Collected from Customers' Building Control Systems

In order to maximize the potential value of the building systems data collected, they must be gathered at the right time and at the right frequency. Cimetrics has developed a special data collection device, called the Infometrics Cache, which gives our analysts excellent control over how information is collected. The Infometrics Cache is connected to the building control system at each customer facility and transmits critical information to Cimetrics via a secure, firewall-friendly internet connection. In most cases, the collection of data consumes a small fraction of the available network bandwidth.

Cimetrics engineers have considerable experience in connecting Infometrics-enabling equipment to different building control systems. BACnet® is the preferred communication protocol—if the control system uses BACnet® as its native network protocol, then the connection of the Infometrics Cache is very simple. If the building control system does not use BACnet®, or if BACnet® is not the primary protocol in use, then Cimetrics may need to arrange for the installation of a hardware or software gateway that will translate the necessary data into a format supported by our device.

The Infometrics Cache temporarily stores the data collected from the facility's building control system, then periodically transmits the information to the Cimetrics Data Center. In most cases, the best way to send the data to Cimetrics is to use an existing internet connection at the customer's facility. The Infometrics Cache transmits information using a secure industry-standard network protocol that is compatible with firewalls.

The Infometrics system has been designed to ensure that the security of customers' systems and data are maintained. Cimetrics I.T. experts are prepared to work with security-conscious customers to ensure that the Infometrics system meets their particular requirements.

Turning Building Data into Actionable Information

Facility managers need concise and accurate information to help them make decisions about how to maximize the performance of building systems. Infometrics was developed specifically to address this need.

Cimetrics delivers actionable information to Infometrics customers using a combination of state-of-the-art technology and analysis by experienced engineers. Infometrics reports include specific prioritized recommendations, most of which can be implemented at low cost, along with estimates of the resulting annual savings.

Infometrics analysis and recommendations are based on industry research in the areas of fault detection, fault diagnosis, building system optimization and commissioning. Cimetrics has used this research to develop algorithms and software tools that allow Infometrics staff to quickly and efficiently analyze the data that are continuously collected from customers' building systems.

An Infometrics analyst is assigned to every Infometrics customer. The analyst is responsible for creating a data collection strategy, analyzing the data, producing the Infometrics reports, reviewing recommendations with the customer, and being available to each customer for consultation when questions arise. Each of the Infometrics analysts has years of experience as an energy, electrical or mechanical engineer. Cimetrics analysts can also consult with staff engineers who are experts on building control systems, HVAC equipment and data analysis.

Cimetrics believes that the Infometrics approach—skilled analysts using state-of-the-art technology to analyze building data—is the best way to deliver actionable information to customers that want to maximize the performance of their building systems.

The following is a brief description of the Infometrics project development process:

Connectivity

The Infometrics system links into a facility's mechanical equipment (heating, ventilation, air conditioning, etc.) through the building's automation system and utility meters, acquiring an ongoing flow of operational data.

Cimetrics collects this information facility-wide from multiple disparate systems through the BACnet® protocol, and transmits it securely via the internet to the Data Center for analysis. This scalable data processing technology is capable of collecting and analyzing information from thousands of buildings worldwide over long periods of time.

Analysis

Cimetrics' energy, electrical and mechanical engineers use proprietary algorithms and software to analyze building efficiency. Infometrics algorithms have been designed based on standard industry techniques and academic research. The Cimetrics team has 100+

years of experience in energy engineering, controls, communications, and software development. Infometrics provides an independent measurement of building system efficiencies. The depth of Cimetrics' expertise enables Infometrics analysts to mine value from data for building owners. Infometrics' analytical approach, examining both static and dynamic data, focuses exclusively on adding value to customers' portfolios.

Reporting

Periodically, Infometrics analysts create a report on facility performance based on data which have been collected and processed by the Infometrics system. This report includes management information on energy consumption and mechanical system performance as well as specific recommendations. Target energy consumption and operational characteristics are identified and variances from predicted results are analyzed, problems identified and appropriate measures for remediation recommended to the owner.

The reports contain a wealth of ongoing unbiased recommendations and management information that enables firms to reduce energy and operational costs, identify potential equipment problems in order to avoid downtime, and enjoy the benefits of a continuously commissioned facility. Cimetrics' engineers work with facility staff and owners' contractors to deliver maximum value to the building owner.

Infometrics Project Timeline
Step 1: Facility Assessment

A facility assessment is intended to investigate the potential for Infometrics to provide energy and operational cost savings, as well as improved performance of facility systems. The assessment typically includes an analysis of two years of the facility's utility bills and a points/equipment list collected from the facility's building automation system. Operating personnel are inter-viewed to obtain general information about the facility (square footage, operating schedules, utility metering systems, energy conservation strategies already implemented, known operational problems, planned system changes, etc.). Based on the results of the facility assessment, Cimetrics develops a proposal for Infometrics services.

Step 2: Infometrics Connectivity

Once Cimetrics and the facility owner have reached an agreement on the scope of Infometrics services to be provided, Cimetrics establishes connectivity with the building systems that are to be monitored. A special Cimetrics device, the Infometrics Cache, connects to the facility's building systems via the BACnet® protocol, and internet connectivity is established. Every building automation point that is needed for Infometrics analysis is entered into the Infometrics database, and each point is assigned a standard name based on its function within the system. Often an Infometrics customer is a building owner with several facilities and varied facility control systems spread over a wide geographic area. Cimetrics' Infometrics solution enables connection and monitoring of all of these locations to produce the raw data that will ultimately be analyzed to produce recommendations.

For example, at American University, Infometrics currently monitors three buildings, totaling 242,000 SF (see case study, page 100).

Step 3: Infometrics Monitoring

Cimetrics monitors the facility owner's electrical and mechanical systems 24 hours a day, and stores the data collected on average every 15 minutes from those systems in a database system located at Cimetrics' secure data center. The data are then trended and studied by Infometrics analysts using Cimetrics' proprietary analysis tools.

For example, in a technique called MicroTrend, a

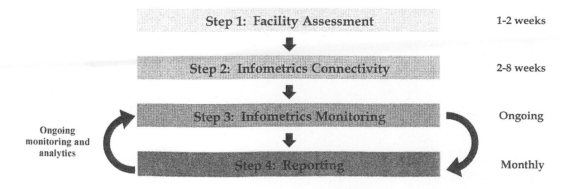

Infometrics Project Timeline

Step 1: Facility Assessment	1-2 weeks
Step 2: Infometrics Connectivity	2-8 weeks
Step 3: Infometrics Monitoring	Ongoing
Step 4: Reporting	Monthly

Ongoing monitoring and analytics

short-duration, high-frequency query is done on all pertinent data regarding a specific system component. For instance, a variable air volume box (computer controller damper) may have up to 16 variables that represent its real time performance. Characteristics that are measured may include: control loop constants, actuator travel, and/or energy consumption per unit of handled air. This technique enables analysts to uncover rapid fluctuations that would not otherwise be apparent.

Step 4: Reporting

Cimetrics provides clients with periodic reports which detail the facilities' targeted energy consumption and operational characteristics in a very usable format for clients, their energy departments, building managers, maintenance personnel, etc. Variances from predicted results are analyzed. Problems are identified, prioritized and reported so that appropriate measures can be taken. An Infometrics Analyst reviews each report with the client during a scheduled conference call.

Technology Overview

The Infometrics system links into a facility's mechanical equipment through the building's automation system and to its utility meters using the firewall-friendly Infometrics Cache device, acquiring a secure and ongoing flow of operational data.

A simplified Infometrics system architecture is shown below. Major hardware components include a data collection device that is connected to each customer's building automation system, the database system,

and the network operations center. A brief description of the components follows:

- **Data Collection Device ("Infometrics Cache").** Acquires data from the customer's building automation system, temporarily stores the data, and periodically transmits the data to the Infometrics database system over the internet using a secure, "firewall friendly" network protocol.

- **The Database System**. Stores configuration information, raw data collected from customer systems, processed data, and issue tracking data. The system is designed to easily scale out as needed.

- **The Network Operations Center ("NOC")**. Monitors the performance of all components of the Infometrics system and critical devices in the customer's building automation system.

Cimetrics has developed a considerable amount of software to support Infometrics, and has also made use of several products that were developed for our BACnet® products business. Infometrics analysts use both commercial packaged software and proprietary software tools.

The Infometrics system collects data from a customer's building automation system by directly connecting to the system's network. If the network does not use the industry-standard BACnet® protocol, then Cimetrics and the customer arrange for the installation and con-

Infometrics System Architecture

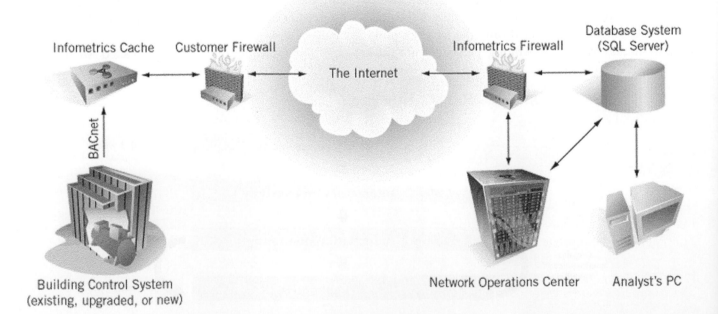

Infometrics Cache   Customer Firewall   The Internet   Infometrics Firewall   Database System (SQL Server)

BACnet

Building Control System (existing, upgraded, or new)

Network Operations Center   Analyst's PC

figuration of a BACnet® gateway. BACnet® gateways are available for most popular building automation systems developed within the last ten to fifteen years. Cimetrics has developed considerable expertise on how to interface to popular building automation systems.

Point naming is of considerable importance in enabling the Infometrics service, and Cimetrics encourages the adoption of standard point names. By using a standardized point naming convention, several automated configuration tools can be used which reduce configuration costs.

Infometrics is fundamentally different from traditional commissioning services in that its value is derived primarily from the analysis of large amounts of building operation data collected over a long duration. The huge volume of data collected from every building each month (1,000 points per building average * data sampling frequencies of 15 minutes or less * 30 or so days per month = over 2 million data points per building per month; multiplied by a growing number of total buildings monitored = hundreds of millions of data points collected per month) must be converted into a high-value decision support tool for the building owner (i.e., a periodic report). Over the past five years, Cimetrics has developed scalable data processing technology capable of collecting and analyzing data from thousands of buildings. The Infometrics system takes advantage of data accessible through the building control system installed in each facility. Building automation systems are connected and points of data relevant to the analysis process are transmitted over the internet to the Cimetrics data center where they are processed using Infometrics algorithms. Cimetrics energy and mechanical engineers work with facility staff and contractors to deliver value to the building owner.

Because of its deep industry experience, Cimetrics is able to provide a complete Infometrics solution from surveying buildings to the full deployment of data connectivity, analysis and service delivery. Because the company helped most manufacturers develop and test their BACnet® products, we have considerable knowledge of the communication capabilities of their products. We also leverage our years of work designing/installing routers and working as the liaison between building automation and IT to transmit data reliably and securely across the internet.

Cimetrics' Infometrics program provides powerful algorithm-based analysis of continuously collected building automation data. Valuable periodic reports enable owners to optimize efficiency and comfort, lower maintenance costs, and effectively manage facility staff and contractors across their portfolio of properties.

CUSTOMER BENEFITS

The benefits of ongoing commissioning have been demonstrated and are becoming better known. Several independent assessments have yielded estimates of cost savings due to ongoing commissioning and optimization (see table below).

Infometrics' success has been driven by four main factors:

1. The growth of internet and broadband wide-area networks, which provide the communications infrastructure necessary for real-time remote monitoring and control;

2. Increasing adoption of BACnet® and other open communication standards for building automation and control, which reduce the complexity and cost of advanced building management systems;

3. The emergence of an enterprise-wide management paradigm; and

4. Rising energy prices, creating increased interest in energy efficiency and cost savings.

Commissioning and Optimization Cost Savings

Study/Agency	Finding
Federal Energy Management Program (FEMPstudy)	20% on average (based on 130 facilities [Texas A&M])[1]
TIAX Report for US Department of Energy (DOE)	5-20% (guidelines)[2]
California Commissioning Market Characterization Study (CCMCS)	15% (existing facilities) 9% (new construction)[3]
National Institute of Standards and Technology (NIST)	US$0.16 per square foot (energy alone)[4]

[1]Liu, Minsheng, David Claridge and Dan Turner. "Continuous Commissioning Guidebook," October, 2002, page v.
[2]TIAX report for the DOE, "Energy Impact of Commercial Building Controls and Performance Diagnostics: Market Characterization, Energy Impact of Building Faults and Energy Savings Potential," November, 2005, page 9-137.
[3]Haasl, Tudi and Rafael Friedmann. "California Commissioning Market Characterization Study," Proceedings of the 9th National Conference on Building Commissioning, Cherry Hill, NJ, May 9-11, 2001.
[4]Chapman, Robert. "The Benefits and Costs of Research: A Case Study of Cybernetic Building Systems," NISTIR 6303, March 1999, page 80.

All of these factors have come into place very quickly over the past five years:

- *The Growth of Internet and Broadband Wide Area Networks*: Until recently, communication between buildings typically required the use of telephone lines, but now low cost wide area network communications systems are available. This enables the cost-effective movement of large amounts of information across multiple locations.

- *The Development of Open Systems*: Just as TCP/IP and other standard protocols were crucial to the development of the internet, open systems are critical to the development and deployment of new technologies in the building industries. The building industry has now adopted BACnet® as the open standard of choice for data communication. BACnet® makes the integration of building systems for various manufacturers and across buildings significantly more straightforward.

- *The Emergence of an Enterprise-wide Management Paradigm*: Major corporations have embraced enterprise-wide management systems in supply chain management, human resources, finance, and manufacturing. Enterprise-wide management systems have now been expanded to building management, enabling building owners and managers to view their energy operations, maintenance and regulatory issues as part of the whole picture and their building automation system as part of the IT infrastructure.

- *The Recent Rise in Energy Prices*: With natural gas prices quadrupling since 2000 and electricity prices rising by as much as twenty percent, building owners are faced with a significant increase in their largest operating cost: energy. The energy problem has become a hot button issue with real estate owners.

The total savings afforded to Infometrics customers are comprised of multiple components, described below:

Summary of Infometrics Benefits
- Reduced energy consumption and energy cost
- Prioritization of equipment maintenance
- Reduced downtime caused by mechanical equipment failure
- Improved facility operations
- Ongoing commissioning of mechanical systems and control systems
- Reduced risk of indoor air quality problems
- Identification of profitable mechanical retrofit opportunities
- Improved occupant comfort
- Knowledge of facility energy consumption patterns and trends
- Integration with information technology systems

- *Reduced energy consumption and energy cost.*—Access to and analysis of energy usage data delivers cost savings resulting from better control system performance, improved energy load management, and smarter energy purchasing.

- *Prioritization of equipment maintenance.*—The information provided in Infometrics reports allows maintenance managers to prioritize maintenance activities, more effectively utilizing both in-house staff and outside service contractors.

- *Reduced downtime caused by mechanical equipment failure.*—The use of Infometrics allows systems and equipment to operate under near-optimal conditions for extended periods of time. In addition, equipment and component malfunctions are diagnosed and remedied before catastrophic failure occurs. As a result, equipment life is extended, fewer replacements are required, and replacement costs decline. Furthermore, better diagnostic information enables support staff to more quickly and effectively repair equipment and components.

- *Improved facility operations.*—Prioritized recommendations allow facility managers to develop proactive maintenance plans. Catastrophic downtime is avoided with ongoing equipment assessment and proper maintenance, and life cycle costs are reduced.

- *Ongoing commissioning of mechanical systems and control systems.*—Systems operate at near-peak efficiency with proper maintenance, minimizing energy waste and extending equipment life. Purchasing and upgrades can be planned well before equipment failures are likely to occur.

- *Reduced risk of indoor air quality problems.*—HVAC system maintenance ensures proper air flow and the correct ratio of outdoor to recirculated air. Occupant health is maintained with maximized air quality and minimized temperature variation.

- *Identification of profitable mechanical retrofit opportunities.*—Equipment performance is continually assessed, highlighting those components which are operating at suboptimal levels and predicting cost savings to be gained by replacing them. Building owners are able to assess ROI for informed decision making on retrofits and replacements.

- *Improved occupant comfort.* Infometrics reports enable improved occupant comfort from enhanced operating performance of HVAC systems. Infometrics gives building operations staff the information needed to provide a more consistent environment which has a significant impact on worker productivity and tenant loyalty.

- *Knowledge of facility energy consumption patterns and trends.* Infometrics reports may enable building owners to aggregate energy buying and predict needs. Energy use trends are shown which can facilitate predictive buying and maintenance opportunities.

- *Integration with information technology systems.*—Infometrics enterprise-wide facility data can easily and cost-effectively integrate into existing information technology systems, providing more centralized control for operational, purchasing, and financial management.

With a low installation cost, Infometrics has achieved a track record of delivering immediate and lasting reductions in a building's operating cost structure. The Infometrics service also gives the customer necessary information for regulatory compliance, occupant comfort/safety and mission-critical monitoring. A sample of actual potential annual savings identified to date at current customer sites is provided in the table below.

Note that these figures only consist of measured energy cost savings and exclude maintenance, operational and regulatory benefits.

Short-term, Infometrics customers have identifiable and tangible savings in energy consumption, repair & maintenance and labor resources. Additionally, the Infometrics service gives the customer necessary information for regulatory compliance matters, occupant comfort/safety and mission-critical monitoring.

In the longer term, building owners and managers as well as utilities and energy providers realize intangible cumulative benefits critical to organizational effectiveness and key to taking advantage of energy deregulation.

CASE STUDY: REMOTE INTELLIGENCE CAPABILITIES HELP AMERICAN UNIVERSITY REDUCE COSTS, IMPROVE EFFICIENCY

Washington, D.C.—Universities large and small are under intense pressure to improve their educational environments and reduce costs to better accommodate faculty and students. American University faces unique challenges for its 84-acre campus in prestigious northwest Washington, D.C., as it transforms itself into "an academically distinctive, intensely engaged and student-centered community," one that provides the ideal balance between financial responsibility and educational priorities. Like most institutions, the university must maximize efficiency without compromising effectiveness.

Client Objectives

The university formulated a 15-point strategic plan that would transform the institution. The plan identified the reduction of costs and increased operational efficiency over three years as keys to the plan's success. In searching for ways to achieve these objectives, the remote analysis and optimization technology that powers Infometrics caught the attention of the university's physical plant management. Infometrics is a comprehensive, ongoing process, performed by industry specialists offsite, that helps institutions resolve operating problems, improve comfort, optimize energy use and

Sample Potential Annual Savings Identified To Date

Customer	Buildings Monitored	Points Monitored	Est. Per Point Savings	Annual Savings Identified
Customer 1	2	8,500	$28.50	$242,000
Customer 2	3	3,000	$22.00	$66,400
Customer 3	3	1,300	$40.00	$52,000
Customer 4	1	10,000	$72.90	$729,000

identify retrofits for existing buildings and central plant facilities.

Solution Overview

A remote analysis and optimization program was created for three of the university's buildings, comprising three components: facility data acquisition, remote expert analysis and reporting. The aim was to give building operations personnel unbiased recommendations and management information so they could reduce costs and enjoy the many benefits of continuously commissioned buildings.

A site survey was conducted with a needs analysis and assessment of all facilities, including gathering relevant, existing performance data and site histories. Cimetrics' Infometrics team also ensured that the university's existing building automation system had the tools and capabilities to facilitate the collection and transmission of large amounts of real-time data.

Non-intrusive, secure BACnet® connectivity was established between the existing building automation system and the remote database to mine and transmit continuous, real time facility data. The Infometrics team installed and configured a firewall-friendly BACnet® routing device as the communications interface for 3,000 points.

The analysis and optimization program provides timely report recommendations designed to assist systems engineers with fault detection, troubleshooting and problem solving while prioritizing maintenance issues and reducing downtime.

With the intelligence provided in the analytical reports, the university is given unbiased energy, maintenance and operational recommendations that offer opportunities to reduce costs and optimize equipment for reliable operation.

Client Results

The university does not have the metering to verify building-by-building savings, but for the three monitored buildings, resolutions to the types of problems found could lead to annual savings in the range of $125,000 in energy alone.

Physical plant operations had an initial increase in the number of repair work orders in the monitored buildings related to the problems identified through the monitoring process. Most of the problems identified had the potential for, and some were actually having, a direct impact on occupant comfort. The increased ability to find and fix these problems before they resulted in an occupant comfort call is in line with the physical plant's strategic direction and is leading to fewer occupant comfort calls.

The remote analysis uncovers faults and produces value for the university—through the eyes of expert, unbiased professionals—that cannot otherwise be reasonably detected or uncovered from a one-shot survey of the buildings or addressed with an off-the-shelf software product.

"We're so busy handling day-to-day symptoms that we don't have time to dig deeper into the root causes of the problems. The remote analysis and optimization service gives us the big picture of how our facilities behave and what we should be doing to address the larger issues."

—*Willy Suter, Director of Physican Plant Operations*

Chapter 11

Enterprise Energy Management System Installation Case Study at a Food Processing Plant

Naoya Motegi, David S. Watson and Aimee T. McKane
Lawrence Berkeley National Laboratory

INTRODUCTION

ENTERPRISE ENERGY MANAGEMENT (EEM) System is a combination of software, data acquisition hardware, and communication systems to collect, analyze and display building information to aid commercial building energy managers, facility managers, financial managers and electric utilities in reducing energy use and costs in buildings. This technology helps perform key energy management functions such as organizing energy use data, identifying energy consumption anomalies, managing energy costs, and automating demand response (DR) strategies. Compared to other data archive and visualization systems, EEM is more tied-in to business enterprise information such as; facilitating energy benchmarking, optimizing utility procurement, and managing overall energy costs. Figure 11-1 illustrates concept of EEM system architecture.

During recent years, numerous developers and vendors of EEM have been deploying these products in a highly competitive market. EEM offer various software applications and services for a variety of purposes. Costs for such systems vary greatly depending on the system's capabilities and how they are marketed.

In spring 2004, Del Monte Foods received funding from State Technologies Advancement Collaborative (STAC) to install an EEM system at one of their food processing plants. Del Monte asked Lawrence Berke-

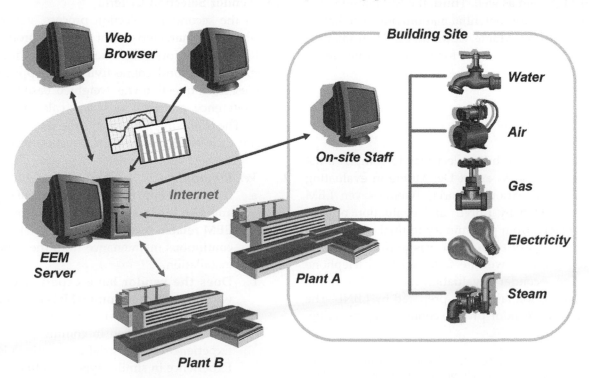

Figure 11-1. Enterprise Energy Management System

101

ley National Laboratory to evaluate the EEM vendors. LBNL has been conducting survey and research on EEM (Motegi, 2003). Role of LBNL in this project is to develop a specification framework of EEM and its selection criteria and procedures, through the food processing plant's case study.

SITE CHARACTERISTICS

Del Monte chose a food processing plant in California central valley to install the EEM. The plant has 1.2 million ft^2 of total building area. Plant operations experience seasonal peaks during the summer months. The maximum demand is approximately 6.5 MW.

Requirement for EEM

Addition to the common EEM data management, visualization and analysis features, Del Monte requested several key features which are critical to their plant operation. First, the EEM had to be capable of managing data from a range of supporting systems, including water, compressed air, gas, electricity and steam, also referenced as "WAGES." Second, the interoperability between the EEM and existing systems had to be established. The plant already had an Enterprise Asset Management System (EAM)* and forklift battery charger system. The data from these systems needed to be fed into the EEM system as well. Third, the EEM system had to be expandable for potential nationwide installation and networking. Del Monte views this EEM installation to this plant as a pilot project for future nationwide installation.

VENDOR SELECTION

Role of LBNL in this project was to compile a list of EEM vendors and to assist Del Monte in evaluating their products by various criteria. Twenty-seven EEM vendors were listed by LBNL, and evaluation process took several steps from preliminary to final.

The preliminary evaluation was performed by reviewing the vendors' website followed by a telephone interview. The secondary evaluation was to ask vendors to answer a questionnaire form prepared by LBNL. The questions include vendor profile, connectivity, metering,

---

*Enterprise asset management (EAM) is the organized and systematic tracking system of an organization's physical assets including plant, equipment and facilities for productivity enhancement, maximizing asset life cycle, and minimizing total cost.

integration, data security, data visualization and analysis, graphical display, demand response, utility procurement management, financial analysis, and reporting capability. The filtering criteria were based on Del Monte's requirements and LBNL expertise. Many detailed questions were asked on features of data analysis web-base tools, and experience and strength in industrial grade metering and control products. Upon completion of the secondary evaluation, the vendors were filtered down to four candidates.

For the final evaluation process, Del Monte sent a request for proposal (RFP) to the four vendors. Prior to the RFP, web demonstrations were also conducted. LBNL helped Del Monte evaluate the proposals. Through the RFP evaluation process, one vendor couldn't provide requested features within the budgetary allowance. The web demonstrations of two other vendors were somewhat unfocused on EEM features as a pre-developed tool. The winning vendor's demonstration covered most of what Del Monte requested, showed flexibility for future software feature expansion, and was well-developed. Interestingly, two of the vendor finalists decided to partner with each other for this project, and the winning vendor company was purchased by the fourth finalist vendor company during the evaluation process. This shows the dynamic changes in the current EEM industry.

EEM Vendor Selection Criteria

In the secondary selection process, various questions about the characteristics of the EEM products were asked of the vendors. The questions included vendor profile, metering and connectivity, and application capabilities. Each question was weighted based on the client's preference, and total score was calculated for each vendor. The following list presents the questions used to evaluate each vendor during the selection process.

1. *Vendor profile—*
    a. **Does the vendor have several years of experience and a stable business history in the particular EEM related services?** It is important to receive continuous maintenance from the vendor after installation.
    b. **Does the vendor have experience working with industrial customers?** It is more important to assure precise installation and operation in industrial sites than in commercial buildings. Experience in industrial applications is critical. Experience in similar type of industrial facility, for instance food processing or bio-tech, will be plus.

c. **Is the vendor capable of single source responsibility?** Make sure that the vendor manages from hardware installation to application customization as a turn-key contractor.

d. **Does the vendor have a regional office in the state?** A regional office nearby is useful to get hands-on support from the vendor during and after the installation.

2. *Connectivity—*

a. **Does the vendor have capability for two-way communication flow (monitoring and control)?** All EEMs should have monitoring capability, while some of them may offer remote control capability as well.

b. **What types of data input are supported by the EEM?** Types include: Pulse, 0~10 Volt DC, 4~20 mAmp, Digital, etc.

c. **What types of metering/communication protocol are supported by the EEM?** Types include: TCP/IP, BACnet, LonTalk, Modbus, Profibus, etc. If your facility uses some specific protocol or has some native equipment for the specific protocol, compatibility between the system and the EEM will enhance their performance.

3. *Metering—*

a. **What types of data are the EEM designed to monitor/archive?** Although most of EEM products can be customized to archive any kind of data, some EEMs may focus their functionalities on specific data types such as whole building electricity, gas or other utilities. If you are planning to perform more detailed diagnoses, built-in analysis functions for electric sub-meter, pressure, or temperature data will be helpful. Calculated data points and the capability to create a virtual data point from real measurement point values, are also helpful.

b. **What is the system response speed?** In general, communication response speed for industrial grade system is 200 ~ 500 ms or less, and commercial grade is 1 ~ 2 seconds. For only monitoring purpose, a slower response speed is acceptable, while industrial control requires faster response speed than commercial.

4. *Database—*

a. What is the database compatibility for the EEM?: Is it compatible with the following systems:

   i. Web-service client with eXtensible Markup Language (XML)

   ii. OLE for Process Control (OPC)

   iii. Open Database Connectivity (ODBC) compliant to interface 3rd party software application

   iv. Structured Query Language (SQL)

   v. Application Program Interface (API) to communicate with specific field devises such as hand-held equipment

b. Is the database and the data transfer through the Internet encrypted?

5. *Application capabilities—*

What package of web application tools is available to visualize and analyze the trend data? Most EEM vendors customize their application to meet the users' needs, while they usually have built-in standard applications which are well-tailored and easy-to-use. Even if a vendor claims that they can offer any type of analysis tool as their customized solution, the customized features could be immaturely designed or costly. LBNL published a report which describes some of the application capabilities in detail (Motegi et al, 2001). Listed below are the application features which were found in existing EEM products as their built-in application.

a. **Graphical display**: shows geographical location of site or equipment in a site.

   i. Geographical site/building display: Display building locations, site drawings, etc so that remote users can get better sense of the site.

   ii. Graphical equipment monitoring: Display mechanical system diagrams and show real-time equipment data (temperature, pressure, status, etc) in the graphics.

   iii. Summary view of equipment status: Summarize key equipment parameters on one screen.

b. **Data visualization/analysis**: visualize the trend data in meaningful fashion by using Excel-like charts and graphics.

   i. Daily profile: Time-series daily load profiles are displayed with time, in intervals of an hour or less, along the horizontal axis and load along the vertical axis.

   ii. Day overlay: Overlay plots display multiple daily profiles on a single 24-hour time-series graph.

   iii. Multi-point overlay: Allows viewing of multiple time series data points on the same graph.

   iv. 3D surface chart: Three-dimensional surface charts often display the time of day, date, and variable for study.

    v. <u>Calendar profile</u>: View up to an entire month of consumption profiles on a single screen as one long time series.

    vi. <u>X-Y scatter plots</u>: X-Y scatter plots are useful for visualizing correlations between two variables.

    vii. <u>Basic statistical analysis</u>: Perform statistical calculation, such as mean, median, standard deviation, correlation, ANOVA, and regression.

    viii. <u>Benchmarking</u>: Benchmark against building energy standards/codes, or public database such as EnergyStar.

    ix. <u>Intra/inter-facility comparisons</u>: Benchmark against the building's historical data, or across multiple buildings in the enterprise.

    x. <u>Aggregation</u>: Aggregate data among multiple data points. Integrate different energy units using energy conversions (ex; kWh, Therm, etc. into Btu).

    xi. <u>Data mining (data slice/dice/drill-down)</u>: Sum-up/drill-down time-series data by monthly, weekly, daily, hourly, or trended interval.

    xii. <u>Normalization</u>: Normalize energy usage or demand by some factors such as building area, number of occupants, outside air temperature (OAT), and cooling or heating degree-days (CDD, HDD), to make a fair comparison between buildings.

    xiii. <u>Hierarchical summary</u>: Summarize usage and cost information by different levels. For example, starting from equipment energy cost, individual building energy cost, site energy cost, to regional energy cost.

c. **Advanced analysis**: offers more specific analysis than the "data visualization/analysis" mentioned above. This type of analysis often takes multiple data points and use more complicated algorithms.

    i. <u>Power quality analysis</u>: Monitor the voltage or current phases for conditions that could have adverse affect on electrical equipment.

    ii. <u>Steam charts</u>: Calculate temperature, pressure, specific volume, and enthalpy for saturated steam and water.

    iii. <u>Refrigerant charts</u>: Create psychometric chart for refrigerant type dehumidification application. Used for refrigeration diagnostics.

    iv. <u>Forecasting</u>: Forecast future trend by historical data and related parameters. Briefly explain the forecasting method (ex; regression

model, neural network).

    v. <u>Validation, editing, estimation (VEE)</u>: A process performed to ensure quantities (kWh, kW, kVar, etc.) retrieved from meters or interval data recorders (IDR) are correct. The process includes validation of data within acceptable error tolerances, editing or correcting erroneous data and estimating missing data.

    vi. <u>Equipment fault detection diagnostics</u>: Detect equipment failure or degradation based on customized algorithm and parameters.

d. **Utility cost analysis**: provides tools to analyze their utility bills by cutting-and-dicing in various manners.

    i. <u>Invoice verification (bill validation)</u>: Utility bills are compared to meter readings to validate accuracy of bills.

    ii. <u>Energy cost drilldown</u>: Using energy tariff and usage data, calculate daily or hourly energy cost breakdown, while usually cost can only be seen in monthly total.

    iii. <u>Real-time cost tracking</u>: Calculates electricity costs every day or hour using real-time meter reading and rate tariffs.

    iv. <u>End-use cost allocation</u>: According to user-defined parameters and algorithms, estimates end-use energy consumption from whole building energy. Generally used for cost allocation to building tenants. A common parameter definition is energy use per square foot.

e. **Procurement management**: provides utility procurement information and analysis tools to develop better procurement plan and save utility bills.

    i. <u>Rate engine/utility tariff integration</u>: Software contains a series of rate tariffs to fit to clients' utility rates. Users can compare energy procurement alternatives. The rate tariffs are; manual input, pre-programmed, or online tariff data acquisition.

    ii. <u>WAGES cost forecasting</u>: Forecast future WAGES cost based on future usage forecast and energy cost trend. Performed by component level and/or whole building level.

    iii. <u>Real-time pricing functionalities</u>: Integrate real-time pricing information into procurement process.

    iv. <u>WAGES cost what-if analysis</u>: Compare current and alternative energy procurement scenario by applying different WAGES tariff/forecast to past, current or future WAGES usage.

    v. <u>Buy/wait suggestions</u>: Based on energy wholesale market trend, weather forecast, and onsite stock, suggest when and which utilities to purchase energy.

f. **Reporting**: provides data exporting and alarming functionalities.

    i. <u>Data export</u>: Export data to manually analyze the trend data or generate a ready-to-use report. Data outputs include; Excel (raw data), Access (database), Word (report), and PDF (report).

    ii. <u>Customized alarm module</u>: Detect anomaly data and send alarm via e-mail, pager, mobile, etc.

    iii. <u>Dashboard functionality</u>: Display selected parameters in small window application which stays on PC screen or task tray, so that operators can notice an alarm quickly.

6. **Demand response**: Needs for demand response capabilities are increasing in both commercial and industrial facilities in recent years. Demand response functionalities include communication with utilities, automated/remote control, and event analysis.

a. **DR communication**: Enable communication with utilities or DR aggregators including notification of DR event via website, e-mail, phone, pager, cell phone, etc, and bidding to DR event via website.

b. **DR control**: Execute demand-shedding operation with; remote manual operation, semi-automated control by time-schedule, or fully-automated control by parameter/external signal.

c. **DR event analysis**: Analyze DR event by using web application tools.

    i. <u>Event report</u>: Track and report kW savings and cost for all events.

    ii. <u>Baseline calculation engine</u>: Calculate baseline based to utility program formula.

    iii. <u>Saving calculation by tariff</u>: Calculate saved demand and program incentive/penalty based on baseline and program tariffs.

FINAL DESIGN OF EEM

Due to the limited budget, Del Monte decided to focus on monitoring three systems for the pilot installation, to be able to partially perform system diagnostics. The targeted systems were the compressed air supply,

Table 11-1. List of EEM Measurement Points

Category	Component	Measured point	Unit
Air	Compressed air	Airflow Air pressure Air temperature	CFM PSI °F
Electricity	Utility main	Power demand Power consumption Power quality	kW kWh
	Air compressors	Power demand	kW
	Forklifts	Power demand Run-time	kW Hour
	Water pumps	Power demand Power consumption	kW kWh
Gas	Utility main	Energy Cost Price	MMBtu $ $/MMBtu
Steam	Boilers	Steam flow Steam temperature Cost	kLbs °F $/kLbs
Water	Water pumps	Water flow Water pressure Water pumped	GPM PSI Gal

the booster house pumps, and the electric fork lift fast charging. They also measured a few key points for electrical supply, gas, and steam, although points will be need to be added later to allow for analyses of systems performance across all WAGES (water, compressed air, gas, electricity, and steam) categories for the plant. Table 1 shows the list of the measurement points included in the initial EEM. This includes existing measurement points for the boiler temperature, pressure, and flow as well as inputs from the forklift charger management system.

The EEM is also designed to interface the existing EAM system. The EEM extracts the production data from the EAM, and makes the data available on the EEM analysis.

Cost Estimate

Figure 11-2 shows cost estimate of the winning vendor's proposal. Total cost is $126,000, includes $66,000 of base system (server and software) and $7,900 per monitoring point (metering hardware, installation, configuration). The base system cost will not significantly increase with the addition of monitoring points. The cost per monitoring point depends on existing meter equipment and the quality of metered data desired.

POST-INSTALLATION ANALYSIS

LBNL conducted post-installation analysis for the use and trend data of the EEM. The analysis is based on the trend data and feedback from the users. From the users' perspective, the EEM is used for 1) energy saving verification and 2) daily system operational decision making.

The main intent of the EEM installation is to identify opportunities for increasing the effectiveness and energy efficiency, while reducing the operating costs, of the monitored systems. Once the energy saving and/or cost saving potential in the operation of a system is identified and quantified, plant personnel can take specific actions with confidence in the outcome. The ability to visually represent both the impact of current practice and the potential for improvement is a major asset in obtaining management approval for system improvements. The EEM can be built out incrementally over time to include measurement and analysis of other systems within the plant. Multiple facilities can also be connected into a corporate network via multi-site EEM installations.

The initial data can be used to establish a time series baseline for operations prior to the any system improvements. Data collected post-improvement may be used to establish a new baseline. Data collected over an extended period post-improvement can be used to trend the energy use and cost to production, as well as to identify opportunities to improve operational efficiency. At the time of this writing, the measurement data are being collected and analyzed to establish a baseline, however preliminary analysis has already identified opportunities to make specific operational improvements.

An example of these operational improvement opportunities was revealed through preliminary analysis of the compressed air system. Currently, the air compressors are operated manually by the facility operator. Prior to the EEM installation, the operator did not have reliable data concerning compressed air output or demand for any specific time period. Therefore, the air compressors were operated in a fixed operational pattern designed to meet production needs without regard to efficient operation. After the EEM installation, the operator was able to check air compressor data on a daily basis, and decide how many compressors to operate to meet the facility's needs. The operator has found that the EEM quite useful in understanding the facility's operation.

Compressed Air System Analysis

We conducted additional data analysis for the compressed air system supply to look at the overall efficiency of the compressed air supply. Compressed air is supplied by eight compressors in a variety of sizes, brands, age, and capacity. As previously mentioned, control is achieved manually rather than through a sequencer or other coordinated control strategy. Two 100 hp rental compressors are added to this configuration during summer production season. Table 2 shows specification for each air compressor. The EEM measures the electric demand of each compressor with the exception of the rental compressors, and total airflow at the header (including the output of the rental compressors during

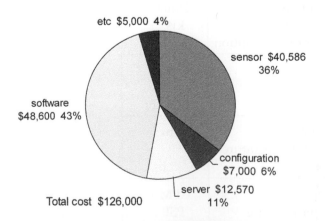

Total cost $126,000

Figure 11-2. Breakdown of EEM Installation Cost

Table 11-2. Air Compressor Farm Specification

	Full Load CFM	Full Load HP	Full Load kW	Full Load Package kW*
Compressor #1	620	125	93	115
Compressor #2	442	100	75	92
Compressor #3	352	75	56	69
Compressor #4	352	75	56	69
Compressor #5	320	75	56	69
Compressor #6	320	75	56	69
Compressor #7	425	100	75	92
Compressor #8	425	100	75	92
Rented Compressor #1	450	100	75	92
Rented Compressor #2	425	100	75	92
Total	3256	725	540	668

*There is a difference between compressor motor nameplate ratings (hp) and actual compressor package kW. Most compressors have motors with 1.15 Service Factor (SF) rating. Also air-cooled compressors have fan motors in the package driven by separate motors, thus increasing the "Package Power" further. In this analysis, the factor of 1.24 which is calculated from Compressor #1's manufacturer specification is applied to all other compressors.

the peak production months). The rental compressors were continuously operated for 24-hours a day at full load from August 1st to September 12th, 2005.

Figure 11-3 plots total compressor power against total airflow with the compressor efficiency curve estimated from the compressor specification. Due to lack of power measurements for the rental compressors, these values were estimated based on the compressors' operational setting during the affected period. In Figure 11-3, the total power demand of the permanent compressor installation is shown in black, and total power demand of both the permanent and rental compressors (with extrapolated values for the rental units) is shown in gray.

The chart indicates that the compressors used 10 to 20% higher electricity demand than their efficiency curve. Another important finding from these data is that the maximum required airflow was around 3,300 cubic feet per minute (cfm), which is almost same as the total capacity of the permanently installed compressors. This indicates that this facility may be able to meet peak demand without the additional cost of the rental compressors. Even allowing for a margin of additional demand, it is unlikely that an extra 200 hp would be required. Since rental compressors are expensive to lease and operate, this could be a significant cost saving opportunity.

The trend data also revealed that Compressor #1, which has the highest efficiency in the compressor supply system, is offline more than 90% of the time. By assigning Compressor #1 to run as base load, the total compressor efficiency can be expected to increase. Thus, capturing compressor operational characteristics may help improve efficiency and reduce the cost of operation.

CONCLUSION

The lack of immediate access to meaningful data is a significant barrier for plant personnel seeking to understand and improve the operational efficiency of plant systems. EEM systems offer great potential to help facility engineers and plant managers run their facilities more efficiently. However, the market for EEM is still immature, and not many products are available. Moreover, the customer awareness and demand for EEM system is low. Promotion of successful installations of EEM will help increase development of EEM market, and will eventually make energy management of industrial sector more efficient.

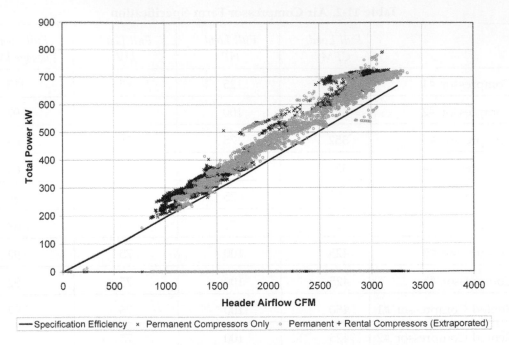

Figure 11-3. Compressor Efficiency Diagnostic Chart

References

Fryer, Lynn. 2002. "Retrocommissioning on Demand: Using Energy Information to Screen Opportunities." Proceedings of National Conference on Building Commissioning 2002.

Motegi, Naoya, Mary Ann Piette, Satkerter Kinny and Karen Herter. 2003. "Web-based Energy Information Systems for Energy Management and Demand Response in Commercial Buildings." High Performance Commercial Building Systems, PIER Program. LBNL#-52510.

Motegi, Naoya, M. A. Piette, S. Kinny, J. Dewey. 2003. "Case Studies of Energy Information Systems and Related Technology: Operational Practices, Costs, and Benefits." International Conference for Enhanced Building Operations, October 13th – 15th, 2003. LBNL#-53406.

Acknowledgment

This article has been prepared by Naoya Motegi, David S. Watson, and Aimee T. McKane of the Lawrence Berkeley National Laboratory (LBNL), as part of the Western United States Food Processing Efficiency Initiative. This initiative, under the auspice of the State Technologies Advancement Collaborative (STAC), was sponsored by the National Association of State Energy Officials (NASEO) and the Oregon Department of Energy, under Contract No. LB05-001006. The information presented in this article draws upon the experience of the authors and other individuals and organization. In particular, the authors wish to acknowledge the contributions of: Anthony Radspieler Jr. of LBNL; Glen Lewis and his colleagues of Del Monte Foods; and many Enterprise Energy Management Vendors, including Emerson, Power Measurement, Rockwell Automation, and Square D, who assisted in this effort.

Chapter 12

Ford Compressor Management System Case Study

Bill Allemon, CEM, Ford Motor Company
Rick Avery, Bay Controls, LLC
Sam Prud'homme, Bay Controls, LLC

INTRODUCTION

MANAGING THE PRODUCTION, distribution and use of compressed air is a frequently misunderstood process and one of the most expensive "products" created at manufacturing facilities. So much so, that compressed air is often referred to as the "Fourth Utility" because of its cost and wide spread use throughout industry.

Compressing air into a usable form can be an energy inefficient process, with up to 90% of the energy consumed by an electric air compressor lost as waste heat, due to mechanical friction and electric motor losses[1]. Some estimate that approximately seven times more energy is required to mechanically compress air, remove moisture, distribute it to the point of use, and convert it back into mechanical energy, than to directly use an electric motor to perform the same task. In response, some industries are converting from compressed air driven tools to direct-drive electric tools when the applicable uses are equivalent. However, there will always remain a market for compressed air driven tools, due to their inherent safety and convenience. The challenge remains: how to generate compressed air efficiently, while meeting the dynamic needs of a manufacturing environment.

In 2003, Ford Motor Company's North American Vehicle Operations Division embarked on a widespread project to significantly reduce compressed air production costs. With the help of the tier one supplier, Bay Controls, a comprehensive energy management system was installed, saving nearly one million dollars in electrical energy consumption within the first year. Vehicle Operations is Ford Motor Company's manufacturing division that includes stamping, body welding, painting and final assembly functions.

This study examines the integrated web-based compressor monitoring system installed as part of an overall control hardware upgrade project, which included Bay's ProTech microprocessor based compressor control devices, local networking and management systems, and BayWatch, a web based monitoring and analysis management system.

More specifically, this case study will focus on the Ford Compressor Management System (FCMS), the Ford application of the BayWatch product. This study will include the following sections: an overview of the capital project scope; a review of the preexisting state of compressed air control in the Ford facilities; the overriding concerns and needs that prompted the installation; the rationale behind selecting the Bay system; a description of how the system functions (both technically and from an end user standpoint); and finally, a survey of the primary economical and operational benefits that the system provides.

EXISTING CONDITIONS

The compressor controls capital project installed new control hardware at 19 manufacturing facilities across North American Vehicle Operations. A total of 132 compressors were modified, consisting of both positive displacement (reciprocating, rotary screw) and dynamic (centrifugal) types at each plant. Mixing compressor types and sizes is a common practice to meet the dynamic air volume needs of manufacturing operations.

Compressed air distribution infrastructures also varied from plant to plant. The majority of plants had centralized air generation with a separate oil-free system in the vehicle painting department. Some plants divided air generation into two systems, while others used a decentralized, point of use strategy. Each plant used common headers to distribute compressed air, with various cross-connections, valves, and back-feed loops for maintenance and redundancy.

Regarding compressor management, there was a mix of prior Bay installations and various competitive systems, the later often based on PLC hardware. A number of plants had no modern control systems at all.

Overall, there lacked a centralized, automated information system to monitor, compare and analyze air generation across the Vehicle Operations division.

The diverse initial conditions between plants meant that compressed air systems were operating with varying levels of efficiency and an assortment of operating strategies. The number and complexity of Ford plants required a solution that was both flexible enough for installation at each location and capable enough to improve energy efficiency in variable conditions.

GOALS AND OBJECTIVES

The primary goal was to reduce the energy consumed in the production of compressed air and provide an application to manage the enterprise at both the plant and divisional levels. Since existing conditions at each plant were unique, the plant-specific deliverables also varied. This required many compressors to be upgraded with newer, more advanced instrumentation. Some compressors needed extensive upgrades to their monitoring and control equipment, while others needed little or no modification. Compressors vary in their complexity, so the number of monitor and control items ranged up to 64 points, and 15 control outputs for each machine. Typical metering points on a compressor include output air pressure and flow, motor (or primary mover) power consumption, and stage pressure, temperature, and vibration.

While most of the energy savings resulted from the new or upgraded ProTech compressor controls, the need for an enterprise energy management system was also identified. This system, which became FCMS, needed to address the following concerns:

- A desire to efficiently and automatically meter and verify energy consumption and savings resulting from the controller installation project.

- How to best extract useful long-term operating data from the built-in monitoring capabilities of the ProTech control systems.

- How to use these data to maintain and improve the operating performance of the compressed air systems.

- Use these data to address compressor system problems before they become serious and affect vehicle production.

- A desire to centrally monitor and benchmark compressed air systems across the Vehicle Operations Division.

PROJECT SOURCING AND FUNDING

Several factors were considered during the selection of the control system vendor for this project: the performance and capabilities of the resulting compressor control and management system; the pros and cons of externally hosting FCMS; and cost effectiveness.

Vendor Selection

Some key aspects in vendor selection included the proven success of Bay Control products at existing plants and Ford's understanding of the product's technical features. At the start of the controller replacement project, 45% of Ford Vehicle Operations plants were operating an earlier version of the Bay Controls system. These controllers would simply require an upgrade of their control hardware; the remaining plants would receive new controllers. Existing long-term installations had proven the reliability of ProTech as a product and Bay as a supplier. They also had proven the product's ability to reduce energy consumption and operate compressors reliably and safely.

System Selection

The FCMS system is based on the Bay Controls BayWatch product, a web based monitoring, data recording and performance reporting software, hardware and engineering service package. Ford Motor Company chose the BayWatch system for several reasons.

BayWatch is designed to integrate seamlessly with the ProTech control system, which made BayWatch easy to install and required minimal custom engineering work. BayWatch is also designed specifically to work with compressed air systems and to manage the complex metering and data recording that is required.

Additionally, the BayWatch system provided these comprehensive enterprise-wide compressed air management features:

- Real time monitoring of plant air compressor systems, as well as the ability to monitor the operating parameters of each individual compressor.

- Comprehensive data recording, reporting and analysis features.

- Straightforward, easy to use web based interface enables anywhere and anytime system management and access.

- Ability to handle multiple facilities and numerous compressors provides centralized monitoring of the entire North American Vehicle Operations compressor system.

- Includes an ongoing engineering service contract, wherein expert operators from Bay Controls perform daily analysis on the connected Ford compressor systems (the scope of this service is detailed later under FCMS Performance Enhancement Service). This, in turn, identifies opportunities for continuous efficiency improvements and preventive maintenance and diagnostics.

These features, combined with the excessive cost that a custom engineered solution would have required, made BayWatch a logical choice for FCMS.

System Hosting

A key decision involved whether to host the FCMS system internally or externally. Ford initially attempted to have the Bay system approved for connection to their internal LAN network, but ran into a number of IT related inhibitors and delays, common in large corporations.

In order to gain approval for internal hosting and connection to the Ford network, the Bay system would have required a lengthy testing and certification process. Corporate IT departments typically provide less support to facilities related systems as compared to systems that support product engineering and manufacturing. Thus, an unacceptable delay would have occurred prior to receiving approval to begin system installation and realize energy savings.

The BayWatch engineering service requires a real-time connection between personnel at the BayWatch Center, located in Maumee, Ohio, and the data hosting system. A connection of this type to internal Ford networks would have required multiple levels of management approval and system tests prior to being fully operational. After launch, management of this connection to comply with dynamic internal standards would have been costly, time consuming, and put the reliability of the data collection system at risk.

Finally, benchmarking verified that external hosting is becoming commonplace for the reasons mentioned above. After final review of all issues, it was decided that externally hosting FCMS would be a less complicated, more economical and generally superior solution.

Project Funding

The FCMS aspect of the project required a separate three year expense contract, justified through energy savings incremental to the overall compressor controls project. A historical incremental return of 20% was used to calculate a savings value and justify a three-year contract. Due to the law of diminishing returns, extending the contract beyond three years could not be justified solely using energy savings. The value of the BayWatch service would be reevaluated prior to contract expiration and either extended across the North American Vehicle Operations plants, continued at select plants, or discontinued due to changing business conditions.

TECHNICAL WORKINGS AND SYSTEM DESCRIPTION

Technical System

BayWatch, the underlying technology behind FCMS, relies on several key integrated hardware and software elements. The ProTech compressor controller is used as the individual compressor monitoring and control device, with a Bay Virtual Gateway connecting groups of controllers at each plant to a remote central server.

Monitoring Capabilities

The functionality of the web based management system is directly related to the monitoring capabilities of the ProTech unit. The controller is an advanced microprocessor based unit designed to work with all makes and models of compressors. Each standard configuration ProTech can support a combination of 64 analog and digital monitoring inputs, which are used to read such values as temperature, vibration, pressure and air flow. In addition, the ProTech controller includes built-in networking functionality, using RS-485 communication hardware to run Bay's proprietary C-Link networking protocol. The networking abilities are used to enable data communication between multiple ProTech controllers, and between a ProTech network and a remote data monitoring system.

In this fashion, the ProTech controller acts as the foundation for a plant wide or enterprise wide compressor management system.

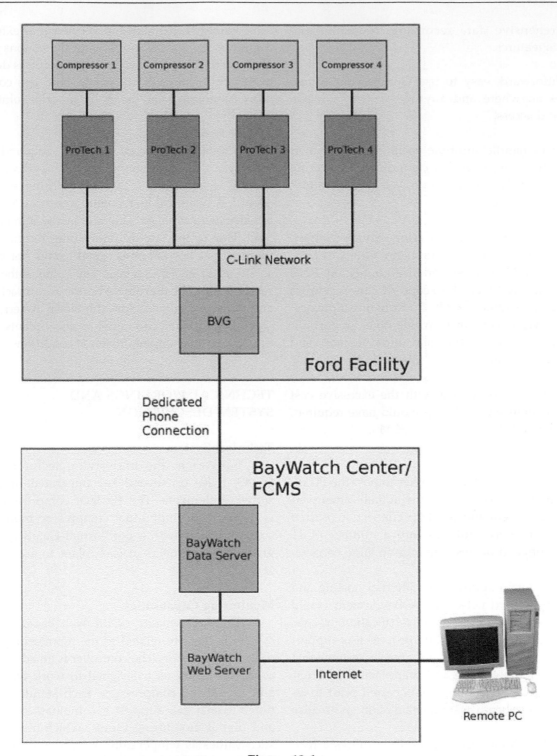

Figure 12-1.

Data Connection

Each plant compressor network is connected to the BayWatch Center via a dedicated phone line or broadband connection. The BayWatch Center is a staffed central monitoring station at the Bay Controls offices (see FCMS Performance Enhancement Service for more information regarding the functions of this team.). Ford approved controls are in place to ensure the security of the internet connection and to prevent unauthorized use by plant personnel. A virtual gateway device is used to interface between the compressor network and the remote connection. In order to maintain continuity of data, each gateway contains internal memory to act as a storage buffer. For security purposes, the virtual gateway establishes the link with the BayWatch Center; it does not accept incoming connections.

BayWatch Servers

Remote data collection services are performed by a dedicated server running a linear, flat file based database. Information is collected for every monitoring and control point from each compressor. The state of the ProTech internal registers are also recorded, which give an additional snapshot of the controller's status. The data are collected continuously throughout the day, approximately every 15 seconds, and are transmitted to the BayWatch Center every few minutes. The database is designed to ensure varied analyses, future expansion, and modification of features.

An additional server supports security functions, maintaining account and log on information. Back up servers duplicate the databases to prevent loss of recorded data. All BayWatch servers use secure shell (SSH) data encryption to ensure the security of transmitted information.

Web-Based User Interface

A web portal provides access to an internet server which displays real-time and historical performance data and supports generation of reports. This server facilitates remote logon to the FCMS system via the public internet, making the system accessible both within and outside of the Ford corporate firewall. Access to this system from any location with internet connectivity has greatly benefited Ford. As time constraints, travel requirements, and workloads increase, Ford requires easy access to information from any location. Ford personnel use the web-based interface to address issues at one plant while being located at another. The system has also been used to address emergency situations during holidays and weekends, without requiring travel to the site.

BAYWATCH USAGE AND INTERFACE

In this section, we will examine how the FCMS system works from an end user standpoint. FCMS is designed to provide several different categories of information, each of which is displayed on a separate screen.

Access to FCMS is password protected. A standard password screen prevents unauthorized users from proceeding to the main site.

System Overview and Facility List

After logging on to FCMS, the user is shown a list of authorized sites to chose from. Selecting a plant displays the system overview screen for that facility.

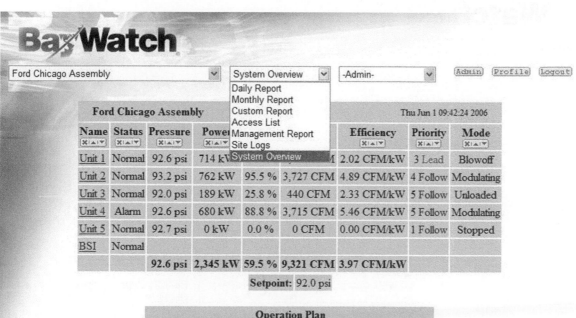

Figure 12-2. System overview shows a summary of the compressor system current operation.

The main system overview screen shows the current compressed air system status of one facility. Displayed on this screen is a tabled list of the air compressors in the plant, along with basic operations data for each compressor: status, pressure, power, load, flow, efficiency, priority and mode. These data points give a general overview of how the system is working, the plant air setpoint, which compressors are producing air, and the efficiency of each compressor. Also on this screen is a drop down list which allows the user to switch from one facility to another. Using FCMS, the entire North American Vehicle Operations compressor system is accessible from this screen, with each facility listed in the drop down list, dependent upon user rights.

The current operation plan for the selected facility is also listed at the bottom of the screen. This table shows the different operating plans for non-production versus production time periods, including changes in system pressure set points and compressor operation.

Compressor View

Selecting the name of an individual compressor displays the individual compressor unit monitoring points screen. Here, every piece of monitored data

routed through the ProTech controller is shown. This view gives a very comprehensive overview of the compressor's state, showing the current readings and alarm status for all monitoring points.

Additional data for the individual compressor are accessed from this page, including a list of control points, ProTech Register values, an operation schedule, and an export screen. The export screen allows a history of selected monitor, control or register points to be saved to a file for use in a spreadsheet.

Operator and Protection Events

Finally, histories of all operator events and protection events for this compressor are also available. Operator events consist of compressor control actions initiated by manual intervention, such as start/stop or setpoint changes. A protection event occurs when an instrument reading exceeds a predetermined level, indicating potential trouble with the compressor. Access to these historical records allows a precise examination of what events and actions occurred prior to any compressor problems; this information can often help resolve the issue more quickly.

BayWatch

Ford Chicago Assembly | Unit 1 | Monitor Points | -History- | Admin | Profile

	Monitor Points - Unit 1								Thu Jun 1 09:48:59 2006	
☒	Description	☒ Tag	☒ Start Low	☒ Trip Low	☒ Alarm Low	☒ Data	☒ Alarm High	☒ Trip High	☒ Start High	☒ Units
1:	System Pressure	PT-101				93.6				psi
2:	Inlet Filter Pres Drop	PDT-107				1.9	10			in H2O
3:	Discharge Pressure	PT-102				99.1				psi
4:	2nd Stage Inlet Temp	TE-103				84	130	140		Deg F
5:	3rd Stage Inlet Temp	TE-104				93	130	140		Deg F
6:	Discharge Temperature	TE-105				191				Deg F
7:	1st Stage Vibration	VT-110				0.27	1.5	2	2.4	mils
8:	2nd Stage Vibration	VT-111				0.36	1.5	2	2.4	mils
9:	3rd Stage Vibration	VT-112				0.41	1.5	2	2.4	mils
10:	Oil Pressure	PT-105	80	80	100	141.3	200			psi
11:	Oil Press Before Filter	PT-108				142.7				psi
12:	Oil Filter Pressure Drop					1.4	4	5		psi
13:	Bearing Pressure	PT-6		300	351	533.2	699			psi
14:	Oil Temperature	TE-101	60	60	70	115	140	150	150	Deg F
15:	Stator Temp A	TE-106				157	330	340		Deg F
16:	Stator Temp B	TE-107				157	330	340		Deg F
17:	Stator Temp C	TE-108				155	330	340		Deg F
18:	IB Motor Bearing Temp	TE-109				164	203	212		Deg F
19:	OB Motor Bearing Temp	TE-110				164	203	212		Deg F
20:	Drive Motor Current	IT-114		10		106.8				Amps
21:	Power Consumption	JT-9				736				kW
22:	Flow Element DP	PDT-113				2.32				in H2O
23:	System Flow	FT-13				3,580				CFM

Figure 12-3. Compressor view.

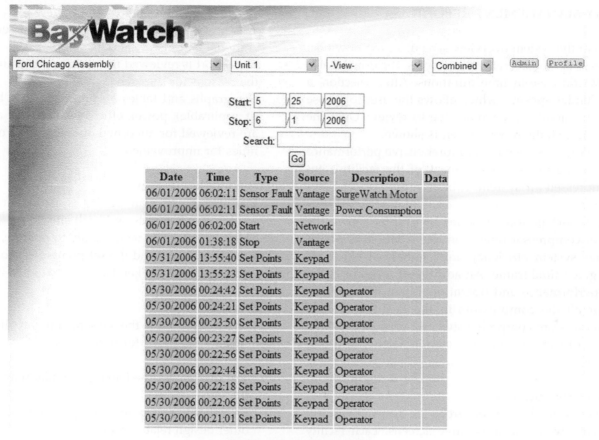

Date	Time	Type	Source	Description	Data
06/01/2006	06:02:11	Sensor Fault	Vantage	SurgeWatch Motor	
06/01/2006	06:02:11	Sensor Fault	Vantage	Power Consumption	
06/01/2006	06:02:00	Start	Network		
06/01/2006	01:38:18	Stop	Vantage		
05/31/2006	13:55:40	Set Points	Keypad		
05/31/2006	13:55:23	Set Points	Keypad		
05/30/2006	00:24:42	Set Points	Keypad	Operator	
05/30/2006	00:24:21	Set Points	Keypad	Operator	
05/30/2006	00:23:50	Set Points	Keypad	Operator	
05/30/2006	00:23:27	Set Points	Keypad	Operator	
05/30/2006	00:22:56	Set Points	Keypad	Operator	
05/30/2006	00:22:44	Set Points	Keypad	Operator	
05/30/2006	00:22:18	Set Points	Keypad	Operator	
05/30/2006	00:22:06	Set Points	Keypad	Operator	
05/30/2006	00:21:01	Set Points	Keypad	Operator	

Figure 12-4. Events log.

Energy Management Report

Ford Chicago Assembly

Wednesday May 31st, 2006

Operational savings of 7,070 kWh (13.3%), $417 over a 1 day period. The energy usage for the period was 49,951 kWh, at a cost of $2,947. The baseline energy usage adjusted for the period was 57,021 kWh, a cost of $3,364.

Independent of operational savings, changes in ambient temperature resulted in an increase in energy usage of 4,064 kWh (7.7%), a cost of $240 for the period.

Changes in system pressure resulted in a decrease in energy usage of 1,591 kWh (3.0%), a savings of $94 for the period. Changes in system pressure caused a decrease in the net amount of leak flow, resulting in a decrease in energy usage of 1,545 kWh (2.9%), a savings of $91 for the period. Changes in the net amount of compressor blowoff resulted in a decrease in energy usage of 4,955 kWh (9.3%), a savings of $292 for the period.

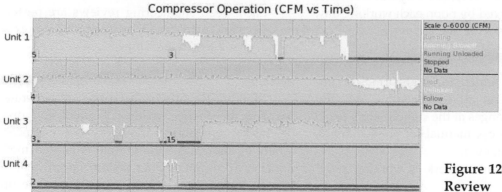

Figure 12-5. Energy Management Review

ENERGY MANAGEMENT REPORT

From the system overview screen, a drop down list provides links to several other reports, including daily, monthly, and custom time durations. After selection, a small calendar appears, which allows the user to choose the day or month they would like to review. Once the date is selected, the report screen is shown.

The report screen is a comprehensive performance and operating report for summary of the facility's entire compressed air system. A brief summary describes any changes in system energy costs that occurred during the report period. A series of graphs shows the details of compressor operation, plant pressure, power and flow, system efficiency and isothermal efficiency for the given time frame. An additional series of tables shows performance and operational data in numerical form for all the compressors in the facility's system. These reports are comprehensive and give an excellent overview of compressed air system efficiency and performance.

Administration Screens

Also available as a link from the system overview screen are several administrative screens. Each facility has a separate list of authorized users; those for the current facility are shown in the access list section. The management report screen documents new issues discovered during the daily performance review. Any exceptions or notes about the facility operations, such as monitoring issues or abnormal energy savings, are shown in this section. Finally, the site logs screen allows the user to select a range of dates to display a list of all history and protection events that occurred for the entire facility.

FCMS Performance Enhancement Service

A description of FCMS would not be complete without examining how the system is used on a daily basis. An integral component of FCMS is the performance enhancement service performed by Bay Controls. Monitored data are reviewed by noon each working day, which consists of the following steps:

System/Log Overview

- Review of alarms, faults, trips, abnormal or unexpected cfm/kW on all compressors, and any manual operator changes at the compressor's panel (i.e., manual start/stop, manual set point changes, unlink a compressor, etc.).
- Verify communication network is working properly.

- Generate the daily report and review
- This report is generated based on most recent operational and historical data.
- The report is reviewed for operational changes and the reasons for these changes.
- The graphs and tables on pressure, flow, blow off (if applicable), power, efficiency, cfm/kW and cost are reviewed for abnormal operation and opportunities for improvement.

Operational Plan Review

The operational plan created is a collaborative effort between Bay Controls and each site.

This plan specifies which units to operate, the run priority of those units and the set points for production, non-production and weekends.

Event Analysis

Items discovered in the daily reviews are discussed, prioritized and assigned for follow-up:

i. Technical Support—Direct immediate contact with the site.
ii. Engineering—Changes to operational plan and other design related changes.

A day end review is performed of important open items to make sure critical events or actions have been resolved or an adequate course of action is being pursued.

Documentation

Management reports are updated for each BayWatch site to include/document new issues discovered during the daily review. This information is accessible online for each BayWatch site under the same pull down menu as the daily and monthly reports.

Additionally, the following weekly and monthly services are also completed:

Weekly

Each week engineering reviews are performed of the operation plan based on identified improvements and/or plant schedule changes.

Monthly

Monthly Reports for each site are generated summarizing the energy savings obtained and documenting the key operating parameters. Bay engineers use a variety of analytical, quantitative tools and methods to model each compressed air system's performance under different operating scenarios, and determine optimum operating plans.

These services could only be performed with the use of an enterprise-wide system, which makes it possible for remote engineers to see a comprehensive view of each compressor and air generation system. The main purposes behind the daily reviews are to quickly identify issues for immediate response, identify incremental energy saving opportunities and monitor progress.

IMPLEMENTATION PROBLEMS

With a project this extensive, encompassing a large number of facilities in locations across North America, implementation difficulties were expected. For this project, the primary issues were technical conflicts with preexisting control systems, connectivity problems, and management issues with system operation.

A number of plants had preexisting system integration solutions that connected an existing compressor control system to a customized head-end computer. Most of these installations were fairly old and in various stages of disrepair. Although the installations were dated, plant personnel were accustomed to and comfortable with the antiquated front-end application. Fear of change and loss of functionality were key concerns from operations personnel. These challenges were slowly overcome through education, product demonstration, and other change management techniques. At plants that truly used the existing system integration solution for local monitoring and control, project underrun funds were used to install the Bay Controls BayView system integration software. This application restored local monitoring and control in a manner that married perfectly with the Bay hardware and web-based system. At locations that used the integration solution only for monitoring and not control, the plant was migrated to the web-based solution.

Internet connectivity issues occurred due to the geographic locations of the plants and central data collector. A few plants located in Mexico experienced difficulty obtaining reliable internet connectivity in a timely fashion. Other plants which had compressors located at a centralized powerhouse were frequently located far from the centralized main phone board. These locations either required a new broadband line installed from the nearest main drop or had to settle for using a dial-up connection and modem.

Management issues included gaining approval from corporate and plant IT departments to install a remote monitoring system. Concerns centered around these primary categories: preventing connection to the Ford local area network, business value of the data being transmitted, fail-safe capabilities of the system, and network protection strategy. Formal review of each issue with corporate, divisional, and plant IT contacts ensured that all concerns were addressed and documented.

ECONOMIC, TECHNICAL AND OPERATIONAL BENEFITS OF USING FCMS

Industrial manufacturers, such as Ford and other automotive companies, prefer to make data-driven business decisions. However, having access to accurate, consistently reliable data is often challenging or not possible. Thus, the ability to use actual data from field instrumentation has been extremely useful to Ford and has added credibility to compressor system analyses. Easily accessible performance reports and analysis from FCMS provide Ford with the tools necessary to maintain and operate an efficient compressed air system. The following sections illustrate how Ford Motor Company uses FCMS to manage compressed air and further drive energy and operational savings.

Efficiency Improvements

The FCMS system has enabled BayWatch engineers to incrementally improve compressor system efficiency by identifying beneficial operational modifications. The BayWatch service is able to recognize the most energy efficient compressors and what is the optimal mix of compressors for a given facility during production or non production periods.

Figure 12-6 shows an example of a compressor system that was operating with different system settings for each shift; the facility had not settled on a most effective, standardized operating plan. The FCMS engineers were able to identify the missed energy saving opportunities and developed a more efficient approach to running the air system.

Verify Current System Efficiency and Efficiency Gains

The ability of FCMS to accurately and consistently calculate the cost of compressed air at each facility has enabled the ability to benchmark compressed air efficiency and production between plants and against other manufacturers. As Ford decides to make future investments in compressed air systems, FCMS will be used to justify capital expenditures using hard data.

An interesting note is that the Vehicle Operations project used field verification and paper records to de-

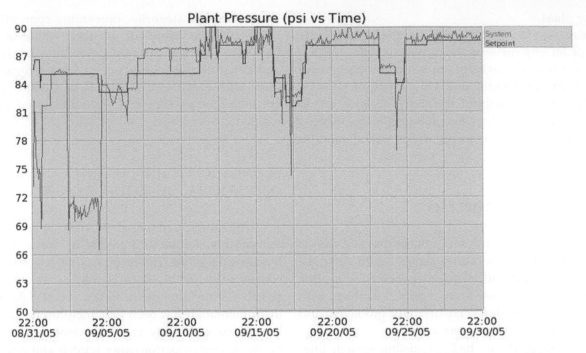

Figure 12-6. This chart shows numerous pressure setpoint changes, identifying a possible area for efficiency improvements.

termine pre-project energy use baselines. This was due to a lack of both historical performance data and time to meter existing conditions. A similar compressor project currently underway at Ford Powertrain Operations is installing BayWatch Virtual Gateways prior to the controller change in order to measure the baseline using the same instruments that the final controller will use. Any resulting efficiency gains will be highly documented thanks to the BayWatch system.

TVM Energy Program

One of the ways that Ford management uses the features of FCMS is in their Team Value Management (TVM) energy program. The TVM program is an ongoing initiative that seeks to maximize energy savings throughout Ford's manufacturing facilities by implementing low-cost and no-cost efficiency actions. One TVM goal is to reduce the non-production consumption of compressed air to 25% of the normal production consumption. Non-production compressed air shutdown performance is measured weekly at Bay using collected data from FCMS, which is then packaged and sent to Ford for internal publication and incorporation into the TVM database. Progress is tracked with the database to determine how close facilities are to meeting targets. What once was a tedious in-plant process of tallying weekly energy expenses and compressor system operating data, is now a simple matter of generating an appropriate report using the web based features of FCMS.

Measurement of Air Leakage

In one example of an efficiency and performance issue, FCMS reporting was used to identify the minimum cfm flow rate during Christmas day. Since it was assumed that there were no activities underway in the Ford plants, the measured flow rate was a close approximation to the leakage rate of the compressed air distribution system. Leakage is a notoriously hard quantity to measure in normal circumstances; the use of FCMS made this measurement relatively easy.

Preventive Monitoring Service

The constant monitoring service performed by Bay has been helpful in identifying and resolving problems with compressors before they become serious. In another example, there was a sensor calibration issue at a plant. BayWatch engineers identified the problem during a normal review and notified the plant before any serious issues with compressor performance or safety could occur. Figures 12-7 and 12-8 show how the power, flow and system efficiency readings were out of their normal ranges, alerting the BayWatch engineers to the problem.

Real Time, Data Driven Troubleshooting

When a problem with a compressor does occur, FCMS data recording and real time monitoring allow Bay engineers to aid with troubleshooting, resulting in a faster resolution to the problem.

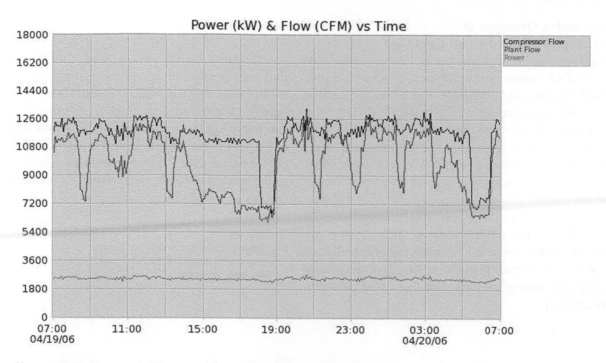

Figure 12-7. Power & Flow vs Time. Starting at about 15:00 hours, the compressor and plant flow readings diverted. The issue was resolved by 19:00 hours.

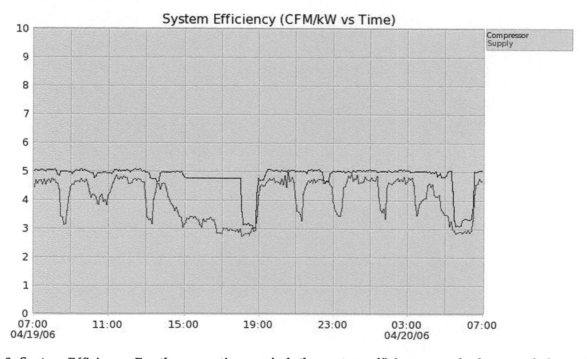

Figure 12-8. System Efficiency. For the same time period, the system efficiency graph also revealed a problem.

In one instance, after a compressor mechanical failure, a plant worked in real time with the BayWatch Center to troubleshoot the machine during restart attempts. This dynamic interaction with off site personnel would be impossible without FCMS capabilities.

Another troubleshooting problem occurred at another plant when a compressor experiencing an increasing number of start/stops. A compressor motor failed and initially the compressor control system was thought to be the root cause. Analysis of the motor windings indicated it was a preexisting condition in the motor. In this case, BayWatch was used to identify the number of start/stops and other performance issues that were taking place before the failure occurred.

**Comprehensive Overview of
Entire Compressed Air System Operation**

Ford has found it invaluable to have a centralized overview of all compressor systems. No other tool provides such an easy and accessible way to see the current status of every compressor on the network.

A soon to be released feature of FCMS is the Tactical Overview, which shows an at-a-glance synopsis of compressor system status at all Ford Vehicle Operations facilities. The columns on the Tactical Overview list the state of the remote network connection (WAN), the state of the local compressor intranet (LAN), the current event status of the system (faults, alarms, shut-offs for any compressors in the system), a running tally of the last 24 hours of system events, positive or negative deviations from projected energy savings, and tasks that currently need resolution.

CONCLUSION

From a broad perspective, this project proved that advanced control and management systems can significantly reduce facility compressed air energy expenses. Additionally, and more specifically, the FCMS component of the project proves the value of web based compressed air management systems. The ongoing performance gains and operational benefits make the underlying BayWatch technology an extremely useful enterprise management tool.

References:
[1] Capehart, B.; Kennedy, W.; Turner, W. Process energy management. In *Guide to Energy Management*; 5th Ed.; The Fairmont Press: Georgia, 2006; 418.

Chapter 13

Technology as a Tool for Continuous Improvement

A building management system serving the Milwaukee campus of Johnson Controls, Inc., delivers information that enables steady advances in occupant comfort, energy efficiency, safety and security.

Ward Komorowski, P.E.

PROPERLY DEPLOYED and operated, a building management system (BMS) is much more than a control device: It is a tool for trending, analyzing, benchmarking, and continuously improving facility performance.

While many BMS function largely as sophisticated time clocks, using only a fraction of their power, the installation serving the Milwaukee, Wis., campus of Johnson Controls, Inc., demonstrates what is possible when a BMS is pushed to its full potential by a creative and energetic staff.

The BMS serves a major corporate center that is home to 1,660 employees. Johnson Controls also uses the system, housed in the Brengel Technology Center, as a showcase to demonstrate the latest in building management capability to visitors from around the world.

The Milwaukee campus includes the Brengel Technology Center and the surrounding downtown complex of six buildings, as well as the corporate headquarters complex in the northern suburb of Glendale, eight miles away. It also includes the Battery Technology Center, a parking structure and the corporate aviation center on the grounds of Milwaukee's General Mitchell International Airport. That facility has hangars and maintenance facilities for two private jets and offices for the pilots.

The BMS integrates HVAC control, energy metering and management, lighting control, electrical distribution, security and access control, and fire and life safety systems. Web-browser-based functionality enables technicians to monitor and control the entire campus from an office in the Brengel Technology Center or from any network-connected desktop computer or portable device with wireless Internet access.

Since its installation in 2000, the system has helped to optimize occupant comfort, energy efficiency, safety and security, while driving down operation and maintenance costs. It has also been used for innovative functions such as:

- Energy consumption benchmarking between buildings and building areas.
- Monitoring of predictive maintenance devices on critical HVAC equipment.
- Control of lighting systems that use daylight harvesting to save energy.
- Environmental discharge control, monitoring and documentation.
- Automatic first responder notification in emergencies.
- Hydraulic shock ("water hammer") mitigation and water conservation.
- Flooding prevention.

Through green-building initiatives and technology applications at the Brengel Center and throughout the headquarters complex, Johnson Controls expects to realize energy and operational savings of more than $4.2 million over 10 years.

CERTIFIED GREEN

As a globally recognized provider of building management technology, Johnson Controls created the Brengel Technology Center not just as an office complex but as a demonstration site for its capabilities. The seven-story, 130,000-square foot building is named for the late Fred Brengel, chief executive of the company from 1967-88.

The structure blends architecturally with the rest of the Johnson Controls Building Efficiency headquarters, which occupies an entire city block and comprises 460,000 square feet of work space.

The Brengel Technology Center was one of the first buildings certified under the U.S. Green Building Council Leadership in Energy and Environmental Design (LEED) program. It later became the first ever re-certified from Silver to Gold under the LEED Existing Buildings (EB) program.

Energy efficiency is a hallmark of the Brengel Technology Center. Its addition increased the size of the downtown complex by 45 percent, yet first-year total energy costs increased by just 17 percent. Overall energy consumption increased by only 15 percent, and when expressed as Btu per square foot, consumption decreased by 20 percent.

With a state-of-the-art lighting system and daylighting control strategies, Brengel Technology Center lighting demands just 0.86 watts per square foot—significantly better than the state energy code requirement of 1.2 watts per square foot. Daylight accounts for 10 percent of the building's potential energy usage.

Compared with 1999 (pre-construction) energy consumption levels, the building has prevented emissions of 2.2 million pounds of carbon dioxide, 13,940 pounds of sulfur dioxide, 5,500 pounds of nitrogen oxides, and 180 pounds of carbon monoxide.

From the Brengel Technology Center, the BMS also seamlessly controls the corporate headquarters complex of three buildings, comprising 165,000 square feet.

TECHNOLOGY LEADERSHIP

In building the Brengel Technology Center, Johnson Controls had objectives beyond providing safe, comfortable, cost-effective work spaces. In large part, the facility was designed as a proving ground for building management technologies, innovative construction methods, and green building concepts, which the company promotes as part of a high-performance green buildings initiative.

The building was constructed in 2000 for just under $17 million, in line with what was then the market-average construction cost of $125 per square foot—demonstrating the concept that green buildings need not cost more. BMS technology was one of many attributes that led to LEED certification for the Brengel Center. Others included:

- Light-colored concrete and roofing and landscaped surfaces to decrease heat islands.
- High-shade energy-efficient windows.
- Water-efficient fixtures that reduce consumption by 20 percent.
- An open courtyard that provides green space for employees.
- Location near bus lines and provision of showers for bicyclists.
- A permanent indoor air quality monitoring system.
- Upgraded copy machines that use zero-ozone-emitting substances.
- Water-efficient landscaping that requires no irrigation system.
- Open ceilings that decrease material usage.
- Aggressive construction waste management and reuse of many existing materials.
- Ensuring that more than half of the building materials contained at least 20 percent recycled content.

A weather station mounted on the Brengel Center roof is tied into the BMS to provide cooling system load forecasting and improve the accuracy of energy applications. Wind, barometric pressure and solar sensors forecast load requirements, allowing staff to adjust energy demands, maximize efficiency, and control energy cost.

Other information-gathering tools include energy sub-metering, load profiling, and cost-report generation. Information from these tools enables management to identify energy-saving opportunities and provides leverage for negotiating with energy suppliers for more favorable pricing structures.

In addition, Brengel Center work spaces are equipped with personal environment control modules that allow individual employees to regulate comfort conditions according to their preferences. Desktop control units incorporated into the HVAC system and integrated with the BMS enable each person to adjust temperature, lighting, air flow and acoustic characteristics as often as necessary to maintain personal comfort.

The modules deliver conditioned air directly to the workstations. All air entering the space is continuously cleaned through an advanced filter, enhancing indoor air quality. The main module contains the filter and an air mixing box. It also generates "white noise" that users can adjust to achieve the desired level of background noise masking. An optional radiant heat panel under the desk provides extra warmth for the feet and legs for people who need it.

The personal modules help minimize energy usage because only occupied offices are conditioned. Each unit includes an occupancy sensor that automatically turns off the fan, lighting and equipment if an office is unoccupied for more than about 15 minutes. In addition, open areas, like hallways and file areas, require less thermal conditioning.

The integration of the modules to the BMS enables additional energy-saving strategies, including morning warm-up and cool-down, and workstation shutdown for long-term absences.

For the facilities staff, the personal modules virtually eliminate hot and cold calls and allow technicians to focus on other issues. The time-saving nature of personal space control is one reason Johnson Controls did not need to add any facility maintenance staff to serve the new Brengel Center.

Personal environmental control and green building attributes placed the Brengel Center on a sound foundation for long-term high performance. BMS technology is the key to continuously improving performance and to ensuring high reliability, low energy costs, and comfortable conditions across the Milwaukee campus as a whole.

FLEXIBLE DESIGN

The BMS for the Johnson Controls campus uses open system standards that accommodate a wide range of control devices and interoperate with legacy devices, protecting existing infrastructure investment.

The core of the system is a network automation engine that uses the standard communication technologies of the building automation industry, including BACnet® and LonWorks® protocols and the N2 Bus (Figure 13-1 shows the basic architecture). This combination of tech-

nologies enables monitoring and supervision of HVAC equipment, lighting, security and fire control systems.

The system can interoperate with more than 1,000 devices from some 125 manufacturers in more than 50 applications categories and is scalable to expand as the controlled infrastructure grows. A single control engine or a network of multiple devices provide alarm and event management, trending, energy management, data exchange and archiving, scheduling and communication.

Internet protocols and IT standards are built in, providing full access from any desktop PC or laptop

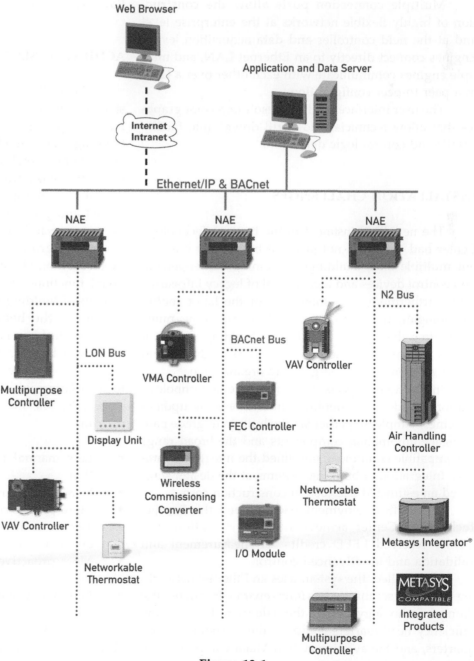

Figure 13-1.

computer with a standard web browser. In addition, the system delivers alerts by way of cell phone, pager, and handheld computing devices. Standard system security and encoding protocols protect against unauthorized access to data and control systems.

The engine supports IT standards and formats, including Internet Protocol (IP), Hypertext Transfer Protocol (HTTP), Simple Network Time Protocol (SNTP), Simple Mail Transfer Protocol (SMTP), Simple Network Management Protocol (SNMP), Hypertext Markup Language (HTML), and Extensible Markup Language (XML). It supports firewall technologies for protection against unauthorized access.

Multiple connection ports allow the construction of highly flexible networks at the enterprise level and at the field controller and data acquisition levels. Engines connect directly to an Ethernet LAN, and multiple engines communicate with each other over a LAN in a peer-to-peer configuration.

The user interface has high-resolution color graphics that allow technicians to "drill down" into facility details and control logic definition.

INSTALLATION CHALLENGES

The new BMS installed in the Brengel Technology Center had to control not just that brand new building but multiple older buildings containing earlier-generation control devices and a great deal of legacy infrastructure. Furthermore, as a showcase for the latest control technologies, it had to be capable of updating much more often than a BMS in a typical building. Finally, it had to accommodate the frequent moves, additions and changes inherent in a fast-growing organization.

In all these respects, the open-systems technology platform made initial deployment and ongoing updates extremely simple. In particular, the built-in upgrade path in legacy automation components and the broad range of compatible equipment simplified the integration process. Integration of building systems at the design stage saved an estimated $225,000 in construction costs.

The BMS deployment itself helped the Brengel Technology Center achieve LEED certification. The facility received LEED credits for measurement and validation and for advanced control.

As installed, the system uses an Ethernet networking web browser that serves four separate locations: The Brengel Technology Center, the balance of the Building Efficiency headquarters complex, the Corporate Headquarters, and the aviation center. Major components of the system include:

- One application and data server.
- 15 Network Integration Engines/Network Automation Engine devices.
- 1,110 N2/LonWorks field controllers.
- 36 network controllers.
- 22 field controllers.
- 4 IFC fire systems.
- One card access system.

All told, the system controls 31,277 different objects and is integrated with products from a wide variety of vendors, including Veris, AMX, Strionair Air, GE, Microlite, Lutron, Liebert, YORK, Cutler-Hammer, PowerWare, and Cummins.

ACHIEVING MAXIMUM BENEFIT

The BMS has enabled the 10-member facilities staff at Johnson Controls to achieve steady improvements in performance among the Milwaukee campus buildings. Personnel can access the system from the office, home, hotel, trade show floor, or airport and, if necessary, make adjustments.

It is common in the industry for BMS to be installed and programmed, then left largely unchanged for long periods even as the facilities they control change and grow older. In other cases, the BMS is heavily focused on HVAC, lighting control and other traditional functions. These approaches neglect significant benefits of building management technology.

Since the beginning, the goal of the Johnson Controls staff has been to use the BMS to achieve the maximum benefit. The basic approach is to ask, for each device deployed in the facilities: Is there a benefit to integrating with the BMS? In numerous cases, the answer is: Yes. In general, tie-ins with the BMS typically mean:

- Less manual reading and recording of information.
- Less need for manual equipment adjustments.
- More precise control over equipment operation.
- Accurate and unassailable documentation of events for regulatory and other purposes.
- More proactive maintenance and operating practices.
- Less worry and stress burden on staff.

In addition, trending and benchmarking tools enable technicians to identify areas for improvement on a regular basis. Just as a medical patient in a hospital is wired with diagnostic devices to track temperature, blood pressure, heart rate and other vital signs, the Johnson Controls buildings are under constant watch. Even when away from the office, technicians check in with the buildings regularly to find out how they "feel."

While this particular system is designed in part to be a showcase for building management technology, control capabilities are not added simply because they are possible. The staff is required to cost-justify all applications, just as any other customer would.

The overall, traditional benefits of the BMS at the Milwaukee campus are outlined above. Here are a number of examples of how technicians use BMS technology to maintain a culture of continuous improvement in building performance.

Benchmarking

It is critical for facilities managers to know where to invest limited capital for the maximum return. To that end, the BMS constantly tracks usage of electricity, fuels, chilled water, and steam, and reports the cost of facility operation, site by site, building by building, department by department. This capability enables comparison of energy performance.

One could assume that the corporate headquarters complex, built in the late 1960s, would consume more energy than the Brengel Technology Center, a certified green building completed in 2000. But how much more energy? And what would be the payoff on energy-saving improvements? Benchmarking provides answers.

The BMS separately reports electrical load for HVAC, power receptacles, lighting, and miscellaneous loads (such as elevators). In spring of 2005, benchmarking showed the corporate headquarters with a total electric power demand exceeding that 8 w/ft^2. (For comparison, the Brengel Center required 2.6 w/ft^2.)

A year later, demand at the corporate headquarters had dropped to 6.25 w/ft^2, thanks to energy-saving projects that included replacement of older chillers with modern variable-speed units and a major lighting retrofit designed to take advantage of natural lighting from largely glass exteriors. If power demand from the corporate data center is subtracted, the headquarters operates at 4.75 w/ft^2. In any case, benchmarking documented savings of roughly 2 w/ft^2.

Trending

Like benchmarking, observation of facility trends supports sound and timely decision making. The Johnson Controls BMS tracks the functions of critical power backup systems—emergency generators and un-

interruptible power supplies (UPS)—around the clock. Any malfunction triggers an alarm, sent automatically by text message to designated facility staff members' handheld computers.

In late 2003, the corporate data center installed a new 400 kW UPS (power factor 0.9, net capacity 360 kW). To monitor growth in demand on that system, the staff set an alarm limit on the demand meter at 160 kW. After repeated alarms, that limit was raised to 180 kW in June 2004, then to 190 kW in August 2004.

Concerned that growing demand would soon consume the capacity of the UPS—a major capital investment—facility staff alerted data center managers. They responded with computer consolidation and other energy-saving measures. Demand declined and then essentially stabilized; in mid-2006 demand stood at 166 kW.

Critical Equipment Protection

Proactive monitoring of critical equipment can help to minimize capital investment. The Brengel Technology Center operates with a single 400-ton variable-speed chiller and a single cooling tower. No redundant backups are needed because the devices are continuously monitored with an online vibration analysis system integrated to the BMS. The fans on the main air-handling unit that serves the majority of the building are also monitored around the clock.

Vibration analysis detects abnormal operating signatures that indicate potential component failure. A vibration analysis spectrum of the chiller installed in the Brengel Technology Center (Figure 13-2) shows a healthy signature indicating that failure is unlikely. In case of anomalies, facility staff is automatically notified, and diagnostics and repairs are scheduled as needed.

Lighting Improvements

Lighting retrofits have been a substantial source of energy savings across the Milwaukee campus. The BMS enables facility staff to calculate energy usage for lighting in a given space, model the expected benefits of a lighting retrofit against utility rates, then document the actual benefits when the project is complete.

At the corporate headquarters, a building that houses the cafeteria and data center has an exterior that is largely glass. A new lighting control system, completed in March 2006, enables daylight harvesting. Foot-candle meters in the offices, tied into the BMS, measure light levels. The BMS then brightens or dims the lights as daylight levels change, maintaining the optimum lighting levels in the work spaces.

At the Brengel Center, modeling showed that day-

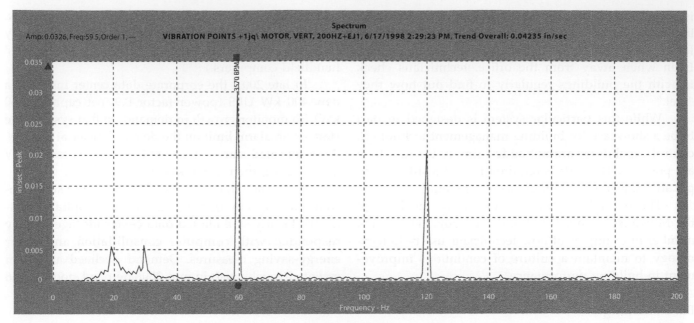

Figure 13-2.

light harvesting would not be cost-effective. Instead, the staff chose light fixtures with two lamps, enabling three levels of lighting: fully on when employees are working, 50 percent on when cleaning crews are present, fully off when the space is unoccupied. Levels are controlled by the BMS.

Water Conservation

A BMS is not typically seen as a device for solving plumbing problems, but it played that role in 2005 at the Johnson Controls Corporate Headquarters. Hydraulic shock ("water hammer") had become a persistent problem. Facility staff traced the problem to sets of four sanitary fixtures on single valves tied into a time clock and flushed every seven minutes. The obvious remedy was to replace those flush valves with motion-sensor flush valves on individual fixtures at a cost of some $30,000.

Instead, staff looked for a lower-cost solution that would correct the hydraulic shock while also conserving water. Because the BMS monitored the water meter, it was easy to trend water usage patterns and identify the peak hours—typically 9 to 9:30 a.m., 11:30 a.m. to 1 p.m., and 2 to 2:15 p.m.

At a cost of $500, the staff tied the flush valves into the BMS and adjusted the flush frequency—more often during peak hours, less often during slower periods. These measures saved more than $2,000 annually in water charges. In addition, timing of the flushes was staggered so that multiple units would not flush at the same time. This significantly reduced hydraulic shock.

Personal Safety

Building automation also can improve responsiveness to emergencies. Johnson Controls has deployed defibrillators around its campus in wall-mounted cabinets. Door contacts on the cabinets are wired to the BMS. If a cabinet is opened, the system automatically notifies first responders on the premises and alerts facility staff.

In another application, research scientists who may have to work alone in the Brengel Center after hours wear an alert device on their wrists. In a medical emergency, a person would push a button on the device, sending a wireless signal to a transmitter, which in turn automatically contacts outside emergency personnel. Facility personnel are also notified, and security guards are dispatched to the proper posts so that when paramedics arrive they can be escorted directly to the scene of the incident.

Environmental Control

The BMS also supports environmental compliance. A research facility at the corporate headquarters complex supports the Johnson Controls automotive battery business. The facility handles battery acid that periodically must be disposed of. To comply with environmental regulations, the acid must be neutralized before disposal, and each disposal must be recorded.

Historically, the process was monitored with an on-site chart recorder that graphed the solution pH (acidity level) on paper disks. Personnel had to visit the site to change the disks, which then had to be filed and stored. Today, the acid neutralization system is integrated with the BMS. A pH meter monitors the neutralization tank. When the solution inside reaches a neutral pH, a solenoid is actuated to open the valve that drains the tank. The BMS automatically records the pH of the solution and the date and time of the discharge.

Security Integration

The BMS at the Brengel Technology Center monitors an extensive security and access control system that safeguards personnel and property. That same system also contributes to energy savings and occupant comfort.

Access control is based on a system of card readers at designated access points. The system is configured so that when an employee "badges in" at a checkpoint, his or her picture appears on the attendant's computer screen. The system also is designed for fast and effective response in emergencies.

Employees in the main lobby and other high-traffic areas have access to emergency switches by which they can trigger an alarm in case of a disturbance. That alarm is routed to the BMS, which automatically focuses security cameras on the area and dispatches security personnel.

The access control system has the capability to play a role in conditioning employee work spaces through integration with HVAC and lighting controls. For example, when an employee "badges in," the BMS automatically turns on the office lights in the person's work area and changes the variable air volume (VAV) controller from the unoccupied to occupied mode. The procedure operates in reverse when the employee "badges out."

Facility Protection

The corporate headquarters is surrounded by a manmade pond that empties into a spillway, which in turn flows to the Milwaukee River. The building is situated slightly below the surrounding grade. In extremely heavy rainfalls, it is possible for the river level to rise higher than the level of the manmade pond, causing water to flow in reverse and potentially flood the building's first floor.

This almost occurred in August 1986, after several inches of rain fell in a few hours. To prevent this in the future, a dam with a movable gate was installed to prevent reverse flow. Today, that gate is controlled by the BMS. Instruments monitor the water levels on both sides of the dam. If the river water rises above the level of the pond water, the gate automatically closes. This means that during rain events, technicians do not have to physically check the water levels and manually operate the gate to protect the building.

Nuisance Alarm Prevention

While alarms are highly beneficial in facility management, it is desirable to have a "quiet" system, as free of nuisance alarms as possible. Reports generated in the BMS enable identification of systems that produce multiple alarms.

Those reports also help identify systems in need of attention. In summer 2005, one VAV controller in the Brengel Center became starved for air and issued an alarm each time the outside air reach 85 degrees F. The BMS recorded a total of 32 such alarms from that controller during the summer. This information alerted facility staff to perform diagnostics and correct the problem.

Anomaly Detection

The BMS can alert facility staff to a wide variety of abnormal conditions. Lighting at the Brengel Center is monitored to report how long lights in a given area are on during a given month. In general, the system alerts facility staff if lights are on for more than 300 hours.

In January 2006, facility staff received a page from the BMS reporting that uplighting on the building exterior had been on for more than 300 hours. A technician visited the roof and found a photocell for a switch that turned the lights off in the daytime was covered by three feet of snow. If the problem had not been corrected, those lights would have burned around the clock.

Here, the primary concern was not wasted energy but the need to change exterior lamps that burn out—a costly process that requires a manlift, blockage of the street, and deployment of a flag person to direct traffic.

TAPPING THE FULL POTENTIAL

A building management system is much more than a lighting and HVAC timing device. When used to the fullest, it serves as the hands, eyes and ears of a skilled facility staff. It is a living system that changes and grows with the buildings, constantly providing information and new capabilities that enable performance improvements: in energy efficiency, occupant comfort, security and safety. The applications of modern BMS technology are limited only by the user's imagination.

Johnson Controls Inc.— Milwaukee Campus

	Number of Buildings	*Size*	*Employees*
Downtown Complex (Includes Brengel Technology Center)	7	460,000 ft^2	1,200
Corporate Headquarters	3	165,000 ft^2	350
Aviation Center	1	35,000 ft^2	10
Battery Technology Center	1	60,000 ft^2	100
Parking Structure	1	500 stalls	
TOTAL	**10**	**720,000 ft^2**	**1,660**

Chapter 14

A Case Study of the Jefferson Health System's Enterprise Energy Management System

Randolph Haines, CEM, Energy Manager
Thomas Jefferson University

INTRODUCTION

THIS IS A CASE STUDY on the installation of an enterprise metering and energy management system in a multi-site healthcare and university group of buildings. I will explain how we went from manually reading meters (the Stone Age) to automatically reading them every 10 minutes (the Information Age), and generating reports that save tens of thousands of dollars per year.

BACKGROUND

The Jefferson Health System (JHS) is a group of hospitals and Thomas Jefferson University, a teaching university comprising 32 large buildings and hundreds of small offices, in the five-county greater Philadelphia area. The system comprises four primary healthcare networks that came together over a period of a few years. At the same time, Pennsylvania deregulated the electricity market. Each network is financially and contractually independent and is part of the JHS alliance to help obtain better rates on insurance, energy, supplies, and other services. The networks each have their own facility manager, with one energy manager who covers the entire system. In total, JHS has more than 10 million square feet of property, with more than 22,000 employees, and has an electrical peak close to 60 megawatts. We consume more than 350,000,000 kWh, 700,000 MM Btu of natural gas, and 400,000,000 lbs. of steam per year at a total cost of $45 million.

IN THE OLD DAYS...

Before advanced metering, the energy manger had to walk to every meter on the main campus and manually record kWh of electricity, pounds of steam, Ccf of gas, and Ccf of water. Recording this information would take 8 hours each month. Due to the distances of sites, many meters were not recorded. The problem with this is that we could not record the peak demand for the month, power factor, or simultaneous demand readings. There was no way to correlate weather conditions with the meter readings. Another problem was by the time the data were read, posted, and distributed; it would be at least two weeks. Those were old data and not very valuable. Who really remembers the weather and other operational details from 4 to 6 weeks ago? Those data were also used to allocate costs to each building and were crude, at best. We had no idea, which buildings were creating our peak demand, or at what time they occurred. Our local utility has very high demand charges and annual ratchet charges. The summer peak sets our ratchet charges throughout the winter. For example, if our summer peak for 1/2-hour hits 20 megawatts, we have to pay a demand charge of 16 megawatts (80% of summer peak) for the entire winter even if we use only 12 megawatts.

With the cost of electricity rising and the amount of companies participating in the Pennsylvania market declining, the previously negotiated three-year, fixed price for electricity was going to take a hit. With a true aggregated metered load, JHS could now go to the marketplace as an educated consumer.

JHS personnel had been talking and trying to get funding for advanced metering for about 10 years. With capital drying up in the healthcare and university industries, a decision was made to select an energy services company (ESCO) to provide energy auditing, funding and project management for energy projects. After completing the selection process, Alliant Energy (recently purchased by Constellation Energy) was chosen to implement approximately $10 million in projects. One project was to install an enterprise energy management system.

With funding in place, we chose the Stark System.

WHY WE CHOSE STARK

During the information-gathering phase of the energy management selection process, we looked at a few different systems. We initially put in place a small pilot program with a software company that dialed into a modem and gathered our data on a nightly basis and posted our information on their web site. After evaluating this, we decided that this wouldn't be adequate for our needs because the information is expensive to gather ($/meter/month), the reports were not very sophisticated and the data were a day behind real time. A lot better than what we had but not quite there. We then met with our building automation system provider and investigated what they offered. They were able to demonstrate their ability to gather data on a real time basis but we had to basically build our own reports. I didn't have the expertise and time for that to happen. Another hurdle was that not all of our buildings in the health system used this one particular building automation system.

We then met with Stark. They demonstrated their ability to gather data on our own ethernet system near real time and offered hundreds of pre-written reports. The company was started in England who more than 10 years prior went though deregulation before the United States and it became obvious that they were ahead of many of their competitors. Also, their pricing beat their competitors. We would own all of the data (stored on our data server) and had the ability to easily modify the canned reports. Their system didn't depend on any particular brand of meter and we knew we were not going to outgrow the system.

With the ability to bring the metering in real time with the software trending, equipment running out of tolerance can be detected quickly. Other benefits include the use of monitoring, customized reporting, verification of utility bills, and remote site monitoring. JHS also decided to use a web-based client to save money because of many users at different locations. This is intuitive software and required little operator training.

The savings were estimated to be $100,000 per year. With the total installed costs including retrofitting the existing meters at $250,000, this provided a simple payback of 2-1/2 years.

HOW THE STARK SYSTEM WORKS

To gather metering data into the Stark System, the first thing that has to happen is the meter has to have the ability to provide a pulse for X number of kilowatts or pounds of steam or cubic feet of gas, etc. Some of the newer meters have this built into them. With our older mechanical meters, we retrofitted the meters by having a meter service company come in and add a pulse generator to each of the meters that didn't already have pulses. The way this works is by painting a white spot on the spinning meter disc and then having an eye pickup the spot every time the disc revolves one revolution. This was cheap to install and very reliable.

After the meters were retrofitted, we installed a data logger and terminal server near the communications closets on our campus (Figure 14-1). Each logger holds up to eight points (meters). Each point can be modified to accept a digital input (i.e., pulse), a 0-10 VDC or a 4-20 milliamp input.

A terminal server was mounted next to the data loggers and assigned an internet protocol address (TCP/IP). This terminal server was connected directly into our ethernet system through a wall jack tied into a switch.

A cable was then installed (shielded, twisted pair) from each meter to the data logger. These cables can be up to 2000 feet from the meter to the logger. Each point on the data logger is assigned a multiplier value on the application software.

We then installed three servers in our data center: data, application and web. We considered having these servers in the energy managers office but decided to place them in the data center due to the UPS backup and daily data backup and other data maintenance services they provide.

Every ten minutes the application server, running Windows 2000 Server software, via the ethernet, goes out to each data logger, collects the information, and stores it on the data server (Figure 14-2). The data server uses Oracle database, version 9.2.0.4 to manage the data. We have had the system in place for more than five years and do not have a problem with storage capacity. The information being stored is a very small packet of information: date, time, identification and a value. We have

Figure 14-1.

been storing meter data for 300 meters every ten minutes for the past five years and to date we have used 4.4 gigabytes since installation. The system continually grows due to the addition of buildings and the desire to learn more about how our buildings are using energy. There is no concern about out growing the Stark System.

The energy manager and a backup person access the information via a client computer and all other users (currently 25) use the web server via Internet Explorer. See Figure 14-3.

INSTALLATION CHALLENGES ENCOUNTERED

To get started, I performed a walkthrough of all buildings to locate all meters and data racks, and identify location of data loggers. After installing a few pilot phone modems, we chose to use the University Ethernet system. Due to the size of the health system, the installation was broken down into three phases. Phase 1 installed all revenue metering in Thomas Jefferson University. Phase 2 installed revenue metering throughout the Jefferson Health System. Phase 3 retrofitted and installed meters on all electrical sub meters and incoming steam and natural gas meters. The revenue and feeder meters captured both kWh and power factor, since JHS pays a penalty on all loads below 95% power factor.

First, the system works great, but a few obstacles delayed installation and added to costs. **This is a very detailed, complex project that should not be taken lightly.** The expertise of many people is required to

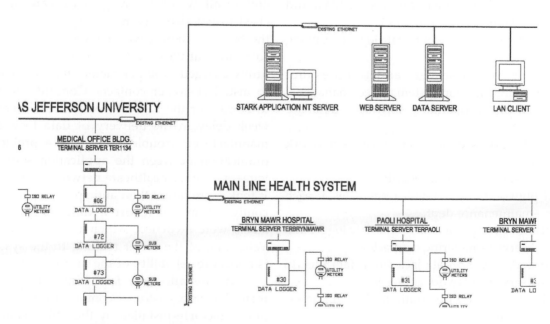

Figure 14-2. Metering System network architecture (partial)

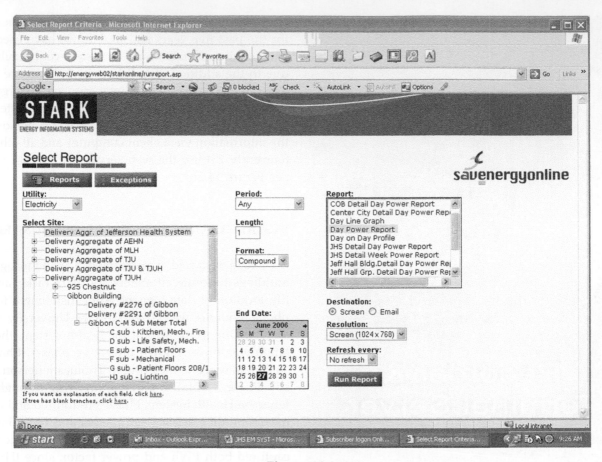

Figure 14-3.

make this project successful. The following is a partial list of project components:

1. Install external customer contacts for kWh and KQ—PECO (our utility company)
2. Run cables from each meter to data logger—cabling contractor
3. Select and install servers and troubleshoot problems—information system project manager
4. Select and install operating software—database administrator
5. Assign IP addresses to data loggers—network engineer
6. Security issues—network security
7. Install plywood and 120-volt receptacles at each meter—maintenance dept.
8. Mount and wire up isolation relays, data logger and terminal server—metering contractor
9. Mount and wire up data jack from data rack to terminal server—network engineer
10. Install software, set up configuration, reports—software engineer
11. Make sure it all works—me!

The following issues were discovered along the way to installation. First, there was much discussion on what servers to purchase to ensure some redundancy and to allow for future growth; care must be exercised to not size servers so large that they would kill the budget. Then it was agreed to use NT and Oracle software, but these cost $1,200 for a 5-user license. The utility charged $750 per meter (and now charge $1300) to install customer contacts. Conduit and wiring had to be run for the 120-volt receptacles. There was a six-week delay in the delivery of data loggers from the manufacturer. Troubleshooting of a problem in communication between the application server and data loggers in other healthcare networks was also delayed. It took more hours than anticipated to set up metering configuration and report writing. A few meter wiring labels were switched, which wasn't readily apparent. Validation of the meter reading to the system reading is critical to reliability of the information.

We installed the following equipment at each meter location: customer contacts for kWh and KQ (for power factor)—installed by the utility company; isolation contacts, data logger/recorder; terminal server and data jack; cabling (CAT5ETR) between each meter (including all subs) and the data logger. A data server, application server, and web server were purchased and installed in the computer center. Stark software was installed on both the application server and the manager's personal computer. A metering configuration was then set up, allowing reports to be automatically sent out via e-mail.

ECONOMIC, TECHNICAL AND OPERATIONAL RESULTS

The benefits in having a full-blown enterprise energy management system have been tremendous. We have saved way more money than the system cost. In fact, in one purchasing deal of the electrical commodity, we paid for the entire metering system. We know exactly how much energy we are using simultaneously in all of our buildings. This is extremely valuable when negotiating purchasing of electricity, natural gas and steam. We now have more than 250 real meters and 50 virtual meters. A virtual meter is the real time total of a group of meters (for example; to make up a total load of a build-

ing). Then a group of buildings are added together for a total load for each health system. Than these totals are added together to obtain one value for the entire health system (see Figure 14-4).

By measuring our energy use and getting the information to operational supervision, we save energy every day. We catch mechanical and electrical equipment running out of tolerance much quicker than in the past. For example, air handlers in one building that were being ramped down each night were accidentally left in operator override and the metering system caught it the next day, saving hundreds of dollars per day. In the past, this may have run like this for weeks or months. We also know how much energy we use based on heating and cooling degree-days. Our buildings have a high correlation to outside weather conditions. We now trend energy use by weather and raise alerts if they get out of tolerance. Having the system in place and getting the information to our operational supervision saves from 2-5% because of increased awareness. Energy related repairs get fixed faster due to this knowledge.

We automatically collect, calculate and e-mail invoices to more than 25 external (retail) users each month

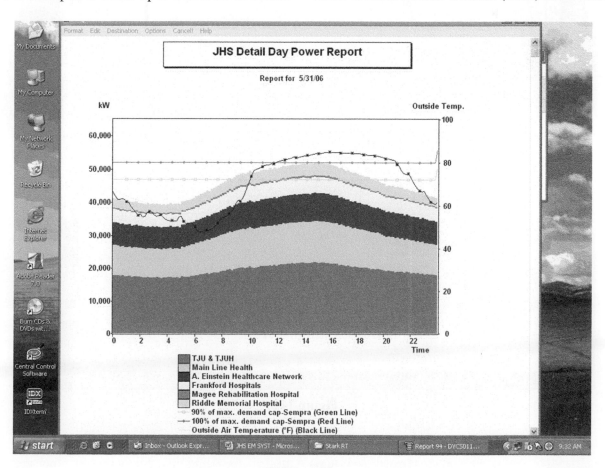

Figure 14-4.

which are more accurate which increases our collection rate (see sample report, Figure 14-5). With the new system, we have a much better calculation of our energy use in buildings we own and operate, which improves the accounting, which in turn improves our collection rate on research grants.

With accurate metering, we verify the value of our energy saving projects. We now save more than 42 million kilowatt hours of electricity per year over what we would have used if we did not install any energy saving projects.

Recently, we started selling real time and day-ahead power on the PJM (our regional transmission operator) open market. By having real time power use, I can bid confidently in the market. In the first six months of the program, I have received more than $30,000. I intend to improve the value of this by implementing a load-shed strategy when prices get very high in the summer months.

CONCLUSIONS

Although this project was a lot of work to install, the results have been tremendous. JHS has moved from the Stone Age into the Information Age in energy management and measurement. Like cars, JHS now has a speedometer and an odometer on its utilities; providing data on how fast the "car" is going and how far it's been. The system has paid for itself many times over. It is fairly developed but will always continue to be expanded as it is used and grows. JHS sees real advantages to gathering real time energy data on one screen in terms of block purchasing and awareness of system problems as they happen. As the system matures, the possibilities for its use are endless. I continue to look at energy use and how we can continue cutting its use or selling its value in the market place. The only thing holding me back is my imagination!

This has been a wonderful investment to improve the future for the Jefferson Health System.

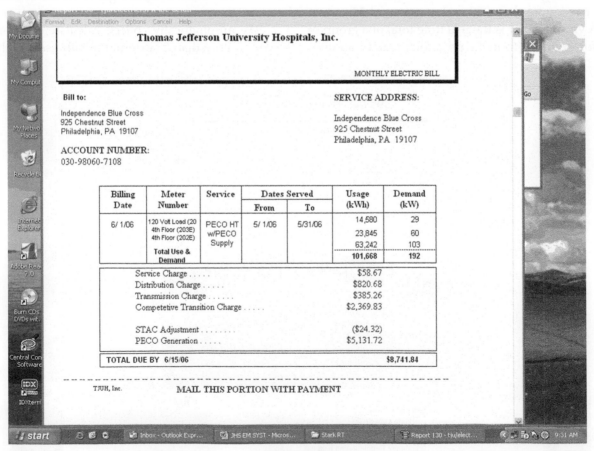

Figure 14-5.

Chapter 15

Integrated Energy Management:

A Proven Approach to a Successful Energy Management Program

Richard Rogan and James Peedin
Honeywell International Inc.

ABSTRACT

IN AN EFFORT TO PROVIDE direction for organizations struggling to manage volatile energy costs, this chapter will dissect two successful integrated energy management (IEM) programs—one in the federal sector (Fort Bragg) and one in the commercial sector (Smart and Final Stores, Inc.). The two programs share a common technology platform and approach which have contributed to more than six years of successful energy cost management.

BACKGROUND

Fort Bragg

Fort Bragg, located in Fayetteville, NC, is the largest U.S. Army post and home of the 18th Airborne Corps. The Fort Bragg Directorate of Public Works (DPW) began a rigorous energy management program in 1999. In 2005, its efforts reduced energy costs by $15 million. To date, the integrated energy approach has delivered $74 million in energy and operational savings.

Additional details about the IEM program at Fort Bragg are available in the following publications:

- HPAC Engineering, August 2005, Innovative Public/Private Partnerships Yield Energy Savings for Army Base, Honeywell

- Engineered Systems, October 2005, Fort Bragg's Arsenal of Efficiency, Joanna R. Turpin

- Energy & Power Management, August 2005, Fort Bragg Attacks Energy Costs, Kristen Anderson, Honeywell

- Distributed Energy, May/June 2004, The New Tri-generation Players: Integrated Cooling, Heating, and Power Systems Are Here, David Engle

Smart and Final Stores

In 2000, Smart and Final Stores, Inc., a California-based warehouse food and foodservice store operator, began an IEM program. Energy savings alone amounted to $6.25 million in 2005, and the program resulted in a $15 million reduction in energy and operational costs over 5 years. The details and results of the program are highlighted in the following publication:

- *Web Based Energy Information and Control Systems* published by The Fairmount Press, Inc. 2005, Compiled and Edited by Barney L. Capehart and Lynne C. Capehart, Smart and Final Food Stores: A Case Study in Web Based Information and Collection, Written by Richard Rogan and John Marden, Honeywell

Although differing significantly in mission and energy usage patterns, these two case studies share a common approach. This unique energy management approach integrates energy supply management with energy demand management to help manage energy costs, consumption and risk. The IEM approach is facilitated by an energy information and control system (EICS). The EICS integrates energy metering, building management, load management and energy analysis tools in order to facilitate identification, development, measurement and verification of cost savings associated with the energy management program.

The EICS for each program utilize technology, software and solutions from a broad array of industry providers. This chapter will provide a high-level overview of the programmatic IEM approach, as well as a vendor independent overview of the major components, functions and features of the EICS which have enabled to the ongoing success of these programs.

THE APPROACH—
INTEGRATED ENERGY MANAGEMENT

The IEM approach treats energy management as a continuous process rather that a series of events. It integrates energy asset management and energy supply management into an active demand management program through the use of enabling services and technologies. The synergistic benefits of the resulting energy management program are significantly greater than the sum of the individual energy conservation measures (ECMs).

IEM provides the process and tools which enable energy managers to drive economic and environmental benefits by utilizing and actively managing the following three components:

- **Energy Asset Management:** The use of high-efficiency equipment and modifications to building equipment in order to reduce energy consumption. The selection and application of energy assets which support multi-fuel options, such as on-site generation (combined heat and power, peaking plants, steam turbines) and alternative fuels (biomass, photovoltaic).

- **Energy Supply Management:** Developing energy supply structures which are designed to access lower-cost wholesale products by leveraging load control capabilities created with the assets listed above. Selection of energy feedstock sources (electric, natural gas, oil, biomass, LP) which provide flexibility in meeting load requirements. Performing ongoing procurement and risk management activities.

- **Active Demand Management:** Real-time integration of energy assets, energy supply structure and energy data via an EICS that supports data-centric decision making, proactive solution development, automated control, and performance reporting across the entire energy chain.

THE ENABLER—ENERGY INFORMATION
AND CONTROL SYSTEM

A robust EICS is core to the successful implementation of an IEM program in that it provides the knowledge of the enterprise energy demand, consumption and

Figure 15-1. Integrated Energy Management
Actively Manage Energy Assets Utilizing an EMCS to Optimize Energy Supply Structures

trends, which will allow the energy manager to increase efficiencies, identify trends, benchmark consumption, purchase and control energy more intelligently.

The EICS is a web-based information system, which integrates utility metering data, equipment monitoring, building control systems and the tools necessary to manage all aspects of the energy chain. An EICS is broadly composed of the following components:

- Input/Output (I/O) Devices
- Building Control Systems (BCSs)
- Building Network Integration (BNI) Platform
- Enterprise Network Integration (ENI) Platform
- Energy Management Applications (EMAs)
- Energy Management Suite/Energy Management Services (EMSs)

Figure 15-2 provides a general overview of a EICS system components, their features and functions. Depending upon manufacturer, the EICS components may combine features and functionality at any level. The components of the EICS utilize various communication media and protocols in order to maintain real-time or near real-time management. The communications media are dictated by site conditions, existing infrastructure and EICS equipment selection, and typically communicate via the organization's local area network (LAN)/wide area network (WAN).

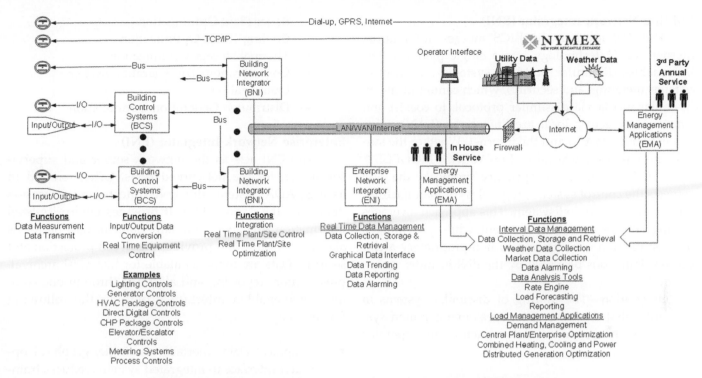

Figure 15-2. Energy Information and Control System Architecture

Input/Output (I/O) Devices

I/O devices are the hardware, transducers, transmitters and relays, such as temperature sensors, flow and power meters, necessary to interface between field equipment and BCSs. They convert ambient information such as temperature, equipment status or power reading into signals which may be interpreted by the BCS. Signals typically used by the BCSs include current, voltage, resistance and contact closure. New "smart" I/O devices contain on-board microprocessors, which convert the sensed data to digital format and allow communication of data directly with higher-level systems.

Building Control Systems (BCS)

BCSs are local microprocessor-based computers, which gather data about the building operations (i.e., temperatures, meter data, smoke levels) and utilize that data to make predetermined changes in the operation of the building assets (i.e., start chiller, run generator, open valve, sound alarm). Some common examples of control systems are building automation, fire alarm, card access, energy metering and video systems.

The installation and modifications of control systems can create energy savings by optimizing the operation of facility mechanical, electrical and water systems based on such parameters as time of day, temperatures, occupancy schedules and operating parameters. Operational savings are realized due to remote system access capabilities, reductions in manpower requirements and reduced wear on building systems.

Energy metering and building automation systems provide the end-to-end control capabilities which enable an active demand management program to close the loop on energy. Pre-programmed control sequences continually manage the efficient operation of the energy assets while supervisory control and optimization routines are provided via the higher level EICS components.

Some of the important functions associated with an IEM program facilitated by BCSs include:

- I/O Data Conversion—Traditional data sensors produce an electronic signal, such as 4-20mA, 2-10 Volts, or contact closure, which represents the condition of the medium they are sensing (temperature, flow, status). The BCS converts these signals into digital information (37°F, 100 FPM, On), which may be stored for future use, read by an operators interface and utilized for real-time control sequences.

- Real-Time Equipment Control—Energy assets such as generators, chillers, and boilers are typically supplied with package controls which ensure safety and manage the stand-alone operation of the equipment. Plant-level direct digital controls (DDCs) are often utilized to provide coordination amongst multiple energy assets and distribution systems (pumps, fans).

Building Network Integrator (BNI)

The BNI level of the EICS merges automation, internet and IT technology into a single platform. The BNI enables dissimilar building systems (i.e., control systems, metering, and security), which communicate on dissimilar media via dissimilar protocol, to coexist and communicate across local area networks (LAN), wide area networks (WAN) and the internet. Whether the systems are based on BACnet®, LonWorks®, MODBUS®, SNMP, OPC, or legacy proprietary protocols, the BNI connects the control systems, regardless of manufacturer into a seamless, unified system. This approach leverages existing control system investment while providing a low-cost method of migrating to non-proprietary architectures. Functions provided by the BNI include:

- Integration—The uniting of dissimilar systems in order that they may operate as one common system with a common data structure and operator interface.

- Real Time Plant/Site Control—The BNI essentially allows direct digital control strategies to operate across disparate control systems. In doing so, existing control system infrastructure may be leveraged to decrease cost and optimize operations. Examples of integrated control opportunities include:
  - Sensors utilized by package controllers are available for plant-wide control strategies in lieu of installing duplicate DDC sensors.
  - Load management applications utilize metering system data as the control input to dispatch generators, manage loads on lighting systems and shed loads via DDC systems.
  - Utilization of digital video systems and remote monitoring of control systems to decrease the number of operators required to manage complex plants and systems.

- Plant Optimization—Plant optimization strategies manage the local mix of energy assets in real time so as to maintain peak system operating efficiency based on current load requirements, equipment availability, fuel options, emission limits, and energy costs. Optimization routines include calculation of plant and equipment operating efficiencies for use in analysis, strategy development and performance reporting. The plant optimization strategies may receive supervisory direction from the enterprise energy management applications. Plant level optimization strategies should include:
  - Chiller Plant Optimization

  - Boiler Plant Optimization
  - Cooling Tower Optimization
  - Thermal Storage Optimization
  - Co-generation/Tri-generation System Optimization
  - Distributed Generation Optimization

Enterprise Network Integrator (ENI)

The ENI acts as the network server and supervisor station for the networked BNIs. Often referred to as a SCADA system, the ENI collects and processes data from the various BCS. It provides enterprise-level information exchange, standard data format, performs alarm processing, and provides password protection and security. Data are made available for historical archival, presentation, reporting and alarming through one common web-enabled interface providing the following functionality:

- Graphical Data Interface—A single, graphical operator interface to integrated systems reduces training costs and increases operator efficiencies. Monitoring and trouble shooting of operating conditions can occur at a single location, thus reducing the need to interrogate multiple operators terminals and machine interfaces to analyze problems.

- Data Trending—The management and presentation of historical data from the various subsystems. Historical trend data are useful in evaluating operational problems. Real-time trends are available for tuning of control loops and analysis of anomalies.

- Data Alarming—The annunciation of abnormal conditions through operator interface, email and text messaging systems reduces the need for real-time monitoring of systems thus allowing more time for preventive maintenance, data analysis and optimization. Typical alarms include; high/low measured value, deviation from setpoint, incorrect equipment status and manual override of control. Although data alarming is a valuable tool, care must be taken to determine the correct application so as to avoid nuisance alarms which may cause operators indifference.

Energy Management Applications (EMA)

The energy management applications (EMAs) provide the tools and functionality necessary to actively measure, manage and control energy assets across the EICS. The EMAs require a level of expertise which is not always resident in an organization. For this reason, many vendors offer the EMA functionality as an annual service plan (ASP) versus a licensed hardware/software solution. Whether they are provided as an ASP or an on-site solution, the EMAs must provide the following basic functionality:

- Interval Data Management—The gathering, management, storage and presentation of interval based energy related data. Interval data which is necessary to an IEM program includes energy consumption, energy production, historical and forecasted weather data and energy market data. The data are made available for presentation, reporting and alarming through a web-enabled interface. Additionally, the interval data are utilized by the data analysis tools and load management applications as described below.

- Data Alarming—The annunciation of abnormal conditions through operator interface, email and text messaging systems. Typical interval data alarm types include; high energy demand, interval consumption greater than forecast and equipment output lower than expected.

- Data Analysis Tools—The tools which are necessary for an organization to manage an active energy program include:
  - Rate Engine—The ability to model utility tariffs and energy contracts so as to be able to calculate the cost of energy. The rate engine is utilized in utility bill validation, load management and optimization applications as well as energy cost budgeting and reporting. The rate engine provides a method for evaluating and reporting the cost associated with energy usage. Supply tariffs or contract determinants are modeled and applied to a load profile in order to determine cost. The rate engine is utilized in contract negotiations, budgeting, reporting, and in load management decisions.
  - Load Forecasting—The ability to predict future energy loads based on historical weather and load data in conjunction with forecasted weather data. Load forecasts are utilized in energy procurement and operational decision support as well as being utilized by the load management applications described below.
  - Reporting—Valuable reporting features including a range of standard on-call reports as well as the ability to create custom reports which draw from real time and interval data, load forecasts, rate engine calculations, benchmarks, etc. Ad-hoc report generation supports operational analysis, while custom reports support management reporting and ongoing operational monitoring.

- Load Management Applications—The software which provides supervisory decision support and automated control of energy assets in order to minimize cost and risk. The applications incorporate real time energy cost, load forecasts, energy asset data (capacity, availability, and operating cost), contract determinants and real time energy demand in order to determine the optimum operation of energy assets. Outputs from these management applications include equipment schedules, load requirements, equipment operating profiles and supervisory direction to control strategies which are distributed across the EICS. Figure 15-3 above provides an overview of an optimization program. Load management strategies include:

  - Demand Management—Shifting and/or shedding of load (fuel or electric) in response to time of use or price signals such as wholesale market prices in deregulated geographies, demand response signals, demand based tariff management, or real time price based tariffs. Electric demand management is accomplished via strategies such as generation dispatch, thermal storage systems, fuel switching, and traditional load shedding. Fuel management is enabled by such strategies as bio-mass fuels, propane air mixing plants and dual fuel appliances.

  - Enterprise Optimization—Managing the operation of multiple, distributed energy plants so as to minimize energy and operating costs in response to load requirements and variable energy prices. The enterprise plant optimization routines provide management direction in the way of mode selection, load requirements, and scheduling to the real-time plant optimization strategies located in the BNIs.

Energy Management Services (EMS)

EMSs are the duties which must be performed in order to ensure the success of an IEM program. As with the proper usage of the EMAs, these services require expertise which may not be resident in the organization. However, these services, whether performed by the organization, consultants or contractors, are necessary to the successful design, implementation and maintenance

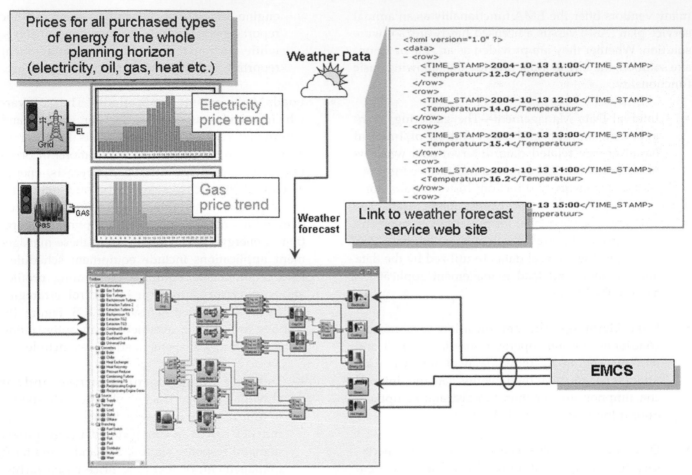

Figure 15-3. Load Optimization Software Utilizes Energy Pricing and Load Forecasts along with a Model of Plant Equipment to Provide Equipment Scheduling via EMCS

of an integrated energy management program. They are aligned with the IEM program components as outlined below:

Energy Asset Management Services

Including the evaluation, design, and application of high-efficiency equipment and modifications to building equipment in order to reduce energy consumption. The selection and application of energy assets which support multi-fuel options, such as on-site generation (combined heat and power, peaking plants, steam turbines) and alternative fuels (biomass, photovoltaic).

- Traditional Demand ECMs—Analysis and design of energy efficiency projects which reduce energy consumption. This includes traditional demand-side measures, such as lighting retrofit, water retrofit and HVAC replacements.

- On-site Generation—Evaluate the feasibility of on-site generation options, such as paralleling switchgear on emergency generators, combined heating

and power systems and steam turbine generators, in conjunction with supply contract options and emission limits in order to provide the options necessary for electric load management.

- Fuel Switching Options—Evaluate the use of biomass fuels, propane air mixing systems, absorption chillers and multi-fuel appliances in conjunction with load requirements and supply contract options in order to provide the options necessary for fuel load management.

- Load Management Strategies—Design the control strategies that manage on-site generation, fuel options and load shifting (thermal storage and load shedding options) in response to market determinants and energy contracts in order to minimize operating costs and risk.

Energy Supply Management Services

Developing energy supply structures designed to access lower-cost wholesale products by leveraging load control capabilities created with the assets listed above. Selection of energy feedstock sources (electric, natural gas, oil, biomass, LP) which provide flexibility in meeting load requirements. Performing ongoing procurement and risk management functions.

Regardless of the location of a facility or its energy sources, many supply structure options exist that can create significant value when effectively coupled with the right on-site energy assets and real-time control. Energy supply management, in the context of an Integrated Energy approach, means three things as follows:

- Contract, Tariff and Rate Analysis—Evaluation of existing energy procurement policies, contracts and tariffs to determine procurement options. In regulated areas, options include tariff changes, load management, and participation in demand response programs. Options in deregulated areas include structuring procurement RFQs with terms which manage risk by leveraging load management options to respond to price signals such as hourly wholesale market prices.

- Develop Risk Management Strategies—Risk management is the process by which one evaluates the supply procurement options and removes the price and volatility risks. Due to the upheaval of the supply markets over the last five years, the management of supply has become an ever increasingly complicated function for customers. Prices have skyrocketed, volatility has increased and customers are looking to energy service providers for expertise to help guide them through these changes, and provide them with tools to manage the risk through the instability.

- Energy Procurement Support—Analyze site load requirements (electricity, fuel, CHW, steam) and designing energy supply structures that are responsive to load management capabilities and supportive of the risk management strategies of the organization. Procurement support includes development of the solicitation, evaluation of responses, recommendation of vendor and management of on-going management of the supply agreements.

Active Demand Management (ADM) Services

Real-time integration of energy assets, energy supply structure and energy data via an EICS that supports data-centric decision making, proactive solution devel-

opment, automated control, and performance reporting across the entire energy chain. Required service functions include:

- Interval Energy Data Analysis—Utilization of the data analysis tools to monitor and analysis of the building and site energy data (electric, CHW, steam, fuel) in order to identify anomalies which may be counter productive to the program. Identification of energy spikes, which may contribute to high energy and/or operational costs, and implementation of changes including operational and reporting/alarming features to continuously monitor the condition.

- Load Management Control Sequences—The design and implementation of control strategies utilizing the energy management applications and the EICS to actively manage energy assets in response to market indicators and energy supply contracts. Load management techniques may include dispatching on-site generation, fuel switching, shedding and shifting loads.

- Equipment and Controls Monitoring & Management—The on-going monitoring of control system functions, alarms and reports in order to assure optimum performance. Implementation of alarm scenarios which provide immediate and accurate indication of equipment and system malfunctions.

- Utility Bill Validation and Error Resolution—Utilize metered data and modeled energy rates or tariffs in order to calculate monthly utility bill via the EICS. Perform a monthly review of metered consumption and calculated cost verses supplier bill consumption and cost. Contact utility supplier to resolve any billing errors.

CONCLUSION

There are three factors which have converged to create a complex energy cost management problem for organizations, those factors are that: 1) Energy prices are substantially higher and are accompanied by price volatility; 2) Building energy infrastructures are aging and have significant inefficiencies; 3) Building loads are typically served by electricity and/or natural gas and have little flexibility.

The IEM approach provides a *proven* outline of the process and the tools which are necessary for organizations to systematically drive economic benefits. By leveraging the three components of IEM; energy assets, energy supply, and control technology, the energy manager will be able to demonstrate solid results in the way of:

• Increased budget control and reduced energy risk associated with volatile energy markets

• Reduced energy costs through the use of alternative, lower cost sources

• Decreased operating costs through the application of energy efficient equipment

• Increased reliability of energy assets through automation and increased flexibility

Frost and Sullivan, a global growth consulting company that has been partnering with Energy Managers to support the development of innovative strategies for 40 years, has awarded Honeywell International, the *2006 Industry and Advancement Award* for this industry leading approach to energy management.

Section III

Web Based Enterprise Energy and Facility Management System Applications

Chapter 16

Automated Commissioning for Lower-cost, Widely Deployed Building Commissioning of the Future*

Michael R. Brambley (michael.brambley@pnl.gov) and
Srinivas Katipamula (srinivas.katipamula@pnl.gov)
Pacific Northwest National Laboratory†
Richland, WA 99352

HOW IS COMMISSIONING likely to be performed 20 or 30 years from now? Will technology advance to the state where building systems and equipment are automatically commissioned on an ongoing basis as conditions demand? The authors of this chapter believe it will and that technologies emerging from research and development today—wireless communication, automated fault detection and diagnostics, advanced controls, and inexpensive smart sensors—are the seeds for the technology that will provide these capabilities in the future.

This chapter takes a brief look at the benefits of commissioning and describes a vision of the future where most of the objectives of commissioning will be accomplished automatically by capabilities built into the building systems themselves. Commissioning will become an activity that is performed continuously rather than periodically, and only repairs requiring replacement or overhaul of equipment will require manual intervention. This chapter then identifies some of the technologies that will be needed to realize this vision and ends with a call for all involved in the enterprise of building commissioning and automation to embrace and dedicate themselves to a future of automated commissioning.

THE VALUE OF COMMISSIONING

Retro-commissioning of existing buildings reduces energy use by anywhere from a few percent to over 60% of pre-retro-commissioning consumption, with most sav-

ings falling into the range of 10% to 30%. [3]–[6] These savings result from detecting and correcting problems such as:

- incorrectly installed equipment (e.g., backwards fans) and sensors,

- incorrectly implemented control algorithms (e.g., economizing cycles),

- inefficient set points,

- equipment and lighting operating during unscheduled times,

- equipment operating at degraded performance (e.g., low refrigerant charges in vapor-compression equipment) and

- missing, failed and uncalibrated sensors.

Energy savings are more difficult to establish for new construction, but by using building simulation, the savings associated with commissioning can be estimated. A meta-analysis found median cost savings on energy of 15% and payback periods of 0.7 years for retro-commissioning of existing buildings, while for commissioning of new buildings, the study found a median payback period of 4.8 years. [7]

While these savings are impressive, penetration of commissioning* in the buildings market remains small; in the U.S., less than 5% of new buildings and less than 0.03% of existing buildings have been commissioned.

---

*This is an update and expansion of earlier versions of this chapter published as References [1] and [2].
†Operated by Battelle Memorial Institute for the U.S. Department of Energy.

*Throughout this chapter, the authors use the term "commissioning" to represent generically all forms of the process applied to both new buildings during design and construction and existing buildings already in operation. The context establishes which use is intended.

[8] A 2003 report on commissioning in public buildings states "While the concept of commissioning is increasingly accepted, there are still barriers—particularly with regard to cost—to implementation of the kind of thorough, independent third-party commissioning that is necessary for the full benefits of commissioning to be realized." [9] Even when performed, pressures exist to keep costs down, which in some cases limits the thoroughness to which the commissioning is performed.

The two factors that likely play the most significant roles in limiting the widespread adoption of commissioning are: 1) lack of knowledge of the benefits of commissioning by key decision makers and 2) the cost of commissioning. This chapter focuses on the second of these by using automation of commissioning processes to reduce labor requirements and, as a result, decrease the cost of commissioning substantially. Fortunately, reducing cost also reduces barriers in decision making, making acceptance by decision makers easier even in the absence of greater information and knowledge. Changes to how commissioning is executed that reduce its cost and make it a routine part of how building systems are designed, installed, and operated could in the long run better promote the objectives of commissioning than keeping it a distinct practice, if by doing so activities producing equivalent results to commissioning penetrate the buildings segment more rapidly and more completely. Key to this is reducing the labor intensity of commissioning by automating as many of the processes involved as possible. Compared to the cost of labor, automation technology is inexpensive.

HYPOTHESIS—THERE'S A BETTER WAY

Commissioning provides important benefits to both new and existing buildings, but there may be a better way to achieve these benefits. In the long-term future (say 20 or 30 years from now), most (but not all) of the objectives of commissioning could be provided using automated processes, reducing barriers that exist today for commissioning and impacting a much larger portion of the building stock. Furthermore, doing so could increase consistency in the commissioning process, improve the reliability of building systems, and make the process of assessing performance, which is a critical part of retro-commissioning, a truly continuous process.

How might this be done? First, identify processes that produce the desired outcomes that could be done automatically. Then develop building equipment, systems, control systems, and tools that implement these processes automatically. Gaps between the automated

processes could be filled by manual procedures, ensuring that the activities that absolutely require human intervention and tie the automated processes together are executed efficiently and cost effectively.

In the next two sections, we first provide a vision for commissioning in the distant future after the automated processes are fully developed and implemented, then take a critical look at commissioning to identify the processes that require direct human involvement and processes that do not and could be automated. We then identify technologies that are key to realizing the automated processes and conclude by identifying research and development that will be essential to accomplishing our vision for commissioning of the future.

In doing this, we are not suggesting that commissioning and efforts to promote it be suddenly terminated today. On the contrary, commissioning as conducted today is important as a transition process. It provides important benefits, but as with most things, it will benefit from improvement over time. Our vision is for the long term. It will not be realized tomorrow, but the vision can help guide research and development (R&D) and product development decisions so they lead us to a future where the benefits of commissioning permeate the entire building enterprise. Likewise, as this technology and tools embodying it emerge, commissioning should adapt and change to assimilate these new capabilities, making the commissioning process faster, less expensive, more thorough, more consistent, more reliable, more cost effective, more continuous, and appealing to more of the market.

COMMISSIONING FOR THE FUTURE—
HIGHLY AUTOMATED AND EFFICIENT

When operation of a new building or new piece of equipment or system in a building is started in the future, with the push of a "start" button, equipment and systems should all test themselves, identify any installation or configuration problems, automatically fix problems amenable to "soft" solutions, and report the need for "hard" solutions requiring replacement or installation of hardware. A report on the performance of all building systems and equipment should be automatically generated, delivered to key recipients, and stored electronically for future reference and updating. During initial operation (e.g., for the first year of a building's life) and continuing over its lifetime, the system should optimize itself, integrating its behavior with external constraints, such as occupancy levels, occupant behavior and feedback, energy prices, energy demand charges, and weather. Although most of the optimization should take place during this initial period of say a year, the building systems should continue to optimize themselves as prices change, spaces are converted to different uses, tenants change, and building equipment itself ages and wears. Using prognostics, systems should automatically inform building staff regarding expected lives and recommended service times for equipment plus the costs of waiting to service the equipment until some time in the future. Automated diagnostics should detect degradation of system performance and failure, make "soft" fixes when possible (alerting building staff to changes made and electronically documenting changes automatically made for future reference), and alert staff to performance degradation, impending failures and required maintenance to prevent them.

Realization of this vision will present unique technological challenges, but methods under development (some examples of which are provided in the next section) are addressing these. Technology will evolve over time and change the practice of commissioning, bringing benefits to building owners and occupants. Ours is a vision for buildings of the future that automatically will perform many of the actions required to meet the objectives of commissioning, where technology will provide the cornerstone for achieving this future.

WHAT PARTS OF COMMISSIONING COULD BE AUTOMATED?

The major tasks composing commissioning of new buildings and retro-commissioning of existing buildings are shown in Table 16-1 by phase of the project. Some processes are critical to accomplishing the objectives of commissioning. Other processes exist specifically to support commissioning as it is performed today. If the overall process of commissioning were changed, some of these processes might become unnecessary or would be modified considerably. The same overall objectives might be achieved without an identical set of processes. Table 16-1 includes designations of whether activities would be manual (M) or automated (A) in a highly automated future for commissioning. Commissioning objectives would be established largely by standards and specialized to specific projects by detailed objectives being automatically inherited by associating objectives with generic types of commissioning projects. For retro-commissioning, most activities would become a routine part of building operation and maintenance and might not require explicit development of objectives. Similarly, objectives associated with equipment startup would be unneeded because all equipment would be automati-

cally started up and tested with standard automated start-up routines satisfying standard objectives.

Commissioning plans similarly would be developed automatically based on information obtained from design documents, and only limited information would be input manually (e.g., special constraints on schedules). Design information would be automatically stored as developed and shared throughout the life of the building. This would include the objectives and the intent behind the design. Automatic storage and universal data sharing protocols would eliminate the need to manually take off information from drawings or re-input information developed in earlier phases, which is labor intensive today. Ultimately, even existing buildings that were designed before automated data storage was routinely used, will possess systems that will automatically detect all aspects of the building, systems and equipment installed, generate the equivalent design documents, and evaluate the design.

Designs will be automatically evaluated with respect to meeting design intents as well as energy and other standards. For many years, researchers have studied the design process and developed methods for automating both design generation and evaluation. A sampling of issues and advancements in design automation can be found in References [10]-[15]. Research in these fields will provide the basis for automation of the review and the revision of designs performed as part of commissioning.

To the extent that commissioning specifications are still required in bid documents far in the future, most of the required language might be generated automatically, reviewed manually, and revised manually in special cases where required. Eventually, though, all documents will be reviewed automatically. Computer-based tools will parse text, "interpret" the meaning, and evaluate the design with respect to needs and design criteria. Given information about the equipment and systems in a building or specified by the design, checklists (to the extent they are still needed) could be generated automatically. With some exceptions, most checklists would be eliminated because checks would be performed automatically. Lists might remain only as an organizational form for presenting results to human users. Just-in-time facility documentation [16] may become the basis for operation and even parts of commissioning and retro-commissioning. Even proper installation of equipment (e.g., whether any fans are installed backwards) could initially be checked automatically. Some problems might require visual inspection after initial automatic detection, but the labor for this would be highly targeted to problem situations, limiting labor requirements and costs.

Table 16-1. Major Commissioning and Retro-commissioning Activities

New Construction Commissioning	*Retro-commissioning Existing Buildings*

1. Conceptual or pre-design phase
   a. Develop commissioning objectives (A)
   b. Hire commissioning provider (M)
   c. Develop design phase commissioning requirements (A)
   d. Choose the design team (M)
	1. Planning phase
	   a. Develop commissioning objectives (A)
	   b. Hire commissioning provider (M)
	   c. Review available documentation and obtain historical utility data (A)
	   d. Develop retro-commissioning plan (A)

2. Design phase
   a. Commissioning review of design intent (A)
   b. Write commissioning specifications for documents (A)
   c. Award job to contractor (M)
	(No design phase activities)

3. Construction/installation phase
   a. Gather and review documentation (A)
   b. Hold commissioning scoping meeting and finalize plan (M)
   c. Develop pre-test checklists (A)
   d. Start up equipment or perform pre-test checklists to ensure readiness for functional testing during acceptance (A)
	2. Investigation phase
	   a. Perform site assessment (M/A)
	   b. Obtain or develop missing documentation (A)
	   c. Develop and execute diagnostic monitoring and test plans (A)
	   d. Develop and execute functional test plans (A)
	   e. Analyze results (A)
	   f. Develop master list of deficiencies and improvements (A)
	   g. Recommend most cost-effective improvements for implementation (A)

4. Acceptance phase
   a. Execute functional tests and diagnostics (A)
   b. Fix deficiencies (M)
   c. Re-test and monitor as needed (A)
   d. Verify operator training (A)
   e. Review O&M manuals (A)
   f. Building/retrofit accepted by owner (M)
	3. Implementation phase
	   a. Implement repairs and improvements (M)
	   b. Re-test and re-monitor for results (A)
	   c. Fine-tune improvements if needed (A)
	   d. Revise estimated energy savings calculations (A)

5. Post-acceptance phase
   a. Prepare and submit final report (M/A)
   b. Perform deferred tests (if needed) (A)
   c. Develop re-commissioning plan/schedule (A)
	4. Project hand-off and integration phase
	   a. Prepare and submit final report (M/A)
	   b. Perform deferred tests (if needed) (A)
	   c. Develop re-commissioning plan/schedule (A)

Source of original table without M and A designations: Haasl and Sharp 1999 [4]

All testing, data collection, analysis, and interpretation of results would be performed automatically. Examples of how some tests could be executed automatically today are given by references [17]-[19]. These capabilities are based on research and development in the fields of automated fault detection, diagnostics and prognostics. (See references [20] and [21] for a comprehensive review of fault detection, diagnostic, and prognostic methods and reference [22] for a review of early fault detection and diagnostic tools.) "Fixing deficiencies" and "implementing repairs and improvements" are designated in Table 16-1 as being done manually; however, only repairs and improvements requiring physical repair, re-

placement, or reinstallation require human intervention. As shown in references [17] and [18], some repairs, such as revising control code, changing set points, and recalibrating sensors, might be done automatically with no human intervention except to read a short report from the computerized system regarding actions it took. A process used to automatically detect, diagnose, and correct sensor bias faults is given as an example in the Appendix. Automatically re-tuning of control algorithms is also possible today for some applications, and most tuning will be done automatically in the long-term future.

As indicated in Table 16-1, most commissioning activities will be done automatically at some time in the future. People will still need to coordinate the processes and ensure that reporting to owners and management is appropriate, but many of the commissioning activities executed manually today will become automatic. This transformation will reduce the labor, time, and cost of commissioning and help overcome some of the key barriers that widespread application of commissioning faces today. Reaching that future, however, will require advances in key enabling technologies and then application of them to building systems. Table 16-2 provides a list of key technologies needed to achieve this future and the capabilities for commissioning that each might provide.

Wireless data communication will eliminate many of the wires required today to collect data or transmit control signals to device actuators. Wires can represent a significant fraction of the cost of a sensor or control point. As a result, wireless communication for sensors and controls will enable more ubiquitous use of sens-

ing, increasing information on the operating state of systems and equipment available at any point in time and enabling better control and maintenance [23]. Plug and play controls and equipment will enable quicker installation and set up of physical systems and controls. Controls will ultimately become self-writing, given some input on the performance objectives for the building and equipment characteristics. Small, embedded, networked processors will distribute control to a greater degree than today's control systems, leading to better, higher resolution, system response while coordinating through networking with other subsystems and components to achieve building-level objectives.

Automated fault detection and diagnostics will lead to greater awareness of system conditions throughout buildings on a continuous basis (see Figure 16-1). Corrective actions will be enacted automatically by "aware" agents capable of correcting faults in some cases (e.g., correcting a control schedule or fixing an

Table 16-2. Technologies Needed for Highly Automated Commissioning

Technology	*Potential Applications*
Wireless sensing, data acquisition, and control	Cost effective sensing and data collection Condition monitoring
Plug and play building equipment and controls	Self-identifying equipment and automatic system design recognition Rapid automatic self-configuration of controls Automatic control algorithm selection and application
Embedded networked sensing and processing	Highly distributed processing of information with local control capabilities coordinated to meet system and building level objectives
Automated fault detection, diagnostics, and prognostics	Automatic detection and diagnosis of operation, equipment, and control faults Automatic detection and diagnosis of designs and hardware installations Anticipation of system and equipment degradation based on condition monitoring Automatic generation of maintenance plans Condition-based maintenance
Automated proactive testing	Automated start up and functional tests, analysis of data, and interpretation of results Continual automated monitoring and testing
Automatic records management and data exchange protocols	Automatic generation of plans and reports Automatic storage of data Automated asset tracking Automatic project management assistance

incorrect set point). In cases where automatic fault correction is not possible, notifications will be provided to building staff and management regarding faults and their costs. No longer will faults go unrecognized or will an engineer need to study data patterns to detect them. The operating state of building systems will be known, along with the performance and cost impacts of problems, so priorities for operation and maintenance can be made with complete information. Prognostic techniques will automatically predict the remaining serviceable life of equipment and suggest condition-based maintenance actions. Automated proactive testing will be the basis for short-term functional testing. These tests allow a wide range of conditions to be simulated over a relatively short period of time so that problems can be detected faster than if only passive observation of routine operation is used. Proactive testing will enable consistent performance of functional tests automatically during initial commissioning and then at regular periods or when needed throughout the life of the building.

Fault detection and diagnostic methods will have applications in design review in addition to use on physical components. Diagnosis of design is similar to diagnosis of a physical device. First a problem or fault is detected with the design. Evaluation of the design indicates that it does not satisfy some design criterion (requirement). This is analogous to fault detection. Then the reason for the fault (its cause) is identified or isolated, which is analogous to fault diagnosis or isolation. The design then needs to be revised to correct the deficiency, which parallels fault correction. When this entire process is automated, it will provide continuous review and evaluation of designs as they evolve. This will likely be done by automated agents (software processes whose purpose is to execute part of the design review and report the results), each of which is responsible for evaluation with respect to a small subdomain. Some of these agents will specifically handle evaluations from the perspectives of commissioning.

Data exchange protocols will provide the basis for sharing data among automated agents as well as commissioning professionals, operating staff, and facility management. Radio frequency identification tags will also play a role in tracking assets as well as enabling easy, automatic identification of each piece of equipment and component, enabling automatic checking for consistency with specifications as equipment arrives on the construction site, and assessing its installation. Tags may also provide physical and performance characteristics from manufacturer tests, which then will become available to processes that evaluate the correctness of

installation, develop control algorithms, evaluate functional test results, and monitor performance. Geographic information systems (GIS) as well as localization algorithms using wireless communications information (see, for example, reference [24]) will be used to determine locations of equipment. Together these technologies will enable realization of highly automated commissioning and operation.

THE PATH TO THIS FUTURE

The impediments to realizing a future where building commissioning and retro-commissioning are largely automated are technological, social and institutional. Without the technology, however, the vision is not possible. With it, automated capabilities for executing all but the repair, replacement, and some management activities of retro-commissioning could be delivered as parts of equipment packages and control systems. With the addition of design tools, this could be extended to commissioning of new buildings. Efforts already underway are beginning to develop tools that automate parts of the commissioning process or provide assistance with parts of commissioning. [17]-[19],[25]-[27] These are initial attempts at using automation to improve commissioning. To realize the full benefits of automated commissioning, advances in each of the technologies identified in Table 16-2 will be needed.

Because the buildings industries are highly fragmented, public R&D organizations will need to provide leadership to produce this technology. Even then, a market demand will need to develop to drive the creation of new equipment and control systems with automated commissioning capabilities. The building commissioning industry will evolve, gaining market share over time as energy and electric power prices increase and more burden for management of the electric power grid is pushed to end users (see, for example, http://gridwise.pnl.gov/ for a vision of the future electric power grid in which "customers" play an active role). Penetrating the market will require improved cost effectiveness for commissioning, as well as education of building owners and operators regarding the benefits of commissioning. Commissioning will need to change in ways that reduce cost while preserving or even enhancing the returns on it. The practice of commissioning will likely change gradually over time with the introduction of new tools that automate parts of the process. Enabling this, however, will require investment in research and development of new automated capabilities.

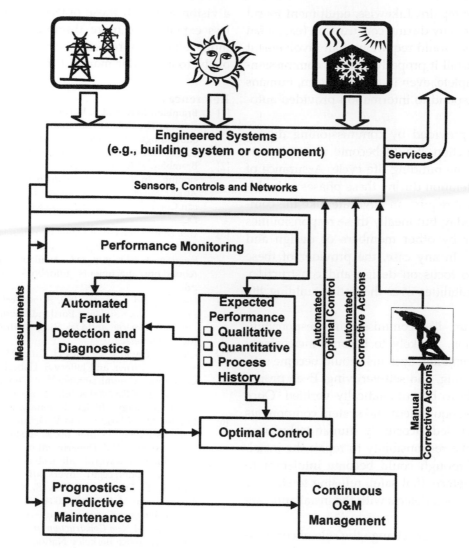

Figure 16-1. Advanced Automated Retro-Commissioning Process

Market transformation programs at the federal, regional, state and local levels can help spread the application of commissioning for the public good. Research, development, deployment and market transformation programs will be needed to accelerate the introduction of automated capabilities and the spread of commissioning, improving the performance of the building stock and bringing energy and environmental benefits. Still, the willingness of the commissioning profession to accept and embrace these technologies will be critical to determining their rate of penetration. Resistance won't stop the introduction of the technologies, only delay their application, but earlier acceptance will help accelerate capture of the benefits associated with high-quality, widespread commissioning even if the mechanism of delivery changes.

CONCLUSIONS—A CHALLENGE FOR THE COMMISSIONING COMMUNITY

Automation could change the nature of the commissioning process for both new and existing buildings. The services performed today as part of commissioning of existing buildings should become part of routine operation and maintenance with automated monitoring, testing, and diagnosis continually performed by the building systems and equipment themselves, taking much of the responsibility off humans.

Still, repairs and replacement of hardware will continue to require human intervention. Deteriorating bearings in pumps, failed windings in fan motors, and leaking valves will need humans to repair or replace them. Automation can only prompt repair technicians to take action to make repairs. Likewise, equipment found to be installed incorrectly during construction (e.g., a fan installed backwards) would require human involvement to remove and reinstall it properly. For the commissioning cycle to be complete, even in the long term, humans will still need to respond to information provided automatically.

The services provided by commissioning during design and construction should become integral parts of those phases of the building life cycle. Assurance of their proper consideration during these phases of building projects may still require an advocate, like the commissioning agent today, but ideally these responsibilities will be taken over by other members of design and construction teams. In any case, the provider of these services is likely to focus on design and construction, rather than responsibilities over the entire building life cycle.

System start up, like commissioning responsibilities during operation, is likely to become increasingly automated. Equipment and systems should become self-configuring, self-testing, and self-verifying. Even proper installation is likely to be automatically verified. Once again, though, when equipment and system components are found to be installed incorrectly, human technicians will still need to take responsibility to repair the installation. Equipment though could become intolerant to some problems it detects (fail safe), refusing to start up until all such problems it detects with the installation are corrected.

Automation will likely change the role of commissioning over time and in 20 or 30 years, its objectives may be met completely differently than they are today. These changes will not occur overnight or even in a few years, but rather over many years, but they should lead to more cost effective delivery of the outcomes promoted by commissioning to a much broader segment of the commercial buildings market. Change is inevitable and will bring benefits. As with use of automation in design [28], detractors will find objections to greater use of automation in commissioning; proponents will grasp increased automation as an opportunity. The authors recommend that the building commissioning community embrace the opportunities posed by new technology and employ them to deliver better services.

Research and development will be required to achieve the benefits of greater automation in commissioning but so will adoption by the various players in the commissioning and broader buildings communities. Researchers and providers of services alike have the opportunity to transform the delivery of commissioning's objectives by working in concert to pursue a vision in which those objectives are delivered faster, less expensively, more thoroughly, more consistently, more reliably, more cost effectively, and more continuously, to a broader market through automation.

References

[1]   Brambley, M.R. and S. Katipamula. 2005. "Beyond Commissioning: The Role of Automation." Published by Automatedbuildings.com. Available on the worldwide web at www.automatedbuildings.com (February 2005).

[2]   Brambley, M.R. and S. Katipamula. 2004. "Beyond Commissioning." In *Breaking Out of the Box, Proceedings of the 2004 ACEEE Summer Study on Energy Efficiency in Buildings*. American Council for an Energy Efficient Economy, Washington, DC.

[3]   U.S. DOE. Undated. *Building Commissioning—The Key to Quality Assurance*. U.S. Department of Energy, Rebuild America Program, Washington, D.C. Available online: http://www.rebuild.org/attachments/guidebooks/commissioningguide.pdf.

[4]   Haasl, T. and T. Sharp. 1999. *A Practical Guide for Commissioning Existing Building*. Portland Energy Conservation Inc., Portland, Oregon, and Oak Ridge National Laboratory, Oak Ridge, Tennessee.

[5]   Claridge, David E, Charles H. Culp, Mengsheng Liu, S. Deng, Wayne D. Turner, and Jeffery S. Haberl. 2000. "Campus-Wide Continuous Commissioning^{SM} of University Buildings." In *Proceedings of the 2000 ACEEE Summer Study*. American Council for an Energy Efficient Economy, Washington, D.C.

[6]   Liu, M., D.E. Claridge, and W.D. Turner. 2002. "Continuous Commissioning^{SM} Guide Book." Federal Energy Management Program, U.S. Department of Energy, Washington, D.C. Available on the worldwide web at: http://www.eere.energy.gov/femp/pdfs/ccg01_covers.pdf.

[7]   Mills, E., H. Friedman, T. Powell, N. Bourassa, D. Claridge, T. Haasl, and M.A. Piette. 2004. *The Cost-Effectiveness of Commercial Buildings Commissioning—A Meta-Analysis of Existing Buildings and New Construction in the United States*. LBNL-56637. Lawrence Berkeley National Laboratory, Berkeley, California. Available online: http://eetd.lbl.gov/emills/PUBS/Cx-Costs-Benefits.html.

[8]   Castro, N.S. and D. Choineire. 2006. "Cost Effective Commissioning for Existing and Low Energy Buildings—A New IEA ECBCS Research Project." In *Proceedings: National Conference on Building Commissioning 2006*. Portland Energy Conservation Inc., Portland, Oregon.

[9]   Quantum Consulting. 2003. *Market Progress Evaluation Report—Commissioning In Public Buildings Project, No. 3*. Report #E03-107. Northwest Energy Efficiency Alliance, Portland, Oregon. Available online: http://www.nwalliance.org/resources/reports/107.pdf.

[10] Gero, J.S. 2002. "Advances in IT for Building Design." In *Advances in Building Technology*, M. Anson, J. Ko and E. Lam (eds.), pp. 47-54. Elsevier, Amsterdam.

[11] Gero, J.S. 2000. "Developments in Computer-Aided Design." In *INCITE 2000*, H. Li, Q. Shen, D. Scott and P. Love (eds.), pp. 16-24. HKPU Press, Hung Hom, Kowloon, Hong Kong.

[12] Caldas, L.G. and L.K. Norford. 2002. "A Design Optimization Tool Based on a Genetic Algorithm." *Automation in Construction* 11(2):173-184.

[13] Iliescu, S., P. Fazio and K. Gowri. 2000. "Similarity Assessment in a Case-Based Reasoning Framework for Building Envelope Design." In *INCITE 2000*, H. Li, Q. Shen, D. Scott and P. Love (eds.), pp. 697-713. HKPU Press, Hung Hom, Kowloon, Hong Kong.

[14]   Fleming, U. and R. Waterbury. 1995. "Software Environment to Support Early Phases in Building Design (SEED): Overview." *Journal of Architectural Engineering* 1(4):147-152.

[15]   Fleming, U. and Z. Aygen. 2001. "A Hybrid Representation of Architectural Precedents." *Automation in Construction* 10(6):687-699.

[16]   Song, Y., M.J. Clayton, and R.E. Johnson. 2002. "Anticipating Reuse: Documenting Buildings for Operations Using Web Technology." *Automation in Construction* 11(2):185-197.

[17]   Katipamula, S., M.R. Brambley, and L. Luskay. 2003. "Automated Proactive Techniques for Commissioning Air-Handling Units." *ASME Journal of Solar Energy Engineering, Transactions of the ASME*, Special Issue on Emerging Trends in Building Design, Diagnosis and Operation 125(1):282-291.

[18]   Portland Energy Conservation, Inc. (PECI) and Battelle Northwest Division. 2003. *Methods for Automated and Continuous Commissioning of Building Systems.* Final Report. ARTI-21CR/610-30040-01. Air- Conditioning & Refrigeration Technology Institute, Washington, D.C. Available online: www.arti-21cr.org/research/completed/finalreports/30040-final.pdf.

[19]   Brambley, M.R. and S. Katipamula. 2003. "Automating Commissioning Activities: Update with Examples." *In Proceedings of the 11th National Conference on Building Commissioning*, May 20-22, 2003. Portland Energy Conservation Inc., Portland, Oregon.

[20]   Katipamula, S. and M.R. Brambley. 2005. "Methods for Fault Detection, Diagnostics and Prognostics for Building Systems—A Review Part 1." *International Journal of Heating, Ventilating, Air-Conditioning and Refrigerating Research* 11(1):3-25.

[21]   Katipamula, S. and M.R. Brambley. 2005. "Methods for Fault Detection, Diagnostics and Prognostics for Building Systems—A Review Part 2." *International Journal of Heating, Ventilating, Air-Conditioning and Refrigerating Research* 11(2):169-187.

[22]   Friedman, H. and M.A. Piette. 2001. *Comparative Guide to Emerging Diagnostic Tools for Large Commercial HVAC Systems.* LBNL 48629. Lawrence Berkeley National Laboratory, Berkeley, California. Available online: http://www.peci.org/library/PECI_DxToolsGuide1_1002.pdf.

[23]   Brambley M.R., M. Kintner-Meyer, S. Katipamula, and P. O'Neill. 2005. "Wireless Sensor Applications for Building Operation and Management." Chapter 27 in *Information Technology for Energy Managers, Volume II—Web Based Energy Information and Control Systems Case Studies and Applications*, B.L. Capehart and L.C. Capehart (eds.), pp. 341-367. Fairmont Press/CRC Press, Lilburn, Georgia.

[24]   Savvides, A., M. Srivastava, L. Girod, and D. Estrin. 2004. Chapter 15 in *Wireless Sensor Networks*, C. S. Raghavendra, K. M. Sivalingam and T. Znati (eds.), pp. 327-349. Kluwer Academic Publishers, Boston, Massachusetts.

[25]   Turkaslan-Bulbul, M.T. and O. Akin. 2006. "Computational Support for Building Evaluation: Embedded Commissioning Model." *Automation in Construction* 15(4):438-447.

[26]   Castro, N.S. and H. Vaezi-Nejad. 2005. "CITE-AHU: An Automated Commissioning Tool for Air Handling Units." In *Proceedings: National Conference on Building Commissioning 2005*, Portland Energy Conservation Inc., Portland, Oregon.

[27]   Salsbury, T.I. and R.C. Diamond. 1999. "Automated Testing of HVAC Systems for Commissioning." In *Proceedings of the National Conference on Building Commissioning 1999*, Portland Energy Conservation Inc., Portland, Oregon.

[28]   Chastain, T., Y.E. Kalay and C. Peri. 2002. "Square Peg in a Round Hole or Horseless Carriage? Reflections on the Use of Computing in Architecture." *Automation in Construction* 11(2):237-248.

APPENDIX: PROCESS FOR AUTOMATED RETRO-COMMISSIONING OF SENSORS [17]

Isolation of Outdoor-, Return- and Mixed-Air Temperature Sensor Problems

The process described in this appendix follows identification of a problem with the outdoor-, return-, or mixed-air temperature sensor in an air-handling unit (AHU) by routine passive (observational) automated monitoring and fault detection during operation.* One of these sensors is faulty but which specific one is not known from the passive process. The active process described here would be executed automatically immediately following detection of the problem or at some later time (e.g., overnight while the building is unoccupied) to isolate the cause of the fault and to automatically implement a (possibly temporary) corrective action.

In an AHU, the return- and outdoor-air streams are mixed and the resulting air stream is called the mixed-air stream. Therefore, the fundamental equations for sensible energy balance along with positioning of the return-air and the outdoor-air dampers can be used to isolate the fault. Placing the dampers at specific positions in this case provides analytical redundancy, which provides additional information.

As shown in Figure 16-2, the first step in the proactive diagnostic process is to close the outdoor-air damper completely and wait for the conditions to reach steady-state, which usually occurs within a few minutes. While keeping the outdoor-air damper fully closed, the return-air and mixed-air temperatures are sampled for a few minutes. With 100% of the return-air recirculated, the average mixed-air temperature should nearly equal the return-air temperature. If this is found, then the return-air and mixed-air temperature sensors are consistent with one another and, because one of the three sensors has failed, the outdoor-air temperature sensor must be faulty.

If the return-air and the mixed-air temperatures are not approximately equal, command the outdoor-air dampers to open fully and wait until steady-state conditions are achieved. When the outdoor-air damper is fully open, no return air recirculates and the average mixed-air temperature should approximately equal the average outdoor-air temperature during the sampling period. If this condition is found, then the outdoor-air and mixed-air temperature sensors are consistent with one another, and the return-air temperature sensor is

---

*See references [17] and [18] for detailed information on the passive fault detection and diagnosis process.

faulty. If the measured mixed-air temperature does not equal the measured outdoor-air temperature, then the mixed-air temperature sensor is faulty (because earlier the return-air temperature sensor was found fault-free).

After isolating the faulty sensor, further diagnosis can identify the underlying cause or nature of the problem. In contrast to relative humidity, air flow, fluid flow, and pressure sensors, temperature sensors are more reliable, but they do exhibit erratic behavior occasionally. In addition to random noise, temperature sensors commonly drift and acquire bias over time. A process for detecting and estimating bias in temperature measurements

is described in the next subsection. The ability to detect the drift over time does not require proactive testing; it can be detected using passive methods (see PECI and Battelle [18]).

Some notes of caution are appropriate for users of the process described here because tolerances of mechanical components can vary widely and change over time. All dampers possess seals to prevent leakage when they are fully closed. Some leakage, however, occurs around the seals, and as the AHU ages, the seals deteriorate, increasing the leakage. Under these conditions, when the return-air dampers are closed, the mixed

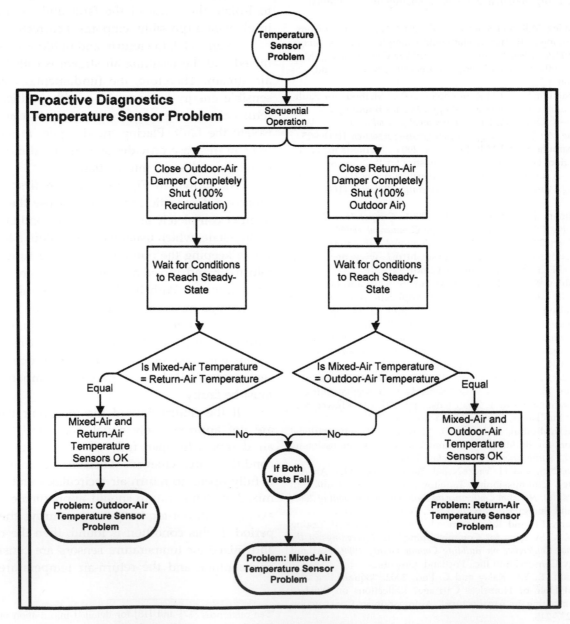

Figure 16-2. Decision Tree of Process to Isolate the Faulty Temperature Sensor

air consists mostly of outdoor air but mixed with some leaked return air. As a result, the mixed-air temperature may not equal the outdoor-air temperature precisely. Therefore, in addition to allowing for measurement inaccuracies of the sensors, the equality tests in Figure 16-2 should also account for damper leakage. Compensation for these sources of uncertainty can be accomplished by relaxing (i.e., increasing) the tolerances on the equality tests. This may sometimes lead to incorrect identification of a faulty mixed-air sensor even when the outdoor-air or the return-air temperature sensor is slightly biased (because it is the least resistive path on the flow chart in Figure 16-2). These sorts of trade-offs between sensitivity of diagnosis to detect problems and the potential for false alarms or false diagnoses are best determined through field tests and experience.

Stratification of air in the mixing box leads to another potential source of error. The measured mixed-air temperature may vary significantly across the duct cross-section. As a result, the mixed-air temperature measured at a single point may differ significantly from the average mixed-air temperature and lead to misleading diagnoses. To prevent this, the mixed-air temperature should always be measured across the duct and averaged using an averaging sensor.

Determining the Bias in an Outdoor-Air Temperature Sensor

In this section, we present an approach for classifying the nature of the fault found in an outdoor-air temperature sensor as an example of a method that can be applied to air-temperature sensors (see reference [18] for detailed schemes for other sensors).

Once a specific temperature sensor has been identified as faulty, further classification of the fault is possible. This section and the next describe a process for estimating the bias and reconfiguring the controls to compensate for it.

The first step in this proactive diagnostic process (see Figure 16-3) is to fully open the outdoor-air damper and wait for conditions to reach steady-state. In this case, values of the mixed-air temperature can be used to identify when steady-state conditions are attained, because at this point in the diagnostic process, we know that the mixed-air temperature sensor is good. One form of steady-state filter is based on the rate of change of the mixed-air temperature. If the rate of change is zero or below a predefined threshold, steady-state conditions have been achieved. After steady-state conditions are achieved, compute the difference between the outdoor-air and the mixed-air temperatures and store the result for further analysis.

The frequency of sampling and the duration of the proactive test depend on field conditions. A sampling rate of a minute or less and total test duration of 15 minutes should be sufficient in most cases. In some cases, the test may have to be performed at different times of the day to ensure that the bias is consistent at all hours of the day. In some cases, something as simple as positioning of the sensor may affect its readings. For example, an outdoor-air temperature sensor positioned so it is exposed to sunlight part of the day may read a few degrees high for those hours of the day, the amount depending on the position of the sun, but may otherwise read normal. This type of bias or problem is difficult to detect, unless the proactive test is repeated several times at different hours of the day and then correlated with other observations, such as solar position. An outdoor-air temperature sensor showing bias during certain hours of the day each day for many days in a row (but not at other hours) would indicate such a problem. As with uncertainty mentioned earlier, field tests are required to better understand these issues.

After the difference between the outdoor-air and the mixed-air temperatures is computed for the duration of the test at a desired sampling rate, the next step is the analysis of the stored data to confirm whether the difference is nearly constant over the entire test period. Commonly-used statistical tests such as the mean and the standard deviation of the sample are recommended. The mean provides the central tendency of the sampled data (the estimate of the bias), while the standard deviation provides the dispersion (how tightly the data are clustered around the mean).

In order for the test to be true (i.e., the difference nearly equal over the test period), the mean must be greater than the tolerance or the accuracy of the temperature sensors and the standard deviation should be reasonable. Another statistical metric called the coefficient of variation can be used to check whether the standard deviation is reasonable compared to the sample mean and the sensor tolerance. The coefficient of variation measures the relative scatter in data with respect to the mean; it is computed as the ratio of the standard deviation to the mean. A threshold for the coefficient of variation must be selected. Below this threshold, the standard deviation would be acceptable and the bias considered constant. Previous studies that used field data to develop empirical models have concluded that a coefficient of variation of about 15% is reasonable.

Reconfiguration of Controls

The final step in the automated proactive commissioning process involves reconfiguring the control algorithms to account for a bias in the outdoor-air temperature sensor. If the previous test concludes that the bias is constant, the value of the bias can be subtracted from the measured value of the outdoor-air temperature to obtain a correct value. If the test concludes that the temperature difference (bias) is not constant, then the controls can be reconfigured to use another properly functioning outdoor-air temperature sensor. Buildings often have several outdoor-air temperature sensors, and substitution of measurements from another outdoor-air temperature sensor should provide a reasonable value for control of the air handler with the faulty sensor. This obviously would not be possible for sensors measuring the return-air or mixed-air temperature, so when a variable bias is detected with one of them, the sensor must be replaced.

Any time controls have been reconfigured as the result of a proactive test, a report should be generated to notify the building manager or the building operator of this change. Then, when the sensor is repaired or replaced, this report will alert the manager or operator that the outdoor-air sensor used for control should be re-configured, removing any software corrections to measured values.

Proactive procedures similar to the one presented in this section can be developed for return-air, mixed-air, and supply-air temperatures (see reference [18]).

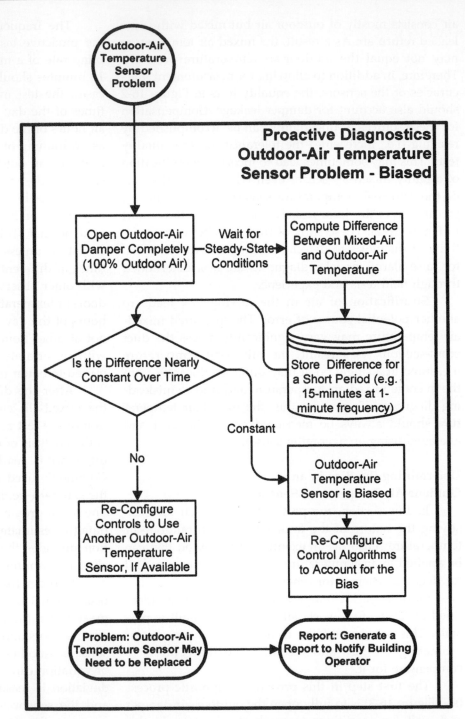

Figure 16-3. Decision Tree for the Process to Check Whether the Outdoor-Air Temperature Sensor is Biased and Implement a Temporary Correction.

Chapter 17

Monitoring-based Commissioning for Facilities

Keith E. Gipson, CTO Impact Facility Solutions, Inc.
Thomas S. Riley, P.E., President—Cogent Energy

INTRODUCTION

THE PURPOSE OF THIS CHAPTER is to introduce the concept of monitoring-based commissioning for facilities and feature an actual customer project utilizing the MBCx process. MBCx is the next evolutionary step of commissioning processes including retrocommissioning, new construction and others. The MBCx process takes advantage of the existing computerized facility or energy management systems. The goal is to use robust, automatic data acquisition tools (gateways) based on IT and web/internet technology for the improvement of facility operation and energy efficiency. Impact Facility Solutions provided the data collection infrastructure and services for Cogent Energy's MBCx tools and processes.

MONITORING-BASED COMMISSIONING

An important aspect of MBCx process is the evaluation of the programmed sequences of operation within the facility management system (FMS). This critical process allows the identification of problems with building performance, and can suggest improvements that will reduce building energy consumption and in many cases, improve building comfort. The evaluation of sequences can be as simple as ensuring that the FMSs programmed time-of-day schedules are aligned with building occupancy, or as complex as validating that a PID loop controlling an economizer cycle is properly tuned.

A challenge in our industry has been collecting FMS trend data in an efficient and meaningful manner. Dozens of energy management system manufacturers, each with a myriad of product generations, are used in facilities throughout the country. In fact, a single University campus could easily have a dozen of FMS types.

Most existing FMSs have limited trending capabilities and even when trending is available, the trends are optimized for display on the proprietary FMSs "front-end" and there is little or no facility to export the data for third-party use. This often results in the commissioning provider developing a custom data collection process for each building, typically a time consuming and inefficient process. Further, based on the FMS system installed, the time interval in which the data are collected often vary, leading to further inefficiencies in by having to re-align the "timestamps" of the collected data from these disparate systems. Only then can it be evaluated and analyzed in a meaningful way.

Monitoring-based commissioning offers a solution for the efficient collection, normalization and evaluation of FMS trend data, regardless of the facility or energy management system manufacturer or product generation.

CASE STUDY—
A LARGE UNIVERSITY CUSTOMER

The customer is a large university in California which serves over 20,000 students. The campus has a legacy facility management system which was utilized for the MBCx process by connecting a communications gateway to two of the buildings, the Science and Physical Education buildings, respectively. The data points for the two buildings were monitored at one-minute intervals. It was necessary to have a relatively high rate of sampling because of the desire to troubleshoot and diagnose issues with the control loops and processes contained in the facility management system controllers. The data were then transported automatically at periodic intervals to Cogent Energy for storage, normalization and analysis.

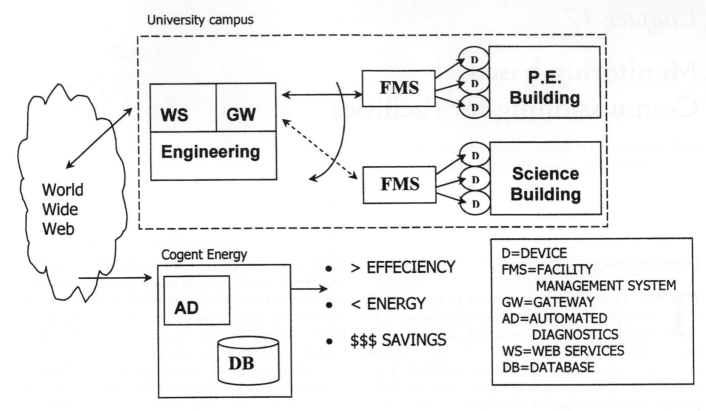

Figure 17-1. System Architecture

PROJECT GOALS

Obtain Cost Effective Energy Savings

Customers must have some reasonable amount of assurance that the savings achieved by any commissioning program can be realized through cost effective means. One of the most cost effective and least labor-intensive methods to monitor the programmed sequences of operation within the facility management system (FMS) is to connect to the FMS using a communications gateway. This approach leverages the existing control system, field devices and sensors; and avoids installing stand-alone, "shadow" devices at a much greater up-front cost than system integration alone.

Measure and Document Energy Savings

Savings realized through increased performance and higher efficiencies is one of the main drivers for any customer to implement a MBCx program. This measurement and validation of energy savings provides the ROI documentation necessary many times to support the MBCx program. Additionally, because M&V (measurement and verification) is at the heart of the MBCx process, the MBCx program generally can be sponsored by utilities, performance contractors and other firms as an energy efficiency project.

Solve Building Environmental Issues

Some building environmental issues are very hard to resolve, especially if the sensors actually acting as inputs to a building automation controller are not the same ones the commissioning professional is looking at. As a former field service technician, I frequently saw problems caused by intermittent errors and problems. By monitoring all of the relevant points at a data rate sufficient enough to "catch" these types of issues through the correlation of the input and output data, correct system operation can be achieved and legacy environmental problems can be resolved.

Facilitate On-going Recommissioning to Ensure Persistence of Savings

This is perhaps the most important aspect of the MBCx process. Unless the results are _**persistent**_, then the savings initially realized can be lost through the natural process of entropy inherent in facility systems. Ongoing monitoring-based commissioning ensures that the initial results persist through seasonal changes, different modes of operation, varying loads and other factors which can contribute to decreased performance or improper operation. A couple of additional benefits of this MBCx process is improved occupant comfort and decreased wear and tear on equipment due to excessive cycling.

FUNCTIONAL TESTING—
CENTRAL AIR HANDLING UNITS

Figure 17-2.

A variety of issues were detected during the MBCx process. The system was first monitored for several months to establish a "baseline" of the operation to compare against the post-MBCx process data. By using the data provided, the following was activities were performed:

• Sensors were calibrated where necessary
• "Point-to-point tests of the facility management system outputs were performed
• Issues were identified and corrected
  — Inoperable relays and mechanical actuators
  — Broken or stuck damper linkages were fixed

Next, the automation sequences of operation were tested and validated using custom field tests and trended data from the facility management system.

ISSUE DISCOVERY THROUGH TREND DATA ANALYSIS: "ALTERNATING" HEATING AND COOLING—PRE-MBCX

Notice the alternating of the air handling system between heating and cooling. Ironically, this type of problem can never be diagnosed by the control system alone, because it originates from within the control logic

of the control system itself! The MBCx process catches this type of systemic, control system anomaly.

ISSUE DISCOVERY THROUGH TREND DATA ANALYSIS: CONTROL LOOP "HUNTING"—PRE-MBCX

This control loop "hunting" issue is typical of a problem with the loop tuning parameters of the control loop which exhibits itself during off-hours and/or partial load situations. This wears out mechanical equipment prematurely and increases cost and maintenance, while decreasing comfort and consistent operation of the system at the desired setpoint.

VERIFICATION OF FIXES THROUGH TREND DATA ANALYSIS POST-MBCX DISCHARGE AIR TEMPERATURE CONTROL

After the monitoring-based commissioning process was completed and the analysis of the trend data was performed, fixes to the control system were verified. This control plot is representative of the kind of even, mid-range control that was exhibited from the systems post-MBCx.

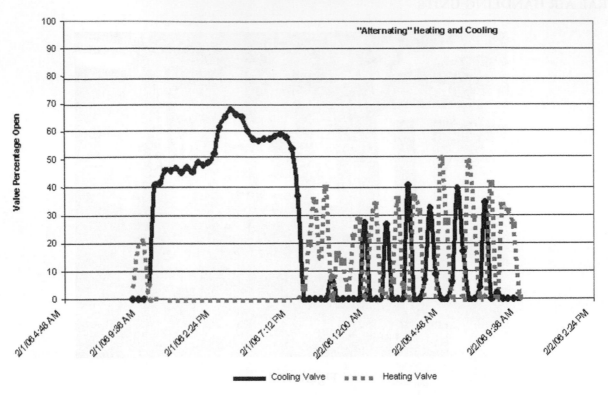

Figure 17-3.

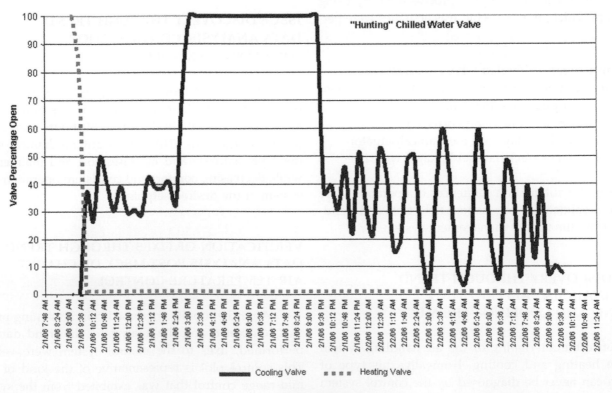

Figure 17-4.

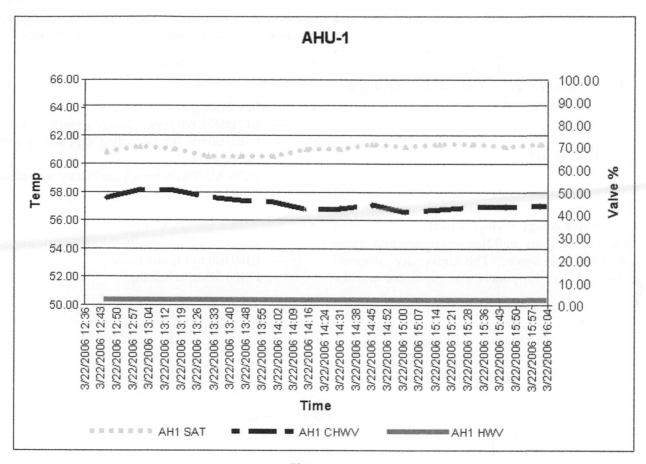

Figure 17-5.

Wear and tear on the mechanical equipment is minimized, cost and maintenance is reduced and comfort and the amount of consistent operation is increased.

SEVERAL CONTROL AND OTHER MEASURES WERE IMPLEMENTED IN THE LABORATORIES AS A RESULTS OF THE MBCX PROCESS:

- Reduced minimum air change rate from 12 to 6 air changes per hour
- Identified and replaced failed controller boards
- Reconnected hood "sash" sensors
- Identified and replaced bad power supplies
- Identified and repaired alarm notification systems
- Identified and replaced pneumatic air solenoids
- Identified and replaced failed pressure switches
- Re-engineered rooms where hoods had been removed
- Identified and replace failed valve linkages

CENTRAL AIR/WATER MEASURE IMPLEMENTED:

- Optimized discharge air temperature sequences to eliminate "alternating" heating and cooling
- Tuned PID control loops to eliminate "hunting" control of valves
- Matched variable frequency drive (VFD) speeds to FMS command
- Identified requirement for and added missing high static pressure shut-down sequences to air handling units (AHUs)
- Added high/low discharge air temperature alarms
- Identified and replaced inoperable differential pressure sensors
- Identified and calibrated temperature sensors where necessary
- Sealed leaking duct work
- Identified and repaired reversed control air signal "normally-open" chilled water valve

- Eliminated irregular cycling of exhaust fans and chilled water pumps
- Fully opened throttling valves to enable VFDs to properly control chilled water flow
- Identified requirement for and added missing differential pressure reset

ACCOMPLISHMENTS

The results of the monitoring-based commissioning process were impressive. The customer was able to obtain cost-effective energy savings which were measured and documented. Many building environmental issues were identified and solved. The University customer was so pleased that the project was expanded and the monitoring-based commissioning process is on-going to ensure persistence of both operation and savings.

- Energy Cost Reduction = $190,000
- Simple Payback Period < 1 year

- Baseline Energy Usage
  — 4,125,000 kWh/year (building electric load)
  — 1,550,000 ton-hours cooling supplied by central chiller plant
  — 18,700 MMBtu heating supplied by central steam plant

- Post-MBCx Energy Usage
  — 3,500,000 kWh/year (building electric load)
  — 1,100,000 ton-hours cooling
  — 11,100 MMBtu heating

PRE AND POST-MBCX CHILLED WATER USAGE

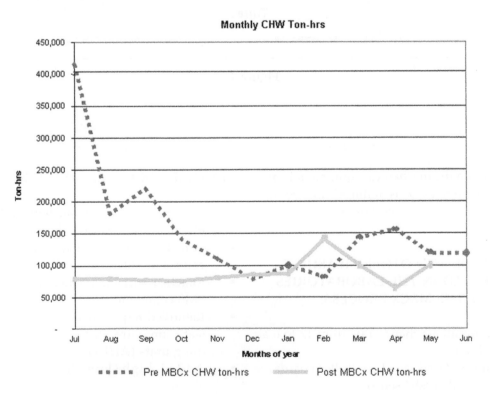

Figure 17-6.

PRE AND POST-MBCX STEAM USAGE

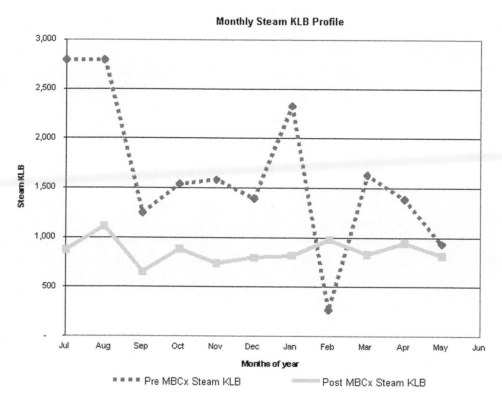

Figure 17-7.

CONCLUSION

Applying advancements such as monitoring-based commissioning and enterprise facility management and data acquisition using facility and energy management systems requires the innovative skills of many people in both the IT and the energy management fields. I would like to thank Tom Riley and the team at Cogent Energy for selecting us to work with them on these advanced applications in the facility and energy management arena and for providing the bulk of the material contained in this case study.

Chapter 18

Effectiveness of Energy Management Systems: What the Experts Say and Case Studies Reveal*

Tom Webster
UC Berkeley, Center for the Built Environment
Lawrence Berkeley National Laboratory
Berkeley, CA

INTRODUCTION AND BACKGROUND

Introduction

I T IS COMMON TO HEAR these seemingly contradictory statements:

- A high proportion of larger federal buildings have some type of building management system (BMS) that is used to operate the building and manage energy use.†

- Virtually all building practitioners feel it is necessary for a building to have one of these systems.

- It is widely believed that BMSs are not as effective as they could be, especially in terms of saving energy, indicating some deficiency in the product offerings.

These are only three of a number of statements we have collected from anecdotal information that raise questions about the effectiveness of installed BCS/EMCS systems.

In the previous articles in this series we focused on the state of practice of building control systems (BCS) and the underlying technology including assessments of newer offerings that can be classified as energy management, control, and information systems (EMCIS). The

purpose of this report, which is the last in this series, is to address the issue of BCS effectiveness in federal buildings by evaluating the authenticity of statements like those above.

Through a variety of techniques outlined below, we gather information from federal facility managers and knowledgeable industry experts to assess these perceptions. We believe that the conclusions reached in this study will convey to facility managers (and indeed all practitioners involved) that maximizing the potential of BMS technology requires adopting a comprehensive view of the entire environment within which the deployment of these systems exists. It is not just an issue of technological capabilities; design, installation, commissioning, operations, the building management structure and the interactions between all of these must be considered as well. Aligning all of these elements into a well-integrated team supportive of the overall goals is essential to achieving a highly effective system.

Background

A BMS is used to manage operations (i.e., physical operation/command and control, maintenance of mechanical and electrical systems, and management of occupant complaints, etc.) and energy use, as well as to control HVAC and other facility functions. Recent trends in BMS offerings appear to emphasize the building management functions of these systems more than energy; energy management more and more is relegated to energy information systems (EIS). [Motegi et al. 2002] Current systems do retain, however, their direct digital control (DDC) functions as well as some ability to integrate with other functions such as facility access and lighting control. In this article we shift our focus from technical issues to understanding how existing BMSs are actually used, if their actual use is consistent with their capabilities, and what types of problems, complaints,

---

*LBNL—57772 (November 2005): This work was supported by the Assistant Secretary for Energy Efficiency and Renewable Energy, Federal Energy Management Program, of the U.S. Department of Energy under Contract No. DE-AC02-05CH11231.
†Here and throughout the report, we use BMS in a somewhat generalized way to represent a broad range of systems of various classifications such as building control systems (BCS), building automation systems (BAS), facility automation systems (FAS), energy management and control systems (EMCS), energy management systems (EMS).

and frustrations operators encounter when using these systems; i.e., how effective these systems are when actually installed in a building.

Organization of the Report

We have organized the remainder of the report into five sections. The first section is a description of our approach and study methodology. The second contains an empirical, broadly stated, definition of effectiveness. The third section contains a summary of the results derived from the study and discusses the findings from expert interviews and case studies. The fourth section provides an analysis and discussion of the results, and the fifth section a summary and conclusions.

STUDY DESCRIPTION AND APPROACH

This study was accomplished using a combination of the following methods for organizing, collecting, and evaluating information:

Identify Accepted Beliefs

Based on author's experience, literature sources, and discussions with experts create a list of the most salient accepted beliefs or claims about effectiveness of BMS technology for systems installed in federal buildings.

Assess Situation

Review demographics of BMS deployment in federal facilities using CBECS and DOE/BTS data. Review past research and experience to identify primary factors that set the context of the study.

Define Effectiveness Criteria

Determine criteria for evaluating effectiveness.

Collect Information

Conduct literature reviews (see references) and interview knowledgeable industry experts (nine industry experts were interviewed) to obtain anecdotal, experienced-based perspectives on existing systems. To determine potential for improvement, conduct interviews with researchers and developers working on new approaches to overcoming outstanding issues. Review results from previous work where GSA building occupants were surveyed.

Conduct Case Studies

Conduct "mini" case studies by questioning operations personnel at selected federal building sites. Since a broad based survey was beyond the scope of this project, the case studies are used to form the perspective developed from the expert interviews. Five sites in GSA Region 9 were contacted based on building size (all greater than 100,000 gross square feet, or GSF), age, use/activity, characteristics of the BMS (e.g., a mix of vendors was sought as well as, vintage, number of points, network architecture) and willingness of personnel to cooperate.

Analyze Data and Evaluate Accepted Beliefs

Using the various sources of data identified above, evaluate the authenticity of the accepted beliefs identified in the first task above.

WHAT IS EFFECTIVENESS?

It is difficult if not impossible to come up with a simple, quantifiable definition of effectiveness. The best we can do is look at the combination of various interrelated factors that together provide a perspective about how well these systems perform needed facility functions. It is not only the technical capabilities that matter, but also the organizational issues that affect how the system was designed, installed, and operated. Based on our experience and research for our previous papers, we developed the following definitions to help organize and categorize the primary factors that allow us to gain a perspective about BMS effectiveness. Our previous research indicates that most of the systems offered today have the capability to perform basic DDC functions so we will not be addressing this factor (although having the capability does not ensure that it will always work properly).

Technical Effectiveness
- How capable is the system in terms of:
- Communications, networking and control
- User interface
- Data monitoring
- Functionality/features tailored to facility needs i.e., ability to transform monitored *data* into *information* in a form that is useful to operations and management
- Support for integration and interoperability

Operational Effectiveness

The degree to which the BMS is effective in supporting facility operations and maintenance (O&M):
1. Occupants:
   - Supports maintaining and improving occupant satisfaction and productivity

- Supports complaint/trouble call diagnosis and resolution
2. Facility maintenance and repair (M&R) and operational efficiency:
   - Provides information for building operations personnel
   - Can improve operations by implementing changes in sequences and control algorithms
   - Provides relevant information for building/ property managers
   - Supports scheduling and alarming functions
   - Allows for remote access
   - Provides relevant information for planning and building management
3. Energy performance:
   - Tracks equipment energy performance
   - Tracks energy use and/or savings
   - Facilitates changes to improve energy performance

Organizational Effectiveness

Those institutional and industry factors that influence the ability to achieve high levels of operational effectiveness using a BMS:
1. Design and deployment:
   - Designed and specified according to owner's intent*
   - Procured, installed, and commissioned successfully
2. Building/property management:
   - Has proactive philosophy, and expectations
   - Prioritizes energy management issues,
   - Has technical knowledge
   - Supports operator training
   - Has no budgets/financial constraints
3. Industry issues:
   - Cooperation between vendors, engineers, contractors, and building management and operations personnel

STUDY RESULTS

Accepted Beliefs

Based on our experience and discussions with numerous building design and operations personnel, we prepared the following list that summarizes our perception of generally accepted beliefs about the overall state of practice of BMS implementations. Our goal is

to investigate these and determine, to the extent possible using the methods outlined above, the veracity of these assertions and to provide insight into underlying issues.

- Almost all larger federal buildings use a BMS
- These systems are considered key to operating a building well.
- Keeping the occupants satisfied is a primary focus of operations staff
- Keeping systems operational is a primary focus of operations staff. These systems are considered effective in this regard.
- The energy saving potential of BMSs is not fully realized.
- The implementation/deployment process for BMSs is very difficult
- The implementation process is fraught with problems.
- In practice, there is a mismatch between the capabilities and sophistication of BMS technology and the ability to exploit them.
- Vendors tend to hype the potential compared to what is actually delivered.
- Despite many claims and much hype, true interoperability has not been achieved.

Situation Assessment
Demographics

According to the 1999 CBECS [EIA 2002] report there are 63,000 federal non-residential buildings in the US. GSA, the largest civilian properties owner, owns and manages 8400 of these properties, 1600 of which are government owned [GSA 2005]. On a square foot basis, about 40% of these buildings have a BMS installed. In Table 18-1 we compare BMS use in federal and commercial buildings. For this comparison, contrary to Table 18-1 of the first article in this series [Webster 2002], we subtracted the federal buildings totals from CBECS data for all commercial (plus other non-federal governmental) buildings. Furthermore, the CBECS commercial data was recomputed to be consistent with the basis of the 1993 Federal Buildings Supplemental Survey (FBSS) [EIA 1997] and divided into small (10-50k GSF) and large (>50k GSF) categories.* Finally, we applied the FBSS distributions of BMSs in buildings to the CBECS data. The table indicates that except for the number of large buildings, BMS use in the private sector is somewhat greater than in federal buildings.

---

*Since these are federal buildings, we include energy efficiency as one of the owner's intents.

*The FBSS study also excluded warehouses and religious buildings.

**Table 18-1. Demographics of BMS use
(Based on 1999 CBECS and 1993 FBSS criteria)**

	U.S. Stock	Federal Stock
By Total *number of buildings*	**27%**	**21%**
Small buildings*	20%	16%
Large buildings	24%	38%
By Total building *floor area*	**48%**	**39%**
Small buildings	24%	16%
Large buildings	62%	51%

*Small buildings are defined as those in the range of 10-50,000 sf.

Building Services Environment

In recent years, outsourcing of building management services has been a dominant trend in the federal government. While this is not unlike what has occurred in the private commercial sector and the BMS products installed are no different, the scale and unique character of the "owners" and associated service environment dictates that we focus our study on experiences with federal facilities. This service environment is one of the keys to understanding BMS effectiveness because it sets the context in which these systems are installed and used. For the purposes of this study, we have focused on buildings owned and/or operated by the General Services Administration (GSA), the federal government's largest landlord. In the GSA these services are either contracted (mostly at the region level) by GSA, or in some cases, (i.e., leased buildings), by the occupying agency or private owners. The GSA buildings we studied (four of the five buildings studied) all used the former model; government owned with services contracted by GSA.

Literature Review

A brief literature review identifies some of the potential benefits and issues related to deploying and upgrading BMSs. For example, it has been demonstrated that high quality monitoring and appropriate trending capabilities can improve energy performance and reduce O&M costs by providing accurate and reliable information and tools capable of being used by skilled operators to investigate operational issues and identify wasteful conditions. Due to the installation of a prototype high quality monitoring system in a 100k GSF San Francisco office building, energy savings of 20% as well as an estimated 20% reduction in service costs due to a reduction in trouble calls were achieved. [Piette et al. 1999]

The value of commissioning buildings and BMSs has been documented in a large study conducted by

Mills and others [Mills et al. 2004]. This study of sixty-nine existing buildings determined measures implemented and cost and savings derived from the commissioning effort. Average energy savings of 18% were reported with paybacks of less than 5 years. Since many of the measures implemented were operations and controls related, these results indirectly indicate the potential for properly operating (and maintained) BMSs.

A study by LBNL identified energy savings of about 9% associated with EMCS retrofits in a San Francisco federal building. [Diamond et al. 1999]. As a part of that study, operators reported a high degree of satisfaction with a new more capable BMS and in fact staffing was reduced by ~50%. Among the new capabilities were a centralized interface that allowed operators to monitor and change control parameters, and the availability of historical operating data, and alarms to alert when problems occur. However, the authors note that the BMS was used primarily for operations and was used little for energy management. Also along with staffing reductions came reassignment of skilled staff by less experienced personnel.

A prototype central monitoring system, GemNet, was installed in GSA Region 9 in an attempt to improve energy performance and reduce cost of operations. While some success was achieved, it appears that there are still a number of technological and operational issues that need to be resolved before the value of these systems can be demonstrated (see discussion below on Potential Improvements). Among these were high maintenance and cost of the central monitoring interface, slow network response, and unavailability (due to budgetary constraints) of staff technically qualified enough to operate the system as it was implemented. [Piette et al. 2002, Levi 2005]

Another view of BMS effectiveness, in terms of savings potential and operator satisfaction, is provided by GSA surveys conducted by the Center for the Built Environment (CBE). An occupant satisfaction survey was conducted in 2004 on 72 GSA buildings with a total of 14,737 respondents. This survey measures the occupant's satisfaction with various aspects of the building they work in and includes a self-reporting productivity scale. A summary of the results are shown in Figure 18-1. [Abbaszadeh Fard et al. 2004] Note that out of all the categories, besides acoustics, thermal comfort and air quality received the lowest mean scores.* This indicates there is potential for improved control, and possibly better HVAC solutions aimed at improving comfort and ventilation.

---

*Each category was scored on a 7-point scale that ranges from -3 (very dissatisfied) to +3 (very satisfied)

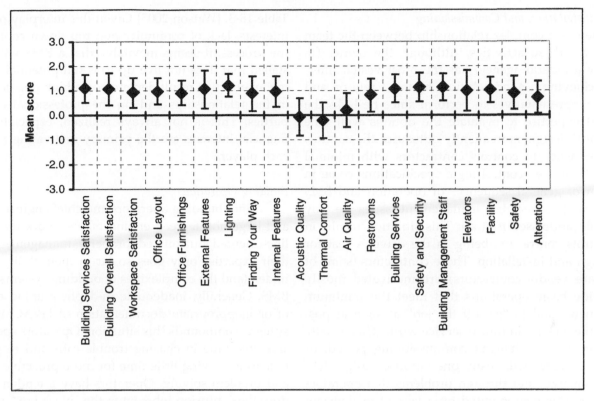

Figure 18-1. Occupant satisfaction survey results for a sample of ~72 GSA buildings (The mean score is the mean of the category means for each building; the error bars show the standard deviation for the sample of ~72 buildings.)

A similar survey (not shown here) was conducted with building operations personnel (one per building) on 72 GSA courthouse buildings. Building management personnel were asked about their satisfaction with the BMS, among other things. These results indicate that, by and large, operators are satisfied with the BMS in terms of the ease of use for operating and maintaining the building, but were not satisfied with availability of control capabilities (i.e., some critical control functions were not present). [Teitjen 2005]

A study of 11 buildings in New England [Fryer-Stein et al. 2004] found that, on average, the BMS systems were producing less than half of the expected savings due to a failure to use even common energy saving techniques like outdoor reset and optimal start/stop.

Other studies, for example a 2002 survey of DOD facilities [Barksdale et al. 2004] it was found, among other things, that 75% of equipment was operating in manual mode bypassing the time clock functions of the BMS. Other problems were attributed to lack of data archiving so historical data is not available for benchmarking to allow comparisons of performance over time, and ineffective management structures due to conflicts and cross-purposes between technicians, engineers, and energy managers. | An economic feasibility study by

Brown [Brown et al. 2000] noted that the Facility Energy Decision System (FEDS) program does not consider savings attributed to EMCSs in economic analyses due to the perception that they are unreliable in delivering savings mainly because of a lack of skilled personnel to operate them.

Expert Interviews

We contacted nine people known by the authors to have considerable experience in some aspect of BMS technology; this experience ranged from application/deployment and commissioning, to research. They provide a "higher level" view different from that obtained from operations personnel in the case studies. We solicited their views individually in open-ended discussions but used the same interview guide that we used for operations personnel as a general guide in conjunction with our working hypothesis that these systems could be more effectively used. These interviews identified a long list of issues that limit the effectiveness of these systems and there was considerable consistency between the respondents. We organized the consensus opinions into categories and have listed them in Appendix A. These results focus on why a BMS is not fully utilized and are generalized in the following summaries:

Design, Installation, and Commissioning

There is a complex relationship between the team players for these activities. Although the owner (in this case a federal agency) has the most vested interest in achieving a good solution, he must rely on and trust his representatives to perform since he lacks the technical expertise to evaluate the options well. The mechanical designers frequently outsource the controls and BMS work to controls contractors with minimal specifications; i.e., control logic specifications come in the form of general sequences of operation and BMS instrumentation and data management issues are only minimally addressed if at all. This situation results in the controls contractors being in the driver's seat for the design and installation. These contractors (some of whom are vendor-contractors) are motivated mostly to provide basic operations that meet the minimum requirements and to "get off the job" as soon as possible while locking-in future service work. This usually results in a short-changed commissioning period, or possibly an extended, costly one because design deficiencies emerge later through problems that operators encounter. This compounded by a lack of or deficient commissioning specifications.

Personal experience of the author confirms difficulties in this area. In a recent project, the system and BMS commissioning was delayed by almost three years due largely to lack of accountability by several players in meeting the spirit and letter of the commissioning intent and the unwillingness to take responsibility for mistakes. Only by continual prodding by researchers, trying to conduct an energy study was the work completed.

A *worst case* view of BMS implementations can be represented by the attitudes and attributes indicated in Table 18-3. [Watson 2005] Given this interplay of vested interests, lack of continuity and top down control over the process, it seems inevitable that a BMS will get installed that does not live up to is full potential. Achieving an effective result depends on a high degree of trust in the relationships and on the professionalism of all parties. This is not always realized, especially when financial interests are the primary motivating factor for each player.

Building Operations

Since building operators and chief engineers are by and large not degreed engineers, and work for bottom-line-oriented outsourced property management firms, their expertise only goes so far in their ability to fully understand the complexities of building systems and the BMS. Generally, inadequate in-depth training and lack of or inappropriate documentation of HVAC and BMS systems compounds this situation. Operators spend considerable time in chasing trouble calls and responding to alarms, leaving little time for more proactive analysis and problem solving. Operators have a tendency to reduce the confusion inherent in the "black box" nature of the inner workings of the BMS, and assert their control, by using manual overrides of automatic functions.

Building Management

Management functions can be characterized by a reactive vs. proactive orientation. At all levels—agency, property management, and on-site—there is a lack of knowledge and understanding as well as little prioritization of getting the most out of the BMS since these operations and energy issues do not seem to be highly valued in terms of bottom line interests.

Table 18-3. Industry hierarchy for BMS design and deployment

Player	Issues and attitudes
Owner/architect	Do not understand controls and BMS technology
Mechanical engineer	Inadequate resources (and/or controls knowledge) to specify quality solutions
BMS vendors	Sell the potential, lock-in proprietary solutions and on-going service
Mechanical and controls contractors	Incentive is to meet minimum requirements and get off the job
Facilities services group	No clout; limited influence or involvement in decision making about BMS solutions

Energy Performance

There is a large untapped potential for improved energy performance. In general, operations staffs pay little attention to energy issues. BMS capabilities that support energy saving features or sequences are little used. However, these capabilities are sometimes used by third party consultants contracted specifically to address energy performance or to conduct re-commissioning projects; the bulk of energy improvement come through this avenue.

BMS Capabilities

Although perceived to be more sensitive and less bullet-proof than older systems, they are indeed sophisticated and capable in terms of controlling equipment and supporting operations. In most cases the primary problem is not the capabilities of the BMS but the fact that many of the capabilities are underutilized. This results from systems being too complex for inadequately trained operators to understand, or the lack of implementation and/or documentation of sophisticated control and management functions. Some other capabilities are best served by other systems (e.g., reporting and operations logging with computerized maintenance management systems (CMMS)). Energy functions appear to

be more and more relegated to EIS systems rather than being incorporated into the BMS. Existing systems do appear to be underpowered for supporting advanced applications such as centralized monitoring or analysis functions that require robust data monitoring and data management capabilities.

"Mini" Case Studies

To provide further insight into real world performance of BMS, we obtained information from operations management personnel using the guide shown in Appendix B, Tables B1-B5. These discussions attempt to capture an operations view with respect to how the system is actually used, what barriers to greater utilization exist, how successful the initial deployment was, and what the management environment is like. We contacted five building operators/managers in GSA Region 9 and attempted to quantify the results by applying scoring to the information collected. Although limited (due to the small sample size) in the degree to which these results can be considered "representative" for all federal buildings, they do provide interesting insights into a number of relevant issues, and are a rich source of anecdotal information.

Table 18-2 shows a categorical summary of results. A high score indicates better effectiveness. We arbitrarily

	Avg Score, % of maximum	SD, % of avg. score
Utilization		
Occupant satisfaction	78%	23%
M&R	62%	28%
Energy management	37%	41%
Barriers	77%	30%
Deployment		
Deployment success	60%	51%
Functionally appropriate	61%	13%
Management		
Owner	34%	52%
Property firm	57%	45%
Building score	**67%**	**24%**

Table 18-2. Case study summary of results by category totals

assume that average scores below 50% and/or standard deviations above 40% indicate a low scoring category. Therefore, energy management, deployment success, and owner management categories score low. All other categories score relatively high indicating that the systems are moderately effective in terms of functionality. There are differences across buildings for the items within a category that tend to get averaged in the results; no weighting was used on the items.

However, when the individual items in each category are reviewed, it is apparent that some items consistently receive low scores (50% or less) or exhibited high variance across buildings. Table 18-3 shows the average score and variance (i.e., standard deviation (SD)) for those items.* These results are in general agreement with the opinions expressed by experts.

The results shown in Table 18-2 are somewhat mixed and cannot be considered definitive due to the small sample size. However, relatively high scores were received for certain topics (e.g., (perceived) occupant satisfaction, M&R, barriers, and appropriate functionality) to indicate that BMSs studied are reasonably effective in keeping a building operating smoothly but the level of effectiveness varies building to building and there are a number of areas where effectiveness could be improved (See Appendix B for complete detailed tables of results).

One case stands out and bears special mention. The San Francisco building received the highest scores of all. However, this building was unique in several ways. An active retro-commissioning effort corrected many problems with both the mechanical systems and the BMS. In addition, the management firm is a newer firm that only performs technical management of operations and repair and the lead operator is a degreed mechanical engineer. This team also receives good support from building management. Since this team is technically well informed and has good support, it has been able to exploit the capabilities of the BMS to improve operations and comfort performance. Further study could illuminate this better.

Potential for Improvement

Currently there are several development projects being conducted that are aimed at improving the effectiveness of BMS systems. The BMS of today would most likely serve as a platform for some of these, but others would require an upgrade in technology.

---

*Individual items were scored on a scale of 1 (low effectiveness) to 5 (high effectiveness).

	Avg Score, % of maximum	SD, % of avg. score
Utilization		
Improving occupant comfort	56%	64%
Create/program graphical displays	44%	75%
Programming closed loop control of processes	36%	99%
Programming of sequences of operations	36%	99%
Diagnostics/troubleshooting with spreadsheets (and/or other software)	30%	67%
Documentation and logging of O&M activities	25%	40%
Report generation	28%	64%
Equipment performance tracking/monitoring	40%	71%
Track/document energy use and savings	32%	34%
Use BMS to improve energy use and cost	40%	50%
Barriers		
Operator training	64%	64%
Software configured to perform functions needed	48%	70%
Operations resources available	52%	84%
Documentation of equipment, systems, control sequences	64%	41%
Deployment		
Procured, installed, and commissioned successfully	56%	64%
System can be replicated easily (multi-building/site, not custom)	56%	53%
Supports integration and interoperability	40%	61%
Management		
Management is knowledge about BMS technology	27%	43%
Management has a proactive vs. reactive attitude about BMSs	27%	43%

Table 18-3. Low scoring items (< 50% average score or > 40% SD)

- *Centralized remote monitoring and analysis.* (GEMnet) This technology relies on an on-site BMS to deliver data from a number of sites to remote database servers where it can be analyzed by expert practitioners (usually part of an applications/energy service provider (A/ESP)) using sophisticated analysis tools for diagnostics, equipment performance tracking, and energy savings tracking. GEMnet is one early example of this technology, currently others are emerging with various levels of capability [Cimetrics 2005]. While central monitoring and analysis appears to have great potential, the underlying BMS technology needs to be robust and capable enough to support it (i.e., communications bandwidth must be adequate to support acquisition of large amounts of monitored data). The central monitoring tools must be robust enough to make analysis efficient (i.e., they must support high quality data management, presentation, and reporting capabilities). It is not clear, given the current cost structure, that the value proposition is ultimately realizable.

- *Energy information systems (EIS)/Enhanced Automation (EA)/Enterprise Energy Management (EEM).* These systems are similar to central monitoring systems but focus primarily on energy efficiency and demand responsiveness (DR). They can be web-based and therefore operated remotely to support ASP services. They can provide a variety of functions such as load profiling, billing analysis, forecasting, and load shifting and demand response. While there were many vendors who developed and sold this technology in the past, only a few exist today. DR has become the focus of development efforts in this area, primarily sponsored by the CEC. [Motegi et al. 2002] Some of these EIS functions can be easily added to or combined with an existing BMS platform to provide for load shedding and energy tracking capabilities without the need for sophisticated analysis tools. These generally fall in the Enhanced Automation category. [CEC 2002]

- *Advanced applications.* There are a number of new applications being developed that use the BMS to provide added capabilities such as data visualization, fault diagnostics, and occupant feedback. These are thoroughly reviewed in our 4th article of this series [Yee 2004].

- *Commissioning.* As evidenced by the many studies, conferences, and commissioning agents now available, it is clear that commissioning is becoming more emphasized in the buildings industry and in federal buildings construction. Although work still needs to be done to improve the outcomes from these efforts, these trends are in the right direction to improving overall building system performance and BMS effectiveness.

- *Advanced monitoring specifications.* The DOE and CEC have jointly sponsored the development of [Hitchcock 2005] advanced monitor specifications for BMS systems. This project attempts to address many of the outstanding issues discussed in this article by developing requirements not only for instrumenta-

tion specifications, but also for data visualization, performance metrics, training, and commissioning. The realization of these objectives would go a long way toward improving those specific aspects that scored low in our results and toward correcting the issues that experts cite as being significant constraints. However, it is an open question as to how much impact these specifications would have because of the joint constraints of operations staff expertise and management orientation. For example, without a concomitant increase in the qualifications of operating staff and management knowledge, or a major improvement in "user friendliness" from more sophisticated BMS monitoring and analysis tools, the same constraints that now exist are likely to continue. Namely, operators have limited time or expertise to use these tools effectively and the business interests of the industry players will continue to dominate the implementation.

ANALYSIS AND DISCUSSION

Evaluation of Accepted Beliefs

The following summarizes the degree to which the results of our study substantiate the accepted beliefs.

1. *Almost all larger federal buildings use a BMS; these systems are considered key to operating a building well.*
    - Table 18-1 indicates that this is somewhat true; for buildings over 50k GSF 38% of buildings and 51% of total building floor area have BMSs installed.
    - However, building personnel and experts agree that operating large buildings without a BMS is difficult.

2. *Keeping the occupants satisfied and systems operational is the primary focus of operations staff. These systems are considered effective in this regard.*
    - Case study utilization results show that a major use of the BMS is to help manage occupant complaints. However, the occupant survey results suggest that operators are unaware of the generally low ratings that comfort gets.
    - Experts confirm this, emphasizing that operators are so busy responding to trouble calls that they have little time left for other tasks such as evaluating and improving comfort and energy performance.

3. *The energy saving potential of BMSs is not fully realized.*

- Case study results show a very low (37% score) for utilization of BMS for energy management functions.
- Expert and operations interviews indicate that these functions are not a routine part of the operators work, and that when it is third party experts more often perform it.

4. *The implementation/deployment process for BMSs is very difficult and fraught with problems.*
   - The case studies results indicate a mixed view of this issue. While deployment success and appropriate functionality on average are rated moderately high at ~60%, deployment success variability is high (SD = 50%).
   - Experts, however, maintain the process is chaotic and difficult due to the overlapping and competing interests of the providers of services and equipment.

5. *In practice, there is a mismatch between the capabilities and sophistication of BMS technology and the ability to exploit it.*
   - Operators indicate that the more complex capabilities of the BMS are little used. The results of discussions with them also indicate that barriers to more effective use are due to lack of training, time availability, and documentation rather than the operator's ability to use the BMS.
   - The two issues of operator expertise and inability to use complex BMS capabilities go hand-in-hand. Experts confirm that the mismatch has to do with operator's expertise; complex problems and analyses are often left to third party experts to solve.

6. *Vendors tend to hype the potential compared to what is actually delivered.*
   - This has not been confirmed; technology is robust enough for many functions simple and complex at least to the degree that operators use them. Even for experts, these systems are capable of supporting some level of analyses with trends, plotting, etc. In general, the operators indicate that functionality delivered is appropriate for their needs.
   - However, experts indicate that existing systems are lacking in terms of support for robust data management necessary for higher end analysis such as benchmarking, central monitoring, and continuous commissioning functions.

7. *Despite many claims and much hype, true interoperability has not been achieved.*
   - Confirmed by everybody; the availability of BACnet and other technologies still does not translate into true interoperable (i.e., direct communication links between systems of multiple vendors that provide for coordination of control and management functions) solutions. There still is a tendency for vendors to make the systems proprietary to lock-in the customers. *Integration* (i.e., the amalgamation of various functional components under the umbrella of facility automation) into the BMS of lighting and access control, on the other hand, appears to be relatively common.

Revised Perspective

Besides generally confirming most of the accepted beliefs above, our study revealed several other perspectives about effectiveness. The following summarizes our revised perspective of the state of practice of BMS implementations based on the results of this study.

- *Demographics:* A significant portion of large federal buildings use a BMS of some sort to operate the building; it is standard practice to use DDC technology and BMS functions to service buildings. But there is still significant potential for deployment of BMS technology in these buildings.

- *Overall effectiveness:* Effectiveness evaluations must be broad based; simply focusing on the technical aspects of the BMS is not enough because these systems exist in a complex design, services, and management environment. In the following, we have annotated the three categories of effectiveness that we have used to capture this broader view to summarize overall effectiveness:

  — *Technical effectiveness is moderate*—Newer systems have adequate technical capabilities to support primary service functions and even more functions that are sophisticated if the expertise is available to exploit them.

  — *Operational effectiveness is high*—Operational effectiveness is high for general and routine building management functions and keeping occupants satisfied (e.g., minimizing trouble calls and supporting low level diagnosis) but is low in terms of proactive improvements in energy and comfort performance. Operators appear unaware of the generally low comfort performance of these systems. The level of operator expertise is relatively low compared to the complexity and capabilities of the HVAC and BMS systems, but lack of training and inadequate time devoted to using the BMS also militate against more effective use. Perhaps technological solutions could be developed that would help overcome the lack of operator expertise.

— *Organizational effectiveness is low*—This factor has a strong impact on constraining use of the full range of capabilities. Inadequate management structures result in overburdened operators who lack training, or do not have the required background expertise. There are few higher-level practitioners (e.g., HVAC/ mechanical engineers, or highly skilled operating engineers) to solve complex problems and actively exploit the functionality available, therefore these tasks are generally outsourced.

• *Distribution of functions:* Third party specialists, on an as needed basis, accomplish most analysis, commissioning, performance improvements, and control sequences programming. Likewise, reporting and logging functions are performed in CMMS applications

• *Design and deployment:* Good design and installation is crucial in getting high performance systems installed but the process is fraught with industry related problems.

• *Energy savings:* Most BMSs are not used actively to manage energy; this function is being accomplished more and more by EIS systems or is the result of third party studies or various types of commissioning.

• *Integration and interoperability:* These are not robust in the industry at large. Integration occurs more frequently, but true ("plug and play") interoperability is not being practiced and in fact is being resisted by vendors intent on maintaining vendor lock-in.

• *Improvements coming:* Various research and business concerns are taking steps to improve the performance of these systems. Targeted efforts aim at the primary issues of better operator tools and training,

commissioning, and high quality instrumentation, and robust data monitoring and management.

SUMMARY AND CONCLUSIONS

An evaluation of effectiveness of BMS focused just on *technology* is too narrow a view; in previous reports in this series we studied these technology factors. In this study, we took a broader look at other non-technological issues and considered how the systems are used in a services environment made up of building operators and managers and industry players such as vendors, contractors, and engineers and architects—all with different levels of expertise and stakes. A full assessment must consider these issues and their interrelationship because they all influence BMS effectiveness. We have organized these issues into three key categories: Technical, operational and organizational. We have evaluated these key areas by collecting opinions from building technology experts, reviewing pertinent literature, and talking to operations personnel. Considered together the interviews, literature review, and the case studies results provide some perspective about what is useful about these systems, what areas are underutilized and what the constraints are that limit them from being used to their full potential.

The operator discussions indicate that, although individual buildings vary, overall they rate moderately in all the key effectiveness categories. However, these results mask those viewed on an individual item basis. Combining these results and the category results, with the interview and literature review as well as our own experience, we found that.

1. The results from this study indicate that effectiveness is limited more by operational and organizational problems than by technical capabilities, although improvements could help here also. The technology appears to be adequate for the level at which a BMS is commonly used, but for more demanding applications, such as central monitoring, further development is needed.

2. On the other hand, some available functions are underutilized. One factor appears to be a lack of the required expertise and/or training to program sequences or fully understand the consequences to the system of sequences they might develop by the operators. Systems tend to be used at the level of expertise of operators; they use the BMS primarily to address faults and trouble calls/complaints

and for scheduling and manual control of equipment. Sometimes outside consultants use the BMS to conduct diagnostics and analyses that are more comprehensive. Programming, when required, is done mostly by the vendor or independent specialists. Finally, there seems to be a disconnect between what operators believe about occupant comfort compared to what the occupants think; thermal comfort and air quality receive the lowest scores of all building indoor environmental factors except acoustics.

3. Decision makers in the building management structure (both agency personnel and outsourced services management) are, in general, not very knowledgeable about BMS technology. Controls and building operations are not a highly valued function so resources (training and level of expertise) are not well matched with complexity and capabilities of available BMS technology. These managers are more interested in other asset management functions and are more reactive than proactive when it comes to BMS issues. This has resulted in increased outsourcing for the more complex functions of energy, equipment performance analysis, programming improvements, and difficult problem solving.

4. Despite the fact that case study results indicate that the systems installed turn out to meet the owners' specifications and the functionality is generally appropriate to the site needs, the design and installation process is such that many systems are inadequately procured, installed and commissioned. This results in poorly functioning equipment, inadequate documentation to support operators, and more costly repairs and upgrades in the future. However, commissioning is becoming more commonly accepted so that deficiencies are being corrected in many older systems during retro-commissioning projects.

We believe that the conclusions reached in this study will convey to facility managers (and indeed all practitioners involved) that maximizing the potential of BMS technology requires adopting a comprehensive view of the entire environment within which the deployment of these systems exists. Although improvement in technology can help, it is not just an issue of technological capabilities; design, installation, commissioning, operations, the building management structure and the interactions between all of these must be considered

as well. Aligning all of these elements into a well-integrated team supportive of the overall goals of energy efficiency, occupant comfort, and low cost O&M is essential to achieving a highly effective system.

ACKNOWLEDGEMENTS

The author gratefully acknowledges Bill Carroll (LBNL) for his support and guidance of this project. We also extend our gratitude to the following individuals for their contributions through expert interviews. Steve Bushby, Leader, Mechanical Systems and Controls Group, NIST; Ken Gillespie, Technologist, Pacific Gas and Electric Co.; Phil Haves, Senior Scientist, Environmental Energy Technologies Division, LBNL. We would like to specially acknowledge Mark Levi, GSA for his assistance in locating case study buildings and in the development of the interview guide, and his candid comments about these issues. In addition, the building operators and managers deserve special mention for their valuable contributions to this study.

References

Abbaszadeh Fard, Sahar, L. Zagreus and C. Huizenga, 2004. "Using the CBE Occupant Survey To Assess Facility Management Performance in GSA Facilities, May-July 2004 Delegated Buildings." Center for the Built Environment, University of California, Berkeley, CA, October.

Barksdale, G.G., and M.D. Smith. 2004. "How to Get the Most Out of Your EIS. A New Organizational Management Paradigm." Chapter for *Information Technology for Energy Managers,* eds. Barney L. Capehart. pp.75-85

Brown, D.R., J.A. Dirks, and D.M. Hunt. 2000. "Economic Energy Savings Potential in Federal Buildings." PNNL-13332, Pacific Northwest National Laboratory, Richland, WA, September.

CEC. 2002. "Enhanced Automation. Technical Options Guidebook." California Energy Commission, May.

Cimetrics. 2005. Infometrics web-site. http://www.cimetrics.com/home/infometrics/

Claridge, D.E., W.D. Turner, M. Liu, S. Deng, G. Wei, C. Culp, H. Chen, and S. Cho. 2002. "Is Commissioning Once Enough?" *Proceedings of the 25th World Energy Engineering Congress,* October 9-11, Atlanta, Georgia.

Diamond, R., T. Salsbury, G. Bell, J. Huang, and O. Sezgen, R. Mazzucchi, J. Romberger. 1999. "EMCS Retrofit Analysis Interim Report," LBNL-43256, Lawrence Berkeley National Laboratory, Berkeley, CA, March.

EIA, 1997. "Federal Buildings Supplemental Survey 1993." SR/EMEU/95-02, Energy Information Administration, USDOE, August.

EIA, 2002. "CBECS. Commercial Buildings Characteristics 1999." Energy Information Administration (EIA), USDOE, May 2002. http://www.eia.doe.gov/emeu/cbecs/detailed_tables_1999.html

Fryer-Stein, L., B. Kiernan, L. Keller, and R. Reiss. 2004. "What Network Building Control Can Do for End Users." Chapter for *Information Technology for Energy Managers,* eds. Barney L. Capehart. pp.41-47

GSA. 2005. GSA Public Buildings Service Website. http://www.gsa.

gov/Portal/gsa/ep/channelView.do?pageTypeId=8199&chan
nelPage=%2Fep%2Fchannel%2FgsaOverview.jsp&channelId=-
13303. General Services Administration, Washington DC.

Hitchcock, R. 2005. Personal Communications, Senior Scientist, Envi-
ornmental Energy Technologies Division, LBNL.

Mills E., H. Friedman, T. Powell, N. Bourassa, D. Claridge, T. Haasl,
M. A. Piette. 2004. "The cost-effectiveness of commercial-
buildings commissioning." LBNL 56637, Lawrence Berkeley
National Laboratory, Berkeley CA, December.

Motegi, N. and M.A. Piette. 2002 "Web-based Energy Information
Systems for large Commercial Buildings." LBNL 49977, Lawrence Berkeley National Laboratory, Berkeley, CA, May.

Piette, M.A, S. Kinney, M., Levi, D. McBride and S. May. 2002. "GEM-
net Status and Accomplishments. GSA's Energy and Maintenance Network." *Proceedings of the 2002 ACEEE Summer Study
Proceedings*, Pacific Grove, CA, 18-23 August.

Piette, M.A., S. Khalsa, P. Haves, P. Rumsey, K.L. Kinney, E.L. Lee,
A. Sebald, C. Shockman. 1999. "Performance Assessment and
Adoption Processes of an Information Monitoring Diagnostics
System Prototype." LBNL 44453, Lawrence Berkeley National
Laboratory, Berkeley, CA, October.

Shockman, C. and M.A. Piette. 2000. "Innovation Adoption Processes
for Third Party Property Management Companies." LBNL-46376,
Lawrence Berkeley National Laboratory, Berkeley, CA, July.

Teitjen, R. 2005. Personal Communication, Senior Architect, Office of
Applied Science, GSA.

Watson, D. 2005. Personal Communications, Senior Scientist, Environmental Energy Technologies Division, LBNL.

Webster, T, 2002. "Trends Affecting Building Control System (BCS)
Development." LBNL 47650, Lawrence Berkeley National
Laboratory, Berkeley, CA, 15 pp.

Yee, G., and T. Webster. 2004 "Review of Advanced Applications
in Energy Management, Control, and Information Systems."
LBNL 53546, Lawrence Berkeley National Laboratory, Berkeley,
CA, 33 pp.

Bibliography

Motegi, N, Piette, M.A., Kinney, S. and Dewey, J. 2003. "Case Studies of Energy Information Systems and Related Technologies.
Operational Practice, Costs, and Benefits." *Proceedings of International Conference for Enhanced Building Operations.* LBNL Report
#53406.

Piette, M.A. 2005. Personal Communications, Senior Scientist, LBNL

Santos, J. 2005. Personal Communications, Principal, Facility Dynamics.

Webster, T. 2002. "BCS Integration Technologies—Open Communications Networking." LBNL 47358, Lawrence Berkeley National
Laboratory, Berkeley, CA

Yee, G., and T. Webster. 2004. "State of Practice of Energy Management,
Control, and Information Systems." LBNL 53545, Lawrence
Berkeley National Laboratory, Berkeley, CA, 23 pp.

APPENDIX A. EXPERTS INTERVIEWS DETAIL*

Design, Installation, and Commissioning

- Design engineers lack controls knowledge; tend to "farm out" controls work to controls contractors.

- Controls contractors tend to treat control systems like a commodity, and work to minimize their costs

*None of these statements is universally true, there will always be exceptions.

which results in meeting minimal requirements and "getting off the job" as soon as possible.

- Controls contractors/vendors work to "lock-in" service contracts by holding access keys, programmed logic, and other technical expertise. However, this may be important to maintaining "version control" of control software since operations staff appear to have little understanding (possible reflecting lack of training) or time or motivation to maintain documentation of this sort. Lock-in is furthered when the system requires parts only the BMS vendor can provide.

- Controls contractors are more rewarded by low bids than good job performance.

- Controls contractors are not generally independent; they represent a select few vendors or are part of a vendor's local field organization.

- Little continuity in implementation from design through commissioning; the owner is not sufficiently capable of understanding technical issues to be able to evaluate success. Owners (or owner's agent) must prove deficiency rather than contractors proving success.

- Commissioning stops when the money runs out. However, commissioning is more frequently being accepted and budgeted for in recent years.

- In some instances there is a conflict of interest between contractors. For example, when the mechanical contractor or general hires the test and balance (TAB) contractor, it is unlikely that the owner's interest will be best served.

- Since owners are under fiscal constraints or architectural issues dominate concerns and budget, systems are first cost driven and therefore designer's resources are so limited that first class solutions are compromised.

Building Operations

- Operating engineers are not really professional engineers and therefore lack the knowledge to understand intricacies of HVAC processes and BMS technology.

- Technical managers are not engineers but are capable operating engineers that have moved up to corporate responsibilities.

- Most buildings operations are outsourced to for-profit companies. There is very little in-house quality control expertise left. Only minor repairs and upgrades are performed by on-site operations staffs, larger ones are outsourced to contractors.

- A limited number of operators are highly capable so they become overburdened with problems since they are the only ones with enough expertise to solve complex issues.

- Lack of training of operations personnel is endemic to the industry. Property management firms are not motivated to invest because their interest is on larger fiscal issues and profit margins. [Shockman et al. 2000]

- Due to the limited expertise and lack of operator training BMSs tend to be viewed as "black boxes." This results in more "brute force" approaches to solving problems or disabling/overriding controls logic in an attempt to achieve understandable performance; i.e., reducing the system operations to the operators level of understanding.

- Lack of BMS and HVAC system documentation and control sequences in operator friendly format is a severe problem. Details of control sequences are largely unknown by operators or are embedded in large tomes of programming minutiae. However, lack of operator expertise in this area tends to confound the problem.

- Operators do not use documentation and logging capabilities typically included in a modern BMS. Sometimes paper logs are used and/or a CMMS; without these logs there is a loss of "institutional memory" (especially if the operator leaves) that makes problem solving more difficult.

Building Management

- There are two levels of management that influence BMS issues. Owner management i.e., GSA or other federal agencies consists of on-site or regional level building managers; they usually are not technical people and thus have little understanding of BMS and HVAC issues. Property management firms to which building operations have been outsourced consist of on-site operating engineers and maintenance staff. Firm management are typically technical managers that were previously operating engineers [Shockman et al. 2000]

- Contracts are highly structured and negotiated to minimize management firm and agency risk. Large upgrades or BMS commissioning, programming, complex diagnostics and system modifications are usually outsourced to third party contractors and consultants.

- Upgrades are relegated to "routine technologies" i.e., known and understood additions that are embellishments that perform the same functions in a new way as opposed to those that perform new and different functions. [Shockman et al. 2000] Innovations come from new buildings, not upgrades to existing.

- Buildings operations and management orientation is predominately reactive as opposed to proactive.

Energy Performance

- Potential is large; most (approximately 80%) of energy savings are controls related. [Claridge et al. 2002]

- High quality monitoring can result in significant energy savings [Piette et al. 1999]

- Energy cost is not a driver; compared to other facility costs and keeping occupants happy, it is a low priority. There is little real incentive to use less energy and few benchmarking procedures for comparisons.

- Equipment performance tracking, i.e., chiller kW/ton, is virtually non-existent.

- Active energy performance tracking is negligible, even when data is available.

- Best opportunity for good energy performance is initial design and installation; after that it becomes difficult and expensive via retro-commissioning activities.

BMS Capabilities

- Modern BMS technology is more capable, complex and sensitive; older systems are more "bullet proof."

- Technical effectiveness/capabilities for most buildings operations are available in modern product offerings. They typically support the needs of routine operations well; it is hard to imagine running a building today without one.

- Off-the-shelf system capabilities are not quite ready for advanced applications such as centralized monitoring; networking and software technology is not robust enough and is too expensive. (See Potential Improvement Section)

- Reporting functions are often not used; nobody is interested in the content.

- No true interoperability, not like IT industry, mostly data exchange/monitoring. Problems can be increased if BMS control sequences attempt to override or interfere with equipment unit controller functions.

- Security issues are not a serious issue for dedicated networks, but may become more critical when combined with IT.

- Operators rarely program either logic or graphics screens; these functions are normally done by outside third party contractors or vendor field offices.

APPENDIX B. OPERATORS DISCUSSION GUIDE

The topics listed below were used to guide the conversation with interviewees. In this way, we were able to collect information in a systematic manner.

Table 18-B1. Building Characteristics

Building characteristics
- Location/City
- Description
- Type/use
- Vintage
- Size, GSF

HVAC Systems
- Description
- Type(s)
- Vintage
- Schedules

BMS
- Description
- Brand
- Vintage
- # of points
- Network architecture
- Energy information system (EIS)?

Table 18-B2. Utilization

Occupant satisfaction
- Responding to complaints/service calls
- Improving occupant comfort

Maintenance and repair
- Monitoring of operations (with or without graphics)
- Change setpoints, schedules, manual control
- Use graphical displays
- Create/program graphical displays
- Programming closed loop control of processes
- Programming of sequences of operations
- Diagnostics/troubleshooting with trend logs
- Diagnostics/troubleshooting with spreadsheets (and/or other software tools)
- Alarms management
- Safeties (software) setting management
- Documentation and logging of O&M activities
- Report generation
- Remote access for alarms or monitoring (via phone or otherwise)

Energy & equipment performance
- Track and monitor equipment performance
- Track/document energy use and savings
- To improve energy use and cost

Table 18-B3. Barriers to Effective Use

Lack of operator training
Operator distrust of system to perform properly
Operator resistance
Interference from others
System reliability (do components and front-ends fail frequently)
Slow response
Erroneous information (false readings/scaling, sensor calibration)
Nuisance alarms
Requires constant "tweaking"
Difficult or complicated to setup and/or use
Software cannot be configured to perform functions needed
Difficult to understand operations via information available
Data access and management (lack of tools, access, storage, archiving)
Limited resources, no time to use system effectively (too many fires)
Lack of documentation on equipment, systems, control sequences
System was oversold or misrepresented by vendor
Other comments

Table 18-B4. Deployment Success and Functionality

Deployment
    Designed and specified according to owners intent
    Procured, installed, and commissioned successfully
Functionality
    System is functionality appropriate for site needs
    System is flexible, adaptable
    System can be replicated easily (multi-building/
        site, not custom)
    Supports integration and interoperability

Table 18-B5. Management responsiveness

Management is knowledgeable about BMS technology
Management makes BMS issues a priority
Management supports operator training
Management has adequate financial resources to support BMS infrastructure
Management has a proactive vs. reactive attitude about BMSs

Location	North CA		North CA		Arizona		North CA		South CA		Totals		
Building size/type	336k GSF, offices		500k GSF Labs, offices		130k GSF, offices		1,500k GSF, office/courts		216k GSF, offices				
BMS, vintage (points)	JCI Metasys, 1998 (2000)		JCI NCMS, ~1990 (3000)		JCI Metasys, 1995 (350)		Alerton, ~1995 (~50-70k)		Delta, 2000 (245)				
	Scores	%	Scores	%	Scores	%	Scores	%	Scores	%	Avg score	Avg %	SD
Utilization													
Occupant satisfaction	6	60%	6	60%	9	90%	10	100%	8	80%	7.8	78%	
M&R	32	58%	35	64%	24	40%	51	93%	33	55%	35	62%	
Energy management	4	27%	3	20%	8	53%	8	53%	5	33%	5.6	37%	
Barriers	61	76%	31	39%	67	84%	78	98%	72	90%	61.8	77%	
Deployment													
Deployment success	3	30%	3	30%	8	80%	6	60%	10	100%	6	60%	
Functionally appropriate	12	60%	12	60%	11	55%	11	55%	15	75%	12.2	61%	
Management													
Owner	2	8%	10	40%	12	48%	10	40%			8.5	34%	
Property firm	6	30%			17	68%	19	76%	11	55%	13.25	57%	
Building score	**126**	**60%**	**100**	**45%**	**156**	**73%**	**193**	**88%**	**154**	**72%**	**145.8**	**67%**	

Table 18-B6. Case Study results, by category totals

Percentages are a better indicator of the results because the maximum score for each section is not necessarily the same for all buildings since some capabilities were not available on some systems (and were therefore not scored; i.e., percentages were based on questions answered).

	Avg Score	Avg %	SD	SD%
Utilization				
Improving occupant comfort	2.8	56%	1.79	64%
Create/program graphical displays	2.2	44%	1.64	75%
Programming closed loop control of processes	1.8	36%	1.79	99%
Programming of sequences of operations	1.8	36%	1.79	99%
Diagnostics/troubleshooting with spreadsheets (and/or other software)	1.5	30%	1.00	67%
Documentation and logging of O&M activities	1.3	25%	0.50	40%
Report generation	1.4	28%	0.89	64%
Equipment performance tracking/monitoring	2.0	40%	1.41	71%
Track/document energy use and savings	1.6	32%	0.55	34%
Use BMS to improve energy use and cost	2.0	40%	1.00	50%
Barriers				
Operator training	3.2	64%	2.05	64%
Software configured to perform functions needed	2.4	48%	1.67	70%
Operations resources available	2.6	52%	2.19	84%
Documentation of equipment, systems, control sequences	3.2	64%	1.30	41%
Deployment				
Procured, installed, and commissioned successfully	2.8	56%	1.79	64%
System can be replicated easily (multi-building/site, not custom)	2.8	56%	1.48	53%
Supports integration and interoperability	2.0	40%	1.22	61%
Management				
Management is knowledge about BMS technology	1.3	27%	0.58	43%
Management has a proactive vs. reactive attitude about BMSs	1.3	27%	0.58	43%

Table 18-B7. Case Study results, low scoring items (low score (< 50%), high standard deviation (SD) (>40%) or both) based on a scale of 1 (low) to 5 (high)*

---

*In the Barriers section the topics have been revised to be consistent with the scoring method and are therefore presented differently here than how they were asked in the survey itself.

	Avg Score	Avg %	SD	SD%
Utilization				
Improving occupant comfort	2.8	56%	1.79	64%
Centralized graphical displays	2.2	44%	1.51	75%
Programming closed-loop control of processes	1.8	36%	1.79	99%
Programming of sequences of operations	2.8	28%	1.79	99%
Diagnostics/monitoring with remote/onsite remote/offsite service	1.5	30%	1.05	67%
Documentation and logging of O&M activities	1.3	28%	0.50	40%
Report generation	1.4	28%	0.89	64%
Equipment performance tracking/monitoring	2.0	40%	1.41	71%
Track/document energy use and savings	1.8	32%	0.50	34%
Use BMS to improve energy use and cost	2.0	40%	1.00	50%
Barriers				
Operator training	3.2	64%	2.05	95%
Software configured to perform functions needed	2.4	43%	1.57	70%
Sufficient resources available	2.8	43%	2.19	88%
Documentation of equipment, systems, control sequences	3.2	49%	1.90	41%
Deployment				
Procured, installed, and commissioned successfully	2.5	56%	1.75	64%
System can be replicated easily (multi-building, not custom)	2.5	56%	1.55	50%
Support, integration, and interoperability	2.0	40%	1.21	67%
Management				
Management is knowledgeable about BMS technology	1.3	27%	0.58	45%
Management has a proactive vs. reactive attitude about BMS	1.3	27%	0.58	43%

Table 16-B7. Case study results; low scoring items (low avg (<50%)), high standard deviation (SD) (>30%) or both based on a scale of 0 (low) to 5 (high).

Chapter 19

The Revolution of Internet Enabled OEM Controls

Travis Short

ABSTRACT

THIS CHAPTER IS DEVOTED to the revolution of mounting internet ready, protocol agnostic control solutions, provided to suppliers of HVAC and electrical equipment on an OEM basis. The evolution of technology will revolutionize how smart buildings will interact with intelligent pieces of equipment. Items that will be discussed in this chapter include; what exactly is meant by internet ready & protocol agnostic solutions in respect to OEM equipment, how these solutions will change the face of the Intelligent building, and finally how these solutions are implemented.

INTRODUCTION

The evolution of the internet has changed how the day to day operation of business is accomplished. This same technology rapidly moved to the building automation industry with solution providers offering web enabled front end solutions to the end user. Now building monitoring was brought to the internet, but there is still a compromise. The main concession is the physical monitoring and control of building systems is still accomplished via the use of varied building automation protocols while the internet aspects are performed by converting this data and displaying it on web pages. While this advance of internet technology has changed the way the building automation industry installs systems, users are typically bound to one protocol and thus one temperature control contractor.

Global energy management can now be accomplished via a direct interface with building systems and not through expensive gateway. Pricey gateways are not a practical application for small building solutions such as gas stations, restaurants, and chain stores. While these facilities could benefit the most from global energy management the current cost of installing a full scale web enabled building automation system (BAS) is not practi-

cal. This is where web enabled building systems such as web enabled roof top units and lighting control systems become a feasible cost effective solution. Now for example global set point and occupancy schedule changes can be controlled from one central location. In addition due to the protocol agnostic characteristics of the equipment building energy monitoring can easily tie into the system via the Modbus protocol and serve up real time energy consumption data over the internet. This energy information can now be used for local demand limiting and global energy forecasting at one central location.

It is now time to discuss taking the awesome power of the internet from the BAS and place this technology at the controlled equipment. You may ask why this concept has taken so long to happen. The answer like most things revolves around cost. Finally the concept can be realized due to advances in technology which have reduced component costs and size. The final piece of the puzzle centers on the use of a protocol agnostic controller. A protocol agnostic controller gives the equipment the ability to be integrated by any control protocol thus eliminating the strangle hold that building automation providers have on facilities.

INTERNET READY, PROTOCOL AGNOSTIC CONTROLS ON EQUIPMENT

Bringing the internet to life at the equipment level will change the way not only building automation is accomplished but how facilities manage their equipment. Some of these radical changes at the equipment level include:

1. The use of archaic LCD displays will no longer be required. Now any device that is internet enabled will be able to connect to, monitor, and perform system set up from any Ethernet connection. Reference Figure 19-1 Connection Options for a sample of different connection options.

2. Data shared between controllers has the capability to be instantaneous and is not bottle necked by the slower communications brought about by serial networks.

3. The onset of internet enabled equipment makes it possible for equipment to email maintenance personnel that a component is about to fail or that preventative maintenance is required.

4. Building automation systems are no longer required for the storage of trend data and system diagnostics. Equipment controllers can now house history data on selected points. As an example data from the space temperature sensor can be stored on the controller and this data can be transmitted to a SQL database running on one of the facilities servers.

One of the inherent benefits brought about by developing controllers based upon the internet is the future proof aspects of the system. Technological advances

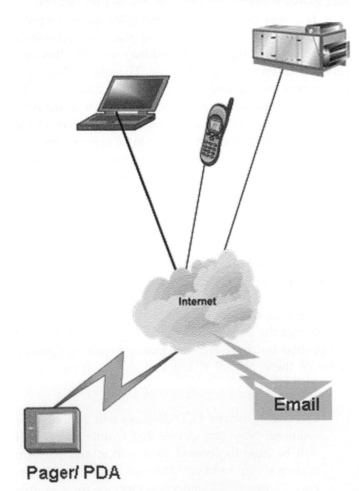

Figure 19-1. Connection Options

happen at such a rapid rate that it is almost impossible to predict what the future will bring. With applications developed using java and a database developed in XML the ability to develop new software platforms to meet the ever changing technology is readily available. The use of java as a development environment frees up the creation of applications from specialized programming environments to universal development environments. This increases not only the speed of development and deployment of new products and applications but also the openness of the systems.

The use of XML as the backend database makes the concept of enterprise level connectivity of diverse building systems a reality. A real world example of such enterprise connectivity is a national hotel chain. In this example the Hotel check in system is tied into the room's controls. Once a room has an occupant checked in the rooms controls change from a non-checked in mode to a checked in mode. The room systems will now use different temperature set points and lighting control schemes. These slight changes will ensure that while a room is not in use lights will be off and room temperature controls will control to maintain global unoccupied set points. While this may seem like a fairly simple example the ramifications on energy consumption are astronomical. No longer will lights be left on in unused hotel rooms and room temperature controls are no longer solely dependent upon the hotel staff setting the temperature set point when a room is not in use.

Now on to the term protocol agnostic and what this means at the equipment level. Protocol agnostic means that the controller is capable of integrating with any field buss protocol in the industry. Of course if we have a controller that is capable of tying directly into the internet why do we need any protocol support at all? There are two main reasons for the support of third party protocols, and they are:

1. The install of new equipment in an existing building with a BAS already installed.
    a. In this installation example the building already has a BAS with a HMI (Human Machine Interface) that the building staff is already familiar with. This is where the protocol agnostic features come into play. The equipment supplier does not have to worry about having support for the existing BAS control system protocol. The same controller and embedded software can be used at any time. This also provides for an ease of integrating the new equipment into the existing BAS, and provides standardization at the equipment level.

2. The install of a small building solution. Small building solutions as mentioned above include those buildings that do not have the install of an expensive BAS system in the budget.

  a. The use of a protocol agnostic solution in this case provides the ability to integrate additional systems into one complete package. The example above discussed integrating a Modbus Power Meter into a small building solution. This is of course just one example. As installed pieces of equipment become microprocessor based the ability to gather real-time operational data becomes a reality. Imagine integrating the Kitchens equipment into the small building solution. Information such as average oven temperature, freezer temperatures, and device failures can be served up over the internet to a central operating location. An equipment failure could be annunciated and a service technician sent to the site before the problem was even noticed locally. As you can see the positive implications to business operations would be noticed immediately.

The ability to serve up real-time data using a building's IT backbone is an incredible resource to facility managers and business managers. The repercussions on small buildings and large facilities will soon be realized as more and more internet ready, protocol agnostic solutions get installed. Consequently the future of the BAS contractor is going to morph from a simple temperature control contractor to that of a whole building system integrator. This evolution is discussed in the next section.

THE CHANGING FACE OF INTELLIGENT BUILDINGS

The previous section was devoted primarily to discussing a revolutionary new hardware and software platform for equipment/system control. This new platform will bring about tremendous changes in the way equipment is controlled and buildings are managed. These changes will also affect the building controls industry and change the way today's buildings will operate. Equipment will have the ability to alert maintenance personnel of potential problems, alarms, or maintenance issues. Changes on a global level will be accomplished from one central location in lieu of being performed locally. Energy consumption and building assets can be monitored and controlled with or without the use of a BAS. This section will explain how the interaction of

divers systems on a M2M (Machine to Machine) basis will increase overall building performance, and how the existing model of current BAS architectures will transform into a model of completely integrated building solutions.

First we will discuss the reality of M2M communication with a couple of practical examples detailing the use of this technology. With equipment and system controllers no longer dependent upon predefined protocols systems that were once diverse, and incapable of sharing data, will now communicate and share information as one complete system. The bridging of different systems and protocols with the use of M2M communications will increase building performance which will have a profound effect on building energy consumption. To better quantify the reality of M2M communications the following two examples will detail the communications of very diverse systems.

Example #1—School HVAC and Lighting Controls

The interaction between a school's HVAC and lighting control systems is the first example. Typically there has been no real interaction between the HVAC systems which maintain space comfort levels and lighting controls which have the main function of turning lights on and off. Control systems designed for the HVAC market have now been around for quite some time and have gone through various changes throughout the years. However the concept of lighting controls is one that is much newer. As the energy costs associated with leaving lights on in spaces that are not in use becomes a reality the control of a buildings lighting systems becomes a necessity. These two systems while so very different both ultimately have the same goal, occupant comfort level. Whether that comfort level is the physical comfort level of a space or is centered on a visual comfort level in the same space, the outcome is still the same. If the space is two warm or cold there is discomfort, likewise if the lights don't turn on work in the space stops.

Now that we've examined to diverse systems we will look at how M2M communications can take these two separate systems and make them work together as one system. Until the advent of new technology as discussed in the first section the use of M2M was a great concept but not a reality. Picture a school using controllers that control these two different systems, but have the inherent ability to communicate with each other as one complete system. This communication can either be over the internet or by using the same communications protocol that the BAS system has. Now onto the practical part of this application, the answer to the main question of M2M communications, what is the benefit of sharing data between these two very separate systems?

Most well designed lighting control applications will utilize devices such as motion sensors and photo cells to enhance lighting control scenarios. The data gathered from these devices when shared with the HVAC controls will enable better energy management and comfort controls. If the motion sensor in the given space is unoccupied the lighting controls shut off the lights for the space, then this information is fed to the HVAC systems feeding that space. Space temperature control is dynamically adjusted to control to unoccupied space set points. This very basic M2M data sharing will ensure that energy savings are realized without affecting day to day operations. Reference Figure 19-2 Advanced HVAC Controls and Figure 19-3 Advanced Lighting Controls for a visual representation of these two systems.

Note: As a disclaimer JENEsys™ is a system developed around the ideals discussed throughout this chapter. JENEsys™ was developed by Lynxspring, Inc. and is powered by the Niagara AX Framework.

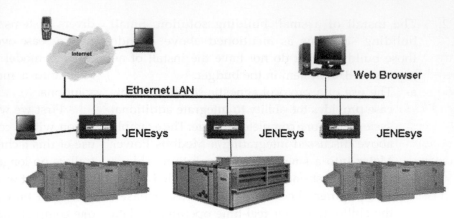

Figure 19-2. Advanced HVAC Controls

Example #2—Retail Store Complete System Interaction

This second example takes the concept of M2M communications and makes it a reality. The first part of this example will discuss a typical installation in a ordinary store as is today with simple disparate systems that control independent of each other. While, the second part of this example, will detail this same exact store with web enabled systems utilizing M2M communications. The dramatic implications of M2M in buildings will become

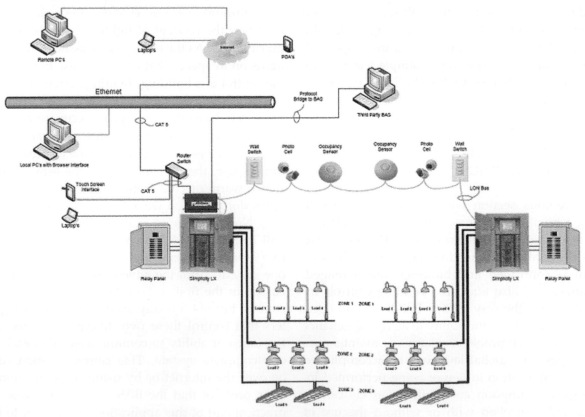

Figure 19-3. Advanced Lighting Controls

even more evident at the end of this discussion.

The typical retail installation includes wall mounted thermostats for control of roof top units and wall mounted light switches for lighting control. These two systems do not have the ability to share data on a M2M level nor can they communicate data to a NOC (network operating center) or to a central data allocation center. The actual performance of the building is maintained locally and is dependent upon store managers to be responsible. For instance space temperature set points cannot be regulated, thus a store manager can place the cooling set point at 68 degrees and waste a tremendous amount of energy. In addition if a store has more than one roof top unit these units can fight each other with one trying to cool the space while the other is heating. In terms of Lighting even greater energy can be wasted. The store is now dependent upon the store manager to turn the lights in the store on and off. More than often the lights will not be controlled properly. Extreme amounts of energy are wasted on a daily basis, thus increasing the operating costs associated with running the store.

Now let's take this same retail store and add in intelligent control systems for HVAC, lighting, and power monitoring, reference Figure 19-4 Retail Store Architecture for a sample system architecture. Instead of using a simple wall mounted thermostat the roof top units will be controlled by internet enabled controllers and use a virtual thermostat. Lighting will in addition be controlled by an internet enabled controller and use networked wall switches, motion sensors, and photo cells. Finally the actual power consumed by the retail store will be monitored, and energy savings schemes are used. The space set points are now set globally from a NOC and are determined based upon company policy. Since data are shared between multiple roof top units they will no longer fight each other to maintain optimum space conditions. Lighting will now be controlled based upon global occupancy schedules and daylight harvesting control schemes are implemented based upon light level readings from the photo cell. Power consumption schemes are devised to perform simple load shedding by dynamically modifying space set points to still main-

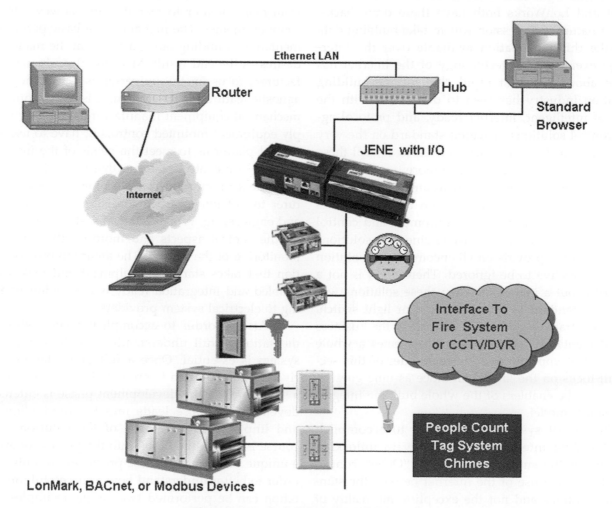

Figure 19-4. Retail Store Architecture

tain comfortable space conditions yet conserve the most amount of energy. Load shedding lighting schemes are devised to shut off predefined lights while not compromising the stores ability to operate. All of the data is now stored in an easily searchable SQL database at central location and global changes are changed from this same central location.

The Transformation of the Building Automation Industry

The past has seen the building automation contractor with the primary role of providing temperature control systems. This role has served companies well over the years and turned several of them into industry giants. However the standard methodology included the use of proprietary protocols and the use of the all too infamous "black box." While being extremely profitable the end user was ultimately affected by hirer and hirer install costs and a lack of selection. This lack of flexibility and ever rising install costs brought about the "Open" protocol wars of the late 90's and early 2000's. While intentions were good the two winners BACnet and LonWorks both have there draw backs. The next natural progression was to take building data and make this information available over the internet. The ever evolving technology of the internet has brought about the most rapid changes the building automation industry has seen to date. Now with the advent of intelligent, internet ready, and protocol agnostic control solutions provided standard on the very equipment that these contractors used to control these contractors must now evolve once more.

The progression of equipment and system manufacturers to utilize smart controls on their systems is only natural. The protection of equipment and control strategies via the use of in-house engineering solutions and the ability to provide an all encompassing solution is all too attractive to be ignored. Therefore it is not a matter of if but a matter of when these solutions will become as standard as the thermostat or light switch. To some contractors this poses a threat to the business as usual mentality, but to others this creates a whole new world of opportunity. The remainder of this section will focus on the latter half since as time goes on they will be the enablers of the whole building integration business model.

The use of systems integration to accomplish whole building integration is the key to unlocking the future of the smart building. As "Open" control protocols and the use of the internet become the standard in buildings and not the exception the reality of creating one common platform from multiple systems

and various manufacturers becomes a true possibility. The protocol agnostic nature of equipment using smart controls makes it possible for the systems integrator to pick their protocol of choice. This will facilitate an ease of integration and ultimately will increase a facilities operational efficiency and energy management.

The future systems integrator will now focus on building integration and leave the control of equipment to the manufacturers. This model will ensure proper equipment operation due to the fact that the manufacturer who designed the system is now in charge of the control schemes associated with the equipment. The integrator has morphed from a simple temperature controls contractor into a manager of all building systems. The integrator will manage alarm data and facilitate the M2M communications between the separate building systems.

IMPLEMENTATION IN THE OEM MARKET

The actual implementation of these solutions varies from manufacturer to manufacturer however there are common themes. The first and most basic part of implementation is finding out exactly what the manufacturer desires as the end result. Mechanical equipment manufacturers focus on the universal aspect of the protocol agnostic features of the system. It is not uncommon for mechanical equipment manufacturers to either not supply equipment mounted controls or have to use several control platforms to meet the needs of the install base. The use of a control platform that is consistent and not dependent upon a given protocol enables the manufacturer to put into practice consistency in programming and engineering. While electrical systems will focus less on the control aspects and more on the system level monitoring of the system. The ability to provide a solution that takes standard equipment and makes it web enabled and integration ready is tremendously appealing to electrical system providers.

Also in order to accomplish a successful implementation, a full understanding of the equipment or system is essential. Once a full understanding is derived then a solution is presented. Upon acceptance of a solution then the development phase is entered. The development phase leads into the production phase and implementation phase of the solution. This of course is a simplified look into the process of applying a unique solution, but these processes are universal. In order to better understand how a successful implementation can be performed two simple examples will be explored.

Example #1—Mechanical Roof Top Unit

As discussed in previous sections the ability for a roof top unit to perform self diagnostics, hold on board trend data, email alerts, and easily incorporate into an integrated building brings your standard unit into the future. When looking at a solution of this nature key components for the manufacturer are the ability to standardize on solutions that are not dependent on control protocols and protect the internal safeties of the unit. Another key to success is the ability to automate the whole process. Automation makes it possible for units to be programmed by simply reading in a model number. Now the requirement for a room full of programmers is eliminated and replaced with computers capable of generating unique applications.

The standard roof top unit transforms from a simple piece of equipment to an integral, self sustaining piece of equipment that is integrator friendly and part of whole building solution. Another consideration is the future proof aspect of the solution. Mechanical equipment has a lifetime of 25 to 30 years while control solutions have a lifetime of 3 to 5 years. Advances in technology happen at such a rapid pace that what is today's solution or protocol of choice is yesterday's archaic technology. That is what makes the use of java and XML so appealing to equipment manufacturers. With a flexible development platform that is not tied to standard control system programs the ability to adapt and change with technology and not against it. This unique ability will increase the lifetime of the investment in the control strategy implemented.

Example #2—Electrical Lighting Control System

The distinctive difference between a lighting control system and the example of the roof top unit is that the lighting control system is a self contained control strategy that has some of the same needs as the roof top unit. The main focus of a lighting control solution is the ability to configure the complete system via the internet and serve up monitoring and control information to the integrated building. This information will then enable the facility to enhance control strategies and increase overall building energy efficiency. The ability to enable a installing contractor to configure lighting control schemes and set up integration points with the simple click of the mouse on a web page makes the use of a controller of this magnitude not only feasible but required.

The above two examples show albeit a simple window into the future of two diverse systems, but also a very exciting glimpse of the future. Buildings of the near future will be complex mesh of networked systems with the ability to monitor and control all aspects of a building from anywhere in the world. The enhanced diagnostics and ease of integration will lead the way into the future of buildings.

CONCLUSION

The use of internet ready, protocol agnostic control solutions will revolutionize the way equipment and buildings operate. The seamless integration of very diverse systems will make M2M communications and fully functional HMIs (human machine interface) a reality and not a dream. The integrated building will ensure facilities perform at there peak operating efficiency and maximize energy usage. The face of the building automation industry will no doubt change as the focus turns from temperature control to systems integration. The future is here and will be realized as more equipment comes from the factory integration ready and web enabled.

Chapter 20

Web Based Wireless Controls for Commercial Building Energy Management

Clifford Federspiel, Ph.D., PE
Federspiel Controls, LLC

ACKNOWLEDGEMENTS

THE PROJECT DESCRIBED in this chapter was funded by direct and in-kind support from the California Energy Commission's Public Interest Energy Research (PIER) Program and the Iowa Energy Center Energy Resource Station. Martha Brook and John House provided project oversight. John House and Curt Klaassen provided input on the design of the experiments. Joe Zhou provided valuable technical support throughout the project. Kurt Federspiel was the software development engineer for this project.

INTRODUCTION

This case study describes a web-based, wireless, supervisory control system for commercial-building heating, ventilating, and air-conditioning (HVAC) systems that has been designed to convert constant air volume (CAV) HVAC systems to variable air volume (VAV) operation in a way that is non-intrusive (no terminal retrofits or static pressure controls), inexpensive, and easily maintainable. The system design avoids the need for asbestos abatement, which is commonly required for buildings that have CAV HVAC systems.

There are three common types of CAV systems that serve multiple zones: single-duct systems with terminal reheat, dual-duct systems, and multi-zone systems. Single-duct systems deliver cooled air to each zone then re-heat it as needed to keep the zone space temperature close to a setpoint. Dual-duct systems supply heated and cooled air to each zone, then mix the two to maintain the space temperature. Multizone systems are a special case of a dual-duct system where the mixing dampers are part of the air-handling unit rather than located at each zone.

CAV systems of the types described above are inefficient. In states with strict energy codes, such as California, they are prohibited in new construction. They are also prohibited by ASHRAE Standard 90.1. For HVAC systems that serve multiple zones, it is now common to use variable-air-volume (VAV) systems.

VAV systems have variable-speed fans and terminal dampers that are controlled so that the amount of simultaneous heating and cooling or re-heating is significantly reduced. There are two common kinds of VAV systems: single-duct and dual-duct. Single-duct VAV systems supply cooled air to each zone terminal unit, where it is metered with a control damper when cooling is required or re-heated when heating is required. When heating, the amount of cooled air is reduced to a low level by the terminal controls, so there is much less wasted re-heat energy than a single-duct CAV system. Dual-duct systems deliver heated air and cooled air all the way to each zone terminal unit with separate air ducts. Dual-duct VAV terminal units have independent dampers that modulate the hot airflow rate to heat a zone and modulate the cold airflow rate to cool a zone. Unlike the dual-duct CAV system, the dual-duct VAV system does very little mixing. Most of the time it supplies a variable amount of hot air when heating and a variable amount of cooled air when cooling. It only mixes air when the zone load is small so that adequate ventilation air is provided.

Although CAV systems are less common in new construction, there is still a large installed base. According to the Energy Information Agency (Boedecker, 2005), CAV systems that serve multiple zones condition 3.5 billion square feet of commercial building floor space in the U.S. Since they are inefficient, retrofit strategies have been developed to modify their design and operation in order to make them more efficient. These strategies require mechanical modifications to the HVAC system.

Mechanical modifications are disruptive to the commercial operations in the building and they will always require asbestos abatement if asbestos is present.

Existing Solutions

For single-duct CAV systems, Liu et al. (2002) recommend adding a VFD to the fan to reduce the fan speed during after-hours operation. During occupied hours the fan is operated at full speed. This strategy does not save energy for systems that are shut off after hours. Even when there is after-hours operation, this method is not cost effective unless the system is large because the energy savings are limited.

For dual-duct CAV systems, Liu and Claridge (1999) describe a means for improving energy performance without retrofitting terminal units. They add a damper to the hot duct and use it to control the pressure in the hot duct. This strategy still requires a mechanical modification, which is intrusive and requires that the system be shut down. It also requires the installation of pressure sensors in the hot air duct and cold air duct.

For multi-zone CAV systems, Liu et al. (2002) describe a means for improving the energy performance by adding a VFD to the supply fan and controlling the supply fan speed so that the most-open mixing damper is 95% open to the hot deck in the heating season. In the cooling season their strategy controls the fan speed so that the most-open mixing damper is 95% open to the cold deck. They do not describe how the strategy works in swing seasons when the unit could be heating some zones while cooling others. The command to the VFD comes from a Proportional-Integral-Derivative (PID) controller that takes the most-open damper position as input. This strategy requires that position sensors be added to the mixing dampers. Position sensors are expensive and difficult to install. Resistive position sensors are prone to vibration-induced pre-mature failure. This strategy cannot be applied to single-duct CAV systems because they do not have mixing dampers.

Johnson (1984) describes a case where a single-duct re-heat system with cooling-only operation was modified for VAV operation by eliminating the zoning, regulating the discharge air temperature, and modulating the average zone temperature by adjusting the fan speed. This strategy yielded large annual energy savings (46.5% reduction in HVAC energy for one unit and 53.9% for another), but could not have worked in a system with heating and most likely had a negative impact on thermal comfort since the zoning was eliminated (i.e., the system was operated as a large single-zone system after the retrofit).

Existing products that are used to convert CAV systems include VAV retrofit terminals and VAV diffusers. VAV retrofit terminals are VAV boxes installed in existing supply ducts where CAV terminals (re-heat units or mixing boxes) are located in a CAV system. VAV retrofit terminals are sold by many manufacturers. Examples of VAV retrofit terminal units are shown in Figure 20-1. When CAV systems are retrofit with VAV retrofit terminal units, the system must be shut down and workers near the terminal location must move because the retrofit requires significant mechanical and electrical modifications to the existing HVAC system.

VAV diffusers are variable-area diffusers that have an actuation mechanism combined with temperature feedback so that the open area of the diffuser is modulated to maintain the local temperature. Modern VAV diffusers can accommodate switchover from heating to cooling mode, and some come with embedded DDC controls. VAV diffusers are sold by Acutherm, Price, Titus, and others. Figure 20-2 shows examples of VAV diffusers. Retrofitting a CAV system to VAV with VAV diffusers is expensive because every diffuser must be retrofit, and most zones have air supplied by several diffusers.

APPROACH

The project involved building and testing a new solution for converting CAV systems to VAV operation.

Figure 20-1. VAV retrofit terminal units.

We designed a web-enabled, wireless control system that included a new control application called Discharge Air Regulation Technique (DART) to achieve the following goals:

1. Short installation time
2. Minimal disruption of occupants
3. Flexible configuration
4. No need for asbestos abatement, should asbestos be present
5. Standalone operation if necessary
6. Long battery life
7. Remote monitoring and alarming
8. Browser-based human-machine interface (HMI)
9. Modular design so that it can be used to deploy other applications

Goals 1-5, when combined with the large energy savings from CAV to VAV retrofits, result in a short payback period. Goals 6-8 yield a system that is easy to maintain. Goals 1-5 are facilitated by the use of low-power wireless sensing and control modules that utilize self-healing, mesh networking. The particular wireless technology that we selected uses a time-synchronized mesh network that enables extended battery life (several years on AA lithium batteries) and mesh networking capability for every node in the network. Both the gateway for the wireless network and the supervisory control system are web-enabled devices. Both of these devices communicate using XML-RPC, both run a web server, and both are configurable via a browser. Additionally, the supervisory controller can send alarms via the internet to an email address or a pager, and it can display time series data. The design of both the hardware and software is modular. We could use a different wireless network, yet still use the FSC and the software running on it. This modular design allows us to use this same platform to deliver other supervisory energy management applications in addition to DART.

DART Application

DART works by reducing the supply fan and return fan speeds at part-load conditions. When there is no load, the discharge air temperature is equal to the zone temperature, so reducing the fan speeds has no impact on the room temperature. At part-load conditions, the discharge air temperatures are somewhat higher or lower than the zone temperature, depending on whether the zone is being heated or cooled. Reducing the fan speeds at part-load conditions causes the discharge air temperatures to increase or decrease, depending on whether the zone is being heated or cooled, so that the heat transfer rate to the zone doesn't change. DART maintains the highest discharge air temperature close to a high-temperature setpoint or the lowest discharge air temperature close to a low-temperature setpoint, which has the effect of keeping the fan speed low when the load is low but causing it to increase as the load increases. Lowering the fan speed not only reduces fan energy consumption, but it also reduces the amount of mechanical cooling, and reduces the amount of (re-) heating.

Control System Architecture

The control system architecture consists of the components shown in Figure 20-3. A supervisory control computer, called an FSC, is connected via a LAN/WAN to at least one wireless network gateway, which is called a manager. The FSC and manager do not need to be co-located. The FSC could be located in a data center in one city while the manager is located in a building in another city. Additionally, systems can be configured with a remote FSC and multiple managers located in different parts of a single building or located in different buildings. The FSC is an embedded, fanless computer with a x86 architecture and a 600 MHz clock. The FSC runs a web server for its HMI. The HMI supports configuration tables, dynamic data tables, and time series graphics. The manager is a web-enabled, embedded, fanless computer with ethernet and serial ports. The manager soft-

Figure 20-2. VAV diffusers.

ware includes a web server that can be used to remotely configure the wireless network. Having the manager and the FSC both web-enabled makes the system highly flexible and easy to manage and service remotely. Sensor inputs and control command outputs are provided by wireless modules, each of which is equipped with analog and digital I/O. To enable the modules to measure temperature with an external probe, a thermistor circuit was added to the HD-15 connector on the wireless modules. To allow the modules to produce 0-10 VDC control commands, we designed a digital-to-analog converter (DAC) circuit that connected to the HD-15 connector of the wireless modules.

The wireless modules and the wireless network manager automatically form a mesh network such as the one depicted by Figure 20-4. The wireless networking hardware we selected operates in the 902-928 MHz ISM band. The modules can use any one of 50 channels (frequencies) in this band. Individual modules in a single network may use different channels. The modules change their channel and their routing parents dynamically in search of a clear channel to avoid interference. These features help improve the reliability of the

wireless communications.

For the DART application, we used wireless temperature sensors to measure zone temperatures, discharge air temperatures (Figure 20-5), supply air temperature, and heating hot water temperature. The wireless modules have an internal temperature sensor that was used for zone temperature measurement instead of the external probe shown in Figure 20-5. The zone temperature sensors were used primarily for monitoring purposes. The supply air temperature sensor and the heating hot water temperature sensor were used to allow the low-temperature and high-temperature setpoints of the DART application to follow the resets on the supply air temperature and heating hot water temperature. Using these two sensors this way eliminated the need to program these reset schedules into DART. By installing the discharge air temperatures in the ceiling plane as shown in Figure 20-5, we could avoid asbestos that might be used to insulate supply ducts or structural components of an older building above the ceiling plane.

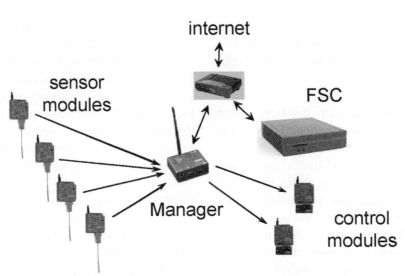

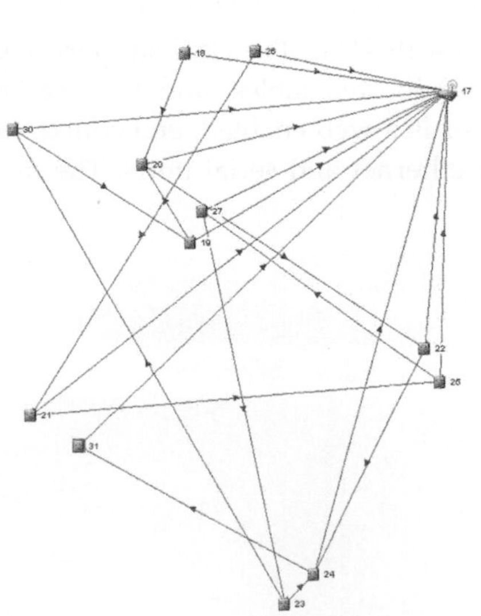

Figure 20-3. Control system components.

The wireless sensors operate on two AA lithium batteries. The wireless network uses a time-synchronized mesh protocol that allows the sensor to be in a deep sleep mode that uses very little power most of the time. Battery life is dependent on the network configuration and how much routing a module performs. Battery life of up to eight years is achievable for modules that do not re-route.

Figure 20-6 shows the wireless control modules. The control modules are line powered. For this application, they were powered from the 24 VDC power supply provided by the variable frequency drive (VFD). They returned a 0-10 VDC speed control signal back to the analog input of the VFD.

The FSC and the manager communicate using XML-RPC. XML (eXtensible markup language) is a standard for creating markup languages that describe the structure of data. It is not a fixed set of elements like HTML, but is a meta-language, or a language for describing languages. XML enables authors to define

Figure 20-4. Mesh network at the ERS

Figure 20-5. Discharge air temperature sensor.

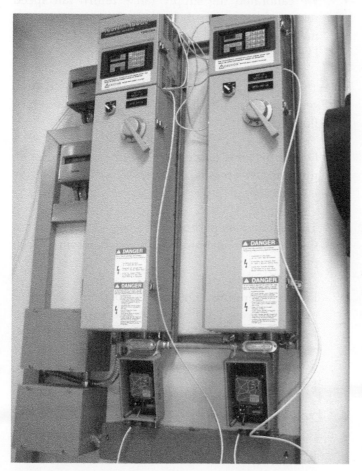

Figure 20-6. Control modules installed beneath the VFDs.

their own tags. XML is a formal specification of the World Wide Web Consortium. XML-RPC is a protocol that allows software running on disparate operating systems, in different environments to make procedure calls over the internet. It is remote procedure call using HTTP as the transport and XML as the encoding. XML-RPC is designed to be as simple as possible, while allowing complex data structures to be transmitted, processed and returned. The FSC uses XML-RPC to configure wireless network settings, to retrieve certain kinds of network data such as battery life and module status (connected or unreachable), and to set the output voltages of the control modules. Sensor data (e.g., voltages measured by the sensor modules that correspond to temperature) are pushed by the manager to the FSC asynchronously. The FSC parses the incoming XML document and places the sensor data in a database.

The FSC has five basic software components. They include the XML-RPC driver for the manager, application control software, a database, a web server, and HMI software (web interface). The XML-RPC driver and the application control software are written in C++, which results in fast, compact code that is beneficial for running on an embedded platform. The driver and application control software are object-oriented, which simplifies software maintenance and upgrades. The application software includes web-based alarms in addition to the DART application software. The system is programmed to send a message to an email address or pager when certain events such as loss of communication between the FSC and the manager or an unreachable control module occur. This feature proved useful during the first week of testing, when a network glitch caused the system to stop. We received the alarm in California, contacted the ERS staff and had them reset the system, so that the controls were quickly operational again. We selected MySQL for the database, Apache for the web server, and PHP for the HMI software because they are all open-source software components, but are also widely used and highly reliable. For this case study, the FSC used the Windows XP operating system, but all of the five software components could run under the Linux operating system.

Energy Resource Station

The wireless control system was installed and tested at the Iowa Energy Center's Energy Resource Station (ERS). The ERS has been designed for side-by-side tests of competing HVAC control technologies. The building has two nominally identical HVAC systems that each serve four zones in the building (three perimeter zones and one internal zone). The building is oriented with the long axis north-south so that the zones of each system nominally have the same solar exposure. Figure 20-7 shows a floor plan of the ERS. The supply and return ducts for the test rooms are color-coded. The numbered circles show the locations of the wireless control devices during the first week of testing. Each number is that device's network ID. Node 17 is the wireless network gateway. It and the supervisory control computer were located in a telecom closet. Both were connected to the internet via a switch, and each used a static IP address provided by the ERS.

The ERS has submeters on all energy consuming HVAC loads including all pumps motors, fan motors, chillers, and the heating hot water boiler. The ERS has approximately 800 monitoring points that are trended every minute for temperatures, humidities, flows, pressures, lighting conditions, and weather conditions. The sensors are calibrated regularly. The ubiquitous and accurate sensing at the ERS enhance the ability to accurately measure the energy performance difference between competing control strategies and technologies.

Additional details about the ERS can be found at http://www.energy.iastate.edu/ers/.

Test Conditions

During the first week of testing, the following configuration was used:

- The A system was controlled with DART, while the B system was run as a CAV reheat system.

- The design flow to the perimeter rooms was 650 CFM (2.4 CFM/sf), while the design flow to the interior room was 300 CFM (1.1 CFM/sf).

- The VAV box dampers were fixed to deliver the design flow to each room at 100% (60 Hz) fan speed.

- We calibrated the supply fan – return fan speed relationship so that the supply flow was approximately equal to the return flow and configured DART to use this relationship.

- The return fan speed of the CAV system (B) was set so that the supply flow was equal to the return flow.

- Discharge air was delivered with one diffuser in the perimeter rooms, but with two diffusers in the interior room.

- The blinds in all perimeter rooms were lowered but the slats were maintained in the horizontal position.

- Low-temperature and high-temperature setpoints of DART were reset based on the supply air temperature and the heating hot water temperature, respectively. These setpoints determine the maximum absolute difference between the zone temperatures and the discharge air temperatures.

- Zone temperature controls of the Metasys system used proportional plus integral (PI) control to modulate reheat coil valves.

- The supply air temperature setpoint was manually reset each morning based on the high-temperature

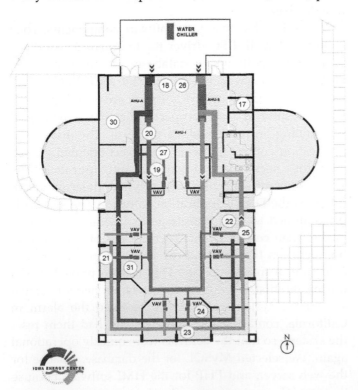

Figure 20-7. Floor plan of the ERS showing mote locations during the first week of testing.

outdoor air temperature forecast for that day. The schedule was 55 to 60 degF as the maximum outdoor air temperature forecast ranged from 50 to 20 degF, respectively.

- The systems were operated 24/7.

- The test ran from February 11, 2006 through February 19, 2006

- The minimum supply fan speed was 40%.

Figure 20-8 shows the false load per room at the beginning of the test. The loads were reduced by 100 Watts (base load turned off for the remainder of the test) at 6 p.m. on February 13 because the peak loads significantly exceeded the capacity of the system. Beginning February 14, the supply air temperature reset schedule was changed to the 52-57 degF range so that the cooling capacity would match the peak loads better. The peak load density (3.7 W/sf at the beginning of the test and 3.4 W/sf after these changes were made) is significantly higher than average. Wilkins and McGaffin (1994) reported load densities from office equipment ranging from 0.48 W/sf to 1.08 W/sf, with an average value of 0.81 W/sf. With the same diversity factor for occupants that was observed for the office equipment, the occupant load density would be 0.53 W/sf, which would give an average load density of 1.34 W/sf. The peak load density in this test was, therefore, 2.5 times higher than average.

Figure 20-9 shows the lighting schedule for each room each day. The peak lighting load was 2.2 W/sf, which is higher than the 1.5 W/sf limit set by modern codes and standards.

During the second week of testing, the following changes were made to the test configuration:

- The B system was controlled with DART, while the A system was run as a CAV reheat system.

- We re-calibrated the supply fan – return fan speed relationship so that the supply flow of AHU-B was approximately equal to the return flow and configured DART to use this relationship. The relationship for this week was not the same as the first week (A and B unit fan characteristics were not the same).

- Low-temperature and high-temperature setpoints of DART were fixed at 60 degF and 90 degF, respectively.

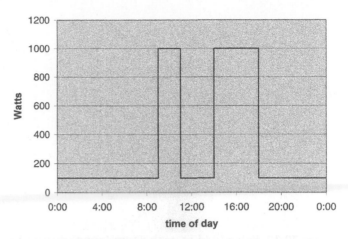

Figure 20-8. False load in each room (zone) during the first week of testing.

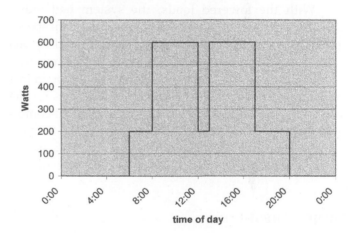

Figure 20-9. Lighting load in each room during the tests.

- Zone temperature controls of the Metasys system used proportional-only control with a 3 degF proportional band to emulate the operation of pneumatic controls.

- The supply air temperature setpoint was fixed at 55 degF at all times.

- The test ran from March 4, 2006 through March 12, 2006

Figure 20-10 shows the false load per room during the second week of testing. The loads were lowered to 1.12 W/sf so that they would more closely represent actual loads in buildings. The base load was provided by a desktop computer because it is increasingly common to keep computers on 24/7 for after-hours maintenance.

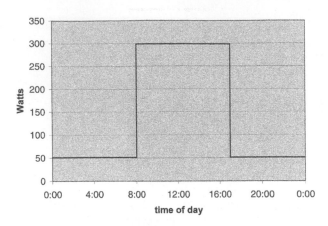

Figure 20-10. False load in each room (zone) during the second week of testing.

With the lowered loads, the system had many daytime operating hours at the minimum fan speed. To exercise the system, the loads were increased by 900 Watts per room on two days (March 8 and 9) from 9:30 a.m. to 11 a.m. and again from 1:30 p.m. to 3 p.m. in just the internal rooms. We picked the internal rooms for the increased loads to emulate the use of an internal conference room.

The lighting loads during the second week of testing were the same as during the first week of testing.

PROJECT RESULTS

Energy Performance Comparison

Table 20-1 summarizes the average energy performance from 6 a.m. to 6 p.m. each day for the first week.

Table 20-1. Energy consumption and savings during the first week of testing (from 6 a.m. to 6 p.m.).

	DART	CAV	% saved
Supply fan kWh/day	15.9	23.5	32.1
Return fan kWh/day	4.31	4.92	12.3
Therms/day	3.16	3.86	18.2

Based on the 12-hour schedule (6 a.m. to 6 p.m.) and the average weekly operating hours reported in the Energy Information Agency's commercial building energy consumption survey (CBECS), which is 61 hours/week, we estimate that DART would save 2.0 kWh/sf/yr and 0.17 therms/sf/yr under this load condition.

Table 20-2 summarizes the average energy performance from 6 a.m. to 6 p.m. for the second week.

Table 20-2. Energy consumption and savings during the second week of testing (from 6 a.m. to 6 p.m.).

	DART	CAV	% saved
Supply fan kWh/day	6.82	23.5	71.0
Return fan kWh/day	2.15	6.24	65.5
Therms/day	2.04	4.07	49.8
Chiller kWh/day	1.44	2.3	37.1

Based on the 12-hour schedule (6 a.m. to 6 p.m.) and the average weekly operating hours reported in the CBECS (61 hours/week), we estimate that DART would save 5.2 kWh/sf/yr and 0.49 therms/sf/yr under this load condition.

If we estimate typical energy savings by averaging the savings from the two weeks, then we get an estimated annual energy savings of 3.7 kWh/sf/yr and 0.34 therms/sf/yr. Using average energy costs reported by EIA for California ($0.1198/kWh and $1.08/therm for 2005), this equates to an energy cost savings of $0.81/sf/yr in California. For Iowa, where the utility costs are $0.0697/kWh and $1.066/therm, the annual energy cost savings should be $0.62/sf/yr. For the U.S. as a whole, where average utility rates are $0.0867/kWh and $1.157/therm, the annual energy cost savings should be $0.71/sf/yr. At these saving rates, the payback period of the entire system, including the VFDs, should be less than two years.

The savings figures are higher than anticipated based on published savings from conventional CAV to VAV retrofits. The high electrical energy savings are probably due to the fact that conventional retrofits don't normally use static pressure reset for the supply fan control, so the supply fan energy isn't reduced as much as possible with a conventional CAV to VAV retrofit. DART operates the supply fan as low as possible while ensuring that the zones are still in control, which yields fan energy savings that should be comparable to VAV operation with static pressure reset. The thermal energy savings may be higher than anticipated because the base case during the second week didn't use supply air temperature reset, and the reset during the first week was modest. Supply air temperature reset is a way for CAV systems with DDC controls on the air-handling unit (AHU) to reduce reheat and mechanical cooling. Since supply air temperature reset doesn't affect fan energy consumption, and since the mechanical cooling savings were small in this demonstration (due to the cool outdoor air temperatures allowing the system to cool entirely with an economizer most of the time), not accounting for supply air temperature reset cannot inflate the potential electrical energy savings. We expect that thermal energy savings for systems that already use a large supply air temperature reset will be about half of the levels observed in these tests.

Temperature Control Performance Comparison

During the first week of testing, the zone temperature control performance of the DART system was more oscillatory than that of the CAV system. The oscillatory behavior was caused by at least the following two factors: 1) the discharge air temperatures were often not stable, 2) the economizer was often not stable. These instabilities were present in both the DART system and the CAV system, but they remain localized in the CAV system, whereas they can become distributed by the DART system because the variable fan speed affects all parts of the system. Additionally, the lower discharge velocities and higher discharge air temperatures under heating conditions may have resulted in a stratification layer that could also result in oscillatory temperature readings at the zone thermostats.

After the first week of testing we made changes to the DART software to make it less sensitive to discharge air and supply air instabilities, switched from Proportional-Integral (PI) to Proportional-Only control for the zone temperatures, and used fixed setpoints for DART. We anticipated that these changes would result in less zone temperature variability. The Proportional-Only zone temperature control emulates the behavior of pneumatic controls, which are still commonly used with legacy CAV systems.

Figure 20-11 shows the maximum absolute deviation of the zone temperatures from the average zone temperature by zone from 8 a.m. to 6 p.m. during the second week. In three of the four zones, the maximum excursion occurred in the system controlled by DART, but the largest excursion occurred in the West zone of the CAV system.

Figure 20-12 shows the average absolute deviation of the zone temperatures from the average zone temperature for the same period on the same scale as Figure 20-11. The average zone temperature variability was highest with the DART system, but it was less than 0.5 degF, so we do not anticipate that DART will result in thermal discomfort due to temperature variability.

We also installed a vertical temperature sensing tree in the East zones to measure stratification. The trees were located half-way between the supply diffuser and the return grill in each room. Figure 20-13 shows the vertical temperature profile in East-A (CAV) and East-B (DART) at 6 a.m. on the coldest morning of the test (27.8 degF outdoor air temperature). At this point in time, the

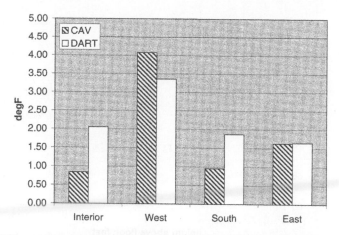

Figure 20-11. Maximum zone temperature excursions during the second week.

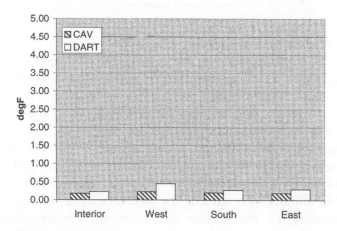

Figure 20-12. Average zone temperature variability during the second week.

discharge air temperature of the East-A (CAV) zone was 71.4 degF, while the discharge air temperature for the East-B (DART) zone was 88 degF with a fan speed of 40%. The CAV system has better mixing, and a nearly uniform vertical temperature profile. The DART system has a clearly increasing vertical temperature profile. The profile is less steep in the occupied zone, and the vertical temperature difference from the ankles to the head (0.5 feet to 5.5 feet) is 5.1 degF, which is less than the 5.4 degF requirement of ASHRAE Standard 55-2004. We conclude that stratification should not be a problem either for thermal comfort or temperature control stability, at least not under load conditions similar to these. In practice there will probably be less temperature stratification because movement of people, mixing between rooms at open doorways, cross-flow in open plan offices (none of which was allowed to occur in these tests) and additional office equipment will mix the air.

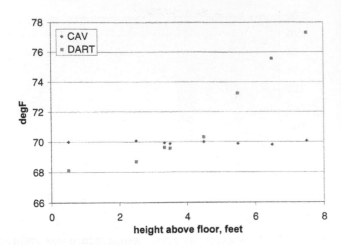

Figure 20-13. Vertical temperature profile at 6 a.m. on the coldest morning of the second week.

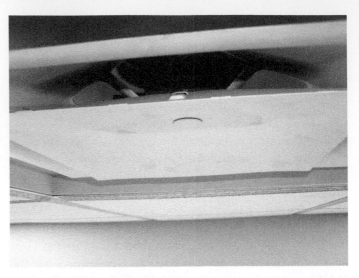

Figure 20-14. Installation of the interior zone discharge air temperature sensor.

Wireless Sensor Network Evaluation

We installed the wireless sensors and controls in locations and orientations that were convenient, and not necessarily ideal for radio transmission. Figure 20-5 shows the installation of one of the discharge air temperature modules. Three of the four zones had this configuration, but in the interior zone we placed the sensor inside the diffuser, as shown in Figure 20-14, because the lower discharge velocities and cross-flow at the diffuser were causing the discharge air temperature readings for the interior zone to be much higher than the other zones when all reheat valves were closed. The diffuser and the connecting ductwork are all constructed of sheet metal, so this module was essentially shrouded in metal. Even in this semi-shielded location, the radio performance was good. The discharge air temperature module installed inside the diffuser (Figure 20-14) provided just 1% fewer packets than the other discharge air temperature modules. We never had an instance where a module was unreachable or where the data were more than three minutes old, which was a criterion that we used for communication failure.

The control modules were installed close to the VFDs (Figure 20-6), which are a potential source of interference. We found that the control modules took longer than the other kinds of modules to join the network, presumably because they had to search longer than the other modules to find a clear channel. However, once they had joined the network, the communications between the control modules and the wireless network manager never failed.

The benefits of wireless communications and mesh networking became quite clear after the installation. The entire network, which consisted of 10 sensor modules, 2 control modules, one manager, and one FSC were installed and operational in less than two hours. By the time we had completed the installation of the sensor modules and control modules and returned to the manager and FSC, the mesh network had configured itself. We did not need to add repeaters to the network or relocate modules to improve the reliability of the wireless communications. Packet loss during the two weeks of testing was negligible.

CONCLUSIONS

We conclude the following from this project:

1.   The energy savings potential from DART is high, even higher than anticipated based on published savings from conventional CAV to VAV retrofits.

2.   The web-enabled manager and FSC, combined with wireless mesh networking make the system easy to install, commission, and maintain.

3.   DART increases the zone temperature variability and heating mode stratification, but the increase in the temperature variability is small and the heating mode stratification is within the bounds of ASHRAE Standard 55-2004.

4.   The reliability of the wireless sensor network technology used for this demonstration was good.

5. The wireless control platform designed for this project could easily be used to provide other supervisory energy management applications.

References

Boedecker, E., 2005, Statistician, Energy Information Agency, U.S. Department of Energy, personal communication.

Johnson, G.A., 1984, "Retrofit of a Constant Volume Air System for Variable Speed Fan Control," *ASHRAE Transactions*, 90(2B), 201-212.

Liu, M. and D.E. Claridge, 1999, "Converting Dual-Duct Constant-Volume Systems to Variable-Volume Systems without Retrofitting the Terminal Boxes," *ASHRAE Transactions*, 105(1), 66-70.

Liu, M., D.E. Claridge, and W.D. Turner, 2002, "Chapter 4: CC Measures for AHU Systems," *Continuous Commissioning Guidebook for Federal Energy Managers*, Federal Energy Management Program, U.S. Department of Energy.

Wilkins, C.K. and N. McGaffin, 1994, "Measuring computer equipment loads in office buildings," *ASHRAE Journal*, 36(8), 21-24.

GLOSSARY

AHU: air-handling unit
ASHRAE: American Society of Heating, Refrigerating, and Air-conditioning Engineers
CAV: constant air volume
CBECS: commercial building energy consumption survey
CFM: cubic feet per minute
DAC: digital to analog converter
DART: discharge air regulation technique
FSC: Federspiel supervisory controller
HMI: human-machine interface
HTML: hyper-text markup language
HVAC: heating, ventilating, and air-conditioning
ISM: industrial, scientific, medical
LAN: local area network
PHP: Hypertext preprocessor (originally called Personal Home Page)
PI: proportional-integral
PID: proportional-integral-derivative
RPC: remote procedure call
SQL: structured query language
VAV: variable air volume
VDC: volts direct current
VFD: variable frequency drive
WAN: wide area network
XML: extensible markup language

Chapter 21

At the Base of the Enterprise Pyramid: Training Building Operators for Digital Monitoring and Control*

Michael Bobker
Building Performance Lab, CUNY Institute for Urban Systems

ABSTRACT

INTRODUCTION OF ENTERPRISE energy information systems strongly suggests the need on the part of facility operators for greater sophistication with digital monitoring and control concepts, equipment, and procedures. A logical model is shown connecting building performance and persistence of savings to operator capabilities in terms of data acquisition, problem recognition, control functions, and corrective response. Training solutions are suggested as a necessary enhancement for organizational transformation in adapting to an EEIS environment, emphasizing hands-on, lab-based work and university collaboration. Need is suggested for new curriculum specifically bridging between physical equipment operations and digital representations of the same operations.

INTRODUCTION

As energy information systems (EIS) are adopted at the enterprise level (EEIS), they will need to grow roots down into the facility and plant level. To a certain extent for data acquisition but more strongly for performance response, digitalized data about system operations will become increasingly important. This projected trend faces a hurdle in having facility operating staff accept and adapt to an increasingly digitalized work environment.

Practical people—mechanics, maintenance workers, tradesmen—live in a world of sights, sounds, smells and sensations associated with the complex machinery of our built-environment, some of which comes from instrumentation but much of which directly reflects equipment operations. These sensory impressions can be quite well defined, highly indicative and finely discriminating. They are a classic case of what has been described as "experiential knowledge" that characterizes expert knowledge in a domain that requires actions (Norman 1993). Re-engineering of building operations that reduces mechanics' regular field exposure risks losing touch with equipment conditions.

Digitalization has penetrated practical work at varying paces in different industries and trades. Machinists routinely program numerically controlled (NC) tools in factories. As drivers we still listen to the symptomatic sounds of our cars, but mechanics must investigate engine problems via computer. Perhaps our buildings lag behind the automotive industry (Capehart 2005) but more and more components integrate electronic chips even if still often remaining as local, equipment-embedded controllers, not tied to a central Building Automation System (BAS). Fire and life-safety systems for large buildings have automated networking to outside emergency responders. Building operators are accustomed to these devices and the chief engineer's office of almost all large commercial buildings today sports a master computer station.

The challenge here is not about the presence of digital components but rather about information utilization. For our building operators, by and large, digitally captured data have not assumed the status of sensory data. Analog gauge readings are recorded on log sheets for periodic scanning and human memory is applied from day to day in processing and comparing observations. Operators know how to talk to each other about what they see, hear and smell in their buildings. The digital record can be at least equally rich and telling but

*An earlier version of this chapter was presented at the ACEEE 2006 Summer Study on Energy and Buildings, Asilomar, CA, under the title "Crossing the Digital Divide: Training Building Operators for Digital Monitoring and Control," co-authored with A. Patel, H. Styles and K. Lenihan.

it is not yet processed and utilized to the same degree by our buildings' human operators.

EEIS, FACILITY PERFORMANCE AND ORGANIZATIONAL TRANSFORMATION

The point of moving towards EEIS is to improve the performance of a portfolio of properties and processes, with performance defined narrowly in the sense of energy but also expanding to a much broader set of concerns. Environmental ideology, energy prices, and even security concerns are all aligned as "push factors." While perhaps not quite at a tipping point, green concepts have been making an impact and gradually transforming market demands upon new buildings. Designs for high performance place new demands on operating personnel. For example, with an increasing adoption of high-performance, green elements for LEED certification of new commercial, governmental, and institutional buildings, the issue facing the market is shifting to whether building operators are properly prepared to understand and operate buildings as intended.

Real estate industry surveys consistently show that thermal discomfort is the predominant complaint in office buildings. Operation to minimize tenant complaint has been the norm and a central criteria for performance. But increasingly our high performance systems operate invisibly, that is without the normal kinds of sensory feedback that indicate performance problems. Conventional monitoring that relies on sensory experience, analog observation, and feedback from the served environment (i.e., complaints)—is rapidly becoming insufficient. If something goes wrong with the power output from a photovoltaic array or almost equally quiet fuel cell, grid electricity will make up the difference *without any visible impact on building services*.

This is equally true for many of the energy system retrofits and upgrades favored by energy efficiency programs. The following list provides a sampling of advanced sub-systems or components whose failure or by-pass does not result in any diminution of building services:

- Variable Speed Drives
- Daylight Dimming
- Economizers
- Outside air controls
- Optimized start-up & shut-down
- Equipment capacity controls
- Heat Recovery
- On-site electrical generation
- BAS functions in manual by-pass

Realization that under-performance is chronic and can easily pass unobserved without specialized monitoring techniques and that, therefore, systems often can dramatically improve performance with just operational improvement, has led to development of existing building commissioning programs. In so far as these programs have adopted an engineering model of "identifying and correcting deficiencies" they fall short on a necessary aspect of organizational transformation: getting operating staff to incorporate monitoring and verification (M&V) approaches to building performance into their normal and standard practice. A logic model of building performance and program designs can help clarify the need to address on-going operational practices.

"SEEING" BUILDING PERFORMANCE, A LOGIC MODEL

System performance deteriorates for a variety of reasons. We know that sensors and controls drift out of calibration for physical reasons as well as for behavioral ones, as they are reset or over-ridden. Valve or damper actuators fail. Belts loosen. Leaks occur. Schedules change. We can think of these all as causes of "performance drift or excursion" from a target state to a sub-optimal state. See Figure 21-1-A. The sub-optimal state or "SOS" can be a "cry for help," if only there are ears educated to hear it.

Maintaining performance (in related contexts also called "persistence of benefits") can be seen to have a central behavioral dimension, as added in Figure 21-1-B. Variance from intended or best practice operation or a degradation of performance is recognized through a testing or monitoring activity and a corrective response is initiated.

Performance, then, depends on some form of regular maintenance awareness and intervention, human, automated or some combination of both. Without such recognition, the sub-optimal state can exist for the long-term and can eventually come to be seen as normal and correct. Performance depends on how quickly recognition and interventions occur to address the development of sub-optimal conditions. The speed and effectiveness of such recognition and response are subject to tools and training. This logical model emphasizes the fundamental importance of a continuing feedback loop between building and building operator for maintaining high performance.

Program Design for Performance Assessment

Retro-commissioning programs to date most com-

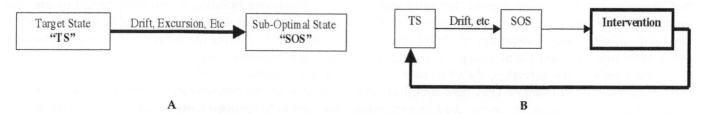

Figure 21-1. Basic Logic Model Schematics for Building Performance

monly emphasize the identification and correction of operating "deficiencies," based on engineering observation and varied forms of analysis, usually by outside consultants. These are quantified as improvement projects and treated much as capital improvement recommendations under the older, familiar model of energy audits. Engineering focus is mobilized, problems found, and solutions implemented. Engineering observation and findings are embodied in a report, generically described as the "Performance Assessment" in Figure 21-2.

The engineering review sits on top of normal maintenance practices, seeking to identify sub-optimal operating conditions and facilitate corrective responses. The model of such program design ignores the repetitive nature of this loop. *The SOS-Intervention linkage cannot for the long term depend on an outside Performance Assessment; this step must be internalized by building operators.*

The EEIS deploys a similar kind of external assessment, conducted at a distance by a corporate central management team. As a continuing corporate function, it is an improvement over one-time program interventions. But the understanding and response of local facility operators is not assured. An EEIS may include various kinds of diagnostic tools or routines but it is an open question as to how much will be passed down to the facility level and, if some is, how effectively operators will be able to use them.

Adding the Training Dimension

Figure 21-3 shows the addition of a training level to the Logic Model to support skills development and process internalization by facility staff. The Performance

Assessment is now structured to identify not only system response needs but also operator learning needs. Identified learning needs are addressed by training that would include key optimization issues for the specific building and also monitoring and diagnostic principles, methods and tools.

This step is critical if the Performance Assessment capability is to be successfully transferred from outside (or corporate) engineering consultants to internal building operators. Appropriate training will enable operators to both recognize "SOS" conditions more readily and to respond more efficiently and effectively. Especially if new diagnostic tools are involved or if an existing automation system has advanced functions that can be better utilized, then training will enable local operators to interface more productively with the enterprise-level system. Ultimately for improved performance, actual equipment operations must be impacted. The operators are the link between the EEIS and the equipment.

With this new "learning needs" element in the Performance Assessment, it becomes significant that there is no formal procedure for identifying operating staff training needs or specifying instructional activities that should occur as part of the commissioning agent's on-site work or in off-site settings. The formal evaluation of learning needs is a discipline outside of engineering. It is relatively easy to say that a learning needs assessment and learning design should become part of retro-commissioning programs. It is much more difficult to expect

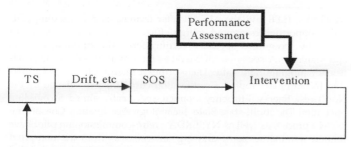

Figure 21-2. Adding an External Performance Assessment to the Logic Model

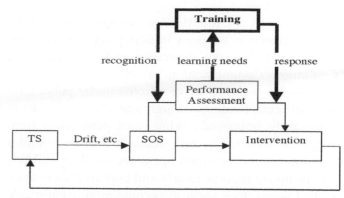

Figure 21-3. Adding Training to the Logic Model

that the retro-commissioning agents, typically staff or consulting engineers, will either readily accept or have the ability to meet such a requirement. Some engineers enjoy working and communicating with operators; given the time and the direction, they can be expected to be reasonably good trainers. This, no doubt, has been occurring relatively informally in the field in successful projects. Consistent program performance on a large scale cannot rely on such undefined procedures. The development of appropriate learning-needs assessment tools needs to be the focus of further work.

SOME BACKGROUND ON BAS UTILIZATION AND DIGITALIZATION IN THE WORKPLACE

That building operator training is needed is also supported by findings about the operation of building control and automation systems. Studies of the effectiveness of BAS utilization find a large degree of underperformance. Two panels of experts asked to categorize performance problems considered "human factors" problems, including operator "interference" and (lack of operator) "awareness," as a leading cause of system underperformance (National Building Controls Information Center, 2002, 2003).

Concepts of "advanced" building monitoring based on data acquired through the BAS have been suggested and developed, up through demonstration projects, including considerations of commercialization, key functionalities and comparisons of commercially available products (Piette, et al., Yee, et al., Haves, et al.). In demonstration projects, highly motivated operators have grasped and utilized the new functionalities, gaining valuable insight into plant operations. Commercial products are increasingly available (see for example Jay Santos and descriptions of the PACRat product) and, to some extent, functions are being incorporated by major BAS vendors but their adoption and utilization remain at an early stage.

Emphasis on monitoring emerges from study of the connection between building performance and monitoring techniques, especially under the rubric of "Continuous Commissioning." Work was done initially to assess the performance of retrofits under public programs in Texas. Findings strongly suggest and document significantly improved savings and persistence from energy-use monitoring and feedback that incorporates operators. Moreover, when operators see their building systems functions more clearly and performance in more detailed terms, they seem to become more comfortable with system optimization changes (Haberl et al.).

Advanced building monitoring functions are intimately bound up with the deeper utilization of digitally acquired data. A large literature on this topic is available from the cognitive sciences, in particular those researchers who examine the impacts and use of computer technologies in the workplace, providing many examples relevant to the process considered in this chapter (Brown 2000, Watson 1993).

TRAINING VIA BUILDING PERFORMANCE LAB, CITY UNIVERSITY OF NEW YORK (CUNY BPL)

Training for energy efficient building operations is readily available in the marketplace. It has typically been offered by non-profit organizations and professional associations, often supported by public sector program funding.* Some university continuing education programs are available, for example at the University of Wisconsin, generally aimed at engineering professionals. More recently community colleges have begun to explore updating the traditional HVAC technician training (Crabtree et al., ATEEC, C. Shockman). Training is also available from equipment manufacturers, albeit for specific products. The "training school" model of some manufacturers is worthy of note.

Undertaking training in the energy efficiency area has generally been a matter of individual decision rather than corporate policy. Roll-out of an EEIS should consider altering this dynamic for key operating personnel. Human Resource managers are usually well versed in training policies and development of training programs. They should be cultivated as valuable allies.

University programs can offer substantial resources and comprehensive programs leading to degrees. A series of focus groups conducted by a CUNY "workforce development initiative" project found that degree credentials were considered very important for career advancement by a cross-section of people in the commercial buildings industry, significantly more so than

---

*ASHRAE, IEEE and the AEE all offer training and continuing education opportunities around energy efficiency; AEE offers a series of nationally recognized specialist certifications. The Building Owners and Managers Association (BOMA) has recently released an energy-efficiency program nationally. The Northeast Energy Efficiency Partnership (NEEP) has adopted a certification curriculum developed by the Northwest Energy Efficiency Council specifically aimed at building operators. The multi-state State Technology Acceleration Consortium (STAC) process as well as NYSERDA's retro-commissioning pilot have funded training sessions, drawing on resources from the non-profit Portland Energy Conservation Inc (PECI). The NYSERDA-supported Building Performance Institute has created certifications for energy efficiency workers in the residential sector.

most certifications.

Through recognition of its own needs for improved building performance, CUNY is recognizing a parallel need in the city's key real estate industry and the workforce that serves it. With support from NYSERDA, CUNY is adopting the model of the Energy Systems Lab at Texas A&M University, through which a combination of performance monitoring, system optimization and practical engineering training will be applied, enhancing the skills available in the city's workforce of building operators.

The BPL has identified two local sectoral training programs with which to collaborate. The Association for Energy Affordability trains operators and energy efficiency workers in multifamily housing. Local 94 of the International Union of Operating Engineers provides staffing for the major commercial office buildings in Manhattan. Both have existing lab facilities used in ongoing programs. CUNY support will help both of these labs develop new training curricula to support adoption and use of next-generation digital controls, emphasizing building performance monitoring.

While neither of these cases represents the corporate environment most likely for EEIS, both demonstrate how a networked community can define its needs and structure training activities.

Connecting Digital to Physical: The AEA Boiler Lab

The Association for Energy Affordability (AEA) has been providing technical assistance to community-based weatherization agencies in New York City for fifteen years through Low-Income Weatherization Assistance, a federally funded program with a history going back to the first oil crises, 1973-77 and the country's first National Energy Plan, under President Jimmy Carter. In urban areas like NYC, the program has an important focus in multifamily housing, with central plant equipment for heating and hot water.

In this case it is interesting to observe how the service delivery mechanism establishes the target audience for training services and topics:

- Community-based agency staff, responsible for initial assessments, owner negotiation, and installation oversight;
- Contractors bidding for work who need to be familiarized with requirements and expectations of the program;
- Building managers, superintendents, and maintenance staff, who must understand intended operation of new equipment and how it relates to performance.

The tasks, needs and level of technical development in these job classifications establish the kinds of training objectives:

for Weatherization and Community-based Energy Program Staff

- Improve diagnosis of existing boiler operations by better understanding of operating sequences, control settings, adjustments
- Test and commission installed work
- Create long-term working relationships with building managers and superintendents as part of a community energy services vision

for Contractor and Service Personnel

- Practice set-up procedures for optimized combustion and equipment cycling so that improved results can be more easily and readily achieved in the field;
- Understand and be better able to implement new generations of controls and capabilities for advanced functions, such as optimized set-back/set-up, reset ratios, firing-rate and lead-lag sequences, oxygen trim, variable speed fan control

for Building Managers and Superintendents

- Understand and recognize various operating patterns and their relationship to building energy performance
- Improve maintenance of efficiency adjustments through better recognition, information, and communication with service firms and mechanics.
- Realize how new sensor technology and GUI-web interfacing can provide data for monitoring building conditions and tracking continuous improvement efforts.

The Boiler Lab Addition

Hands-on aspects of training previously relied on access to operating boiler rooms, the Lab facility, Figure 21-4, greatly improves logistics and, at least as importantly, enables a more systematic approach to hands-on exercises.

Lab exercises, Table 21-1, range from fundamental to relatively complex, with exercises 6-10 benefiting from digital instrumentation to show performance records.

The boilers are overlaid with a LabView data acquisition system, Figure 21-5A. Various pressure, temperature, on/off, and flow points are integrated. The layout of the lab facilitates a training mode interactive between the physical and the digital. A room immediately off the main lab space, originally planned for stor-

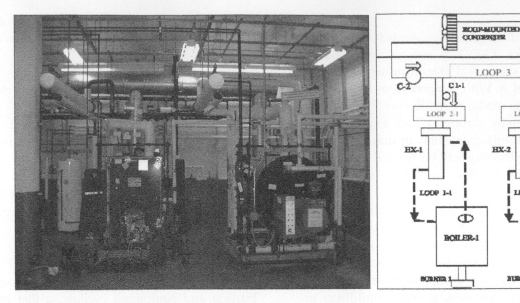

(4A) Left: Physical Boilers (4B) Right: Schematic Layout

Figure 21-4. Boiler Lab at AEA's Bronx Energy Management Training Center
Source: Association for Energy Affordability

Table 21-1. Boiler Lab Exercises

1. Normal boiler start-up and burner firing sequence
2. Opening boilers for inspection, cleaning, leak identification, and re-closing with proper gasketing
3. Low-water cut-off blow-down and switch testing, full boiler blow-down, monitoring of make-up water additions
4. Flame failure safety shutdown, response, and troubleshooting
5. Identification of surging and priming and corrective steps such as water level adjustment, firing rate adjustment, and skimming blow-down
6. Domestic hot water production and mixing valve control at various boiler temperatures and load conditions
7. Combustion efficiency testing and adjustment at various firing rates
8. Pressure control settings and burner firing rate modulation
9. Boiler lead-lag control and cycling in relation to varying load conditions
10. Outdoor temperature reset sequences and adjustments for steam and hot water

age and benchwork is being set up as the "remote data lab." Students can observe a physical sequence of operations on the boilers and then readily review its digital traces. Or teams can work in both rooms simultaneously, communicating changes and seeing how to "tweak"

operations in both directions, digital-to-physical and physical-to-digital. Development of the user-interface associated with lesson-planning is the subject of other work (Huang).

Across the full range of boiler room functions in this size range, the penetration of digital controls is still largely limited to embedded chips with code-based message interface. More advanced digital systems are only beginning to be deployed. Thus operators have very limited experience in looking at large amounts of digital data and connecting them to physical equipment operating patterns. It's at this level that AEA is developing its digital systems training with CUNY.

The first area of expertise to be developed into digital-based curriculum is boiler capacity control, via burner firing-rate modulation, a common mechanism for which is shown in Figure 21-5B, and multiple boiler lead-lad control. Often found sub-optimized in the field, digital data tracks are able to graphically demonstrate load-matching dynamics under various conditions. A set of varying load conditions and operating pattern results under different firing-rate control modes is shown in Table 21-2. Varying the heat rejection loop's temperature and/or flow enables simulation of load conditions. These dynamics reflect the common facility operating conditions in which heating and cooling part-loads lead to significant inefficiencies.

When combined with metered fuel and hydronic flows, students will be able to see the energy use impacts of different operating modes under different load condi-

 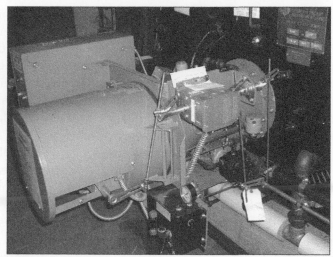

(3A) Left: LabView Data Acquisition point
(3B) Right: Burner showing modulating motor, jackshaft and linkages

Figure 21-5. AEA Boiler Lab, Details
Source: Association for Energy Affordability

Table 21-2. Firing Modes, Load Conditions and Observations

Burner firing mode	Load Condition	Observations
On/Off	Fixed	Cycling pattern, timing
Modulating	Fixed	less cycling, overshoot and anticipation
On/Off	Varying	Increased cycling as load reduces
Modulating	Varying	more modulation action as load reduces

tions. Students will construct energy balances that will be cognitively connected to the experience of setting up the physical boiler operating patterns. The principles of this hands-on training are equally applicable to training for cooling system capacity control.

The latest models of commercial controls for this market are beginning to place digitally acquired data onto websites, where data visualization tools can be applied to accumulated data. Wireless communications is reducing the cost of obtaining apartment temperatures, and early adopters find that reducing overheating can endow significant economic benefits. Market demand for operators who know how to use digital systems to optimize system function and economics will not be far behind.

Introducing Advanced Tools: International Union of Operating Engineers, Local 94 Training Center

The situation with the Operating Engineers union is different from that of apartment building operators. This union staffs engineering operations for most of the city's major office buildings, almost all of which have building automation systems (BAS) in place. Thus most of these building operators have direct experience with integrated digital systems, hierarchical architectures with local control panels, and graphical displays (GUI) at central control consoles. This population already has developed a cognitive connection between physical equipment and digital representations.

Although a rigorous work-study of these engineers and their BAS remains to be done, general observations suggest that these operating engineers will typically

• know how to check the status of major components—air-handlers, chillers—via the BAS console;
• will access to real-time "snap-shot" views of operating conditions; and
• will use, in a limited way, historical logs maintained by the BAS.

Use of the BAS logs is similar to the way logbook

entries of gauge readings have traditionally been used—relatively short-term scanning of variables for status and significant excursions from normal operations. Such rather limited use of available historical data suggests the frontier of current practice for this class of operating engineers. From a combination of what (older) systems offer and what skills operators have (or lack), digitally available data are generally under-utilized, with theory-practice gaps along the following dimensions:

• Use of multivariate data review and data visualization tools

• Trending and weather-adjusted baseline comparisons of energy use as telltale for performance excursions.

• Diagnostic data analysis to suggest functional areas of under-performance

• Documentation for understanding of interactive component and building responses

• Application of AI expert systems for automation or semi-automation of building system optimizations.

We might characterize current management of building systems as highly intuitive at the level of Chief Operating Engineers, with rules-of-thumb rooted in successful past practice governing decisions about how to match equipment operations to conditions. Operating decisions are made with judiciously wide safety margins for maintenance of comfort, with avoidance of complaint as an over-riding performance metric. As a result, for example, confidence in demand response actions is low.

The build-out of the BAS lab at the Local 94 Training Center is just entering its planning stage, which will be undertaken through a collaborative process with senior union members, building owners, and BAS vendors. The opportunity for technology transfer is enormous. The exploration of new tools and techniques can be developed through a well-defined "community of practice."

The Local 94 Training Center is supported by the local real estate industry with a set of training courses that are mandatory for new hires into the workforce. Among various hands-on training set-ups, their facility includes a lab for DDC components, following up on basic training in pneumatic controls and electric relaying. Courses are cross-coordinated with certifications through the Building Owners and Managers Association (BOMA) and BOMA's Building Owners and Managers Institute (BOMI). Advanced, non-mandatory course work is considered key to job advancement in multiple steps reaching to the level of Chief Operating Engineer

("Chief").

The emerging plan for the BPL-supported build-out of an "advanced BAS Lab" at Local 94 is to create a telecommunicated data hub equipped with a large assortment of data analysis, visualization, and diagnostic tools that will be able to work across the range of BMS platforms. A variety of communications design and access issues need to be resolved but the goal is to have capability for data-dumping from a building BAS and/or read-only remote access to the BAS. The initial target training audience will be Chiefs and aspiring Chiefs, who can arrange management approval for secured data access.

To use the lab facility Chiefs and senior engineers, possibly along with service engineers from BAS vendors, will meet in a seminar format in which tools can be learned through application to building data, exploring various system configurations and functions. CUNY academic credit towards a Bachelors degree is anticipated. Learning and technology transfer objectives are several-fold:

• Familiarize target audience (Chiefs) with range of tools, data visualization, and diagnostic methods and case study examples of how tools have been applied;

• Provide exercise in defining building optimization problems and data needs for their solution;

• Identify data accessible through site-specific BAS, data necessary for addressing specified building issues, possible data acquisition gaps and solutions;

• Begin transfer of specific (public domain) monitoring and diagnostic techniques from lab to buildings and/or support exploration of BMS features and/or commercially available tools;

• Enhance understanding of opportunities and needs for existing BMS upgrading in participant facilities;

• Gain feedback to tool developers from highly experienced building operators;

• Gain understanding of monitoring and optimization procedures that can usefully be instituted as part of normal practice for Chiefs and their staffs.

CONCLUSION:
TRAINING FOR EEIS IMPLEMENTATION

Adoption of EEIS as a central corporate function will almost surely drive increasing application of advanced tools for building performance monitoring and diagnostics at the individual facility level. A missing link in effective adoption and utilization of such tools has been training. Our common sense, along with research, experience and a logic model, tells us that training of building operators is an essential element of Organizational Transformation towards new high-performance practices. The real question is what kind of training and how best to structure its delivery. This chapter has argued, we hope convincingly, that committed educational partners can create mechanisms to reach key segments with hands-on activities that will open new vistas of digital information utilization that, in turn, will translate into enhanced building performance.

References

Advanced Technology Environmental Educations Center n.d. *Partnership for Environmental Technology Education (PETE) C4 Program for Energy Management Technicians* materials available at http://www.ATEEC.org/energy

Brown, John Seely and Paul Duguid 2000 *The Social Life of Information* Harvard Business School Press, Boston

Capehart, Barney et al. 2004 "If Buildings Were Built Like Cars—The Potential for Information and Control Systems in Buildings" *Strategic Planning for Energy and the Environment* volume 24 no. 2

Crabtree, Peter et al. 2004 *Developing a Next-Generation Community College Curriculum for Energy-Efficient High-Performance Building Operations* Proceedings of the Summer Study on Energy Efficiency in Buildings American Society for an Energy Efficient Economy (ACEEE) Washington DC

Haberl, Jeff 1996 *An Evaluation of Energy Savings Retrofits from the Texas LoanSTAR Program* Energy Systems Laboratory Technical Report, Department of Mechanical Engineering, Texas A&M University

Haves, Phillip et al., 2006 *A Specifications Guide for Performance Monitoring Systems Spec Guide Draft 007*

Huang, Da-Wei and Michael Bobker 2006 "Seeing by Degrees: Programming Visualization from Sensor Networks" *International Conference on Enhanced Building Operations*, Shanghai (forthcoming)

Norman, Donald 1993 *Things That Make Us Smart: Defending Human Attributes in the Age of the Machine* Perseus Books, Cambridge

National Building Controls Information Program 2002a *"NBCIP Roundtable Summary: What's Wrong with Building Controls?"* Iowa Energy Center

_____ 2002b *"NBCIP Roundtable Summary: What's Wrong with Building Controls? Part 2"* Iowa Energy Center

_____ 2003 *"Summary Report: Characterization of Building Controls and Economizers Using the Commercial Building Energy Consumption Survey"* Iowa Energy Center

Piette, Mary Ann and Chris Shockman 2000 "Innovation Adoption Processes for Third Party Property Management Companies" in *Proceedings of the Summer Study on Energy Efficiency in Buildings* ACEEE, Washington DC

Santos, Jay "Controls Education: More Critical than Ever" n.d. *Networked Controls* (available via internet, http://www.facilitydynamics.com/publications)

Shockman, Christine 2006 personal communication

Yee, Gaymond et al. 2005 "State of Practice of Energy Management, Control, and Information Systems" re-printed in B. Capehart *Web Based Energy Information and Control Systems (Information Systems for Energy Managers, volume 2)* Fairmont Press, Atlanta (originally LBNL Report No. 53545 part 3).

Chapter 22

Facility Scheduling Program (FSP)

David C. Green, Green Management Services, Inc., Fort Myers, Florida
Paul J. Allen, Walt Disney World, Lake Buena Vista, Florida
Robert Proie, Orange County Public Schools, Orlando, Florida

ABSTRACT

THE FACILITY SCHEDULING PROGRAM (FSP) provides an easy-to-use method for Orange County Public Schools (OCPS) staff to request after-school operation of heating, ventilating and air conditioning systems (HVAC) in any OCPS facility. Additionally, the FSP provides a convenient method to display the normal programmed HVAC schedules for each school building for weekdays, weekends and holidays. The FSP is a web-based program accessed via the OCPS intranet using a web browser.

INTRODUCTION

Any large campus of facilities that doesn't operate around the clock can save money by setting back or turning off HVAC equipment when the staff does not occupy the buildings. At Orange County Public Schools the problem has been in managing the occupancy days and times and controlling the equipment accordingly in an environment where after-hours activities are not always consistent. Facility managers would like to minimize the task of collecting occupancy data as well as setting the equipment controls. A web-based database application is well-suited to transfer occupancy days and times in the form of a schedule to facility managers. The database also provides a historical record of the normal schedules and special event requests.

WHY IS THE FSP NEEDED?

HVAC that operates in unoccupied buildings is costly in terms of energy use. Without proper management of the HVAC equipment occupancy schedules, equipment can get re-programmed to run longer than is needed. Periodically evaluating HVAC occupancy schedules and returning them to optimal settings is the proverbial "low hanging fruit" for the energy manager. Documenting and organizing the details of the optimal HVAC equipment schedules is part of this task. Using a web-based system to access this information lets users review the optimal schedules and compare this to how the HVAC equipment is actually operating.

The University of Texas at El Paso Efficiency Task Force reported in the fall of 2003 considerable savings from its "HVAC Operational Clock Management" initiative. According to the initiative, technicians adjusted the schedule of operation by the facilities energy management control systems (EMCS) to match classroom schedules. The savings was $43,000 per year. Administrative costs and costs for maintaining the schedules was $3,000 per year leaving a net cost savings of $40,000 per year. [1]

One of the fundamental features of an energy management system (EMS) is the ability to set-back or turn off HVAC equipment at a particular time in the evening and then turn it back on in the morning. The ideal EMS system would be able to manage the time and set-point schedules for a large campus facility in a master schedule database. Time schedules should be set up to incorporate the facility opening/closing times and dusk/dawn times by day of week. Optimal temperature and humidity set-points should also be documented in this master schedule database. The EMS would be able to reset all time schedules and set-points each day to restore the optimal settings that might have been changed during the day. Additionally, the EMS should also be able to handle special events that occur after the normal open/close time schedules by sending additional time schedules that effectively increase the HVAC/Lighting equipment runtime to accommodate the special event. [3]

However, what typically is found in most large campus facilities today are several EMS manufacturers that do not communicate together. Changes to the time or set-point schedules are handled differently for each EMS. To be able to have the ideal EMS setup described above would require the investment in costly hardware/software and further complicate an already complicated system. Typically, users will invest the time to understand how to change the time and set-point schedules using the particular EMS manufacturers front-end software.

Regardless of how the time and set-point schedules are actually programmed at the EMS, the energy manager should have a separate information system that keeps track of the optimal settings. Ideally, this would be web-based so that users throughout the campus would be able to review their own facility settings. Additionally, the web-based system would provide users an easy-to-use tool to make requests for HVAC and lighting to accommodate special events. The facility manager can then make appropriate changes to the EMS time schedules to match the users requests. The web-based program streamlines the communication process between the building occupants and the facility managers. The web-based program does not attempt to solve the costly task of integrating disparate EMS systems, but instead organizes and manages the equipment scheduling task.

For large campuses, the communication process between facility managers and building occupants to make temporary changes to the EMS schedules can grow into a huge task. Let's imagine how much communication would be required for a school district with 182 schools and over 2,000 buildings or
. Control points are areas within a building or outside a building that the HVAC system controls separately from the building itself, such as parking lot lights. Someone from each school has to notify the facility managers that people will occupy a building after-hours or else the normal energy saving cycle will either set-back the thermostat settings or turn the equipment off completely. For any given week a normal school will have at least one after-hours activity per night. The problem is that it is not always in the same building or on the same night at the same times. Fortunately, some activities always occur in the same building on the same night at the same times. But this still requires a huge number of phone calls or emails to communicate the needs of the schools. Also, on the other end, the facility manager has to keep track of all these after-hours requests. And then let's consider the cancellations that are bound to occur adding more phone calls, emails and confusion. Often time's schools require confirmation that facility managers have scheduled equipment for operation during important events. This adds even more communication traffic.

Kent State University uses a "Building HVAC Scheduling Request Form" to request the equipment be operational after-hours. The form is available online but must be signed and faxed to the appropriate person to make the request complete. This helps to minimize the work of the facility manager in collecting and tracking the occupancy information and serves as a method to verify the legitimacy of the building use by requiring a signature. [2] However, it is a long way from addressing the need to collect and track this information efficiently.

Without the use of a database to hold the information there is risk that it can be lost. Also, a database might be helpful in organizing the information in chronological order and keeping track of who made requests for after-hours activities. Of course, the problem is that information gathered in the ways described previously and stored in a database requires a lot of duplicate effort. When a school calls in the request the facility manager writes it down and enters it into a database. This could be a considerable amount of information. First of all the request must contain the name of the school, the date and time of the activity, the duration of the activity, the purpose of the activity and who is making the request along with the requestors phone and email address. Then in order to verify a request the school has to call in again and the facility manager has to look up the request in the database. The same goes for confirming a time schedule change since the facility manager has to lookup the requestor in the database and phone or email the requestor to confirm that the equipment time schedule change is complete for the appropriate times.

Even though it's proven that scheduling HVAC equipment operation to correspond with building occupancy saves energy and money. All of this administrative effort to collect information on time scheduling is costly and diminishes the returns of cutting back on energy use. Thus, there is a need to minimize this effort by decreasing the communication traffic and eliminating the duplicated tasks of getting the information into a database.

FSP DEVELOPMENT GOALS

The main goal of the FSP is to minimize the effort of the facility manager in collecting and organizing occupancy information for buildings. It needs to do this in a way that is user-friendly to both the requestor and the facility manager. It should involve a secure communication means that tracks who the requestor is automatically but does not require email or phone conversations. The FSP should allow others to access the information if needed and derive reports from it.

As mentioned earlier OCPS has approximately 182 schools with over 2,000 buildings or control points that facility managers may need to schedule each and every day. It's obvious that communicating the occupancy information for these schools could generate a lot of phone calls or emails. Since we want to collect information on a per-building basis transcribing this information onto some other medium like a spreadsheet or a database requires a great deal of additional effort. So, our first step is to eliminate those phone calls and emails by sending the information directly into a database.

Databases are complex software systems not commonly used by everyone yet. The FSP needs to have a user-friendly way to input the information into the database. The most recent advances in this regard are web-based database applications. The question is… is it secure? In this case network security takes away most of the risk since the FSP is only accessible from inside the firewall of the OCPS intranet. The FSP even adds another layer of security since a user must first log in to make a schedule request for an after-hours activity. The database maintains personal information about the user so that the FSP can fill in some of the information pertinent to each request such as the phone number and email address of the requestor automatically.

The idea is for the requestor to use their web browser to log in to the FSP, fill out a form requesting after-hours operation of HVAC equipment on a building by building basis and have the request show up on the web browser of the facility manager. Of course, it is marked in such a way as to show that it is not yet scheduled and thus needs attention. This eliminates the need for a phone call or email request. The facility manager then takes the required action and marks the request as "scheduled." It then shows up on the requestor's browser as "scheduled" and thus eliminates the confirmation phone call or email. The same process applies to canceling a request. Changes made to the database online are immediately available to anyone who has access to the FSP. There is no need to transcribe information once the requestor submits the schedule for an after-hours activity.

This brings up another important goal which is to allow others to view requested schedules. Through filtering and sorting actions performed on the information contained in the database anyone with the proper authority could analyze the data as needed to identify trends or metrics concerning the number of hours buildings are used. Security personnel can monitor the FSP to make sure their responsibilities are covered.

So, a web-based database application is best suited to provide a user-friendly interface to transmit building occupancy information directly from the schools to the facility managers with minimal effort on the part of both. This eliminates traditional communication traffic and transcribing of information. It provides additional advantages in that anyone can analyze the information if needed.

FSP: THE SOLUTION TO FACILITY SCHEDULING

The facility scheduling program (FSP) is a web-based database application that minimizes the effort of collecting and organizing occupancy information for the buildings of Orange County Public Schools (OCPS). It records normal operating schedules as well as schedule for after-hours activities. It is a user-friendly application that allows users to request schedules for multiple buildings quickly and easily. Requests show up simultaneously on the requestor's web browser as well as that of the facility manager. The application has features that allow for analyzing the information as well as maintaining user information and authorizations.

Development Partners:
1. OCPS Facilities: Responsible for providing the data required to populate the building database. They are responsible for administering the operation of the program.

2. OCPS I.T. Department: Responsible for providing the web-server all server-related functions (backups, security, and disk space).

3. Green Management Services: Design and implement the FSP based on the scope of work.

4. Walt Disney World: Assist in the initial scope development and facilitate implementation of program.

The FSP uses a FoxPro database application in conjunction with FoxWeb which allows database queries through a web browser. The application is secure since it is only accessible within the OCPS intranet and not to the general public. The following FoxPro database tables are the basis of the FSP:

Schools Table—A list of all the schools and their type (i.e. elementary, middle, high, etc.).
Schedules Table—A list of the normal schedules for each building.
Events Table—A list of each requested schedule for an after-hours event.
Guidelines Table—A list of operating guidelines as to the number of scheduled hours per day.
Authorizations Table—A list of users and their authorization levels.

The schools and schedules tables populate pull-down menus on the main page of the FSP. These pull-down lists act as filters for the resulting report after the user selects the "Display" button (see Figure 22-1).

FSP users choose **"View Operating Guidelines"** to report the operating guidelines for each school type. (Figure 22-2.)

Users can view the normal operating schedules for each school by selecting the action **"View Regular Class/Event Operating Schedules."** As you can see in Figure 22-3, the hours scheduled are compared to the operating guidelines to show if they fall within or not. Authorized users can edit the regular class operating hours by clicking on the links in the **"Times"** column.

Users view the scheduled events for each school by selecting the action **"View Scheduled Events."** (See Figure 22-4.) The report also shows how much time elapsed before the event was actually scheduled in the energy management system. These requests show up in chronological order with the "unscheduled" requests at the top of the list so the facility manager knows which ones need attention first. Authorized users can edit scheduled events by clicking on the link in the **"Times"** column. This is also how a user may cancel an event if needed.

If a user has selected the **"Log In"** link and logged in with the correct user i.d. and password they can schedule an event for an after-hours activity. (See Figure 22-5.) Once a user has entered an event the user can pick another schedule name and schedule an event for that building for the same date and times. Note that the FSP fills the requestor information into the form automatically.

Users may also choose to **"View Hourly Use."** (See Figure 22-6.) These values represent the total number of scheduled hours by building. Some schools have more buildings than others. The purpose of this report is to show the hours that are used the most.

Users can view and change authorizations in a similar manner. Each user is authorized to either view or edit the actions listed in Figure 22-1. Any of the reports can be filtered by using the pull-down lists or sorted by selecting a heading that is linked to analyze the information in the database as needed. Some reports have links at the bottom of the page to export the data to an Excel spreadsheet.

CONCLUSION

The FSP is a simple easy-to-use example of how web-based automation can aid in facilities management. The result is a tool that can save valuable dollars in

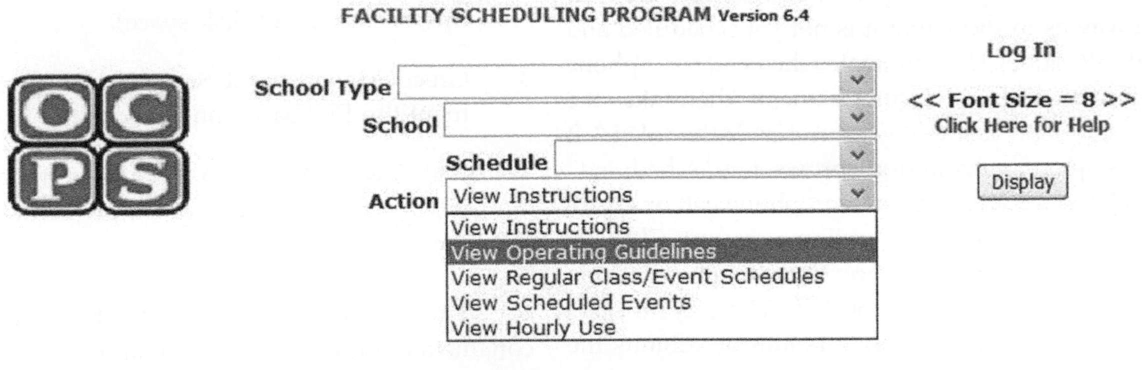

Figure 22-1.

OCPS Operating Guidelines			
SCHOOL TYPE †	OPENING	CLOSING	OPERATING HOURS
ALTERNATIVE EDUCATION	6:00AM	6:00PM	12.00
ELEMENTARY	6:00AM	6:00PM	12.00
HIGH SCHOOL	6:00AM	8:00PM	14.00
MIDDLE SCHOOL	6:00AM	7:00PM	13.00
NINTH GRADE CENTER	6:00AM	7:00PM	13.00
ORANGE TECHNICAL EDUCATION CENTER	5:00AM	11:00PM	18.00

Figure 22-2.

Regular Class/Event Schedules

School Type = NINTH GRADE CENTER
School = APOPKA 9TH GRADE CENTER

SCHEDULE	LOCATION	TIMES	DAYS OF WEEK	SCHEDULED	OPERATING HOURS
ENTIRE CAMPUS	ENTIRE CAMPUS	No Schedule		No	
L1	MAIN OFFICE	6:00AM-4:30PM	Mon, Tues, Wed, Thur, Fri	Yes	10.50
L10	CLASSROOMS	6:45AM-3:30PM	Mon, Tues, Wed, Thur, Fri	Yes	8.75
L11	CLASSROOMS	6:45AM-3:30PM	Mon, Tues, Wed, Thur, Fri	Yes	8.75
L12	CLASSROOMS	6:45AM-8:30PM	Mon, Tues, Wed, Thur	Yes	13.75 Exceeds Operating Guidelines of 13.00

Figure 22-3.

Scheduled Events (as of 06/18/2006 09:01:54 AM)

SCHOOL	SCHOOL TYPE	SCHEDULE	LOCATION	DATES	TIMES	DAYS OF WEEK	REASON	SCHEDULED ↑	HOURS TO SCHEDULE
BOONE HIGH	HIGH SCHOOL	ENTIRE CAMPUS	ENTIRE CAMPUS	05/24/2006-06/30/2006	6:00AM-4:00PM	Mon, Tues, Wed, Thur, Fri	School Sponsored Event	No	
CYPRESS SPRINGS ELEMENTARY	ELEMENTARY	L-3	ART/MUSIC	07/11/2006-07/11/2006	4:00PM-7:30PM	Thur	School Sponsored Event	No	
BOONE HIGH	HIGH SCHOOL	ENTIRE CAMPUS	ENTIRE CAMPUS	08/07/2006-05/22/2007	6:00AM-9:00PM	Mon, Tues, Wed, Thur, Fri	School Sponsored Event	No	
GLENRIDGE MIDDLE	MIDDLE SCHOOL	BLDG. 3	MUSIC	01/12/2006-TBD	6:30PM-8:30PM	Thur	Facility Lease Agreement	Yes	23.7
EDGEWATER HIGH	HIGH SCHOOL	MEDIA	MEDIA	01/21/2006-TBD	6:00AM-9:00PM	Sat	School Sponsored Event	Yes	0.3
CYPRESS SPRINGS ELEMENTARY	ELEMENTARY	L-2	CAFETERIA	01/25/2006-06/30/2006	6:00PM-8:30PM	Wed	Facility Lease Agreement	Yes	16.1
CYPRESS SPRINGS ELEMENTARY	ELEMENTARY	L-3	ART/MUSIC	01/25/2006-06/30/2006	6:00PM-8:30PM	Wed	Facility Lease Agreement	Yes	16.0
CYPRESS SPRINGS ELEMENTARY	ELEMENTARY	L-4	MEDIA	01/25/2006-06/30/2006	6:00PM-9:30PM	Wed	Facility Lease Agreement	Yes	16.1

Figure 22-4.

energy conservation by curtailing HVAC operations during hours that a building is not in use. It also minimizes the amount of effort required to collect and organize the normal operating schedules of buildings as well as after-hours activities. It cuts down on phone and email traffic while making the information available to a wider audience. The database provides an accurate record of each request for further analysis if needed. The FSP allows verification that school has requested a specific operating (or special event) schedule, and further verification

Add a New Event							
SCHOOL	Schedule	LOCATION	DATES	TIMES	DAYS OF WEEK	REASON	SCHEDULED
ALOMA ELEMENTARY	OSS 1	CAFÉ, STAGE, MEDIA, OFFICE	06/18/2006 to 06/25/2006	1 ∨ : 00 ∨ AM ∨ to 1 ∨ : 00 ∨ AM ∨	Mon Tues Wed Thur Fri Sat Sun	School Sponsored Event ∨	No ∨
			Select the **Start** and **End** dates; the schedule will repeat every day for the date range specified	Select the **Start** and **End** times	Select the **Days of the Week** (Hold the CTRL key down to select multiple days)	Select a **Reason** for the event	**FSP Administrator only**
Schedules needed the same day must be submitted by Noon!							
Date Requested	Requested By		Requestor Phone				Save
06/18/2006	DAVID C GREEN		(800)409-3874				

Figure 22-5.

OCPS Hourly Use																								
School Type = ELEMENTARY																								
School = ALOMA ELEMENTARY																								
Schedule = OSS 1																								
	1	2	3	4	5	6	7	8	9	10	11	12	13	14	15	16	17	18	19	20	21	22	23	24
Total Regular Class Schedules Use	0	0	0	0	0	0	1	1	1	1	1	1	1	1	1	1	0	0	0	0	0	0	0	0
Total Event Schedules Use	0	0	0	0	1	1	1	1	0	0	0	0	0	0	0	0	14	14	12	4	0	0	0	0
Note: These values represent the total number of scheduled hours by zone. Some schools have more zones than others. The purpose of this report is to show the hours that are used the most.																								

Figure 22-6.

that any changes requested are actually implemented. It is used to ensure that school schedule change requests are placed into affect in a timely manner.

Once the users enter the information into the database there is no limit to the analysis that the facility manager can perform at a later date if needed. It allows detailed monitoring of the operating schedules of all schools, and by type of school, enabling the district to look for trends and out-of-place data, so that schools may be aware regarding opportunities to trim operating times to further conserve energy where possible.

The FSP also allows for monitoring of after-hours use of facilities to ensure that school joint-use agreements properly account for and bill community organizations for their fair share of cost, ensuring that these joint use agreements are in place where after hours

usage is noted as a cross-check verification. It has been used to respond to concerned community groups that did not understand the cost of operating a facility after-hours through accurate knowledge of facility operating times and calculated costs per hour.

References

[1] "HVAC Operational Clock Management"; internet page, http://admin.utep.edu/Default.aspx?tabid=17377, accessed 6/4/2006; University of Texas at El Paso Efficiency Task Force, El Paso, Texas Paso, TX El Paso, TX

[2] "Building HVAC Scheduling"; internet page, http://www.kent.edu/ceo/Energy/Building-HVAC-Scheduling.cfm, accessed 6/4/2006; Campus Environment and Operations, Kent State University

[3] Capehart, Barney; "Custom Programs Enhance Building Tune-Up Process"; Information Technology for Energy Managers; p.368; Fairmont Press, Inc., Lilburn, Ga. 2004.

Chapter 23

University of North Carolina at Chapel Hill—
A Proactive Approach to Facility Management

Sanjyot Bhusari
David Brooks, P.E.

FACILITIES LEADERS today manage service businesses with budgets in the tens-of-millions of dollars. Typical of most service organizations, there is a demand on the one side to provide excellent customer service by providing a safe, healthy, comfortable environment, while keeping down costs. To help achieve these goals, facilities managers employ various hardware and software technologies to address needs such as comfort management, preventive maintenance, hot/cold call resolution, energy management, and operational budget planning among others. But, has the typical facility manager kept up with the current available technologies, to advance the facility's business beyond the traditional model?

In this chapter the enterprise building management system (EBMS) is introduced. The University of North Carolina at Chapel Hill (UNC-CH) recognized the value in the concept and the approach to enhancing business processes. The UNC-CH project is the first of its kind and represents the latest in both technological approaches and design approaches.

The traditional building automation system (BAS) represents the core technology at most large facilities. The BAS typically represents a multi-million dollar investment, and its operation (or intended operation) can support and/or hinder the goals mentioned above. The BAS industry has evolved from the old days of pneumatic systems to current advances in direct digital control (DDC) systems. Recent trends include web-enablement and the steady migration towards systems that are modeled around IT standards. These evolutions were meant to make facilities management functions simpler and give the facilities manager the tools needed to achieve their goals with the least amount of staff. But has this really happened? Have the internal business processes within a facility (i.e. equipment scheduling, alarm management, preventive maintenance, energy management, and asset management) evolved or changed from the days of pneumatic systems to the current state of DDC

systems? What does a facility manager want from BAS? Why has technology not been accepted? What are the obstacles in the way of embracing technology? How has budget impacted the use of technology? Do facility managers have the right skill set to run the facilities? What are the driving forces that affect a facility manager's decisions today?

To answer these questions, we interviewed several facility managers running a wide range of facilities that included: universities, community colleges, government organizations, financial organizations, hospitals among others. The following list of questions and responses attempts to address the big question "Are we, as an industry, taking full advantage of the technologies available to us and are we looking at new ways to do business?"

Question 1

Have the internal business processes within a facility like scheduling, alarm management, preventive maintenance, and documentation evolved or changed from pneumatic systems to DDC systems?

Most facility managers observed/noted that their facility management functions like scheduling, alarming, preventive maintenance and documentation have not really changed or evolved from pneumatic systems to DDC. One facility manager noted/quipped "I could schedule equipment with one electromechanical time clock, now I have three computers." A lot of facilities are still debating whether it made sense to switch to DDC systems or stick with the pneumatic system. Many senior operators found pneumatic systems to be simple and easy to understand. They find DDC systems complicated and cumbersome. DDC systems have come with their share of pain, with proprietary protocols and control vendor strategies to lock owners into their solution. In that context many facility managers appreciate the open nature of pneumatic systems. Such concerns make us wonder if DDC systems have really lived up to their promise.

Several universities have a structure where departments occupying the building are not responsible for the energy bill. Hence, there is little incentive for the occupants to worry about occupancy schedules. As a result, facility management does not get the required information to schedule units off during an unoccupied mode. Some facility managers would like to change this business process by giving occupants control over their air handling unit occupancy schedule, and making them accountable for their energy bill.

Facilities having a multi-vendor environment have ended up with multiple computer systems running different/proprietary software for different controls vendors. Each controls vendor deals with alarms differently. Many facilities have a single operator trying to manage alarms coming from different systems. Most systems do not have a way to prioritize alarms as well. As a result, operators struggle dealing with alarms and routing them to appropriate personnel.

DDC systems have capability of delivering a lot of information if it is implemented correctly. However a lot of facilities operate under the "put fire out first" mode and have not utilized these capabilities to proactively manage their facilities and give preventive maintenance the importance it deserves.

Question 2

What does a facility manager want from BAS?

There is an overwhelming desire from facility managers for BAS to run like it is intended. Most facility managers feel the need to baby sit their systems. One facility manager wondered "If air planes can run on auto-pilot, and cars can run on cruise control, why can't BAS run on its own?" Overall, facility managers wanted their BAS to be a tool that they could use to:

* help ensure a safe and secure environment
* reduce energy consumption
* manage growing expectations for comfort and service
* reduce operating and maintenance costs
* provide proactive notification
* correct problems remotely (troubleshooting)

A number of facilities currently utilized their BAS as a reporting tool only. However, there is a strong desire to use it as a troubleshooting tool as well. Trends are a powerful way of troubleshooting; however, consulting engineers have traditionally asked for a very few points to be trended. To a large extent these decisions are based on perceived costs. In most systems trends have to be set up after a problem is reported. A facility manager

complained that he asked his tenants to tell him if they were going to get hot or cold a week in advance, so that he could start his trends.

To a large extent, consulting design engineers have not understood maintenance requirements. For example, when they specify a status point, more than likely what is displayed on the screen is the command output rather than the actual status. Canned graphic programs have seldom accurately described the configuration of a system. Warning alarms, when things are starting to get out of acceptable ranges, are great predictive maintenance tools to allow the maintenance staff to be proactive rather than reactive. However, facility managers feel that they have not gotten everything that existing BAS is capable of delivering.

Question 3

Why has technology not been accepted? What are the obstacles in the way of embracing technology?

A common response to this question was that "a system is only as good as the operator operating it." Consulting engineers could design a highly advanced BAS with all the bells and whistles, but facility managers feel that they rarely have adequate resources to utilize the BAS per its design intent. State institutions and universities find it difficult to compete with private institutions and multi-national companies to retain qualified people to run and maintain their systems.

Another key issue with technology is training or lack of it. Though consultants routinely specify a big chuck of training time for new construction projects, they rarely get into details of how this training needs to be delivered. General consensus among facility managers was that detailed training requirements be specified. It would help to split the training program into two parts. The first part introduces the operators to the new system, and then second part gets into technical details. In between part one and two there should be a gap of a couple of months for operators to get more familiar with the system.

Another common issue is of utilizing proprietary systems. Some facility managers prefer to live in the sole source environment that allows them to learn one control company's product line and be self sufficient in maintaining them. In such situations they feel they get locked in with the technology that that particular controls company supports.

In general for open systems, procurement processes usually utilize the low bid approach. Controls companies working under this approach and trying to make profit have found it difficult to invest enough time to implement DDC system solutions correctly.

Overall there is a lack of idea and understanding of how business processes could be improved with current implementation of technology. Technology has changed drastically over the years, leaving facility managers with the difficult task of keeping up with it.

Question 4

How has budget impacted the use of technology?

The intent of building systems is to provide an environment that is safe, secure and comfortable. However, that is seldom the main intent of the building owners. Corporations, universities and hospitals would rather invest in business systems, research, and latest medical equipment than in their BAS. A common problem facility managers face is that of funding. BAS is in most cases a necessary evil.

Working with limited budgets, facility managers struggle to get their people the right amount of training to run the system effectively. Limited budgets are a crucial factor for facilities not been able to hire and retain qualified controls techs. Many facility managers complained of not having adequate resources to cover for staff going for training.

These factors severely restrict the use of new technology.

BAS has rarely been seen from a return-on-investment point of view. If implemented correctly, facility managers feel that BAS could offer return on investment of less than three to four years.

There is a real need for looking at facility management as a profit center and not as a business that gets its profit from saved energy costs. Facilities need more effective maintenance efforts (all too many do not have a PM program because they are always in a crisis mode and do not have time to be proactive).

Question 5

Do facility managers have the right skill set to run the facilities?

With limited budgets and fast-changing technology, facility managers have to struggled to maintain the right skill set to run their facilities. Open protocols and systems have helped to bring competitive prices to the procurement processes; however, they have left facility managers and controls techs with disparate systems to learn, manage and maintain.

With the increased convergence of BAS with information technology, new skill sets are required. Now facility managers not only need staff adequately trained in traditional control systems like controller tuning, programming etc., they also need staff to manage digital communication, web programming and networking.

Question 6

What are the driving forces that affect a facility manager's decisions today?

Facility managers feel that they operate in the "urgent" mode than in the "important" mode. With limited staff and funding, and a growing list of expectations, they are increasingly reacting rather than responding. As one facility manager puts it, "Whoever screams the loudest gets my attention!"

Question 7

Do facility managers get enough management support?

Our survey got a variety of responses to the different questions. However, there was a unified answer to one question: Facility managers whole heartedly agreed that they did not get enough management support. Facility managers felt that management tended to look at their core business as the only priority, neglecting the support resources. Management rarely realizes that the physical condition or "curb appeal" of their business is as important to a potential client as the expertise you have to offer. For example, a facility manager working for a major hospital wondered, "What would your perception be if you were going into a hospital for major surgery and you were told that a renowned doctor was doing the operation—and as you walk in you see dirt on the floors, stained ceiling tile, and walls that need painting? This curb appeal impression would most likely influence or change your perception about how good that doctor is."

The information collected, led to a rather interesting and potentially debatable technology curve representation of the facility management industry spanning the last 40 years.

The following technology gap curve (Figure 23-1) shows an almost flat innovation line for both facility management functions as well as the traditional mechanical, electrical, and plumbing (MEP) design industry. The technology curve followed a similar flat profile until the 80s when DDC systems hit the scene, and were affordable to the medium to large facilities. As noted in Figure 23-1, technology has grown exponentially leaving the facility management industry to carry on "status quo."

It is important to note here that the current technological state of BAS industry has advanced beyond the technical capabilities of the traditional HVAC mechanic. This gap will continue to grow larger until the facility management industry begins to look at new models and take advantage of all the technology available to it. There is plenty of blame to go around but it is our responsibility (facility manager's, consultants and

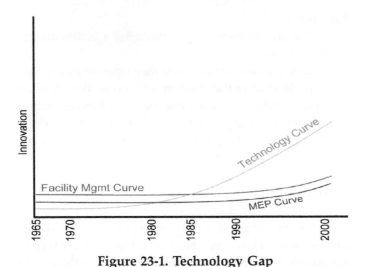

Figure 23-1. Technology Gap

vendors) to begin the long road of catch-up. Facility managers must be willing to consider new processes and business models, consultants need to be educated in the technologies available to their clients, and vendors need to focus more on applications and solutions and less on product throughput.

The enterprise building management system (EBMS) approach is a design concept that addresses this technology gap both from a technical perspective as well as a business process perspective. From a technical perspective, EBMS integrates disparate BASs into a single coherent graphical user interface (GUI) by utilizing information technology (IT) standards referenced in this chapter as "web services." Web services is basically an IT term used to describe a group of tools used to define how systems will communicate and share information. Since web services has become the de-facto standard within the third-party applications business, most enterprise applications such as maintenance management, utility billing, human resources, and event scheduling can be seamlessly integrated into the building management system. It is this ability to marry various applications that drove the University of North Carolina at Chapel Hill (UNC-CH) to pursue the EBMS approach to facility management. From a business process perspective EBMS places information at all levels of the organization, while minimizing the resources needed to manage that information. Facilities leaders can finally manage cost versus comfort, have the data to optimize systems efficiently, and create real change in business processes, leading to better customer service and improved credibility.

Let's Consider a Few Examples

a.   Preventive maintenance is a crucial facility management function. It includes a wide range of activities that include filter changes, belt replacement, coil

cleaning, greasing bearings, etc. At many facilities these functions are carried out in accordance with a fixed time period in lieu of management by runtime usage. An air handler may have a very dirty filter resulting in loss of energy efficiency, yet its filter will get replaced only at some pre-scheduled time. With DDC systems a differential pressure transmitter can alarm the BAS when the condition of the filter exceeds manufacturers' recommendations; however, this feature is rarely used. An EBMS solution to such preventive maintenance issues could be to generate a work management "to do" list for HVAC techs every morning that would include run time information and vital statistics of major equipment. This would allow preventive maintenance to be based on run time conditions. EBMS also allows automation to be taken several steps further. In the above filter example, the EBMS can also check inventory, automatically order filters, and issue a work order.

b.   Scheduling is another key facility management function. When implemented it is also one of the most important energy conservation strategies in the facility manager's arsenal. Schedules have evolved from the days of the electromechanical time clock to the current DDC system's using user friendly calendars. Yet, aggressive scheduling programs are rarely pursued. However, a major problem with DDC systems has been interoperability. There are many reasons for this, but the primary reason in the lack of resources needed to create and manage any scheduling program. Managing different building automation system manufacturers makes this task even more daunting when one considers the many different approaches BAS manufacturers use for scheduling applications. The task is, at a minimum, overwhelming and cumbersome, and as a result, scheduling programs are abandoned or equipment is set for the maximum runtime to account for all possible cases. An EBMS approach to scheduling is to have a common scheduling interface across different BAS manufacturers. Global commands can start and stop equipment. At the same time, further automation is possible through integration of third-party programs, placing many of the scheduling functions in the hands of the end user. For example: A classroom-type building can get scheduled automatically as students register for classes. An on-line form for after hours use can schedule the building for air conditioning and lighting as well as automatically generate billing for its use.

c. Most DDC systems come with a canned alarm management program. The facility manager is stuck with features they may never use or, must struggle to force the canned application to fit the organization's needs. Features could include the ability to snooze, disable, delete, or simply acknowledge an alarm. Such abilities could be assigned depending on operator skill sets. Alarms can be sorted and given priority by assigning them a number as well as moving them on top of the list. Alarms can be color coordinated depending on their status. Specific alarms could be linked with trends of different points associated with the situation leading to an alarm, giving the operator a more global view of the conditions leading to the alarm. EBMS approach leads to "intelligent alarms" that could be set up to diagnose the problem as well as offer recommendations to the operator for possible course of action. Further automation is now possible by linking alarms to work order management software packages and other facility third-party programs (i.e. internal emails, cell phone, IT broadcast messaging, etc.).

d. Facilities often find it difficult to hold onto competent and talented staff. Quality and quantity of staff are seldom at a level where a facility manager wants them. A goal of the EBMS approach is to use technology to make several facility functions easier and automated, and give the facility manager the ability to do more with less. Intelligent alarms, semi-automated schedules, pro-active preventive maintenance strategies, and effective documentation are some of the things that could make facility management functions easier, and refine some of the internal business processes.

e. EBMS is a new approach to multiple building operations that allows the traditional BAS to actively interact with new and existing enterprise systems. This interaction is made possible through the use of a normalized data (web services) package which can be shared with and drawn from existing enterprise programs such as space planning, registration, billing or any number of other enterprise level systems used within the facility environment.

f. The EBMS provides facility managers with the management and analysis tools needed for system diagnostics and pro-active system management (i.e., trending, occupancy, performance analysis, scheduling and alarm management). These and many other application examples will give the HVAC facility services staff the tools necessary to ensure occupant comfort, energy conservation, and assets that are effectively managed in a pro-active instead of reactive mode. This can now be done at the lowest possible cost and with the least number of staff.

g. EBMS allows the traditional BAS to move from a proprietary and private world, little understood by the building occupants, to a system which everyone can actively use to enhance their use of the space they occupy.

CASE STUDY

The University of North Carolina at Chapel Hill (UNC-CH) enterprise building management system (EBMS) project includes more than 140 buildings and approximately 33,000 building automation system (BAS) points. The existing building-level control systems are reporting to eight different BAS front ends using materials and software from six different manufacturers, and all operating with various open and proprietary communication protocols (BACnet, LONWorks, JCI-N2, etc.). BAS capacity is expected to grow proportionally by an additional seven million square feet of new construction currently being constructed or under design. The sheer complexity and the growing point count associated with this period of growth will further limit the ability of the current HVAC facility services staff to effectively manage the many different and incompatible control systems under the current model. This model (similar to most large campus environments) stretches the already limited resources in such a way that limits or, in many cases, completely removes the concepts of energy conservation (EC), preventative maintenance (PM) and pro-active problem (PPR) resolution from the staff's standard operating procedures.

The UNC-CH management requested a system that would integrate the existing BAS vendors into a single coherent system, and follow current information technology (IT) standards of information transfer. There was also an overwhelming need to develop a simplified and intuitive operational environment that would ensure the HVAC facility services group would regain its ability to focus on the value added services described in the previous paragraph (EC, PM, & PPR). The biggest challenge of the project was to develop a method of integration that ensured a normalized data stream at the enterprise (or WAN/LAN level) of the system.

As the BAS is increasingly seen as part of a much larger enterprise structure, the need for meaningful and usable information is growing. Certain IT trends, i.e., lower bandwidth costs and lower memory cost, continue to push the need for data integration at all levels of the building environment, including the BAS. The IT industry has already developed a set of rules describing the way data are shared. The general term used to describe these rules is "web services" and it is used to describe the methods needed to share data computer-to-computer. Web services are web applications that are self-contained and modular. They can be run over the internet and can be integrated into other applications. One of UNC-CH's goals was to transform its wealth of BAS data into an environment that would enable intelligent and real time business decisions. In other words, the EBMS needed to have the ability to import and export data into and out of other third-party applications like maintenance management, human resource planning, utility management/billing, and asset management.

A key component of the EBMS design is the incorporation of what was defined under this project as the enterprise building level translator (EBLT). Since most of the existing BAS data at UNC-CH were part of legacy and/or proprietary systems, some mechanism for data normalization had to be developed. The EBLT "normalizes" or transforms these data into XML/SOAP messages. XML (extensible markup language) and SOAP (simple object access protocol) are the basic components of Web Services. XML is a specification detailing how to encode information and is ubiquitous in the IT infrastructure. XML, however, is not a complete communication system and must be layered on top of a communication component, i.e., SOAP.

The EBLT's web services provide web methods for securely authenticating client applications. Minimum security requirements include user name and password authentication. Further security is provided by utilizing a third-party security application which gives the administrators the ability to assign enterprise-wide access policies, deploy programs to many computers, and apply critical updates to an entire organization. These third-party security applications are then incorporated into the EBMS. The third-party security packaged used at UNC-CH stores information about its users that acts in a similar manner to a phone book. This allows operators to work from any thin client and have the same rights that are assigned to them, regardless of the application. It also gives administrators the ability to remove all rights of an individual with one key stroke, in the event of termination or security threat.

The new EBMS has been specified to utilize scalable vector graphics (SVG) to represent data. SVG is an emerging web standard for two-dimensional graphics. Like HTML, SVG is written in plain text and rendered by the browser, except that in this case, it is not just text that is rendered but also shapes and images, which can be animated and made interactive. SVG is written

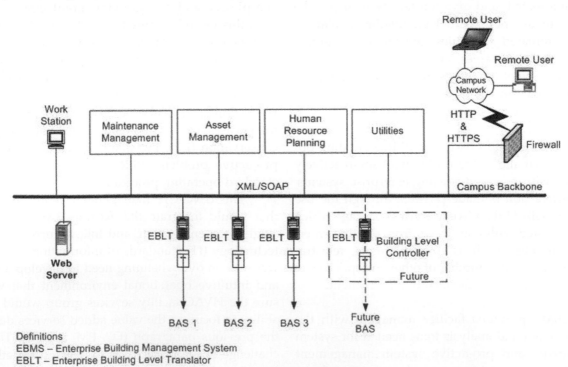

Figure 23-2. EBMS Network Architecture

in XML and developed by the World Wide Web Consortium. Its scalable nature means that similar graphics can be used on 52" plasma monitors, lap tops or PDAs without loosing any of the resolution or functionality. Since SVG is XML based, points/data can be embedded in them so that graphics represent real time information. For example, colors and shapes of graphics representing temperature can be customized to appear different at different temperatures. Different alarms can show graphics differently making it easy for operator to decipher and act on alarms. Most of the data generated from a BAS are never stored and hence never analyzed.

The EBMS employs a robust data historian server application that provides constant and real time data storage. Data storage requirements can become quite large when considering the storage needed to record data for 33,000 points at 15-minute intervals and over a 1- to 2-year span. While the data storage needs can be large, the decreased $cost/MB has made this a non issue. Utilizing web services to access these data makes it very easy to analyze and integrate it with different applications. The primary goal of the historian is to analyze systems to show financial impact of decisions and measure performance relative to established engineering models. For example, at UNC-CH the EBMS will collect utility usage information, real time utility rates, and calculate real time $/sqft usage at the building level. The impact of control strategy modifications and building performance deficiencies can be reviewed in real time.

This integration method also gives traditional building HVAC control systems the ability to have new custom applications developed economically for non-traditional clients by using readily available software development platforms like Microsoft's .NET or JAVA's J2EE. These platforms are easy to use and, hence, give the facility management engineer the ability to create custom applications geared specifically to their organization.

EBMS provides the facilities group at UNC-CH the ability to publish services for other departments. For example, web access to a selected subset of data points can be provided to a researcher to monitor his laboratory conditions remotely. An administrative assistant can set up occupancy schedules and temperature preferences for several dozen conferences simultaneously from their own spreadsheet. The classroom schedules setup by the registrar's office can be automatically incorporated into the HVAC occupancy schedules.

The EBMS allows development of cooperative energy use agreements between utilities and the building occupants. Users may subscribe to various levels of

demand management. Higher levels of demand management could result in preferential rate structures, as compared to lower levels of demand management.

Creation of the EBMS involved a thorough understanding of UNC-CH campus requirements, and the various BAS systems currently installed within the campus. The resultant request for proposal (RFP) package included all bidding requirements, technical specifications, legacy (a term used to describe older systems) point-list, and network architecture requirements. For a systems integrator to bid the project accurately, knowledge of the existing BAS information was critical. An exhaustive survey of the campus buildings was undertaken to document the existing BAS system(s) and develop a detailed legacy point-list. The point-list included information on system type, building level and system level, controller types, manufacturer, software and hardware revisions currently in use, and planned for EBMS integration. In addition an extensive survey was done to confirm current workstation functionality and graphic utilization, ensuring the EBMS would match certain characteristics common to these existing interfaces.

The RFP and subsequent procurement was developed with a two-step process in mind. The goal of the RFP was to identify the best "value" solution and not necessarily be limited to a low-bid option. The rules of procurement in the state of North Carolina allowed for this sort of procurement strategy when purchasing technology-based solutions. The first step required the bidders to submit a detailed technical proposal based on the RFP package pre-qualification requirements. Where a technical requirement could not be met, the bidders were required to submit an exception or alternative solution. The evaluation criteria emphasized the need to

Specifications:	Maximum Points:
Technical Specification Requirements	200
Project Work Plans	50
Training Plan	50
Total Cost of Ownership	300
Financial Information	50
References	125
Corporate Experience	50
Key Personnel Experience	125
Other Value Added Services	50
TOTAL	1,000

Figure 23-3. Sample Evaluation Criteria
Note: Actual numbers may vary from state to state and site to site

meet, or improve on the technical requirements established under the RFP. References and experience of key personnel on the team were also evaluated. As indicated in the scoring sheet described below, the total cost of ownership was given the most weight supporting the goal of the university to find the best first cost and life cycle cost solution. The second step involved subsequent interviews, questions and answers, and actual price proposals. In accordance with the local state procurement office, the "best value" response would be awarded the project.

CONCLUSION

UNC-CH entered this project with the goal of developing the highest level of operational support to the traditional HVAC customers, becoming better stewards of the energy being consumed, and all while the campus continues to expand in size and complexity. Creating such a system based on IT standards will enable the HVAC facility services staff to have a much more significant impact on the operation and management of the campus. At the same time UNC-CH will also have the ability to implement enhanced demand management strategies and analyze building energy consumption in near real-time fashion.

EBMS is a new approach to multiple building operations that allows traditional BAS to actively interact with new and existing enterprise systems being utilized by both business and educational units all across campus. This interaction is made possible through the use of normalized data (XML/SOAP) which can be shared with and drawn from existing enterprise programs such as space planning, registration, billing or any number of other enterprise level systems.

The EBMS will provide UNC-CH with the management and analysis tools needed for system diagnostics and pro-active management (i.e., trending, occupancy scheduling, and alarm management). These and many other application examples will give the HVAC facility services staff the tools necessary to ensure occupant comfort, energy conservation, and assets that are effectively managed. This can now be done at the lowest possible cost and with least number of staff.

At UNC-CH, the EBMS will allow the traditional BAS to move from a proprietary and private world, little understood by the building occupants, to a system which everyone can actively use to enhance their use of the space they occupy.

Chapter 24

Web Based Information Technology— A Supply Chain for Energy Strategy in Hotel Operations

Khaled A. Elfarra, PhD, CEM, DGCP
General Manager, Engineering/Projects Department
National Energy Corporation—Egypt (NECE)
khfarra@link.net

ABSTRACT

Energy information systems (EIS) have been considered the main tool in assessing hotel operations, but they are dependent on the individuals who perform the operations, and based on predetermined operational procedures. This concept is still true in many hotels of Egypt, since financial and human resources are major obstacles against such automation. Development in both the tourism sector, and the commercial sector, competitively attracts international chain hotels management companies to bring new business approaches in operation into their market.

Originally, EIS were manually operated to monitor and organize building energy consumption and related trend data by the energy managers, and to enable the operations budget. Over the past decade, web based technology (WBT), or EIS , has been introduced. This technology helps perform key energy management functions such as organizing energy use data, identifying energy consumption anomalies, managing energy costs, and automating demand response strategies. Moreover, this technology needs infrastructure including hardware, software, human resources and financial capabilities, to properly perform the new energy approach strategy and to ensure its sustainability. In addition to the energy strategy, all operation resources must be included to enhance the performance and market niche of the hotel industry.

This chapter presents the development in hotel operation with respect to WBT and energy managers' operation strategy. Operation strategy and new business approaches are necessary for the implementation of WBT and automated EIS, when considering a project for one of the corporate management companies in Egypt.

INTRODUCTION

In the past, most of a hotel management company's business approach was centralized. All information was gathered in individual hotels and corrective actions were taken by the central management company. This process was suffering from continual monitoring of performance, and from quick response against performance improvement. This management strategy does not enhance roles and responsibilities by utilizing all involved hotel parties, from top management to floor employees.

With the changes in management strategies and the development of innovative technologies for communication and information handling, many hotel management companies are in the process of updating their hotels with new management technologies involving WBT and EIS.

Using WBT and EIS systems, hotel management companies (corporate) are promoting comprehensive operations strategies covering the following aspects:

- Gap Analysis;   Comprehensively assesses the existing situation of the hotel/ individual activities in order to set the baseline of comparison including all lines of operation.

- Planning;   Emphasizes the essential requirement for strategy setup including the necessary hardware, software, human and financial resources.

- Monitoring;   Mainly controls the different aspects of performance via benchmarking and preparedness of corrective actions.

- Linkage Protocols; Set up unified communication protocol between individual activities and corporate company to enhance the overall image of the company in view of customer satisfaction, higher profitability, more secured operations, and higher competitive strengths.

The motivation of the corporate company towards its individual activities will result in higher performance as well as full commitment from employees and managers. Hence, energy managers as a part of company operations will be enrolled in initiating proper energy programs and their performance. In the following sections of this chapter, each component of the operation strategy will be demonstrated in details considering data analysis for corporate hotel management companies in Egypt.

GAP ANALYSIS

The main objective of energy managers is to clearly identify the existing conditions, such as:

- Hotel electricity demand and relevant energy consumption with respect to hotel occupancy levels.

- Installed utilities that serve hotel performance, such as chillers, boilers, heat exchangers, piping

system, insulation level, indoor/outdoor lighting systems, distribution network for electricity and for water systems—either domestic use, chilled water, or hot water.

- Applied control systems, building management systems (BMS), local area network (LAN), and protocols in use.

- Existing maintenance programs and operational procedures.

- Applied international certifications in the hotel and the internal auditing system.

When a hotel energy manager has recorded the readings of consumption, there are many parameters that can be measured to assess performance. In Figure 24-1, hotel consumption is illustrated on a monthly basis over the year with respect to hotel occupancy levels. This figure can easily explain to the energy manager the trend in consumption with respect to the hotel occupancy level. Using this information, the monthly energy consumption can be illustrated versus the hotel occupancy, to obtain the hotel baseload. This baseload reflects the fixed operating expense, irrelevant of hotel occupancy levels. This fixed operating expense can be further investigated by the energy managers, and considered as a potential core project for energy conservation measures.

Referring to Figure 24-1, the energy manager can calculate the baseload, based on a simple approach

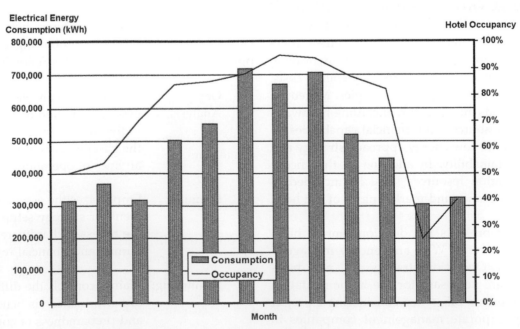

Figure 24-1. Hotel Monthly Energy Consumption and Occupancy Level

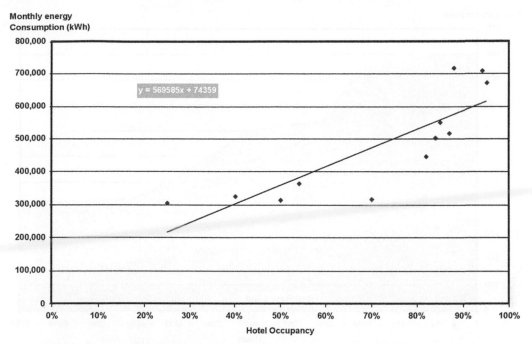

Figure 24-2. Hotel Monthly Energy Consumption Versus Occupancy Level

which computes a formula that represents the hotel's monthly energy consumption with respect to its occupancy level. Figure 24-2 shows the result of this computation. This benchmarking of the hotel performance can be set as the baseline for comparison purposes.

The hotel's monthly energy consumption is presented in a scattered graph; the curve for these data shows the relationship between the consumption and relevant occupancy. The equation (y = 569585x + 74359) shows that the hotel's baseload consumption is about 74,359 kWh/month, or about 15.5% of its total annual energy consumption. The energy manager can simulate the average hotel demand over the year in order to monitor hotel performance, showing the baseload versus the variable load which changes with respect to occupancy. Figure 24-3 illustrates the hotel load duration curve presenting the baseload, which is unrelated to hotel occupancy, and the variable load which is related to hotel occupancy. Moreover, the energy index for this hotel can be computed in order to compare this index with the other indices of the corporate company and the other standardized norms. The hotel energy index is presented in Figure 24-4.

Using this information, energy managers can start to study the deviation in energy index at different occupancy levels, even inside the hotel itself, and can set corrective procedures for normalizing the variable portion of the energy index. The computed energy index is tabulated in Table 24-1.

The computed values show that the values related to the baseload decrease with higher occupancy levels, where the values of the variable load vary according to occupancy levels. This gives an indication to energy managers that there is a potential for energy efficiency measures.

Using the same procedure, water consumption and hydraulic load of the waste water treatment plant of the hotel can be assessed by gathering water consumption and waste water effluents data. Therefore, the hotel energy manager can set the hotel baseline and include such data in an EIS of the corporate company. This helps the energy manager demonstrate the deviation in hotel energy performance with respect to benchmarking and setting the energy efficiency program for such activity.

The next step in hotel gap analysis is to log performance data sheets for all installed utilities. These utilities include boilers, chillers, pumps, air handling units, fan coils, waste water treatment plant, insulation level of internal distribution piping system, lighting system, etc. The performance data sheets will enable energy managers to conduct proper assessment of such utilities and set the deviations that might take place from the first commissioning of the utilities. The gathered database can be considered the core for the hotel/corporate company EIS.

In line with benchmarking and the utilities' database, energy managers must comprehensively evaluate existing maintenance programs and applied control systems. This task will identify the needs for building

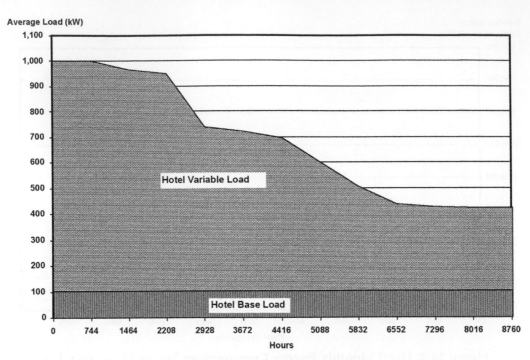

Figure 24-4. Hotel Energy Index

Table 24-1. Hotel Energy Index

Month	Room Nights (RN)	Energy Index (kWh/RN) due to		
		Base Load	Variable Load	Total
August-04	4,185	17.8	57.6	75.4
September-04	4,374	17.0	66.9	83.9
October-04	5,859	12.7	41.7	54.4
November-04	6,804	10.9	63.0	73.9
December-04	7,115	10.5	67.1	77.5
January-05	7,366	10.1	87.6	97.7
February-05	7,182	10.4	83.4	93.8
March-05	7,868	9.5	80.7	90.2
April-05	7,047	10.6	63.1	73.7
May-05	6,863	10.8	54.4	65.2
June-05	2,025	36.7	114.3	151.0
July-05	3,348	22.2	75.4	97.6

capacity, hardware/software upgrades, and the required updates for communication protocols among all individual hotels with their corporate company.

To incorporate WBT into hotel performance, the business strategy of the corporate company must be audited and gaps must be eliminated for enhanced performance. The auditing strategy must cover at least the tabulated items in Table 24-2 for different disciplines of hotel/corporate company operations.

PLANNING

Strategic planning in hotel operations is the keystone for efficient performance. In this regard, energy, and water are part of the hotel operation, so energy managers must include them in their operation budget to the management.

With large-scale WBT/IT availability in the market, corporate company management and its facility energy managers started thinking of planning corporate energy strategy. However, building a corporate energy strategy needs analysis of individual hotels as well as the corporate management company. This may not be a recipe for success, but without it a strategy is much more likely to fail. A sound strategic plan should:

- Serve as a framework for decisions or for securing support/approval.

- Provide a basis for more detailed planning.

- Explain the strategy to others in order to inform, motivate and involve.

- Assist benchmarking and performance monitoring.

- Stimulate change and become a building block for the next plan.

Table 24-2. Hotel/Company Strategy Components

1. Management	2.5 Functionality
Policy	Staffing & recruitment
Vision	Job Descriptions
Objectives & Targets	Outsourcing
Organizational structure & Behavior	Procedures and work instruction
2. Planning	Communication Channels
2.1 Operational	Documentation Control
Baseline & Bench Marketing	Technology Maintenance
Technology Applied Innovation	Reliability, Maintainability & Sustainability
History and Data Manipulation (rate of growth)	3. Implementation
Human Resources	Structure and Responsibilities
QC & QA	Regulations and Legislations
Material & Supply	Internal Auditing
Inventory for Products and Material	Continual Review for Business Plans
Products Categorization	Corrective Actions
R&D Resources	Documentation and Control
2.2 Sales & Marketing	IT & WEB Technology
Sales Force	4. Performance Measurement
Sales Tracking, Monitoring & Resources	Timely Implementation of all Aspects
Distribution network & Promotion	Efficiency
Credit Terms	Effectiveness
Customers Segmentation	Impact
Market Survey	Sustainability
Sales Services	5. Management Review
Market Opportunities	Review Audits Outcome
Contribution Margin	Improve/Review Corrective Actions
Marketing Resources & Competence	Dedicate Required Resources
2.3 Capacity Building	Measure Success
Training Needs	Enhance Strategy
Training Schedules & Programs	
Skills Update	
2.4 Finance and Financing	
Budgeting	
Cash Flow	
Finance Support	
Fund Raising	
Return on Investment	
Financing Controls	

A strategic plan is not the same thing as an operational plan. The former should be visionary, conceptual and directional in contrast to an operational plan which is likely to be shorter term, tactical, focused, implementable and measurable. As an example, compare the process of planning a vacation (where, when, duration, budget, who goes, transportation—all strategic issues) with the final preparations (tasks, deadlines, funding, weather, packing, etc.—all operational matters). A satisfactory strategic plan must be realistic and attainable so facility/energy managers and entrepreneurs can think strategically and act operationally. Hence, on setting up the plan for the corporate energy strategy, all the facility/energy managers of the individual hotels should assess their facilities in order to uncover the strengths, weaknesses, opportunities, and threats (SWOT) for the corporation. For the example project, the SWOT analysis is tabulated in Table 24-3.

Table 24-3 encourages the corporation to improve its weaknesses, and to understand its strengths. This is the responsibility of facility/energy managers at the individual hotels, who must tie in their locations in Egypt and the world-wide hotels of the corporate company. Therefore, in building the company energy strategy, building a capability for all relevant staff is a virtual necessity.

A critical review of past performance by hotels' owners and management, and the preparation of a plan beyond normal budgetary horizons require a certain attitude. Some essential points to keep in mind for the energy strategy during the review and planning process include:

- Relate to the medium term, i.e., 2-4 years.
- Be undertaken by owners/directors.
- Focus on matters of strategic importance.
- Be separate from day-to-day work.
- Be realistic, detached and critical.
- Distinguish between cause and effect.
- Be reviewed periodically.
- Be written down.

This then leads to strategy development covering the following issues:
- Vision
- Mission
- Values
- Objectives
- Goals
- Programs

Figure 24-5 presents the process of building the energy strategy and its main elements.

WBT PLANNING

WBT is an essential part of corporate energy strategy execution. The role of facility/energy managers is to assess current resource needs of their hotels and the extra requirements for building their systems. Most of the hotels of the corporation have old control systems without communication protocols. Therefore, the development plan will request some hardware to incorporate WBT in these hotels. Under close collaboration with hotel energy managers, the system is designed to facilitate the performance and increase the utilization efficiency of available resources. Figure 24-6 shows the proper setup of such a WBT system that serves as a communication tool among all hotels of the corporate company in the initial phase. The next phase will enable communication with all interested parties outside the corporation.

In setting up a WBT system for each individual hotel, the company vision is to unify the system for all hotels. This unification will have the following components:

1. Full assessment of the existing control systems and establishment of the required bill of quantity. The bill of quantity will list the field devices required, the calibration for operative devices, the proposed new controllers with their associated communication protocols and the software required for monitoring and control.

2. Set up the operational strategy for the system so that the software and hardware accommodate the monitoring, tracking, and control functions.

3. Document all operational procedures and forms using support software that facilitate the flow of information among all departmental business activities of the corporation.

4. Set up all required algorithms for data processing and manipulation, to enhance the benefits of gathered readings relevant to energy, water, and utilities performance.

5. Choose the activities that must be added to a WBT system such as energy consuming equipment, operating utilities, security access to guest rooms hotel, closed security monitoring systems, hotel check in/out procedures, etc.

The energy managers' role is to focus on monitoring, measuring, and verifying the consumption of en-

Table 24-3. Existing SWOT Analysis

STRENGTHS	WEAKNESSES
● Wide diversity in operations expertise. ● Distinguished relations with tourism authority. ● Outstanding Links with international institutions, firms, and associations. ● Availability of wide range of services to the customers. ● Long experience in organizing and implementing energy/environmental projects/programs. ● Hotels certifications. ● Lower operations budget without customer dissatisfaction. ● High quality and proficiency in services. ● Clear management policy for operations. ● Robust sales force and marketing.	● Gap in benchmarking. ● Lack of energy projects finance. ● Lack of stipulating data and its correlation with planned demand on energy. ● Obstacles and barriers against the sustainability concept. ● Financing agreement between Management Company and hotels' owners.
OPPORTUNITIES	**THREATS**
● Enhancement of networking and collaboration within the regional countries. ● Exposure to know how and technology transfer. ● Partnership with different stakeholders. ● Enhancement of "win-win" relationships. ● Demonstration for energy efficiency measures of proven technology. ● Build capacities and awareness.	● Regulations and legislations imposed in the region. ● Inadequacy in energy codes and standards. ● Uncertainty of data and information gathered from the region. ● Booming demand on energy and market needs. ● Energy prices, subsidy, and rate structures. ● Energy Policy and strategy in view of globalization and free markets. ● Economic growth rates and reforming.

ergy and water to build the operational indices of their facilities. This role has been implemented for some of the hotels under this project study, and the existing indices are tabulated in the Tables 24-4 through 24-8 covering the monitored water and energy consumption.

For energy managers, Tables 24-4 through 24-8 indicate hotel performance and focus on opportunities for measures of conservation or higher efficiency performance. These data are manually gathered and processed so error deviation is probably high and some decisions might be taken at higher risk considering the investment required. The automation process, even though a high initial investment is required, will avoid errors, accelerate corrective actions, increase company image, and upgrade company market competence.

WBT PERFORMANCE MONITORING

The implementation of WBT in the corporate company provides a virtual basis for improving performance in different disciplines of individual hotels. However, such implementation needs continuous monitoring and evaluation by facility and energy managers to comprehensively maintain system performance at a high level. The monitoring and evaluation matrix is presented in Table 24-9. WBT project monitoring must maintain full involvement of different interested parties of the corporate company. This involvement will help enforce continual improvement, considering the need for commissioning and re-commissioning of these types of technology.

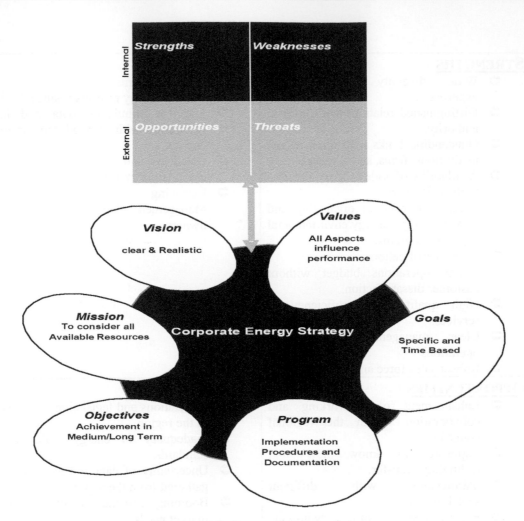

Figure 24-5. Energy Strategy Setup

Table 24-4. Annual Electrical Energy Consumption and Cooling Load

Hotel #	# of Guest rooms	Annual Occupancy	Room Nights (RN)	Installed Cooling Capacity (TOR)[1]	Electrical Energy Consumption (kWh)	Cooling Consumption (TORh)[2]
1	191	65%	45,315	107	2,145,674	625,822
2	195	78%	55,517	400	4,195,231	1,174,665
3	82	75%	22,448	70	2,333,258	482,743
4	300	72%	78,840	109	2,558,390	639,598
5	207	68%	51,377	211	2,260,000	652,889
6	312	74%	84,271	250	5,545,300	1,512,355
7	207	83%	62,711	465	5,000,000	965,517

[1]Tons of Refrigeration   [2]Cooling load consumption in tons of refrigeration hours.

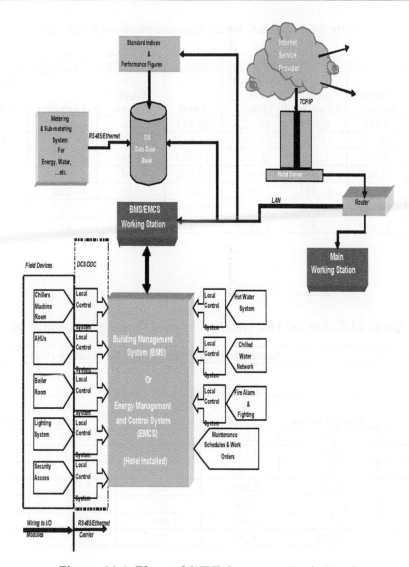

Figure 24-6. Planned WBT System at Each Hotel

Table 24-5. Annual Electrical Energy and Cooling Load Indices

Hotel #	Equivalent Full Load Hours for Chillers Operation	Electrical Energy Consumption (kWh/RN)	Cooling Consumption (TORh/RN)
1	5,849	47.4	13.8
2	2,937	75.6	21.2
3	6,896	103.9	21.5
4	5,868	32.5	8.1
5	3,094	44.0	12.7
6	6,049	65.8	17.9
7	2,076	79.7	15.4

Table 24-6. Annual Water Consumption and Indices

Hotel #	Annual Water Consumption (m³/yr.)				Consumption (lit./ Room Night (RN))			
	Guest Rooms	Laundry	Kitchen & Others	Total	Guest Rooms	Laundry	Kitchen & Others	Total
1	9,485	8,963	10,522	38,971	430	198	232	860
2	30,534	13,740	16,794	61,068	550	248	303	1,100
3	8,867	2,559	5,971	17,397	395	114	266	775
4	33,901	16,951	16,951	67,802	430	215	215	860
5	28,258	11,303	16,955	56,515	550	220	330	1,100
6	46,349	19,930	26,419	92,698	550	237	314	1,100
7	32,610	14,674	17,935	65,219	520	234	286	1,040

Table 24-7. Annual Hot Water and Space Heating Consumption

Hotel #	Annual Energy Consumption for Heating (GJ)				
	Domestic Hot Water				
	Guest Rooms	Laundry	Kitchen & Others	Total	Space Heating
1	1,511	2,054	424	3,989	2,330
2	2,368	3,149	677	6,193	4,374
3	688	586	241	1,515	1,798
4	2,629	3,884	683	7,196	2,382
5	2,192	2,590	683	5,465	2,431
6	3,595	4,567	1,064	9,226	5,632
7	2,529	3,363	723	6,614	3,595

Table 24-8. Annual Hot Water and Space Heating Indices

Hotel #	Annual Energy Consumption for Heating (kJ/RN)				
	Domestic Hot Water				
	Guest Rooms	Laundry	Kitchen & Others	Total	Space Heating
1	33,349	45,325	9,355	88,029	51,429
2	42,656	56,713	12,187	111,557	78,793
3	30,635	26,122	10,717	67,474	80,084
4	33,349	49,266	8,662	91,278	30,210
5	42,656	50,412	13,295	106,363	47,322
6	42,656	54,193	12,631	109,479	66,830
7	40,329	53,620	11,523	105,472	57,334

CONCLUSIONS

This chapter presented the setup of an energy strategy in one of the corporations that manages a group of hotels in Egypt. The setup of the energy strategy is based on WBT and EIS. The main goal of this chapter is to clearly define the real connection between energy strategy and the overall system performance for facility/energy managers. The main indices for the hotel business were presented, as they are clear indicators for continual improvement in facility performance.

The identification of WBT system components and requirements will motivate the proper implementation of a company energy strategy, achieving for the

Table 24-9. Monitoring and Evaluation Matrix

Level	Issue	Monitoring Indicators and/or Criteria	Means of Observing	Freq.	Monitoring Procedure
Activity Level	Timely Implementation of Activities	Number of project activities implemented in time	- Review Contracts - Contractor Assessment Form - Samples Visits	Each Contract	Members use routine Monitoring Form
	Service Provider Performance	%age of approved deliverables in time	- Review Contracts - Contractor Assessment Form - Training Evaluation Sheets - Samples Visits	Each Service Provider	Members use routine Monitoring Form
	Meeting Service Requirements	Number of project indicators being met	- Indicators - Review Contracts - Contractor Assessment Form - Samples project visits	Each Contract	Members use routine Monitoring Form
	Relevance and Quality of Project Design	How the stated objectives correctly address the identified problems or real needs?	- Analysis of Relevance through Routine Project Visits, Interviews, Analysis of Plans	Samples	Members use special Monitoring Form
Activity and Overall Objectives Levels	Efficiency	How well the various activities transformed into the intended results?	- Analysis of Efficiency through Routine Project Visits, Interviews, Analysis of Plans	Samples	Members use special Monitoring Form
	Effectiveness	How far the results used or their potential benefits were realized?	- Analysis of Effectiveness through Routine Project Visits, Interviews, Analysis of Plans	Samples	Members use special Monitoring Form
	Impact	How far the activity objectives helped in realizing the planned overall objectives and Log frame indicators?	- Analysis of Impact through Routine Project Visits, Interviews, Analysis of Log frame, and Plans	Samples	Members use special Monitoring Form
	Sustainability	How far the positive outcomes of the project are likely to continue after the project commissioning?	- Analysis of Sustainability through Routine Project Visits, Interviews, Analysis of Log frame, and Plans	Samples	Members use special Monitoring Form

company: a) policy, b) vision, c) mission, d) values and objectives, and e) performance programs.

Operations automation thru WBT offers a central and accurate database for facility/energy managers. It reduces the effort in cross checking data validity and accuracy, and properly assesses the mandatory requirements for high efficiency performance of the facility.

References

Clark W. Gellings and John H. Chamberlin, "Demand Side Management: Concepts and Methods," Fairmont Press Inc., 1993.

"Annual Reports of Energy Conservation and Environment Project (ECEP)," 1990-1997, Cairo, Egypt.

"Energy Monitoring, Targeting, and Accounting," ECEP Manual, ECEP/DRTPC, Cairo University, 1997.

Albert Thumann and D. Paul Mehta, "Handbook of Energy Engineering," Fairmont Press Inc., 2001.

"Combined Energy and Environmental Audits for Eco-Tourism Sector at Hurghada and Red Sea Area," ECEP/DRTPC, Cairo University, 1996-1997, Cairo, Egypt.

Khaled Elfarra, "Demand and Energy Management for HVAC Load in Egypt," the First International Conference on Electrical Energy in the Syrian Arab Republic, Nov. 27-29, 1995, Damascus, Syria.

"Study on Energy Patterns and Analyses of Commercial Sector at Cairo Governorate," Energy Conservation and Environment Program, Cairo University, 1998, Egypt.

Khaled Elfarra, Adel Khalil, Hindawi Salem and Osama Elbahar, "Energy Efficiency Program in Egypt: Objectives and Implementation," First International Conference on Energy Conservation, February 12-14, 2000, King Fahd University of Petroleum and Minerals, Dhahran, Saudi Arabia.

Section IV

Web Based Enterprise Energy and Facility Management Systems in the Entertainment Park Industry

Chapter 25

How Disney Saves Energy (Hint: It's Not *Magic*)

Paul J. Allen, P.E.
Walt Disney World

ABSTRACT

THE WALT DISNEY WORLD RESORT near Orlando, Fla., is among the most highly visited destinations on earth. Its "campus" consists of hundreds of buildings that include world-class hotel and conferencing centers, exotic ride environments, and precisely controlled spaces for horticulture and animal care.

In addition to a Wall-Street eye on the bottom line, Walt Disney himself encoded the company's DNA with an ethic toward conserving natural resources and the environment that remains to this day as a program called Environmentality. Environmentality is a way of thinking, acting, and doing business in an environmentally conscientious way—from saving energy and water to reducing waste and other environmental impacts.

At Disney, energy is a key to success. Air-conditioning, refrigeration, compressed air, and water-moving systems for buildings, rides, and transportation all run primarily on electricity and natural gas. To maximize energy conservation and efficiency while minimizing costs and environmental concerns, the Walt Disney World Resort has implemented a state-of-the-art energy management program (EMP) that can serve as a role model to owners and administrators of public and private facilities.

This chapter describes the energy management program at the Walt Disney World Resort near Orlando and discusses its results in terms of energy and cost savings. Perhaps in doing so, other facility owners worldwide will develop their own energy management programs and cultivate the economic, energy, and environmental benefits enjoyed by Disney [1][4][6][9][10].

THE ENERGY STAR FOUNDATION

The cornerstone of the Disney EMP is its strong relationship with the U.S. Environmental Protection Agency (EPA) through the EPA Energy Star Buildings program, which has five main components:

- Building tune-up (recommissioning).
- Energy-efficient lighting (Green Lights).
- Load reductions.
- Fan-system upgrades.
- Heating- and cooling-system upgrades.

The relationship between the Walt Disney World Resort and the Energy Star Buildings program was established in 1996, when Disney implemented the EPA Green Lights program across 17 million sq ft of facilities. This was completed in 1998 and resulted in annual electrical savings of 46 million kWh. Also in 1998, the Walt Disney World Resort began the implementation of numerous other cost-effective energy-saving projects.

Disney's projects included:

- Optimizing compressed-air-system controls
- Upgrading hot-water-boiler controls
- Retrofitting variable-speed-drives into air, pumping, and chilled-water systems
- Retrofitting demand-controlled ventilation into convention-center spaces
- Upgrading and integrating energy-management-systems (EMS), including networking one EMS vendor's stand-alone EMS to centralized network-based servers.
- Installing utility-submetering systems in areas operated by non-Disney companies working in Disney facilities for utility cost recovery purposes.

In aggregate, the efforts Disney has undertaken since 1996 have resulted in a 53-percent internal rate of return (IRR) and metered annual reductions of approximately 100 million kWh of electricity and one million therms of natural gas.

THE DISNEY EMP FRAMEWORK

Disney's EMP has three main components: the energy management systems (EMS) that are installed in each building or facility; the energy information system (EIS), which is a suite of information technologies that works with the EMS to provide data and information to energy managers and other stakeholders; and Disney staff (called "cast members"), who collectively participate in the EMP. It's the combination of technology and people that makes Disney's EMP successful and sustainable.

ENERGY MANAGEMENT SYSTEM

Overview

The energy management systems (EMS) used at the Walt Disney World Resort are used to control energy consuming equipment—primarily for heating, ventilating and air conditioning (HVAC) equipment and lighting control. The parameters of greatest interest are temperature and humidity setpoints and equipment operating time schedules.

Over the years, Disney has installed a variety of energy management systems from different vendors, which it continues to operate. One vendor's system controls more than 80 percent of the installed EMS base. This system was upgraded to a centralized server-based system connected to the corporate ethernet-based intranet. This upgrade provided Disney's EMS with a standard and stable hardware and software platform along with the other benefits shown below:

- Review of EMS field panel programming and real-time operation can be made "globally" through any desktop PC on the corporate network.
- The EMS program and data are stored on network servers that are maintained by Disney's Information Services team. Backups are made daily.
- Automatic reset of equipment time and setpoint schedules are made daily from a server-side control program.
- Data collection for both EMS point trends and utility meter data can be collected and used by the energy information system for quick and easy display.
- Maintenance and training of the EMS is simplified.
- Services contracts are minimized or eliminated.
- EMS spare parts inventory is minimized.

Disney EMS Design Strategy

Disney's approach to EMS is simple and straight-forward. By standardizing on one manufacturer's EMS, Disney has achieved the following desirable results:

- Single-Seat User Interface
- Compatible with Existing EMS
- Easy-to-Use
- Easily expandable
- Competitive and Low-cost
- Owner Maintainable. No service contract required

Disney's EMS vendor's system uses a proprietary protocol, which is not "open." Standardizing on one EMS-vendor's system minimizes the importance of the EMS protocol used. Disney considers the ability to design, install, program, create graphics and continuously improve their EMS in-house the most important feature.

With one single-vendor EMS, this begs the question of how Disney keeps procurement competitive. Here's how: During a procurement effort, Disney will:

- Design the EMS controls as part of the construction documents. This includes details on EMS-panel wirelists, EMS communications interconnections, and EMS panel mounting details.
- Prescriptively specify every detail of the EMS (sensors, actuators, wire, etc.) so there is absolutely no doubt what will be installed and how it will be installed.
- Competitively bid the installation of the wire, conduit, sensors and actuators.
- Owner furnish the EMS panel and everything inside (EMS hardware, interface modules and power supply)

During construction of the EMS, there are actually two contractors. The controls contractor scope of work is competitively bid and includes the installation of all the end devices and pulling all the control wires (properly labeled per the wirelist detail drawing) back into the EMS control panels. The EMS contractor performs the wire termination, programming and startup of everything inside the EMS panel. The EMS contractor is very knowledgeable of how Disney's EMS vendor's equipment is installed and configured.

The key to Disney's success is that it is in complete control of the process. The EMS vendor is part of the process, not the master.

ENERGY INFORMATION SYSTEM

Utility Reporting System

The philosophy, "If you can measure it, you can manage it," is critical to a sustainable EMP. Measurement for management is the job of the EIS. The EIS is a suite of programs and computers that take data from the EMS and other data collection sources and churn them into actionable information for use by operators and managers. The EIS measures energy at the facility level and tracks the resulting energy conservation efforts over time.

Continuous feedback on utility performance pinpoints problems in the EMS that need attention. Such feedback also drives Disney's incentive program, which keeps people actively seeking to reduce consumption and expenses without creating new problems.

Disney created their own web-based EIS that uses an off-the shelf database management system to store the vast amount of energy data they collect [5][7][8]. The custom program, called the utility reporting system (URS), resides on a network web server. The URS gathers, stores, and processes monthly utility bill data and hourly meter data from a variety of data collection sources. The URS's reports are created in web-accessible (HTML) formats and can be reached via the Disney intranet.

One popular feature of the URS is a "report-card" format for publishing utility data and historical information. The report card is distributed via e-mail on a monthly basis, with each message containing high-level (summary) information and hyperlinks allowing "point-and-click" access to greater detail. Some links are to graphs that compare current data to data from up to 12 previous months. Also, data can be filtered to compare one Disney area against others. For example, how is Epcot performing relative to Animal Kingdom? Such comparisons stoke a healthy spirit of competition among area managers.

Specialized reports are used to monitor and report utility usage in areas operated by non-Disney companies working in Disney facilities, which helps to keep them aware of their usage rates. By measuring actual energy consumption instead of a square foot allocation, operating participants are motivated to manage their energy usage to keep their utility expenses low.

Disney Goes to School

The Walt Disney World Resort participated in a public/private effort to develop an energy information system, called utility report cards (URC), to help Orange County Public Schools (OCPS) better manage energy costs. The URC program was based on the energy information system methods and techniques developed at the Walt Disney World Resort. The URC is a web-based energy-information system that reports and graphs monthly utility data for schools.

Each month, a web-based report is automatically generated and e-mailed to school principals and staff as encouragement to examine their school's electricity usage (energy efficiency) and to identify schools with high-energy consumption needing further investigation. The URC also is intended for teachers and students to use as an instructional tool to learn about school energy use as a complement to the energy-education materials available through the U.S. Department of Energy's EnergySmart Schools program (ESS). To see how the URC operates, go to http://www.utilityreportcards.com and click on "URC Live."

The URC was created to help OCPS staff understand and, therefore, manage their utility consumption and associated costs. The URC allows school principals to become aware of how their school is performing relative to a projected benchmark and to other schools of similar design and capacity. Giving recognition to schools that improve performance from prior-year levels could create a spirit of competition with the opportunity to recognize success. Those schools identified as high-energy users become the focus of attention to determine the reasons for their consumption level and ultimately to decrease the energy used. All of this is done by using the monthly utility data that are provided electronically at minimal or no cost to the schools by the utilities.

PEOPLE ARE THE REAL ENERGY STARS

Organization

Conservation has always been one of Disney's core values. In a public-service announcement recorded while he was the honorary chairman of National Wildlife Week, Walt Disney defined "conservation" and thereby set a tone for Walt Disney's Environmentality ethic:

"You've probably heard people talk about conservation. Well, conservation isn't just the business of a few people. It's a matter that concerns all of us. It's a science whose principles are written in the oldest code in the world, the laws of nature. The natural resources of our vast continent are not inexhaustible. But if we will use our riches wisely, if we will protect our wildlife and preserve our lakes and streams, these things will last us for generations to come."

Disney's Environmentality program provides the framework behind Disney's resource conservation efforts. Everyone has a role to play. There is a dedicated staff of energy conservation engineers and technicians who orchestrate the energy conservation efforts and keep the program moving forward by refining the EMS, EIS and other program components.

Management supports the EMP by promoting and encouraging energy savings efforts and authorizing budgets sufficient to get meaningful work done. New projects are considered based on their expected internal rate of return (IRR). There may also be other non-financial benefits that weigh in to the energy project funding decision.

Disney also recognizes that cast members need to be involved in the EMP to establish a facility-wide sense of ownership and accountability for energy usage. Through Disney's Environmental Circles of Excellence, Disney cultivates Environmentality instead of dictating it. These local teams meet monthly and work on various resource conservation projects in their respective park, resort or support area. Using the Environmentality motto, "Every little bit makes a BIG difference" lets cast members participate in identifying energy waste no matter how small the detail.

Energy Star Tool Bag

The Energy Star Tool Bag was created as a guide to help cast members look for energy waste.

Overall Building
1. Heating, ventilating and air conditioning (HVAC).
   • Turn off units during unoccupied hours.
   • Adjust temperature and humidity setpoints to minimize unnecessary heating and cooling.

2. Turn off interior and exterior lighting when not required.

3. Perform walk-throughs—look for energy waste
   • Any exterior lighting on during the day?
   • Note "too cold" or "too hot" areas.
   • Note any areas that are "too humid"
   • Close open doors during hot or cold weather
   • Is all non-essential lighting turned off/dimmed down?
   • Are there any PCs left on?
   • Are there any decorative fountains on?
   • Can building facade or other decorative lighting be turned off?

4. Review utility metering reports and look for energy waste.

In the Office
1. Turn your lights off when you leave your office or conference room.
2. Program your PC monitor, printer and copier to "go to sleep" during extended periods of non-activity
3. Turn your computer off completely when you leave to go home.

In the Kitchen
1. Minimize kitchen equipment pre-heat times.
2. Turn cooking equipment down or off during slow periods of the day
3. Eliminate water waste, report leaking faucets.
4. Turn off kitchen hoods after closing.
5. Turn off or reduce lighting levels in dining areas & kitchen after closing.
6. Keep refrigerator/freezer doors closed. Install plastic strip doors on refrigerator/cooler doors.

In Convention Areas
1. Turn off lighting and HVAC equipment during unoccupied hours.

Swimming Pools
1. Adjust pool water heating temperatures to minimize natural gas consumption during winter months.

In Guest Rooms
1. Setback guest room thermostat to low cool.
2. Close drapes in guest rooms.
3. Keep sliding doors closed.
4. Turn off lights in guest rooms.

Energy Star Team

Cast members that participate in the Disney's EMP are part of the Energy Star team. The Energy Star team meets on a monthly basis and is comprised of cast members from parks, resorts and support areas. The meetings provide a great opportunity to share and discuss best practices and learn of vendors energy-saving products. If one team member wants to try a technology in his or her area, the product is installed and metered to measure the results. If it works, it can be rolled out to the entire property. The Energy Star team meetings also provide a good venue for providing on-going training to operators of the EMS and EIS.

The cast members that make up the Energy Star team are typically from maintenance or operations. Since the cast members that participate do so in addition to their normal job responsibilities, the most successful team members are the ones that want to participate. While knowing some technical details of energy management is helpful, more important is the ability to get things accomplished by effective networking within their areas. The dedicated energy management staff can generally provide the technical skills training needed for the team.

Utility Report Card

The challenge for any large scale EMP is to sustain the process and to keep everyone focused on saving energy. Disney uses their monthly utility report card as a scorecard on how each area is doing. By making the Utility Report Card easy to use and delivering it quickly and efficiently via email, management can easily determine how well their area is performing relative to other areas. Areas that have increased consumption are easily identified from the URC report. Management can question the increases and that is generally all that it takes to motivate those responsible individuals to take action. Knowing everyone is looking keeps the focus on finding new ways to saving energy.

Energy Star Awards

The Energy Star Awards program was developed to increase the awareness of energy usage among Disney management and cast members. It works by recognizing and rewarding successful energy conservation efforts and demonstrates that energy conservation can be simple and fun.

Because utility meter information is readily available in the URS, a report was developed to provide feedback on how well each area is doing relative to prior year usage. For example, Magic Kingdom, Epcot and Contemporary are all separately tracked and reported.

A spirit of competition is created by ranking each area based on percent change from the prior year. The areas with the greatest reductions rise to the top of the list. The awards recognize those areas that are at the top and identify those areas on the bottom as being in need of improvement.

Each year, a report is generated to show the award winners. Details on the award winners' accomplishments are highlighted in Disney's corporate Environmentality report, Enviroport. No cast member individually benefits financially from the EMP. Recognition for doing a good job has been the only incentive needed.

THE BUILDING TUNE-UP PROCESS

Overview

The building tune-Up (BTU) [2] or re-commissioning [3] step in the Energy Star Buildings program has been very cost-effective for Disney. The BTU process concentrates on optimizing the operation of the EMS and generally results in a 5-20 percent reduction in utility usage in a very cost-effective manner. Disney's savings attributed to the BTU process have been approximately equal to the savings resulting from its Green Lights program, but at a fraction of the cost (Figure 25-1).

At the start of a BTU effort, teams are formed from the engineering and operations departments to review the building and EMS control devices, programming and settings. The BTU process typically results in the following:

- Reduced utility consumption by optimizing air conditioning and lighting time schedules and setpoints
- Improved EMS performance by improving energy management system programming & documentation
- A list of corrective actions identified by monitoring HVAC system operations
- Measured utility savings using the URS

The BTU process is simple and can be started at anytime. The area Energy Star team member generally organizes the BTU team for his or her area. An energy management engineer facilitates BTU teams with support from facility-engineering and building operations and maintenance departments. Initially, 1-hr meetings are held once a week. This may not seem like much time, but a slow and steady pace is best for this type of work. Most of the detailed work is spent between the weekly meetings in the review and documentation of the EMS control sequences, setpoints, and time schedules. Each building and each HVAC system is evaluated one system at a time until all of the systems have been reviewed and everything is working properly.

Computer Programs

The key to a successful BTU program is to manage the details. To help with that effort, Disney uses several custom software tools that work together to keep EMS settings intact.

The facility time schedule (FTS) program manages equipment time schedules and temperature setpoints in a central database. Because time schedules and setpoints can be changed from their optimal settings and not set back, the FTS was created to reset them automatically on a daily basis.

The building tune-up system (BTUS) is a web-based program that allows users to view EMS control settings without accessing the EMS. Critical parameters of EMS operation are stored in a relational database. A web-accessible program displays the information via a web browser. The BTUS shows:

- HVAC equipment and the area serviced, including color-coded floor plans of the area serviced.
- Equipment time schedules and setpoints.
- Action items for follow-up repair for each HVAC system.
- Links to EMS trend graphs and historical data of utility-consumption.

The energy management tune-Up system (ETS) is a web-based program that lets users search through their energy management systems for equipment not working properly. The ETS is actually two separate programs. The first program reads each EMS point and records those values in a central database. The data are recorded twice a day—during the daytime operational period and during after midnight closing period.

The second program provides the users with a web-based interface to scan though their area data to look for the following EMS problems:

- Chilled water valves that are closed but the supply air temperature is <60F (chilled water valves not closing).
- Fans that are commanded off but the fan status is on (fan is not turning off).
- Temperature and humidity sensors out of range (sensors defective).
- Override forces on control points

PROJECT SUPPORT

New projects and renovations provide a great opportunity to incorporate energy saving products into the design. Even though the incremental cost to install an energy saving project would most likely be lowest if installed during a scheduled facility downtime or as part of new construction, the project budget might not be able to support the increased incremental cost. Estimating the potential internal rate of return (IIR) resulting from the expected annual cost savings helps justify increased project budget. A business case can be prepared that details the scope, shows alternatives,

describes potential risks and rewards of the project. A strong business case and a high IRR will certainly help sell the project to management.

Replacement of heating, ventilating and air conditioning (HVAC) equipment is an opportune time to incorporate more efficient equipment and new energy management system controls. For example, with the Walt Disney World Resort's hot humid climate, the addition of a heat pipe wrapped around the cooling coil in a 100% makeup air unit provides an efficient method to control humidity while minimizing cooling and reheat energy costs [11].

CONCLUSION

Environmentality is part of the way of life at the Walt Disney World Resort. Energy management programs are good for the environment and make good business sense.

The Disney EMP began by working with and learning from the well-established Energy Star Buildings program, which is available to everyone at http://www.energystar.gov.

Disney adopted Energy Star and then tailored it by integrating commercial energy management systems with a custom energy information system. This technology-based solution is used throughout the Disney World Resort organization by administrative managers, engineering, operations and maintenance staff, and cast members. This combination of people and technology has resulted in a sustainable energy management program at Walt Disney World. As the Walt Disney World Resort continues to expand, these programs will continue to play an important role in reducing energy costs in both new and existing facilities.

References

[1] How Disney Saves Energy and Operating Costs, Paul J. Allen, P.E., Heating/Piping/Air Conditioning (HPAC) Engineering, January 2005.
[2] Continuous Commissioning[SM] in Energy Conservation Programs, W. Dan Turner, Ph.D., P.E., Energy Systems Lab, Texas A&M University, Downloaded from http://esl.tamu.edu/cc on Dec. 20, 2004.
[3] ENERGY STAR Buildings Upgrade Manual—Stage 2 Building Tune-Up, US EPA Office of Air and Radiation, 6202J EPA 430-B-97-024B, May 1998
[4] Disney's "Environmentality Program," Paul J. Allen, Brett Rohring, Proceedings of the Energy 2003 Workshop and Exposition, August 17-20, 2003
[5] Information Technology Basics for Energy Managers—How a Web-Based Energy Information System Works, Barney Capehart, Paul J. Allen, Klaus Pawlik, David Green, Proceedings of the 25th World Energy Engineering Congress October 9-11, 2002
[6] Sustainable Energy Management—Walt Disney World's Ap-

proach, Paul J. Allen, Proceedings of the Energy 2002 Workshop and Exposition, June 2-5, 2002

[7] Managing Energy Data Using an Intranet—Walt Disney World's Approach, Paul J. Allen, David C. Green, Proceedings of the Business Energy Solutions Expo November 28-28, 2001

[8] Measuring Utility Performance Through an Intranet-Based Utility Monitoring System, Paul J. Allen, Ed Godwin, Proceedings of the 23rd World Energy Engineering Congress October 25-27, 2000

[9] Walt Disney World's Environmentality Program, Paul J. Allen, Bob Colburn, Proceedings of the Business Energy Solutions Expo December 1-2, 1999

[10] Walt Disney World's Energy Management Program, Paul J. Allen, Ed Godwin, Proceedings of the Business Energy Solutions Expo December 9-10, 1998

[11] Applications of Heat Pipes for HVAC Dehumidification at Walt Disney World, Paul J. Allen, Khanh Dinh, Proceedings of the 15th World Energy Engineering Congress October 27-30, 1992

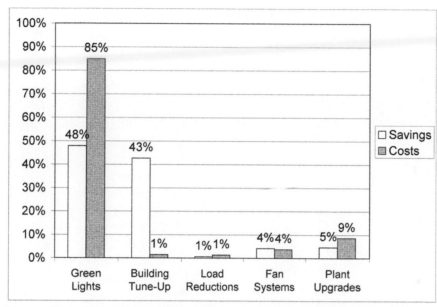

Figure 25-1. Walt Disney World Resort's Energy Star Program Results

Chapter 26

An Owner's Approach to Energy Management System Design

Paul J. Allen, P.E.
Chris Sandberg
Walt Disney World

ABSTRACT

THIS CHAPTER DESCRIBES an owner's approach to design, install and operate an energy management system (EMS). This approach bucks the multi-vendor "interoperability" trend by standardizing on one EMS system (i.e., one manufacturer). Maintaining a competitive procurement process with one EMS vendor might sound impossible, but the owner's EMS design approach makes this happen. The ability to design, install, program, create graphics, and continuously improve the EMS in-house is important to most owners. Costs for training, spare parts and EMS software licenses are minimized by using one EMS. This methodology works well. There are no surprises and in the end, it all works, which is the important thing.

INDUSTRY TRENDS IN ENERGY MANAGEMENT SYSTEMS

There is little doubt that future energy management systems will integrate information technology (IT) Internet standards. These open and standard approaches from IT have set the stage for the EMS vendors to develop systems which interface to the Web. Hopefully this move toward IT standards will help drive down EMS costs and make systems easier to use.

Aside from the impact that IT will have on future EMS, there are some fundamental features which owners have always desired and will continue to desire from these systems:

- Single-Seat User Interface
- Compatible with Existing EMS
- Easy-to-Use
- Easily expandable
- Competitive and Low-cost
- Owner Maintainable. No service contract required

Over the past decade there have been changes made by the EMS manufacturers to meet some of these desires. Each has introduced its own levels of difficulties:

- "Open" Protocols, LonWorks and BacNet
- Overlay Systems which interface with multiple EMS

The owner's EMS design approach is simple and straightforward. By standardizing on one manufacturer's EMS, an owner can obtain all of the above-mentioned desired features. The key to the success is that the owner is in complete control of the process. The EMS vendor is a partner in the process, not the master.

The Traditional EMS Design Approach

Let's take a look at the traditional EMS design process. Although this process has all the right goals and objectives, the results generally fall short of the owner's ultimate desires. Here is the traditional step-by-step process which most companies follow today:

A/E Design

1. The mechanical engineer completes the HVAC design—including chiller and boiler plant, pipe and pumping systems, air handlers, ductwork and temperature controls.
2. The control design is generally the last item to be completed because all of the other components must be designed first to know what is needed for controls.
3. The control drawings are generally schematic in nature and the specifications performance-based.
4. To allow for competitive bidding, several EMS vendors systems are called out as acceptable.
5. In an attempt to standardize the EMS for the owner, the specifications call out for the EMS panels to communicate with an "open" protocol.

The thought is that the system selected could be compatible with the existing EMS installations in some form.

Construction

6. The project is bid and a mechanical contractor is awarded the contract, more often than not having used a rough order of magnitude estimate for the controls based on a percentage of the mechanical contract.

7. The mechanical contractor solicits bids from several controls contractors and generally selects the lowest bidder.

8. The controls contractors scope is all-inclusive—controllers, sensors, actuators, cable, conduit, etc.

9. The temperature controls contractor prepares a submittal which shows the actual EMS design and materials required for the job.

10. The mechanical engineer reviews the submittal to determine compliance with the plans and specifications.

11. The EMS is installed by the controls contractor, programmed to meet the desired sequence of operation, started up and commissioned.

12. The system is turned over to the owner.

Operation

13. The EMS vendor contacts the owner and requests a service contract to maintain the system.

Let's pick this approach apart and see where the weak links are. First, the engineer's controls design is often not much more than a set of "typical" schematics accompanied by a narrative specification which describes the desired results with little or no "how-to." In other words, the engineer delegates the responsibility for the detailed EMS design to the controls contractor. This must be done, because the engineer does not yet know which EMS vendor will be selected. Even if a single EMS vendor was selected, it is very rare that a design engineer would be sufficiently knowledgeable with this system to produce a detailed design. Thus, the resulting EMS design is by nature somewhat vague and entirely performance-based.

Many times the engineer will specify that the EMS protocol be "open" (LonWorks or BacNet) to have the systems be "interoperable." This sounds good, but there is currently little commonality between different EMS vendors' low-level panel programming, and different service tool software is needed for each EMS vendors' system regardless of the "open" protocol. "Interoperable" ends up meaning "talks to the same front end, but

not peer-to-peer with other systems."

In the end, the owner gets a system which meets the original specification, but might not be the same as or even compatible with the existing facility EMS. The new EMS might require the owner to obtain an annual service contract because the system is so difficult to change, re-program, and repair. The owner operates this new system as just one of many disparate EMS systems in their EMS portfolio.

Several companies offer products which will interface to different EMS vendors systems to provide a common user interface and result in a "single seat." This approach comes with additional costs (both hardware and software) and results in an added level of complexity to the existing EMS. Furthermore, training, spare parts, and software licenses required for each EMS vendors system are still required.

The Owner's EMS Design Approach

The owner's approach to EMS design is based on two fundamental principles: (1) to ensure competitive bidding the EMS hardware shall be owner furnished (2) the design documents and specifications shall prescriptively define the EMS design. Let's look at the owner's EMS design process in more detail:

A/E Design

1. The mechanical engineer prepares all the HVAC design—including chiller and boiler plant, pipe and pumping systems, air handlers, ductwork, water and air flow diagrams and temperature controls.

2. The mechanical engineer fills in a temperature controls drawing showing the controls points and the sequence of operation and includes a controls specification based the owners EMS.

3. The design drawings also show all EMS panel locations with 120V power, locations of all space temperature, humidity and CO_2 sensors.

4. The electrical drawings detail all motor and lighting control schematics including control relays. All underground conduits are also shown on the electrical drawings.

EMS Design

5. The energy management engineer prepares the EMS design.

6. An EMS design drawing is prepared that includes the unit control panel (UCP) wirelist showing point numbers and sensor types for all input/output cables. EMS communication block diagram and the UCP enclosure mounting details.

7. A specification for the EMS panel work describes the work to be completed inside the EMS panel (i.e., cable termination standards, UCP programming standards).

Construction

8. The project is bid and a mechanical contractor is awarded the contract.

Owner

9. All of the EMS modules, EMS panel and other components used inside the UCP are owner furnished. The basic EMS programming is also owner furnished.

Controls Contractor—"Outside-UCP" Work

10. The mechanical contractor competitively bids the "outside-the-UCP" work which includes the wire/conduit/sensors/actuators as detailed EMS design drawings and specifications. A controls contractor is hired for this work.

11. The controls contractor prepares a submittal which show the equipment details for the job. This might be as simple as a set of material cut sheets, or they may produce control diagrams for each piece of equipment with the point number and cable information from the EMS design drawing.

12. The mechanical engineer reviews submittal to determine compliance with plans and specifications.

13. The controls contractor installs, labels and terminates all of the field devices and pulls cables back to the EMS panel leaving 10 feet of slack for final termination by owner or the EMS contractor.

14. The controls contractor completes a cable/device checklist to verify proper installation and labeling. A signed copy is submitted to the owner as notification for the EMS contractor to begin.

EMS Contractor—"Inside-the-UCP" Work

15. A separate EMS contractor familiar with the owners EMS is hired by the controls contractor to perform the "Inside-the-UCP" work. This includes wire termination on EMS modules and start-up. In special cases, the owner may self-perform the EMS contractor work.

16. The temperature control contractor and the EMS contractor work together to commission the EMS and verify the operation of each control point.

17. The system is turned over to the owner after building acceptance.

Operation

18. The owner coordinates the connection of the EMS to the corporate ethernet network and uploads all EMS panels to the central EMS server.

19. The owner prepares the EMS graphics and fine-tunes time and setpoint schedules.

With the owner's EMS design approach, the EMS is completely designed as part of the design documents. All details about the EMS are prescriptively defined in the design documents. The owner knows exactly what they are getting and knows that it will seamlessly integrate into their existing EMS.

Competitive bidding is used to select the controls contractor that will comprise the largest portion of the EMS installation cost. The installation of all the wire/conduit/sensors and actuators can be competitively bid because it was included in the design documents and shows each input/output point number for the EMS.

The owner negotiates with the preferred EMS vendor on unit pricing for their EMS control modules. These modules are provided as owner furnished material to the project based on the EMS design. Keep in mind that the success of this process is predicated on the idea of direct parts sale from the preferred EMS vendor to the owner at deeply discounted prices. Some EMS vendors may not be willing to do this.

Once the controls contractor has all of the field wiring properly labeled and pulled into the EMS panel, the EMS contractor completes the final wire termination, programming and startup. The owner's energy management team coordinates the addition of the new panels to the central EMS server as well as prepares the EMS graphical interface.

EMS DESIGN TOOLS

In order to create the EMS design drawings and specifications, the owner must be very knowledgeable of their EMS hardware and software inside and out. They should be able to take an EMS controller out of the box, connect input/output field wiring, program the controller software and add the graphical interface to the EMS server. This level of sophistication lets the owner define their EMS design process.

AC-701 Drawing

The AC-701 drawing is used by the mechanical design engineer to define the EMS control points and basic sequence of operation for the HVAC system. The AC-701 also includes any additional lighting EMS control points designed by the electrical design engineer. The AC-701 format was created to allow the mechanical design engineer to easily specify the control points desired by simply marking the points desired in the matrix. The AC-701 uses a schematic design and simple EMS point table format for quick and easy completion by the design mechanical engineer (see Figures 26-1 through 26-3).

AC-702 Drawing

The AC-702 drawing is used to assign the EMS control points specified on the AC-701 drawing to the actual EMS controller input/output points. Additionally, details on EMS panels communications interconnections and EMS panel mounting details are shown on the AC-702 drawing (see Figures 26-4 through 26-6).

The energy management engineer prepares this drawing for inclusion into the design documents. Since there is only one EMS vendors system used, the EMS design can be tailored exactly to the EMS controller used. Once the AC-702 drawing is completed, all of the EMS hardware is known and defined. This enables the owner to purchase the EMS hardware needed for the job.

Wirelist Program

Without a doubt, the most powerful tool in the EMS design process is a program called the "wirelist." The wirelist is a custom written database program that keeps track of all the input/output points for an EMS panel. The program allows the user to enter sensor code, the point name and description for all points. The software adds in the details on what wire to use and where the input/output wires are landed on the EMS controller. The wirelist provides the EMS panel details that are needed for the EMS design drawings.

The wirelist program is continually being added to and improved on. One of the most recent wirelist additions enabled the basic EMS controller program to be created directly from the wirelist. This saved a tremendous amount of time by eliminating the need to manually re-enter all of the basic configuration data into the EMS controller program. Since the wirelist is a database, it is used to record and document any changes made to EMS panels, well after the initial EMS design. A new wirelist EMS panel schedule is printed and installed in the EMS panel to keep the documentation up-to-date.

Specifications

There are two specifications used to define the EMS installation details. Specification 15950 is used to define the scope of work to be completed by the controls contractor. This is basically all of the work outside of the EMS panel. The 15950 specification prescriptively defines all of the sensors and actuators used in the EMS. This includes all of the sensors/wiring/conduit/actuators used for the EMS. It also defines that all EMS panel components, including the EMS panel itself, are owner furnished.

The 15951 specification is used to define the work completed inside the EMS panel. The controls contractor hires an EMS contractor for EMS panel wire termination and startup. In special cases the owner might use their own technicians to self-perform the EMS contractor work. The 15951 specification defines the approved EMS contractors, EMS panel wire termination standards, EMS software naming conventions and programming standards. The owner provides the basic EMS panel programming to the EMS contractor. The 15951 specification was developed from the many years of experience in wiring and programming EMS panels. The EMS contractor works closely with the controls contractor to assure a smooth start-up of all EMS equipment.

CONCLUSION

Making the EMS design part of the design documents helps an owner become the master of their EMS. This starts from an intimate understanding how their EMS works—both hardware and software. EMS design expertise comes from day-to-day EMS operational experience and from the development of EMS design standards.

The owner's EMS design approach is both simple and straightforward. Some may consider using only one EMS as a limiting factor, reducing competition. Instead, owners need to keep their focus on the end result—an EMS that is both easy-to-use and flexible enough to meet their changing facility needs.

TEMPERATURE CONTROL SCHEDULE

UNIT DESIGNATION	OPTIONS	ANALOG IN	DISCRETE IN	ANALOG OUT*	DISCRETE OUT	REMARKS

REMARKS:
- 19 VAV BOX CNTLRS (SPT,SUPT) (TYP OF 9)
- (TYP OF 15)
- (TYP OF 4)
- INTERIOR LIGHTING CONTROLS
- EXTERIOR LIGHTING CONTROLS

UNIT	R1 MIN. OSA, NO PRECOOL	R2 MIN. OSA WITH PRECOOLING	R3 100% OSA	R4 100% RETURN AIR	C2 SPLIT COOLING COIL	S1 SINGLE ZONE BLOW THROUGH	S2 MULTIZONE BLOW THROUGH	S3 SINGLE ZONE DRAW THROUGH	S4 VAV DRAW THROUGH	① SPACE TEMPERATURE	② SPACE HUMIDITY	③ SPACE CO2	④ RETURN AIR TEMPERATURE	⑤ RETURN AIR HUMIDITY	⑥ RETURN AIR CO2	⑦ SUPPLY AIR TEMPERATURE	⑧ SUPPLY AIR PRESSURE	⑨ OSA COLD DECK TEMP.	⑩ MIXED AIR TEMPERATURE	⑪ COLD DECK TEMPERATURE	⑫ VFD SPEED FEEDBACK	⑬ AIR FLOW	⑭ SETPOINT ADJUSTMENT	⑮ EXHAUST/DUCT AIR TEMP.	㉑ OSA FILTER DIFF. PRESS.	㉒ FILTER DIFF. PRESSURE	㉓ SUPPLY FAN STATUS	㉔ TRANSFER FAN STATUS	㉕ EXHAUST FAN STATUS	㉖ CONDENSATE FLOAT SWITCH	㉗ VFD READY STATUS	㉘ VFD SUMMARY FAULT	㉙ AC UNIT/MISC. STATUS	㉚ LOCAL MANUAL SWITCH	㉛ LIGHTING ZONE STATUS	Ⓐ ZONE VOLUME DAMPER	Ⓑ ZONE F&B DAMPER	Ⓒ COOLING COIL VALVE	Ⓓ OSA COOLING COIL VALVE	Ⓔ MODULATING HEATER CNTRL	Ⓕ MODULATING OSA DAMPER	Ⓖ VFD SPEED CONTROL	Ⓛ SUPPLY FAN START	Ⓜ TRANSFER FAN START	Ⓝ EXHAUST FAN START	Ⓞ STAGED GAS / ELEC. HEAT	Ⓟ STAGED DX COOLING	Ⓠ 2-POSITION OSA DAMPER	Ⓡ 2-POS. ISOLATION DAMPER	Ⓢ LIGHTING ZONE CONTROL
AH-1	X							X		1	1	1	1	1	1	1	1				1						1					1						1		1	1	1	1							
AH2-10	X			X				X		1	1	1				1				1	1						1					1						1		1	1	1	1							
AH11	X			X					X	1						1				1							1											1			1		1							
AH12								X		1						1											1											1					1							
AH13								X		1						1											1																1							
EF1-15																													1																1				
EF16-19																													1																1				
INT LITES																																			12														12
EXT LITES																																			6														6

*NOTE: EACH ANALOG FUNCTION MAY REQUIRE MULTIPLE ACTUATORS.

Figure 26-1. AC-701 Control Point Specification Matrix

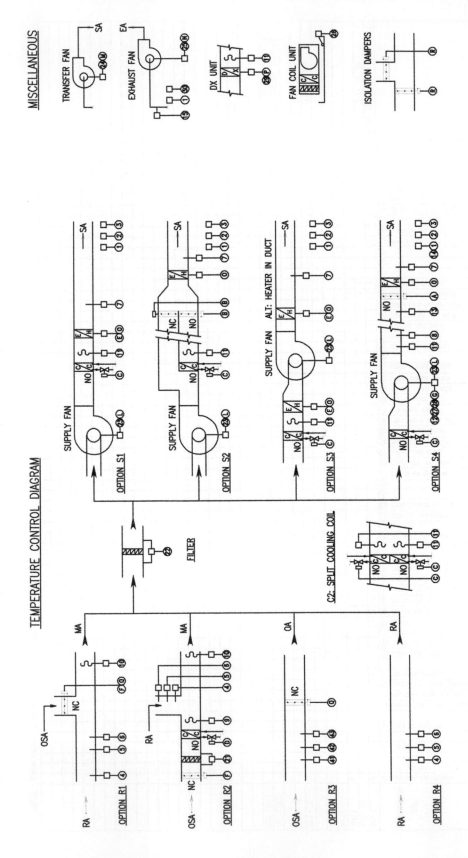

Figure 26-2. AC-701 Air Handler Control Schematic Diagram

TEMPERATURE CONTROL SEQUENCE OF OPERATION

A) GENERAL
1) SHUTDOWN SEQUENCE:
   a) SUPPLY, TRANSFER, AND EXHAUST FANS OFF; HEATERS OFF.
   b) CHILLED WATER VALVES ARE CLOSED; OSA DAMPERS ARE CLOSED.

2) OPERATING HOURS:
   a) SUPPLY, TRANSFER, AND EXHAUST FANS ON; OSA DAMPERS OPEN.
   b) CHILLED WATER VALVES AND HEATERS CONTROLLED PER OPTIONS BELOW.
   c) VOLUME AND FACE/BYPASS DAMPERS ARE CONTROLLED PER OPTIONS BELOW.

3) IN THE EVENT OF A CONTROL POWER FAILURE:
   a) CHILLED WATER VALVES OPEN, FACE/BYPASS DAMPERS OPEN TO COIL FACE.
   b) OSA DAMPERS CLOSE, ZONE VOLUME DAMPERS OPEN.
   c) SUPPLY FANS ON, ELECTRIC HEATERS ARE OFF.

4) SPLIT COOLING COIL--OPTION C2
   a) ALL COILS WITH MULTIPLE CONNECTIONS SHALL BE PROVIDED WITH MULTIPLE CONTROL VALVES AND SHALL OPERATE AS A "SPLIT COIL."
   b) LOWER COIL TO MAINTAIN 50°F TO 55°F WHILE COOLING IS REQUIRED TO ENSURE MINIMUM DEHUMIDIFICATION OCCURS.
   c) UPPER COOLING COIL MODULATES TO MAINTAIN AVERAGE COLD DECK TEMP. TO SATISFY SPACE TEMPERATURE AND HUMIDITY REQUIREMENTS.

5) SYSTEMS UTILIZING DIRECT EXPANSION (DX) COOLING COILS WILL OPERATE SIMILAR TO OPTIONS DESCRIBED HERE WITH "STAGED DX COOLING" SUBSTITUTED FOR "COOLING COIL VALVE".

6) HUMIDITY OVERRIDE MODE. WHEN SPACE HUMIDITY EXCEEDS HUMIDITY SETPOINT:
   a) COOLING VALVE MODULATES TO MAINTAIN 50°F TO 55°F COLD DECK TEMP.
   b) ZONE DAMPERS (WITHOUT ZONE HEATERS) MODULATE TO MAINTAIN SPACE TEMP.
   c) HEATERS (IF USED) MODULATE OR SEQUENCE ON TO MAINTAIN SPACE AT THE COOLING SETPOINT. ZONE DAMPERS (WITH HEATER) OPEN TO FULL COOL POSITION.

7) NIGHT SETBACK: ALL EQUIPMENT USING OPTIONS R1, R2, AND R4 SHALL BE CAPABLE OF OPERATING AT NIGHT WITH THE OSA DAMPERS CLOSED, TRANSFER FANS OFF, AND USING SETBACK (UNOCCUPIED) TEMPERATURE AND HUMIDITY SETPOINTS.

8) ELECTRIC HEATERS SHALL HAVE BUILT-IN PROOF-OF-AIRFLOW HARDWIRE INTERLOCK TO LOCALLY PREVENT OPERATION WITH NO AIRFLOW.

B) AIR HANDLING UNIT OPTIONS (OPERATING HOURS):
R1) MINIMUM OUTSIDE AIR, NO PRECOOLING:
   a) OSA DAMPERS OPEN TO MINIMUM POSITION. AIR HANDLERS SHALL USE MODULATING DAMPER, FAN COILS WILL TYPICALLY USE TWO-POSITION DAMPER.
   b) UNITS WITH OSA DELIVERED FROM A PRIMARY COOLING UNIT MAY REQUIRE OSA DAMPER FOR PROPER NIGHT SETBACK OPERATION.
R2) MINIMUM OUTSIDE AIR WITH PRECOOLING
   a) OSA DAMPERS OPEN TO MINIMUM POSITION.
   b) PRECOOLING COIL VALVE MODULATES TO MAINTAIN PRECOOL DISCHARGE TEMP. BETWEEN 50F AND 55F TO DEHUMIDIFY THE VENTILATION AIR.
R3) 100% OUTSIDE AIR
   a) OSA DAMPER TO FULL OPEN POSITION.
   b) DAMPER ACTUATOR LIMIT SWITCH (BUILT-IN) SHALL BE MONITORED AND WILL TRIGGER SUPPLY FAN TO START VIA SOFTWARE INTERLOCK.
R4) 100% RETURN AIR
   NO DAMPER CONTROLS--TYPICALLY FOR SMALL FAN COILS ONLY.

S1) SINGLE ZONE BLOW THROUGH
   a) COOLING COIL VALVE MODULATES TO MAINTAIN COLD DECK TEMPERATURE AS DETERMINED NECESSARY TO SATISFY SPACE COOLING/DEHUMID. LOAD.
   b) MODULATE/STAGE ELECTRIC HEATER TO MAINTAIN SUPPLY AIR TEMPERATURE AS DETERMINED NECESSARY TO SATISFY SPACE HEATING LOAD.
   c) SUPPLY AIR TEMPERATURE FOR KITCHEN HOOD SHALL BE AT 78°F FOR SUMMER , AND AT 70°F FOR THE WINTER, NO REHEAT.

S2) MULTIZONE BLOW THROUGH
   a) COOLING COIL VALVE MODULATES TO MAINTAIN COLD DECK TEMPERATURE AS DETERMINED NECESSARY TO SATISFY SPACE COOLING/DEHUMID. LOAD.
   b) ZONE DAMPERS MODULATE TO MAINTAIN ZONE SUPPLY AIR TEMPERATURE AS DETERMINED NECESSARY TO SATISFY SPACE CONDITIONS.
   c) MODULATE/STAGE ELECTRIC HEATER TO MAINTAIN SUPPLY AIR TEMPERATURE AS DETERMINED NECESSARY TO SATISFY SPACE HEATING LOAD.

S3) SINGLE ZONE DRAW THROUGH
   OPERATES THE SAME AS OPTION S1 SINGLE ZONE BLOW THROUGH.

S4) DRAW THROUGH VARIABLE AIR VOLUME (VAV)
   a) COOLING COIL VALVE MODULATES TO MAINTAIN COLD DECK TEMP. AS DETERMINED NECESSARY TO SATISFY AVERAGE SPACE CONDITIONS.
   b) VARIABLE FREQUENCY DRIVE (VFD) SHALL BE COMMANDED TO VARY SUPPLY FAN SPEED TO MAINTAIN STATIC PRESSURE SETPOINT IN THE DUCT.
   c) ZONE VOLUME DAMPER MODULATES TO SATISFY THE ZONE COOLING LOAD.
   d) IF ZONE DAMPER IS AT MINIMUM COOLING POSITION AND SPACE TEMP. CONTINUES TO DROP, ZONE HEATER (IF USED) WILL STAGE ON AS NECESSARY TO SATISFY THE ZONE CONDITIONS. VOLUME DAMPER MAY NEED TO OPEN BACK UP TO A NEW POSITION AS REQUIRED FOR MINIMUM HEATER AIRFLOW.

S4.1) WHEN VAV BOX CONTROLLER IS SPECIFIED FOR ZONE CONTROL:
   e) AIRFLOW SENSOR SHALL BE UTILIZED TO MAKE ZONE CONTROL PRESSURE-INDEPENDENT.
   f) SETPOINT ADJUSTMENT "SLIDER" ON TEMPERATURE SENSOR SHALL OFFSET THE ZONE TEMPERATURE SETPOINT +/-2°F BASED ON OCCUPANT INPUT.

C) EXHAUST FANS:
1) AIR BALANCE INTERLOCK: FAN STOPS AT NIGHT BASED ON TIME SCHEDULE BUT ALSO STOPS WHEN ASSOCIATED AIR HANDLER OR FAN COIL STOPS.
2) TEMPERATURE CONTROL: FAN CYCLES TO MAINTAIN SPACE TEMP. SETPOINT (30°C FOR ELECTRICAL ROOMS). MINIMUM ON/OFF TIMES SHALL BE ADJUSTABLE TO PREVENT SHORT CYCLING.
3) PROJECTOR EXHAUST: FAN STARTS WHEN LOCAL SWITCH IS ENABLED OR WHEN TEMPERATURE RISE IS SENSED IN THE PROJECTOR EXHAUST DUCT. FAN STOPS WHEN EXHAUST AIR TEMPERATURE RETURNS TO ROOM TEMPERATURE. ISOLATION DAMPER(S) MAY BE REQUIRED IF SINGLE FAN SERVES MULTIPLE PROJECTORS.

D) REDUCED VENTILATION MODES:
1) OSA DAMPER SHALL BE MODULATED TO A LIMITED DEGREE BELOW THE "MINIMUM" POSITION DURING OFF-PEAK OCCUPANCY PERIODS BASED ON THE CO2 VALUE IN THE SPACE RELATIVE TO THE OUTDOOR AIR CO2.
2) IF A LARGE VENTILATION AMOUNT IS SPECIFIED FOR "OPEN-DOOR" PRESSURIZATION, OSA DAMPER SHALL BE MODULATED BELOW THE "MINIMUM" POSITION TO LIMIT SUPPLY (AND RESULTANT BUILDING) PRESSURE DURING "CLOSED-DOOR" OPERATION. DURATION SHOULD BE LIMITED, AND DAMPER MUST RE-OPEN IF CO2 LEVEL RISES.

Figure 26-3. AC-701 Air Handler Sequence of Operation

UCP-1   FIELD INSTALLATION PANEL SCHEDULE

LOCATION: PRE-FUNCTION MECHANICAL ROOM 1000

PT #	ELEC RUN:	PT TYPE	SENSOR CODE	BLDG NAME	SYSTEM NAME	DEVICE NAME	COMMENTS
01	1P	AI	T55	BU	AH31	TR	AH-1 RETURN AIR TEMP
02	1P	AI	H21	BU	AH31	HR	AH-1 RETURN AIR HUMIDITY
03	4C	AI	CD2	BU	AH31	CD	AH-1 RETURN CO2 SENSOR
04	1P	AI	T49	BU	AH31	TC	AH-1 COLD DECK TEMP
05	1P	AI	P7	BU	AH31	PS	AH-1 DUCT STATIC PRESSURE
06	1P	AI	VM2	BU	AH31	MVD	AH-1 VFD SPEED MONITOR
07	1P	AI	P4	BU	CHW	PD1	CHILLED WATER PRESS DIFF
08	1P	AI	P4	BU	HHW	PD1	HEATING HOT WATER PRESS DIFF
09	4C	AO	EA2	BU	AH31	CC	AH-1 COOLING COIL VALVE
10	4C	AO	EA2	BU	AH31	DE	AH-1 OUTSIDE AIR DAMPER
11	1P	DO	R21	BU	AH31	FS	AH-1 SUPPLY FAN START/STOP
12	1P	DO	R21	BU	AH31	FE42	EXHAUST FAN 42 START/STOP
13	1P	DO	R21	BU	AH31	FE43	EXHAUST FAN 43 START/STOP
14	1P	DO	R21	BU	LTG	Z1	CONTACTOR 13A, RM 2505, EXT
15	1P	DO	R21	BU	LTG	Z2	CONTACTOR 13B/C, RM 2505, INT
16	1P	DO	R21	BU	LTG	Z3	CONTACTOR 13D, RM 2505, INT
25	1P	DO	R21	BU	LTG	Z4	CONTACTOR 13E, RM 2505, INT
26	1P	DO	R21	BU	LTG	Z5	CONTACTOR 14A/B, RM 1525, INT
27	1P	DO	R21	BU	LTG	Z6	CONTACTOR 14C, RM 1525, EXT
28	1P	DO	R21	BU	LTG	Z7	CONTACTOR 14D, RM 1525, INT
29	1P	DO	R21	BU	LTG	Z8	EMERGENCY LTG ELP/EHP, INT
30	1P	AO	VD2	BU	AH31	CVD	AH-1 VFD CONTROL
31	4C	AO	EA2	BU	AH31	DR	AH-1 RETURN AIR DAMPER
67	1P	DI	P23	BU	AH31	AF	AH-1 FILTER ALARM
68	1P	DI	AUX	BU	AH31	SFS	AH-1 SUPPLY FAN STATUS
69	1P	DI	AUX	BU	AH31	AVD	AH-1 VFD ALARM STATUS
70	1P	DI	IR2	BU	AH31	SFE42	EXHAUST FAN 42 STATUS
72	1P	DI	IR2	BU	AH31	SFE43	EXHAUST FAN 43 STATUS
73	1P	DI	AUX	BU	LTG	SZ1	CONTACTOR 13A STATUS
74	1P	DI	AUX	BU	LTG	SZ2	CONTACTOR 13B/C STATUS
75	1P	DI	AUX	BU	LTG	SZ3	CONTACTOR 13D STATUS
77	1P	DI	AUX	BU	LTG	SZ4	CONTACTOR 13E STATUS
78	1P	DI	AUX	BU	LTG	SZ5	CONTACTOR 14A/B STATUS
79	1P	DI	AUX	BU	LTG	SZ6	CONTACTOR 14C STATUS
80	1P	DI	AUX	BU	LTG	SZ7	CONTACTOR 14D STATUS
82	1P	DI	AUX	BU	LTG	SZ8	EMERGENCY LTG ELP/EHP
83	1P	DI	AUX	BU	LTG	SZS	EXT LIGHT ON/OFF

GENERAL NOTES:
FOR "ELEC RUN:", P = TSP #20, BELDEN 8762 OR OWNER APPROVED EQUAL.
4C = 4 CONDUCTOR #18, BELDEN 8489 OR APPROVED EQUAL.
FOR PT TYPE, AI = ANALOG IN, AO = ANALOG OUT, DI = DISCRETE IN, DO = DISCRETE OUT.
LABEL BOTH ENDS OF THE CABLE WITH THE POINT NUMBER (PT #).
LEAVE TEN FEET OF SLACK CABLE IN THE UCP FOR OWNER TERMINATION.
EACH VALVE AND DAMPER MAY REQUIRE MULTIPLE ELECTRONIC ACTUATORS FOR PROPER TORQUE;
ADD A 4C#18 (BELDEN 8489) FOR EACH ADDITIONAL ACTUATOR REQUIRED.

UCP-2   FIELD INSTALLATION PANEL SCHEDULE

LOCATION: MEZZANINE RM 1111

PT #	ELEC RUN:	PT TYPE	SENSOR CODE	BLDG NAME	SYSTEM NAME	DEVICE NAME	COMMENTS
01	1P	AI	T55	BU	AH2	TA	AH-2 SPACE TEMP
02	1P	AI	H21	BU	AH2	HA	AH-2 SPACE HUMIDITY
03	1P	AI	T49	BU	AH2	TC	AH-2 COLD DECK TEMP
04	1P	AI	T42	BU	AH2	TS	AH-2 SUPPLY AIR TEMP
05	1P	AI	VM2	BU	AH2	MVD	AH-2 VFD SPEED MONITOR
06	1P	AI	T55	BU	AH3	TA	AH-3 SPACE TEMP
07	1P	AI	H21	BU	AH3	HA	AH-3 SPACE HUMIDITY
08	4C	AI	CD1	BU	AH3	CD	AH-3 SPACE CO2 SENSOR
09	4C	AO	EA2	BU	AH2	CC	AH-2 COOLING COIL VALVE
10	4C	AO	EA2	BU	AH2	HC	AH-2 HEATING COIL VALVE
11	4C	AO	EA2	BU	AH2	DO	AH-2 OUTSIDE AIR DAMPER
12	4C	AO	EA2	BU	AH2	DR	AH-2 RETURN AIR DAMPER
13	1P	DO	R21	BU	AH2	FS	AH-2 SUPPLY FAN START/STOP
14	4C	AO	EA2	BU	AH3	CC	AH-3 COOLING COIL VALVE
15	4C	AO	EA2	BU	AH3	HC	AH-3 HEATING COIL VALVE
16	4C	AO	EA2	BU	AH3	DO	AH-3 OUTSIDE AIR DAMPER
17	1P	AI	T49	BU	AH3	TC	AH-3 COLD DECK TEMP
18	1P	AI	T42	BU	AH3	TS	AH-3 SUPPLY TEMP
19	1P	AI	VM2	BU	AH3	MVD	AH-3 VFD SPEED MONITOR
25	4C	AO	EA2	BU	AH3	DR	AH-3 RETURN AIR DAMPER
26	1P	DO	R21	BU	AH3	FS	AH-3 SUPPLY FAN START/STOP
36	1P	AI	P4	BU	CHW	PD2	CHW DIFF PRESS
37	1P	AI	P4	BU	HHW	PD2	HHW DIFF PRESS (UCP5 PT23)
41	1P	DO	R21	BU	FE	FE1	EXHAUST FAN 1 START/STOP
42	1P	DO	R21	BU	FE	FE2	EXHAUST FAN 2 START/STOP
43	1P	DO	R21	BU	FE	FE3	EXHAUST FAN 3 START/STOP
44	1P	AO	VD2	BU	AH2	CVD	AH-2 VFD CONTROL
45	1P	AO	VD2	BU	AH3	CVD	AH-3 VFD CONTROL
67	1P	DI	AUX	BU	AH2	SFS	AH-2 SUPPLY FAN STATUS
68	1P	DI	AUX	BU	AH2	AVD	AH-2 VFD ALARM STATUS
69	1P	DI	P23	BU	AH2	AF	AH-2 FILTER ALARM
70	1P	DI	AUX	BU	AH3	SFS	AH-3 SUPPLY FAN STATUS
72	1P	DI	AUX	BU	AH3	AVD	AH-3 VFD ALARM STATUS
73	1P	DI	P23	BU	AH3	AF	AH-3 FILTER ALARM
78	1P	DI	IR1	BU	FE	SFE1	EXHAUST FAN STATUS 1
79	1P	DI	IR1	BU	FE	SFE2	EXHAUST FAN STATUS 2
80	1P	DI	IR1	BU	FE	SFE3	EXHAUST FAN STATUS 3

SENSOR CODES:

T55 = SPACE TEMPERATURE W/PUSH BUTTON
T42 = 18" DUCT TEMPERATURE
T49 = 20' NICKEL RTD AVERAGING
AUX = AUXILLIARY CONTACT
R21 = 24VDC DPDT RELAY
P23 = DIFFERENTIAL PRESSURE SWITCH
H21 = SPACE HUMIDITY
H22 = DUCT HUMIDITY
DSn = AUX'S OR P23'S X n
CD1 = CARBON DIOXIDE SENSOR (SPACE)
CD2 = CARBON DIOXIDE SENSOR (DUCT)

T44 = PIPE TEMPERATURE W/THERMOWELL
P4 = WATER PRESSURE DIFFERENTIAL
FM1 = WATER FLOW METER
EA2 = MODULATING ELECTRONIC ACTUATOR
EA3,4,5,6,7 = MULTIPLE EA2'S
EA1 = 2 POSITION ELECTRONIC ACTUATOR
EA8 = MULTIPLE EA1'S
R4n = R21'S X n
IR2 = CURRENT RELAY
P7 = AIR DUCT STATIC PRESSURE
MVD = VFD MONITOR INPUT (4-20ma)
CVD = VFD CONTROL OUTPUT (4-20ma)

Figure 26-4. AC-702 EMS Panel Wirelist

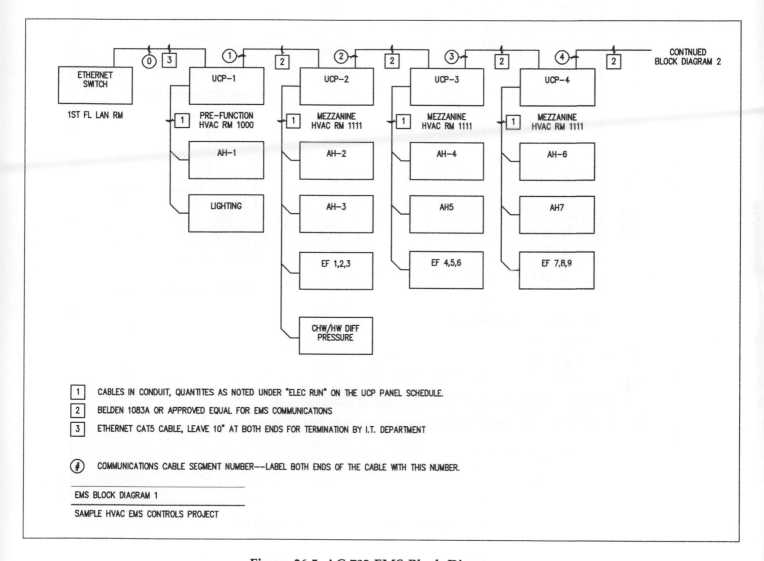

CONTINUED
BLOCK DIAGRAM 2

ETHERNET SWITCH
1ST FL LAN RM

UCP-1
PRE-FUNCTION
HVAC RM 1000
AH-1
LIGHTING

UCP-2
MEZZANINE
HVAC RM 1111
AH-2
AH-3
EF 1,2,3
CHW/HW DIFF PRESSURE

UCP-3
MEZZANINE
HVAC RM 1111
AH-4
AH5
EF 4,5,6

UCP-4
MEZZANINE
HVAC RM 1111
AH-6
AH7
EF 7,8,9

1   CABLES IN CONDUIT, QUANTITES AS NOTED UNDER "ELEC RUN" ON THE UCP PANEL SCHEDULE.

2   BELDEN 1083A OR APPROVED EQUAL FOR EMS COMMUNICATIONS

3   ETHERNET CAT5 CABLE, LEAVE 10" AT BOTH ENDS FOR TERMINATION BY I.T. DEPARTMENT

(#)   COMMUNICATIONS CABLE SEGMENT NUMBER——LABEL BOTH ENDS OF THE CABLE WITH THIS NUMBER.

EMS BLOCK DIAGRAM 1

SAMPLE HVAC EMS CONTROLS PROJECT

Figure 26-5. AC-702 EMS Block Diagram

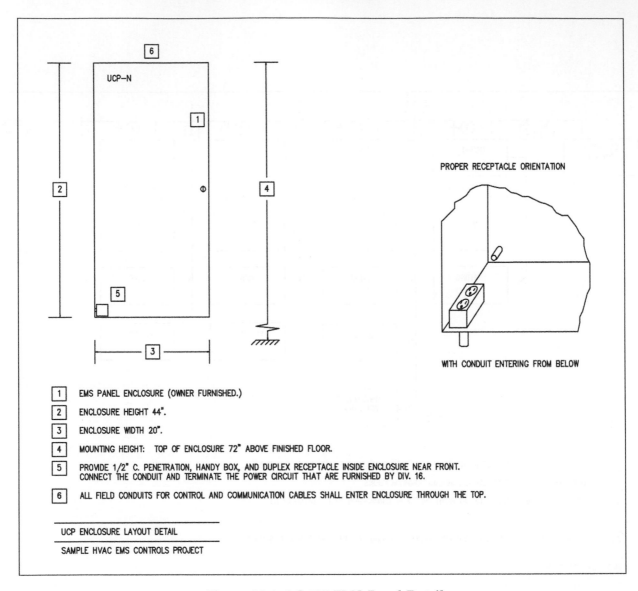

Figure 26-6. AC-702 EMS Panel Details

Chapter 27

Using the Web for Energy Data Acquisition and Analysis

Paul J. Allen, P.E.
David C. Green
Jim Lewis

ABSTRACT

ADVANCES IN NEW EQUIPMENT, new processes and new technology are often the driving forces in improvements in energy management, energy efficiency and energy measurement processes. Of all recent developments affecting energy management, the most powerful new technology to come into use in the last several years has been information technology—or IT. The combination of cheap, high-performance microcomputers together with the emergence of high-capacity communication lines, networks and the internet has produced explosive growth in IT and its application throughout our economy. Energy information systems have been no exception. IT and internet based systems are the wave of the future.

Timely energy information is particularly critical to energy service companies (ESCOs) as this information can be invaluable in establishing baseline energy consumption, commissioning new installations and ongoing monitoring of savings. Web-based energy information systems provide the ability to view time-stamped data on a daily basis which provides both the contractor and the building owner the information to compare actual performance of the energy conservation measures (ECMs) to the projected performance and to make adjustments to correct problems quickly. All of this can be done with just a web browser from the contractor's office, eliminating the need for costly site visits.

This chapter describes the fundamentals of an energy information system (EIS) and presents two case studies that showcase how energy data collection and analysis can be used to quantify energy project results.

ENERGY INFORMATION SYSTEMS

The philosophy, "If you can measure it, you can manage it," is critical to a sustainable energy manage-ment program. Continuous feedback on utility performance is the backbone of an energy information system[1]. A basic definition of an energy information system is:

Energy Information System (EIS): Equipment and computer programs that let users measure, monitor and quantify energy usage of their facilities and help identify energy conservation opportunities.

Everyone has witnessed the growth and development of the internet—the largest computer communications network in the world. Using a web browser, one can access data around the world with a click of a mouse. An EIS should take full advantage of these new tools.

EIS PROCESS

There are two main parts to an EIS: (1) data collection and (2) web publishing. Figure 27-1 shows these two processes in a flow chart format.

DATA COLLECTION PROCESS

The first task in establishing an EIS is to determine the best sources of the energy data. Utility meters monitored by an energy management system or other dedicated utility-monitoring systems are a good source. The metering equipment collects the raw utility data for electric, chilled & hot water, domestic water, natural gas and compressed air. The utility meters communicate to local data storage devices by pre-processed pulse outputs, 0-10V or 4-20ma analog connections, or by digital, network-based protocols.

Data gathered from all of the local data storage devices at a predefined interval (usually on a daily basis) are stored on a server in a relational database (the "data warehouse"). Examples of relational databases are Fox-Pro, SQL and Oracle[a].

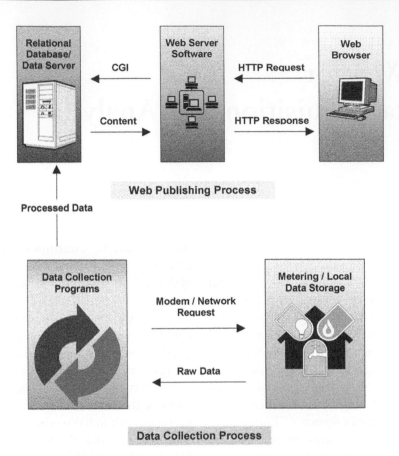

Figure 27-1. Energy Information System Schematic

Using an EMS for Data Collection

Identifying and organizing the best energy data sources is the first step in establishing an EIS. One potential data collection source is from an energy management system (EMS). An EMS typically has a built-in procedure that can produce daily reports on points connected to the system. The process is to program the EMS to collect the desired utility data and then to move this data into the EIS relational database. Shown are the steps required for this to happen:

1.   The EMS PC should be on the corporate LAN.

This might be as simple as installing an ethernet card in the existing EMS PC workstation. However, IT departments are generally very particular about the PC hardware installed on the corporate LAN and will probably require that the existing EMS PC be replaced with one of their standard PC's. This could cause compatibility issues with older EMS software operating on a new PC's operating system (OS) and require the EMS vendor to upgrade their software to a version that is compatible with the newer PC OS. So a seemingly simple task of putting the existing EMS PC on the corporate LAN

might be turn out to be an expensive proposition.

2.   Transfer the EMS reports to the EIS server on a daily basis.

Once a path is established to the EMS PC front-end, the EMS report files need to transferred to the EIS server. There are numerous methods to accomplish this task. This might be simple as mapping a drive to the EMS PC from the EIS server. A DOS-based batch file can be launched at a specific time of day to copy the EMS report files to a subdirectory on the server.

Another method would be to use file transfer protocol (FTP) to transfer files from the EMS PC to the EIS server. Typically, a program on the EIS server is required to coordinate the FTP file transfers.

3.   Capture the data from the EMS reports.

Once the EMS reports have been moved to the server, a custom program is developed to extract the data from the EMS reports and update the EIS relational database. Ideally, this program is developed to read multiple EMS vendors EMS reports.

Since the relational database would have a standard format, the data from different EMS vendors' reports can be reported in a consistent format.

Collecting Data Using a DAS

Another approach to collecting utility data is to use a dedicated data acquisition server (DAS). The DAS allows users to collect utility data from existing and new meters and sensors. On a daily basis the DAS uploads the stored data to the EIS server. Once the data have been transferred to the EIS server, a program reads the DAS data files and updates the data in the EIS relational database for use by the web publishing program.

The AcquiSuite system from Obvius is typical of the emerging solutions and is a Linux based web server which provides three basic functions:

- Communications with existing meters and sensors to allow for data collection on user-selected intervals.

- Non-volatile storage of collected information for several weeks.

- Communication with external server(s) via phone or internet to allow conversion of raw data into graphical information.

The backbone of the system is a specially designed web server. The DAS provides connectivity to new and existing devices either via the on-board analog and digital inputs or the RS 485 port using a Modbus protocol. The analog inputs permit connection to industry standard sensors for temperature, humidity, pressure, etc and the digital inputs provide the ability to connect utility meters with pulse outputs. The serial port communicates with Modbus RTU devices such as electrical meters from Veris, Square D and Power Measurement Ltd.

WEB PUBLISHING

The internet, with the world wide web—or web—has become quickly and easily accessible to all. It has allowed the development of many new opportunities for facility managers to quickly and effectively control and manage their operations. There is no doubt that web-based systems are the wave of the future. The EIS web publishing programs should take full advantage of these web-based technologies.

To publish energy data on the internet or an intranet (a private network that acts like the internet but is only accessible by the organization members or employees), client/server programming is used. The energy data is stored on the EIS server, and waits passively until a user, the client, makes a request for information using a web browser. A web-publishing program retrieves the information from the EIS relational database and sends it to the web server, which then sends it to the client's web-browser that requested the information.

There are many software choices available for the web-publishing process. One method uses a server-side common gateway interface (CGI) program to coordinate the activity between the web-server and the web-publishing program. Using CGI enables conventional programs to run through a web browser.

The web-publishing client/server process for an EIS uses the steps below (See Figure 27-1). This entire process takes only seconds depending on the connection speed of the client's computer to the web.

1. A user requests energy information by using their web browser (client) to send an HTTP (hypertext transfer protocol) request to the web server.

2. The web server activates the CGI program. The CGI program then starts up the web-publishing program.

3. The web-publishing program retrieves the information from the relational database, formats the data in HTML (hypertext markup language) and returns it to the CGI program.

4. The CGI program sends the data as HTML to the web server, which sends the HTML to the web browser requesting the information.

Web Publishing Programming Options

There are many programming alternatives available other than the CGI approach described above. PERL, Active Server Pages (ASP), JavaScript, and VB-Script, Java Applets, Java Server Pages, Java Servlets, ActiveX controls and PHP are a few of the more popular choices available today. Some of these are easier to implement than others. ASP for instance, is a part of IIS so no installation is required. PERL and PHP require installation of their respective programs on the web server machine to run. There are also security issues with some of these approaches. The client's machine downloads Java Applets then executes them from there. Some experts view this as a security risk not worth taking. Javascript and VBScript are somewhat limited in that they are just a subset of the other full fledged programming languages. Most browsers interpret them correctly so no installation is required. Java Server Pages and Java Servlets run on the web server in the same way as ASP but may require some installation depending on the web server used.

Although there are many web servers available to choose from, two are the most popular by far. Microsoft's Internet Information Services (IIS) comes with Windows 2003 server. Apache web server is a good choice for other operating systems. Any web server needs some configuration to produce web content, especially if it is querying a database. The web-publishing task will likely require custom folders, special access permissions and a default page.

After installing the web server, the web-publishing administrator must put a default page in the root directory of the web server. This is the first page users will see in their browser when they type in the web site's internet address. The pages are usually named "default.htm" or "index.htm" but can be anything as long as the web server is configured to treat them as the default page. Next, if CGI is used, the administrator creates a special folder to store the scripts. This is usually called "cgi-bin" or just "scripts." This folder must have permissions specifically allowing the files in the folder to be "executable." In some cases, "write" permissions are required for the folder if the CGI programs write temporary files to it. Other custom folders may be required to organize the web publishing content. Once the web-publishing administrator configures the web server he or she can install and test custom CGI programs and pages. If the CGI program or pages accurately return data from the database then the task of creating custom reports for the energy data can begin.

EIS Implementation Options

Deciding which web server and programming method to use along with configuring and implementing it to create a web publishing system can be quite a task. It really requires an expert in these areas to do a reliable job. Three approaches have evolved to satisfy web-publishing requirements.

1. Use internal resources to accomplish this task. This works well if there are already experienced web programmers available and they have time to work on the project. This makes it easy to customize the web publishing content as needed quickly and cost effectively. Finding time for internal personnel to focus on the project is usually the problem with this option.

2. Hire an outside consultant to do the configuration and programming as needed. This works well if the consultant has a good working relationship with someone internally to facilitate access to the protected systems and help with understanding the data. The consultant must be willing to work for a reasonable rate for this approach to be cost effective. The consultant must also be responsive to requests for support.

3. Purchase and install a somewhat "canned" version of the web publishing software and then customize it to fit the energy data as required. This approach has many possible problems in that the software is usually quite expensive and requires a great deal of customization and support from the outside to make it work well. However, for small simple projects this may be a good fit.

For users, who do not want to invest the time and effort required for this "do-it-yourself" approach, there are numerous companies that provide a complete EIS service for an on-going monthly service fee. The EIS service company provides all of the IT-related functions, including the energy data collection/storage and the web-publishing program. The user accesses the EIS service web site with using a web browser, enters a user ID and password and then uses the available reports/graphs to analyze energy data.

The advantage of this approach is that the user does not get involved with the details and operation of the EIS, but instead is able to work with the EIS service provider to develop the utility data reports most helpful to their operation. The downside to this approach is the on-going monthly service fee that is a function of the amount of data processed—the more meters or bills processed, the higher the monthly fee. There may also be additional costs to customize any reporting from the standard reports already created by the EIS service provider. The Building Manager Online service from Obvius[b] is one of the many choices available to users today.

MEASUREMENT AND VERIFICATION

The majority of energy saving retrofit projects are implemented based on engineering calculations of the projected return on investment [2]. As with any projections of ROI, much of what goes into these calculations are assumptions and estimates that ultimately form the basis for implementation. As the folks at IBM used to say, "garbage in—garbage out," which in the case of energy retrofits means that if any of the assumptions about parameters (run times, set-points, etc.) are wrong, the expected payback can be dramatically in error. The establishment of good baselines (measures of current operations) is the best way to determine the actual payback from investments in energy and sub-metering.

Just as important as building an accurate picture of the current operation is measuring the actual savings realized from an investment. If there is no effective means of isolating the energy used by the modified systems, it may be impossible to determine the value of the investment made. Using monthly utility bills for this analysis is problematic at best since the actual savings achieved can be masked by excessive consumption in non-modified systems.

Consider, for example, a commercial office building whose central chiller plant has an aging mechanical and control structure that provides limited capability for adjusting chilled water temperature. To improve efficiency, the building owner plans to retrofit the system to provide variable speed drives on pumps for the chilled water and condenser water systems along with control upgrades to allow for chilled water set-point changes based on building loads. In the absence of baseline information, all calculations for savings are based on "snapshots" of the system operation and require a variety of assumptions. Once the retrofit is completed, the same process of gathering snapshot data is repeated and hopefully the savings projected are actually realized. If the building tenants either add loads or increase operational hours, it is difficult if not impossible to use utility bills to evaluate the actual savings.

In contrast, the same project could be evaluated with a high degree of accuracy by installing cost-effective monitoring equipment prior to the retrofit to establish a baseline and measure the actual savings. While each installation is necessarily unique, building a good monitoring system would typically require:

- Data acquisition server (DAS) such as the AcquiSuite from Obvius to collect the data, store it and communicate it to a remote file server.

- Electric submeter(s)—the number of meters would vary depending on the electric wiring configuration, but could be as simple as a single submeter (e.g., Enercept meter from Veris Industries) installed on the primary feeds to the chiller plant. If desired, the individual feeds to the cooling tower, compressors, chilled water pumps, etc. could be monitored to provide an even better picture of system performance and payback.

- Temperature sensors (optional): in most installations, this could be accomplished by the installation of two sensors, one for chilled water supply temperature and the other for chilled water return temperature. These sensors do not provide measurement of energy usage, but instead are primarily designed to provide feedback on system performance and efficiency.

- Flow meter (optional)—a new or existing meter can be used to measure the gallons per minute (gpm). By measuring both the chiller input (kW) and the chiller output (tons) the chiller efficiency can be calculated in kW/ton.

The benefits of a system for actually measuring the savings from a retrofit project (as opposed to calculated or stipulated savings) are many:

- The establishment of a baseline over a period of time (as opposed to "snapshots") provides a far more accurate picture of system operation over time.

- Once the baseline is established, ongoing measurement can provide a highly accurate picture of the savings under a variety of conditions and establish a basis for calculating the return on investment (ROI) regardless of other ancillary operations in the building.

- The presence of monitoring equipment not only provides a better picture of ROI, but also provides ongoing feedback on the system operation and will provide for greater savings as efficiency can be fine-tuned.

Viewing and Using the Data

Historically, much of the expense of gathering and using sub-metering data has been in the hardware and software required and the ongoing cost of labor to produce useful reports. Many companies are leveraging existing technologies and systems to dramatically reduce the cost of gathering, displaying and analyzing data from commercial and industrial buildings. The AcquiSuite data acquisition server uses a combination of application specific hardware and software. A standard web browser, such as Microsoft Internet Explorer, provides the user interface.

The AcquiSuite DAS automatically recognizes devices such as meters from Power Measurement Ltd. and Veris Industries, which makes installation cost effective.

The installer simply plugs the meters in the DAS and all configuration and setup is done automatically with only input required being the name of the device and the location of the remote file server. The DAS gathers data from the meters on user-selected intervals (e.g. 15 minutes) and transmits it via phone line or LAN connection to a remote file server where it is stored in a database for access via the internet.

To view the data from one or more buildings, the user simply logs onto a web page (e.g. *www.obvius.com*) and selects the data to view.

The gathering and sorting of data do not provide sufficient energy management guidance unless the data are analyzed, transformed into usable information, and implemented. To help illustrate this point, the following case studies are offered.

CASE STUDY—RETAIL STORE LIGHTING

Background

A retail store chain in the Northeast was approached by an energy services company about converting some of their lighting circuits to a more efficient design. On paper, the retrofit looked very attractive and the company elected to do a pilot project on one store with a goal to implementing the change throughout the entire chain if it proved successful. The retailer decided to implement a measurement and verification (M&V) program to measure the actual savings generated by comparing the usage before the retrofit (the baseline) and after.

Implementation

The store had 12 very similar lighting circuits, all of which were operated on a time schedule from a central control panel in the store. Since the circuits were very similar, it was decided that measuring the impact on one circuit would provide a good indication of the savings from the other circuits. The M&V equipment consisted of the following:

- An electrical sub-meter (see Figure 27-2) was installed on the power lines feeding the lighting circuit;

- A data acquisition server (see Figure 27-3) was installed in the store to record, store and upload time-stamped interval data to a remote server for storage and display. The DAS provides plug and play connectivity to the sub-meter and uses an existing phone line or LAN to send data from the store to a remote server on a daily basis.

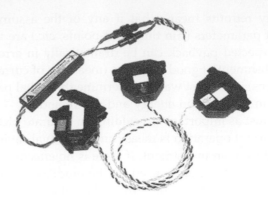

Figure 27-2. Retrofit Electric Sub-meter

- The remote server was used to monitor consumption before the retrofit and to measure the actual savings

Results

Figure 27-4 shows the actual kW usage over roughly 24 days. The left side of the chart shows the kW usage for the first 11 days before the retrofit and the average usage is fairly constant at around 1.45 kW. On Feb. 11, the retrofit was performed, as indicated by the drop to zero kW in the center of the chart. Immediately after the retrofit (the period from Feb 11 to Feb 15, the kW load dropped to around 0.4 kW, a reduction of over 70% from the baseline load in the left of the graph.

The good news for the retailer was that the retrofit performed exactly as expected and the M&V information obtained from monitoring the energy on this circuit provided clear evidence that the paybacks were excellent. The initial good news, however, was tempered somewhat after looking at the chart. It was immediately evident that this lighting circuit (and the other 11 identical circuits) were operating 24 hours per day, seven days a week. The store, however, operated from 10 AM to 9 PM each day and the lighting panel was supposed to be

Figure 27-3. Data Acquisition Server

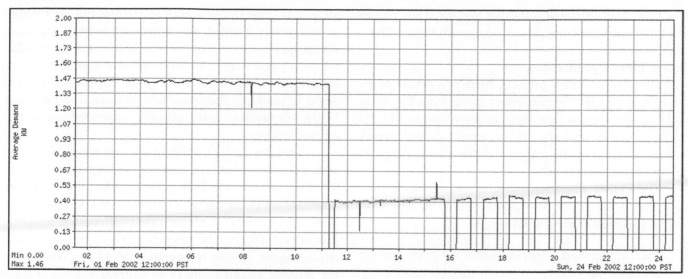

Figure 27-4. kW Loads for a 24-day Period

shutting off the circuits during non-operating hours.

The electrical contractor was called in to look at the system and determined that a contactor in the panel had burned out resulting in continuous operation of the lighting circuits throughout the store. Once the contactor was replaced the operation of the lighting panel was restored so that the lights were only on during operating hours and shut off during the night, as indicated by the right side of the chart.

This simple chart of energy usage provides an excellent example of two uses of energy information:

1. **Measurement and verification of energy savings—** The chart clearly shows the actual energy reduction from the lighting retrofit and the data provided can be used to extrapolate the payback if this same retrofit is applied throughout the chain; and

2. **Use of energy information to fine-tune building operations—**In addition to the M&V benefits of energy information, this example also shows how a very simple review of energy usage can be used to make sure that building systems are operating properly.

CASE STUDY—CHILLER PLANT OPTIMIZATION

Chiller plants consume an enormous amount of energy to produce the air conditioning required for a building. By sub-metering the chiller operation, the operator can directly measure the effects of changing set points and time schedules on energy usage.

Background

This case study is the result of the optimization effort at a new chiller plant built for a large convention center. The mechanical engineer designed the chiller plant as a variable flow primary system. An energy management system controlled the operation of each of these chiller plant operations.

Sub-metering of the chiller plant operation was part of the original design for the convention center. Each chiller came equipped with an on-board electric meter, chilled water supply and return temperature sensors. A chilled water flow meter was added to each chiller to allow for the chilled water tons to be calculated. Additional electric sub-meters were included to measure the condenser and chilled water pump motors and the cooling tower fans. The energy management system collected the energy data, chiller operational parameters and the convention center space temperature and relative humidity values.

Each night, the energy management system produced a report file for the data being trended. This file was stored on the energy management system server. A custom program read this file and reformatted the data and added this to a relational database along with all the other data collected from other projects. This is the data collection process shown in Figure 27-1 previously.

A custom program was developed to display and graph the energy data on a user's PC using a web-browser. Another custom program pushed the energy data report via email to the convention center maintenance personnel. These reports provided the information to the user so they could determine if the chiller plant was operating efficiently. The process to pull the data from the relational database and produce reports was the web-publishing process shown in Figure 27-1.

Chiller Plant Optimization

As with most new projects, there is a period of test and adjustment that results when going from the drawing board into actual operation. In consultation with the design engineer and the chiller manufacturer, several changes resulted in reducing the chiller plant energy usage by approximately 30%.

The chiller plant had several energy saving features ~~that~~ are used to control the operation. The main control parameters are shown below:

- Each chiller had a leaving water set point and operational schedule.

- Chilled water pumps maintained a differential pressure at the farthest air handler unit. Variable speed drives were used to vary the motor speed and the resulting pump flow.

- A chilled water bypass valve is operated to maintain a minimum flow rate through the chillers when the chiller plant flow to the building was low.

- The condenser water pumps were interlocked with the chiller operation.

- Cooling tower fans are controlled by variable speed drives to maintain a condenser water supply temperature set point back to the chillers.

- The air handlers in the convention center exhibit space were set up as single-zone variable air volume systems. The discharge air temperature was fixed and variable speed drives modulate the fan speed to maintain the space temperature set point.

- CO_2 sensors were used to measure the occupancy. If the CO_2 sensor was below the set point, the outside air dampers were kept at their minimum values. As the CO_2 levels increased, the outside air dampers were modulated open to maintain the CO_2 levels at set point.

The first operational issue was the result of the chillers tripping offline from surging. After studying the problem, it was determined that the humidity control programmed for the air handler chilled water valves caused rapid flow rate changes resulting from the chilled water valves opening and closing to maintain humidity

set points. The solution was to change the air handler chilled water control algorithms to maintain a fixed discharge air temperature. Fortunately, each air handler was equipped with a variable speed drive. The control algorithm was changed to ramp the supply fan up and down to maintain the space temperature. The humidity in the convention space actually improved using this control strategy.

Once the chiller operational issues were solved, the next focus was to reduce the energy usage in the facility. The first changes resulted from lowering the minimum flow rate set point for each chiller from 600 gpm to 400 gpm. This resulted in an increased differential temperature at each chiller. The next change was to increase the chilled water leaving temperature to 44 degrees F from the original design value of 40 degrees F. Both of these adjustments combined to result in a 19% energy reduction at the plant.

The next operational change was to shut the entire chiller plant down from midnight to 6 am. Each air handling unit and all exhaust fans were also simultaneously shut down. The temperature and humidity values in the convention center were trended and did not show any significant increases. They quickly recovered in the morning before any convention activity occurred. This change in operation resulted in a further reduction of 15% in the chiller plant energy usage.

An operational problem was detected when the chiller plant daily report data showed a significant increase in the chiller plant electric usage even though there was no convention activity. It turned out that the chiller maintenance company had manually forced the chillers for testing purposes, but failed to put them back into automatic control causing one chiller to run from midnight to 6 am. The problem was quickly resolved and the chiller plant returned to automatic operation.

Chiller Plant Data Analysis Timeline

Figure 27-5 shows the raw data and timeline of events for the chiller plant optimization. The data was measured on a hourly basis and summarized into daily totals. The web-based energy information system captured all of the data and was updated daily. The EIS allowed the users to interrogate the data and measure the impact of operational changes. In making comparisons to determine operational changes, the daily totals could be matched against ambient weather conditions and convention center activities to ensure an apples-to-apples comparison. By also trending the convention temperature and humidity values, operational impacts resulting from the energy conservation strategies could be quantified as well.

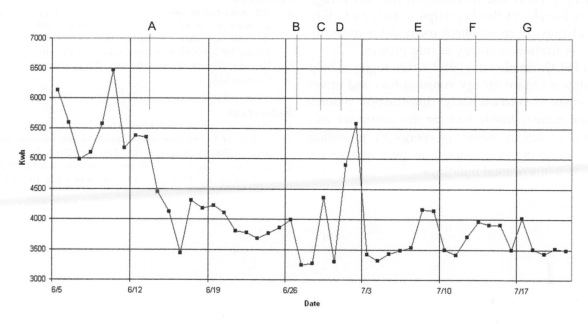

Figure 27-5. Chiller Plant kWh Usage

A. On June 14th, the chilled water set point was raised from 40F to 44F. Additionally, the minimum flow through the chillers was lowered from 600 gpm to 420 gpm.

Event	Outside Temp	Avg kWh	kWh Change	% Change
Before Reset	82.3F	5,192		
After Reset	83.0 F	4,205	-987	-19%

B. Starting on June 27th, the chiller plant was completely turned off from midnight to 6 pm.

Event	Outside Temp	Avg kWh	kWh Change	% Change
Before Shutdown	80.7F	3,873		
After Shutdown	81.1F	3,271	-602	-15%

C. On June 29th, a test and balance company performed some performance tests on the chiller plant that resulted in excessive chiller plant operation.

D. On July 1st and 2nd, the convention center was occupied with a very large event increasing chiller plant demand significantly.

E. On July 8th, the chiller maintenance company mistakenly forced the chillers out of automatic control which caused a chiller to operate from midnight to 6 pm. The mistake was corrected on July 10th.

F. On July 13th, the convention center was occupied with convention activity.

G. On July 17th, the convention center was occupied with convention activity.

By using the electrical submetering data as feedback, adjustments to the chiller plant control strategies were made that resulted in a 30% reduction in energy usage from the origin sequence of operation.

CONCLUSION

Historically, hardware, software and installation of energy information systems has been prohibitively expensive and has limited implementation to those commercial and industrial facilities that could afford to pay for custom systems integration services. These costs have fallen dramatically as companies leverage the enormous investment in the internet to provide tools to the building owner that make do-it-yourself data acquisition a cost effective reality. Hardware and software designed specifically for data acquisition and using available tools such as TCP/IP, HTTP and Modbus put valuable energy information literally at the fingertips of today's facility owners and provide an excellent method for measurement and verification of energy saving projects.

Successful ESCOs employ the latest technologies to help their customers reduce energy consumption and lower operating costs. Web-based energy information systems provide one more valuable tool for the contractor and the building owner in managing energy by providing timely and accurate measurement of performance for individual systems or total buildings.

Footnotes

a. Any reference to specific products or name brands of equipment, software or systems in this chapter is for illustrative purposes and does not necessarily constitute an endorsement implicitly or explicitly by the authors of this chapter or the others in this book.

b   Jim Lewis is one of the authors of this chapter and is the CEO of Obvius, LLC.

References

1. Barney Capehart, Paul Allen, Klaus Pawlik, David Green, *How a Web-based Energy Information System Works*, Information Technology for Energy Managers, The Fairmont Press, Inc., 2004

2. Jim Lewis, *The Case for Energy Information*, Information Technology for Energy Managers, The Fairmont Press, Inc., 2004

Chapter 28

Disney's Enterprise Energy Management Systems

Paul J. Allen, P.E.

ABSTRACT

ISNEY'S ENTERPRISE ENERGY MANAGEMENT systems integrates commercial energy management systems with custom web-based energy information systems. This technology-based solution is used throughout the Walt Disney World Resort organization by administrative managers, engineering, operations and maintenance staff, and cast members. Using Disney's energy information system, each Disney business unit's energy usage is continually measured using a "utility report card." This special report format stokes some healthy competition between the areas by ranking them based on their energy usage target. Disney's energy management system is used to control the energy usage of air conditioning and lighting systems. Building tune-ups are used to continually adjust the energy management systems to match the facility operational requirements. This combination of people and technology has resulted in a sustainable energy management program at the Walt Disney World Resort. The goal of this chapter is to show how these technical systems all work together to drive Disney's energy management efforts.

DISNEY'S ENERGY MANAGEMENT SYSTEM INFRASTRUCTURE

The energy management systems (EMS) used at the Walt Disney World Resort are used to control energy consuming equipment—primarily for heating, ventilating and air conditioning (HVAC) equipment and lighting control.

Over the years, Disney has installed a variety of energy management systems from different vendors, which it continues to operate. One vendor's EMS, the Carrier Comfort Network (CCN)* from Carrier Corporation, controls more than 80 percent of the installed EMS base at the Walt Disney World Resort. ComfortVIEW (formerly called ComfortWORKS) is the Windows-based graphical user interface for the CCN and is based on a client/server network-based system that uses an SQL database on a server and custom client software on the workstations.

Since ComfortVIEW was designed as a client/server based system, it is already designed to accommodate multiple EMS networks using a common SQL server database. Connecting a new attraction or resort to an existing ComfortVIEW server involves the installation of a Carrier CCN-to-ethernet converter. This device provides the connectivity to the ethernet Lan and the CCN RS-485 controller network. Up to 32 CCN networks can be connected to one ComfortVIEW server in this fashion. Using the high-speed ethernet, there is no physical distance limitation of the CCN-to-ethernet converter to the ComfortVIEW server. As an example, the World of Disney Store in New York City is connected to one of the ComfortVIEW servers at the Walt Disney World Resort using a CCN-to-ethernet converter.

There are five ComfortVIEW EMS servers that connect all of the parks, resorts and support facilities together at the Walt Disney World Resort. Likewise, there are ComfortVIEW EMS servers at the Disneyland Resort in California, Disneyland Paris Resort in France and Hong Kong Disneyland. All of these ComfortVIEW EMS servers can be accessed from a user's PC using the ComfortVIEW client software. Logon access control is through a user ID and password that defines the user's access rights. Other features of Disney's EMS infrastructure are shown below:

- Review of EMS field panel programming and real-time operation can be made "globally" through any desktop PC on the corporate network.

---

*Any reference to specific products or name brands of equipment, software or systems in this chapter is for illustrative purposes and does not necessarily constitute an endorsement implicitly or explicitly by the authors of this chapter or the others in this book.

- The EMS program and data are stored on network servers that are maintained by the Walt Disney World Information Services group. Backups are made daily.

- Automatic reset of equipment time and setpoint schedules are made daily from a server-side control program.

- Data collection for both EMS point trends and utility meter data can be collected and used by the energy information system for quick and easy display.

- Maintenance and training of the EMS is simplified.

- Service contracts are minimized or eliminated.

- EMS spare parts inventory is minimized.

The key to Disney's success is that they are able to design, install, program, and operate their EMS. As facilities are added or upgraded, the new HVAC and lighting controls are also upgraded and added to one of the existing ComfortVIEW EMS servers. In the end, everything works; that's the most important thing.

DISNEY'S ENERGY INFORMATION SYSTEM INFRASTRUCTURE

In 1997 Disney developed their first energy information system and chose to go the "do-it-yourself" route, primarily because there were few choices available at the time. Today, there are numerous software and hardware tools available on the market that can be used to create an energy information system. Likewise, there are numerous companies today that provide an EIS for an on-going monthly service fee. Even with these new options, Disney's "do-it-yourself" approach has proven to be the most cost-effective strategy based on the amount of utility data processed.

The hardware and software products selected by the development team were based primarily their familiarity of the products. The first item required was to obtain a server. Disney's I.T. department supplied and maintains the server hardware and operating system. They are responsible for updating the operating system security and anti-virus updates, along with routine backups for the server.

To create programs that display energy data on an intranet or the internet, web server software is re-

quired. Microsoft Internet Information Server (IIS) was used as the web server program since it was already included with the Microsoft XP server operating system supplied with the server.

To "warehouse" the large amount of data collected from meters and energy management systems, a relational database management system is needed. Microsoft Visual Foxpro was used as the database management system. Another program, called FoxWeb, was chosen to interface the Visual Foxpro programs and the web server software. More information about FoxWeb is available at http://www.foxweb.com.

Utility usage graphs are also very useful for charting trends in the utility data. Kavacharts uses Java applets for graphing and can be downloaded for free at http://www.ve.com.

Finally, the Autotask 2000 program was used to schedule when data collection programs are launched during the day. Additional information on Autotask 2000 is available at http://www.cypressnet.com.

UTILITY REPORTING SYSTEM (URS)

The utility reporting system (URS) was Disney's first web-based energy information system and was created in 1997. The URS was developed to report and graph monthly utility data from the Reedy Creek Improvement District (RCID) to track the results of energy saving efforts at the Walt Disney World Resort. [1]

The URS uses off-the-shelf database management system programs to (1) gather the data from all data sources and (2) publish the data on the Disney intranet. The advantage of this approach is that the programs can be customized to collect all utility data—no matter their source—from a variety of existing utility data sources.

The URS was developed to make sub-metering more effective. By continuously "shining a light" on utility usage at each facility, utility costs are minimized by the actions of people who receive timely and informative reports. Continuous feedback on utility performance pinpoints problems in the energy management system that needs attention.

The URS is used every day by users throughout the Walt Disney company. The URS is now used at the Disneyland Resort in California, Disneyland Paris Resort and Hong Kong Disneyland. This is a testament to the URS simplicity and low cost of operation.

URS Program Overview
Each month the RCID utility billing data are out-

put from the RCID utility billing system as a comma delimited file for each meter showing the account number, utility, consumption and cost. By making the RCID billing data available electronically each month, the time required to update the data is reduced significantly and data input errors are eliminated.

A custom Visual Foxpro program was developed to read the RCID monthly data into the database tables used by the URS. Aside from updating the monthly billing data, the program also determines if any new RCID utility accounts were added, and automatically updates the URS meter account definition table.

To speed up web-browser access to the URS, the meter-level monthly billing data are summed to different hierarchical levels. Besides the meter level, which provides the finest level of detail, the utility data are aggregated for each building or group of buildings to produce a subarea-level data table. Likewise, the utility data for each business unit are aggregated to produce an area-level data table.

Besides the monthly billing data, the URS also provides access to a wide range of hourly submetering data. These data can be very useful for determining how energy is used on a near-real time basis. They provide a finer level of detail and help energy managers quantify their energy saving efforts on a hourly/daily basis. Problems can be pinpointed quickly and controls adjusted to keep energy consumption minimized.

The URS updates the submetering data tables on a daily basis. The data are recorded hourly by the respective data collection system and are transmitted to the URS web server on a nightly basis. Visual Foxpro programs read the various raw submetering data files into common submetering database tables. Once the database tables are updated with the prior day's submetering data, they are copied to the appropriate subdirectory on the web server and are then available for viewing using the URS web-publishing program.

Data Collection Programs

A variety of data collection programs are used to pull the utility data into Visual Foxpro data tables. Custom programs read the various data sources and organize the data into common data tables. A separate data collection program was developed to read each utility data source.

Shown below are the various data collection tasks in the URS:

- Monthly Utility Data: The RCID billing system data are downloaded monthly in an ASCII comma delimited format. As new utility accounts are add-

ed, a new record in the account definition table is added.

- Power Monitoring Hourly Data: RCID's supervisory control and data acquisition (SCADA) power monitoring system records max, min and average hourly data for electric meters and outputs these data to an ASCII file each day. A Visual Foxpro program reads these data and reformats them into an hourly data table.

- Energy Management System (EMS) Hourly Data: An EMS can be programmed to produce files that include trends of analog/digital points and consumable data from utility meters connected to the EMS. Data collection programs copy these files from the EMS servers to the URS Server after the EMS creates them each night. A Visual FoxPro program reads the data from these reports, reformats them, and then adds them to an hourly data table

- Acquisuite Data Collection: Hourly data are recorded in local data collection devices called Acquisuites. On a nightly basis, the EnertraxDL data collection program supplied by Obvius automatically uploads data from each Acquisuite into comma separated files (CSV). A Visual FoxPro program pulls the data from these CSV files, reformats them, and then adds them to an hourly data table.

It is important to point out that an enormous amount of data can be generated from hourly utility data. To keep the data manageable, the hourly data are broken up into separate monthly data files.

Once the data collection programs finish updating all of the data from the various hourly data sources, a Visual FoxPro program creates additional files that total the hourly data to sub-area and area levels. This step helps speed up web-browser access to the URS. Finally, the data are copied to the URS Server data directory where they are ready for viewing on the Disney intranet.

URS Email Reports

To make the URS easy to use, it sends HTML-based reports via email on a daily basis (to report on hourly data collected) and a monthly basis (to report on monthly billing data). A daily utility report is created for each business unit and emailed to the business unit distribution list. Using HTML-based email reports allows the tabular report to link to graphs showing daily and monthly utility profiles. Users view the reports using their email program (Microsoft Outlook) and are able to produce graphs by simply clicking on hot-links in the email. Sending email on utility usage helps to increase employee participation in reducing their facility's energy consumption.

Email also increases the likelihood that the User views the utility data. Instead of waiting for the User to visit the URS web site and figure out how to generate the same report, the URS delivers the report via email.

Web Publishing Program

Visual Foxpro is the program language used to generate the web pages for the URS. FoxWeb is another program that is used to interface Visual FoxPro with the Microsoft Internet Information Server web server. The URS uses several reports to view and graph both the monthly utility billing information and the hourly utility data. Kavacharts Java applets, called from the Visual FoxPro programs, generate the graphs used in the URS.

The challenge of producing an effective EIS is to create reports that are both informative and easy to use. The program should be designed so that a User can easily produce reports and graphs with a few clicks of the mouse. The URS makes extensive use of embedded links to sub-reports and graphs. This programming interface makes the URS intuitively easy for the User to navigate.

A new report format, called the utility report card (URC), was recently added to the URS to let users see how their energy saving efforts rank relative to other facilities. Areas that have increased consumption are easily identified from the URC report. Executive management can question the increases and that is generally all that it takes to motivate those responsible individuals to take action. Knowing that everyone is looking keeps the focus on finding new ways to saving energy.

URS Results

The URS, like all EIS programs, does not result directly in energy reductions. Instead, the knowledge, operational insight, and experience gained from utility data can result in operational changes and corresponding energy savings.

The most significant result of the URS has been increased awareness of utility usage. Wasteful practices are corrected and energy-efficient systems are showcased as best practices. Individuals have the URS to track and monitor their energy use and this has translated into lower utility bills.

One other significant result of the URS is its ability to report on utility cost reimbursement to quantify operating participant utility usage. Utility submeters are automatically read each day and a daily report emailed to each operating participant. At the end of the month, the Disney accounting department uses the URS to totalize the utility cost reimbursement for each operating participant. Submetering results in more accurate utility cost recovery compared to a cost per square foot allocation method.

BUILDING TUNE-UP SYSTEM (BTUS)

The building tune-up process is one of the most cost-effective energy management projects available to an energy manager. The actions taken are generally low-cost or no-cost adjustments to an existing EMS and will not only minimize current operating costs but will also lower future maintenance costs. [2]

The building tune-up does not necessarily involve the purchase and installation of new equipment or technology. Instead it requires an investigative-style approach to ensure that the EMS controls are working and controlling the HVAC and lighting systems optimally. The building tune-up process is a systematic approach to fine tuning an energy management system for optimal performance. This effort can be considered one of those proverbial "low hanging fruit" energy projects that all energy managers should focus on.

The non-technical side of the building tune-up focuses on the development of the building tune-up team. It allows both technical and non-technical staff to work together in an on-going continuous improvement process to lower utility costs with minimal capital outlay. This provides an excellent venue to organize energy conservation efforts within an organization. Every team member can play an important role and can contribute to the overall team success.

The technical side of the building tune-up focuses on computer programs that keep the energy management system settings at their optimal state. The building tune-up system (BTUS) is a web-based program that allows all users to view the time schedules and cooling/heating setpoints from their own PCs using web browser software. The purpose of the BTUS is to give building owners a view into their building's heating, ventilating and air conditioning (HVAC) and lighting systems without accessing the EMS directly. The building owners provide the information on how their HVAC systems are controlled by establishing the time and cooling/heating setpoint schedules. Because it is web-based, it also allows a broader audience to access it via its web browser interface and makes it simple for everyone to know how their facilities are operating.

The BTUS keeps track of the following information for each HVAC system:

- Description of area serviced. A color-coded floor plan can be displayed if available.

- Time and setpoint schedules. Includes both desired schedules and a link to look at the most recent schedules broadcast by the FTS program.

- Shows equipment in need of repair.

- HVAC temperature, humidity and status trends can be graphically displayed if available

The main control on the BTUS is a drop-down menu that allows the user to select the area desired. The user is then presented with a list of buildings from which to pick. Once the user selects an individual building, the detailed data for each HVAC system is displayed. Links are available to show the detailed information on each HVAC system.

Another key feature of the BTUS is its ability to display trend data from an EMS. Most EMSs on the market today have some method for collecting data from sensors, control devices and meters attached to the system. The EMS text files produced from these trend reports are huge. There is generally no easy way to graph, sort or filter the data from these reports. The BTUS organizes this voluminous amount of data so they are visualized easily and quickly for each HVAC system. Simply clicking on a link for each HVAC system displays the EMS trend data report and shows the history of the temperatures, humidity and status to let users know what is actually happening.

FACILITY TIME SCHEDULE (FTS) PROGRAM

While the BTUS helps users keep track of their optimal HVAC system settings, the facility time schedule (FTS) program provides a method to prevent EMS degradation by auto-resetting time and setpoint schedules on a daily basis. Without this automatic daily reset feature, the EMS time schedules and setpoints eventually get changed from their optimal settings in response to routine "too hot/too cold" calls.[2]

The FTS program is a custom client/server program that interfaces with the Disney's Carrier Com-

fortVIEW energy management system. The purpose of the FTS program is to provide the energy manager a method to manage the time and heating/cooling setpoint schedules for a large campus facility in a master schedule database. Time schedules can be set up as "relative schedules" that incorporate the facility opening/closing times, dusk/dawn times and day of the week. Each day, the FTS program determines the appropriate open/close/dusk/dawn times and calculates actual time schedules that are broadcast to the Carrier EMS controllers on the Carrier ComfortVIEW Network (CCN).

The FTS program can also handle special events that occur after the normal open/close time schedules by sending additional time schedules that effectively increase the HVAC/Lighting equipment run-time to accommodate the special event. The FTS program provides some additional operational features:

- Users can make local EMS panel adjustments to both time and setpoint schedules to respond to building conditions without worrying that the changes would be permanent. All schedule changes will revert to the master schedule at the programmed download time the next day.

- Setpoint schedules can be grouped into common areas or types and can be programmed with a bias offset. This offset can be used during loadshed conditions to change the setpoint low and high values to a user adjustable level, reducing energy consumption. Schedule groups can also be used to pre-cool or pre-heat an area during special functions. The setpoint bias will be removed from the setpoints after the next automatic download.

- These master schedules are sent automatically to each Carrier EMS controller on a daily basis. For each schedule, the system administrator can choose whether and when to send either or both time and setpoint schedules.

- If the facility open/close times are changed during a given day, the new time schedules are re-calculated and downloaded again to the EMS controllers to reflect these changes.

The key feature of the FTS program is to automatically reset the energy management system time schedules and heating/cooling setpoints to their optimal valves. Keeping the control values optimal ensures that the EMS is operating at peak efficiency.

EMS REPORTS

The data that are reported from an EMS (energy management system) can be very useful in managing system operations. [3] Typical EMS reports include data trending, alarm reports and EMS system activity. Most EMS on the market today do an adequate job of collecting the data but a poor job of publishing and presenting the data to a wide audience. In the past, the EMS was the only system that could access and display the data collected. Only a few people in any one organization could turn these raw data into something meaningful. Sharing EMS data with others, perhaps across the globe, is a reasonable expectation. Simply put, the internet has changed the way we share data.

The amount of data that can be produced from an EMS is huge and overwhelming. To turn these raw EMS data into useful information, the data need to be pulled into a relational database so that they can be reported, filtered, sorted and graphed. If the report files are formatted so they can be read into a relational database, then a custom web-based program can be written to display, search, sort, filter and graph the data using a standard web-browser interface. Access to these data becomes quicker and easier than ever and is also available to everyone on the company intranet instead of the limited few that are EMS users.

The EMS reports program was developed as a custom web-based program to display this information in an intuitive and user-friendly interface that further enhances and broadens the access to this information. The EMS report program has provided the engineering services department with a simple tool to track the activities on their EMS. Knowing what changes were made, by what EMS user and when they were done have proven to be invaluable in troubleshooting EMS issues by continuously tracking all events and alarms.

This approach provides a low-cost method for enhancing an existing EMS by simply capturing existing EMS report data. All of this information helps provide continuous improvement for a successful EMS operation.

EMS TUNE-UP SYSTEM (ETS)

The energy management tune-up system (ETS) was created to provide a means to focus attention on EMS control points that might not be working properly. The ETS identifies defective points that are in need of repair on the Carrier ComfortVIEW energy management system. These defective points in most cases are wasting energy and increasing utility expenses.

The ETS is actually two separate processes. The data collection process uses a feature in Carrier ComfortVIEW that allows the real-time EMS data to be read one EMS panel at a time and recorded into a relational database. Since Disney's EMS is so large, the ETS data collection program is set to record each panels real time data only once per day. Even with this limited scan rate, defective EMS control points can be easily identified for repair.

A separate web-based program was developed to allow users to search the EMS data for specific problems. The ETS web program allows for problems to be identified and corrected in a timely manner. For example, the ETS web program can identify the following problem areas:

1. Chilled water valves that are closed, but the supply temperature is still cold (valve is not shutting).

2. Fans that are commanded off, but the fan status is shown on (fan is not turning off).

3. Temperature and humidity sensors out of range. (defective sensors)

4. EMS points that have been forced. (points not working in automatic control)

To get these problems repaired, users can select the records to have work orders automatically generated through Disney's preventative maintenance program.

OTHER WEB-BASED PROGRAMS

Several smaller web-based database programs were developed to meet specific needs and are described below:

Project tracking system (PTS) was developed to keep track of energy conservation project results. This program was created to prepare executive management reports on the cost-effectiveness of the energy projects. The PTS provides the project scope, the annual energy consumption and dollar savings, the cost of installation and the internal rate of return for the project. Reports can be displayed by business unit or for a particular type of conservation project across all business units.

Building square footage system (BSF) was developed in partnership with the Walt Disney World tax department to display building square footage data for each facility based on county tax records data.

Environmental permit tracking system (EPTS) was developed to keep track of environmental permit reporting due dates. Environmental checklist reports can also be generated to aid in permit compliance verification.

Daily production reporting system (DPRS) was developed to track and report utility production data on Reedy Creek Energy Services operations.

Facility time schedule web program (FTSW) was developed to let users view the park hours and special events times programmed into the FTS programs on each EMS Server.

Central alarm notification (CAN) was developed as a means to notify users of critical alarms generated by the energy management system. The program sends a text page to the group responsible for the equipment, and automatically generates a repair work order through Disney's preventative maintenance program.

CONCLUSION

Disney uses data from its utility meters and energy management systems to drive their energy management program. By using off-the-shelf tools and taking full advantage of low cost web-based programs, Disney has created systems that let users understand their energy usage and continuously keep the focus on reducing energy consumption. Using a monthly utility report card as a scorecard on how each business unit is doing relative to each other is the key and keeps everyone focused on improvement. As the Walt Disney World Resort continues to expand, Disney's enterprise energy management systems will continue to play an important role in lowering energy costs in both new and existing facilities.

Bibliography

[1]   Barney Capehart, Paul Allen, Klaus Pawlik, David Green, *How a Web-based Energy Information System Works, Information Technology for Energy Managers*, The Fairmont Press, Inc., 2004

[2]   Paul Allen, Rich Remke, David Green, *Custom Programs Enhance Building Tune-up Process, Information Technology for Energy Managers*, The Fairmont Press, Inc., 2004

[3]   Paul Allen, David Green, *Creating Web-Based Information Systems From Energy Management System Data, Information Technology for Energy Managers*, The Fairmont Press, Inc., 2004

Section V

Web Based Enterprise Management Systems for Demand Response Applications

Chapter 29

Participation through Automation: Fully Automated* Critical Peak Pricing in Commercial Buildings

Mary Ann Piette, David Watson, Naoya Motegi, Sila Kiliccote
Lawrence Berkeley National Laboratory
Eric Linkugel, Pacific Gas and Electric Company

ABSTRACT

CALIFORNIA ELECTRIC UTILITIES have been exploring the use of dynamic critical peak prices (CPP) and other demand response programs to help reduce peaks in customer electric loads. CPP is a tariff design to promote demand response (DR). Levels of automation in DR can be defined as follows. Manual demand response involves a potentially labor-intensive approach such as manually turning off or changing comfort set points at each equipment switch or controller. Semi-automated demand response involves a pre-programmed demand response strategy initiated by a person via centralized control system. Fully automated demand response does not involve human intervention, but is initiated at a home, building, or facility through receipt of an external communications signal. The receipt of the external signal initiates pre-programmed demand response strategies. We refer to this as auto-DR.

This chapter describes the development, testing, and results from automated CPP (Auto-CPP) as part of a utility project in California. The chapter presents the project description and test methodology. This is followed by a discussion of auto-DR strategies used in the field test buildings. We present a sample auto-CPP load shape case study, and a selection of the auto-CPP response data from September 29, 2005. If all twelve sites reached their maximum saving simultaneously, a total of approximately 2 MW of DR is available from these twelve sites that represent about two million ft^2. The average DR was about half that value, at about 1 MW. These savings translate to about 0.5 to 1.0 W/ft^2 of de-

mand reduction. We are continuing field demonstrations and economic evaluations to pursue increasing penetrations of automated DR that has demonstrated ability to provide a valuable DR resource for California.

BACKGROUND

California electric utilities have been exploring the use of critical peak prices (CPP) and other demand response programs to help reduce peak demands from customer electric loads. CPP is a form of price-responsive demand response. Recent evaluations have shown that customers have limited knowledge of how to operate their facilities to reduce their electricity costs under CPP (Quantum Consulting and Summit Blue, 2004). While lack of knowledge of how to develop and implement DR control strategies is a barrier to participation in DR programs like CPP, another barrier is the lack of automation in DR systems. Most DR activities are manual and require people to first receive emails, phone calls, and pager signals, and second, for people to act on these signals to execute DR strategies.

Levels of automation in DR can be defined as follows. **Manual demand response** involves a labor-intensive approach such as manually turning off or changing comfort set points at each equipment switch or controller. **Semi-automated demand response** involves a pre-programmed demand response strategy initiated by a person via centralized control system. Fully automated demand response does not involve human intervention, but is initiated at a home, building, or facility through receipt of an external communications signal. The receipt of the external signal initiates pre-programmed demand response strategies. We refer to this as auto-DR. One im-

---

*This chapter was previously published in the Proceedings of the American Council for an Energy Efficient Economy's 2006 Summer Study on Energy Efficiency in Buildings.

portant concept in auto-DR is that a homeowner or facility manager should be able to "opt out" or "override" a DR event if the event comes at a time when the reduction in end-use services is not desirable. Participation of more then 30 large facilities in the last three years of demonstrations has shown that the automation can be provided with minimal resistance from facility operators.

The PIER Demand Response Research Center conducted a series of tests during the summers of 2003, 2004, and 2005. The objectives of these tests were two fold. First, we sought to develop and evaluate communications technology to send DR signals to commercial buildings. This was necessary because buildings use controls with diverse protocols and communication capabilities. Second, we sought to understand and evaluate the type of control strategies facility owners and managers would be willing to test in their buildings. During these past three years we have evaluated auto-DR in 28 facilities; the average demand reductions were about 8% over the three to six hour DR events. Many electricity customers have suggested that automation will help them institutionalize and "harden" their electric demand savings, improving overall response and repeatability. The evaluation of the California's 2004 DR programs found that ten to fifteen of the sites that participated in their study could not participate in the DR event because the person in charge of the demand reduction was not in the facility on the day of the event (Quantum Consulting and Summit Blue, 2004).

Table 29-1 shows the number of sites that participated in each year's field tests along with the average and maximum peak demand savings. The electricity savings data are based on weather sensitive baseline models developed for each building that predicts how much electricity each site would have used without the DR strategies. Further details about this research are available in previous reports (Piette et al., 2005a and b). One key distinction between the 2005 and the previous tests is that the 2005 test sites were actually on a CPP tariff, while the 2003 and 2004 tests used fictitious prices

and there was no actual economic incentive for the sites. The "fictitious" test consisted of an actual shed based on fictitious prices. There were no DR economics incentives. The sites were willing to conduct the DR to understand their DR capability and automation infrastructure.

The focus of the rest of this chapter is the design and results from the 2005 auto-CPP field tests, with some additional comments about the previous years' tests. The next section describes the auto-CPP project description and test methodology. This is followed by a discussion of auto-DR strategies used in the field test buildings. We then present a sample auto-CPP load shape case study, and a selection of the auto-CPP DR data from September 29, 2005. The summary section provides an overview of key findings. Since the buildings only participated within the program during the later DR events of the summer 2005, we do not have detailed economics on the impact of CPP. Each site, however, saved money.

Automated CPP Project Description

PG&E's critical peak pricing (CPP) program is a voluntary alternative to traditional time-of-use rates. The CPP program only operates during the summer months (May 1 through October 31). Under the program, PG&E charges program participants' higher prices for power on up to 12 hot afternoons between May 1 and October 31. Manual CPP customers are notified by email and phone by 3 p.m. the previous day that the following day is a CPP day. The customer sees lower electricity costs on non-CPP days. The price of electricity rises on a maximum of 12 hot days, with the DR event triggered by temperature. The additional energy charges for customers on this tariff on CPP operating days are as follows (Figure 29-1):

*   **CPP Moderate-Price Period Usage:** The electricity charge for usage during the CPP moderate-price period was three times the customer's summer part-peak energy rate under their otherwise-applicable rate schedule multiplied by the actual

Table 29-1. Average and Maximum Peak Demand Savings during Automated DR Tests.

Results by Year	# of sites	Duration of Event (Hours)	Average Savings During (%)	Highest Max Hourly Savings (%)
2003	5	3	8	28
2004	18	3	7	56
2005	12*	6	9	38

*Some of the sites recruited were not successful during the 2005 CPP events because of delays with advanced meters and control work, but are expected to be ready for the 2006 tests.

energy usage. The CPP moderate-price period was from 12:00 Noon to 3:00 p.m. on the CPP operating days.

- **CPP High-Price Period Usage**: The total electricity charge for usage during the CPP high-price period was five times the customer's summer on-peak energy rate under their otherwise-applicable rate schedule multiplied by the actual energy usage. The CPP High-Price period was from 3:00 p.m. to 6:00 p.m. on the CPP operating days.

The 2005 auto-DR project design was a collaboration between LBNL, the DRRC, and PG&E. PG&E had offered voluntary critical peak pricing in 2004, with over 250 sites participating. We recruited 15 PG&E customer facilities to participate in fully automated response critical peak pricing. There were three categories of recruits. First, five of the sites had participated in the 2004 auto-DR tests and were willing to move from the fictitious tests to the actual tariff. Second, we worked with the PG&E customer account representatives to recruit two sites that had been on CPP to include them in the auto-CPP tests. Third, eight sites were recruited for the 2005 tests that had not been on CPP or had not participated in the previous auto-DR tests.

Demand Response Automation Server

PG&E sent the critical peak price signals to each participating facility using the demand response automation server developed by LBNL and Akuacom. The automation server communicated via XML with PG&E DR communications system, Interact II. Qualified sites were configured to respond to automated price signals transmitted over the internet using relays and gateways that send standardized signals to the energy management control system (EMCS). A few sites used the day-ahead automation notification for their pre-cooling strategies. Most of the sites used the signal in real time that alerted them at noon on the CPP day that the event was triggered. During the 2005 summer test period, as the electricity price increases during a CPP event, pre-selected electric loads were automatically curtailed based on each facility's control strategy. The automation server uses the public internet and private

corporate and government intranets to communicate CPP event signals that initiate reductions in electric load in commercial buildings. The researchers worked with the facility managers to evaluate the control strategies programmed in the energy management and control systems (EMCS), which executed pre-determined demand response strategies at the appropriate times.

Connectivity was provided by either an internet gateway or internet relay (as shown in Figure 29-2). The internet gateways typically connect the internet communication protocol (TCP/IP) to the protocol of a given EMCS. This means that a different internet gateway type is usually required to communicate with each different EMCS brand or product line. Gateways provide a variety of functions further described in Piette et al. (2005). An internet relay is a device with relay contacts that can be actuated remotely over a LAN, WAN or the internet using internet protocols (IP). The internet is based on a standard protocol (TCP/IP) and all EMCS can sense the state of relay contact closures (regardless of their particular EMCS protocol). Because of this, internet relays can be used on virtually any commercial building that has a standard connection to the internet. Internet connectivity directly to the EMCS is not required.

The four elements of the diagram are as follows:

1. PG&E uses their standard InterAct II system to notify the automation server of an upcoming CPP event (notification occurs day-ahead).

2. The automation server posts two pieces of information on its Web services server:

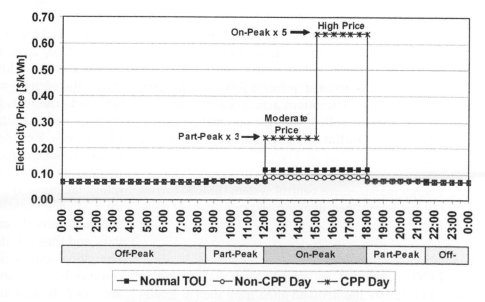

Figure 29-1. Critical Peak Pricing Tariff

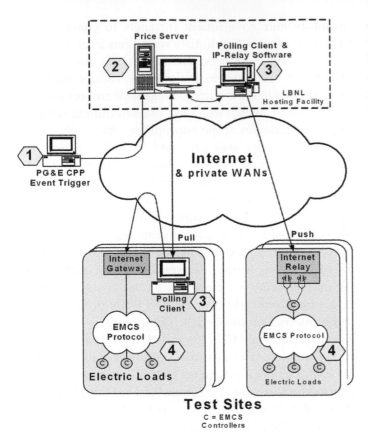

Figure 29-2. Building type, size, year in Auto-DR, and DR control strategy used.

consumption data from Interact for each site. We subtracted the actual metered electric consumption from the baseline-modeled consumption to derive an estimate of demand savings for each 15-minute period. The model is described in previous papers (Piette et al, 2005). PG&E uses a baseline for the CPP evaluation. The demand response strategy was considered effective if in either or both of the moderate price and the high price periods, the average power savings over the 3-hour period was larger than the average of the standard error in the baseline model. For each building we derived the hourly electric load savings, percent savings in whole-building load, and power density reduction (W/ft^2). Sample results for the auto-CPP events are shown below.

The CPP baseline used by PG&E does not include weather data, but is based on the average hourly load shape of 3 highest consumption days in the last 10 working days (excluding holidays). The baseline algorithm considers the site electric consumption from the period of noon to 6 p.m. to choose the highest 3 days. CPP event days are excluded from the reference days. The CPP baseline estimate may be lower than the actual demand if the site's demand is weather-sensitive, since a CPP day typically occurs on a higher temperature day. If the ten previous working days were cooler than the CPP day, the baseline will be lower than weather normalized baseline.

There are a few other features about the project that we do not have space to review in this chapter. The evaluation included post-event surveys to determine how well each strategy performed and if there were any outstanding issues in the DR control strategies. The evaluation also examined the cost to program control strategies in the EMCS and to connect the internet gateways and relays.

— There is a pending event. This is posted immediately upon receipt from PG&E at approximately 3:00 p.m. the day ahead.
— There is an active event of a given level. Moderate-level demand response events are posted between 12:00-3:00 p.m. High-level demand response events are posted between 3:00 p.m.-6:00 p.m. on the day of the event.

3. Polling clients request information each minute. Logic software determines actions based upon latest information polled from the automation server. Actions are initiated based on predetermined logic.

4. Energy management control system (EMCS) carries out predetermined demand response control commands.

Evaluation Methodology

LBNL developed an electric load shape baseline model to estimate the demand shed from the DR strategies for each building. First we collected the electric

Auto-DR Field Test Results

In 2003, 2004, and 2005 we conducted automated DR tests in 28 buildings listed in Table 29-2. Table 29-3 shows the entire list of sites and which years they participated. The tests included numerous building types such as office buildings, a high school, a museum, laboratories, a cafeteria, data centers, a postal facility, a library, retail chains, and a supermarket. The buildings range from large campuses, to small research and laboratory facilities. The table lists the DR control strategies used at each building. The full reports from the auto-DR field tests describe these strategies in greater details, and they are also discussed in Watson et al. (2006). The global zone temperature adjustment was the most commonly used strategy, though 16 other strategies are listed. Nearly all of these strategies were based on direct connections to the EMCS. Further details on pre-

Table 29-2. Building Type, Size, Year in Auto-DR, and DR Control Strategy Used.

	Building use	Total conditioned area	# of bldg	2003	2004	2005	Global temp. adjustment	Fan-coil unit off	SAT reset	Fan VFD limit	Duct static pres. reset	Fan quantity reduction	Electric humidifier off	CHW temp. reset	CHW current limit	Chiller demand limit	Boiler lockout	Pre-cooling	Extended shed period	Slow recovery	Common area light dim	Office area light dim	Anti-sweat heater shed	Fountain pump off	Transfer pump off
300 CapMall	Office	383,000	1		X		X			X		X		X										X	
ACWD	Office, lab	51,200	1		X	X	X		X	X				X	X		X		X						
Albertsons	Supermarket	50,000	1	X																	X	X		
B of A	Office, data center	708,000	4	X	X	X				X	X	X		X	X									
Chabot	Museum	86,000	2		X	X											X							
Cal EPA	Office	950,000	1		X						X									X	X			
CETC	Research facility	18,000	1		X						X	X												
Cisco	Office, tech lab	4,466,000	24		X			X	X							X			X	X			
2530 Arnold	Office	131,000	1	X	X	X												X						
50 Douglas	Office	90,000	1	X	X	X												X						
Echelon	Corporate Headquarter	75,000	1	X	X	X			X		X	X								X	X			
GSA 450 GG	Federal office	1,424,000	1		X	X																		
GSA NARA	Archive storage	202,000	1		X	X																		
GSA Oakland	Federal office	978,000	1	X	X	X																		
Gilead 300	Office	83,000	1		X			X																
Gilead 342	Office, Lab	32,000	1		X	X		X																
Gilead 357	Office, Lab	33,000	1		X	X		X																
Irvington	Highschool	N/A	1		X	X										X								
IKEA	Retail	300,000	1		X	X																		
Kadent	Material process	-	1		X																			X
LBNL OSF	Data center, Office	70,000	1		X	X									X									
Monterey	Office	170,000	1		X															X				
Oracle	Office	100,000	2		X	X		X																
OSIsoft	Office	60,000	1		X	X																		
Roche	Cafeteria, auditorium	192,000	3	X	X						X													
Target	Retail	130,000	1		X						X									X				
UCSB Library	Library	289,000	3	X	X			X	X				X											
USPS	Postal service	390,000	1		X	X								X			X							

cooling research, which may prove to be an important DR control strategy, are presented in Peng et al, (2004 and 2005).

Example of Demand Response from an Office Building

This section provides an example of the DR electric load shape data for the 130,000 ft^2 Contra Costa County office building. The graph shows the electric load shape during an actual auto-CPP event on September 29, 2005. The baseline power peaks around 400 kW, with the weather sensitive LBNL baseline and the PG&E CPP baseline also shown. The vertical line at each baseline power datum point is the standard error of the regression estimate. The vertical lines at noon, 3 p.m., and 6 p.m. indicate price signal changes. The building shed about 20% of the electric loads for six hours by setting up the zone temperatures from 74 to 76 during the first three hours and 76 to 78 F during the second three hours. This strategy reduced the whole-building power density by an average of 0.8 W/ft^2 during the six hours.

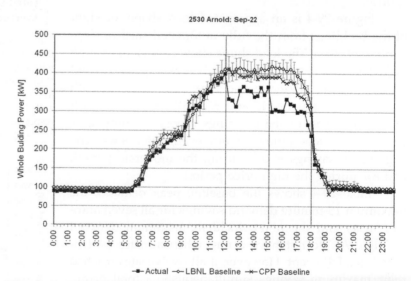

Figure 29-3. Baseline and Office Building Electric Load Shape during Auto-DR Event

Aggregated Automated Demand Response

The auto-CPP tests consisted of seven events that took place from August through November 2005. Configuring many of the sites to participate in the auto-CPP was time consuming because of complications related to control programming and meter installation delays. Thus, several of the CPP events took place before our sites were configured. To account for this delay, we developed a series of fully automated mock-CPP tests that took place in October and November 2005. These days were not as warm as actual CPP days and the DR events show lower electric savings than we observed on warmer days.

Table 29-3 shows sample results from eight buildings that participated in auto-CPP on an actual CPP day. The table lists the average and maximum peak demand savings, whole building percentage savings, and power density savings during the two three-hour price periods: Moderate and High. The average reductions per building ranged from 2 to 184 kW, with maximum savings of 31 to 291 kW. The table shows the total DR (Shed kW), whole-building power reduction (WBP %), and power density reduction (W/ft^2). The columns list both the average and maximum savings for the moderate and high priced CPP periods. The maximum is the fifteen-minute max demand response in the six hour monitoring period. Average percentage reductions ranged from zero (negligible) to 28% savings, and maximum percentage reductions from 3 to 37%. The average power density reductions ranged from 0.02 to 1.95 W ft^2, with maximum demand reductions 0.21 to 4.68 W/ft^2. The Bank of America site dominates the aggregated demand response.

Figure 29-4 is an aggregated load shape for eight of the buildings from the fully automated shed on September 29, 2005. The load shape shows a total of about 8 MW. The automated DR provided an average of 263 and 590 kW in the moderate and high price periods, with maximum savings of 617 and 952 kW, or nearly 1 MW maximum. Most of the buildings report no complaints or comfort issues following our event interviews. The aggregated savings is 3% during the moderate period, and 8% during the high price period.

Table 29-4 shows the baseline peak demand, the maximum 15-minute demand savings for all seven auto-CPP tests and the non-coincident maximum demand savings. We do not have a day when all sites participated in a DR event. However, if all twelve sites reached their maximum savings simultaneously, a total of approximately 2 MW of demand response is available from these twelve sites that represent about two million ft^2. Using the sum of the average demand response for each of the twelve sites shows the average demand response was about 1 MW. These results indicate that 1 to 2 MW of demand response can be expected for two million ft^2 of buildings (0.5 to 1.0 W/ft^2 of demand saving) with this type of automation. As mentioned, following each

Table 29-3. Average Demand Response by Price Period, September 29th

Unit	Site Name	Average		Max	
		Moderate	High	Moderate	High
Shed kW	ACWD	67	57	101	72
	B of A	22	184	132	291
	Chabot	2	32	31	88
	2530 Arnold	34	58	90	89
	Echelon	32	109	42	143
	Gilead 342	45	55	73	75
	Gilead 357	48	62	94	150
	Target	14	33	53	44
Total: Σ(ΔP)		263	590	617	952
WBP %	ACWD	24%	19%	38%	23%
	B of A	0%	4%	3%	6%
	Chabot	0%	3%	10%	28%
	2530 Arnold	8%	14%	21%	21%
	Echelon	9%	28%	12%	37%
	Gilead 342	13%	15%	19%	20%
	Gilead 357	9%	11%	16%	25%
	Target	4%	9%	15%	12%
Total: Σ(ΔP)/Σ(BP)*		3%	8%	8%	12%
Average: Σ(ΔP/BP)/N		9%	13%	17%	21%
W/ft^2	ACWD	1.53	1.29	2.30	1.63
	B of A	0.04	0.30	0.21	0.47
	Chabot	0.02	0.37	0.35	1.02
	2530 Arnold	0.26	0.44	0.69	0.68
	Echelon	0.43	1.45	0.56	1.91
	Gilead 342	1.39	1.72	2.30	2.36
	Gilead 357	1.50	1.95	2.95	4.68
	Target	0.13	0.30	0.48	0.40
Total: Σ(ΔP)/Σ(A)**		0.23	0.52	0.55	0.85
Average: Σ(ΔP/A)/N		0.66	0.98	1.23	1.64

*The average of the individual average whole building response and the average of the maximum individual DR results are shown, along with the aggregated shed compared to the total baseline power.
**The power densities are also shown for the average of the demand intensities (sum all building densities and divide by the sample size) and the sum of the total area and the total aggregated total demand response.

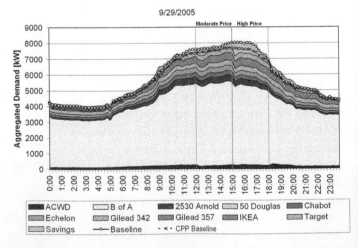

Figure 29-4. Automated CPP Aggregated Demand Saving Results, September 29th

event LBNL interviewed building managers to evaluate if any problems occurred. There were some minor complaints in a few cases. Overall the sites were able to provide good demand response with minimal disruptions. We have begun to explore the costs required to configure the auto-DR communication systems and program DR control strategies within an EMCS. Initial research suggests we can configure auto-CPP systems with the existing financial incentives available as part of California utility DR technical assistance funds. Ideally auto-DR systems would be installed as part of retro-commissioning programs. With their knowledge and skills, today's retro-commissioning engineers may be key players in providing building control tune-ups and developing custom DR strategies during field work (Piette et al, 2006). Installation and configuration of auto-DR systems require a good understanding of HVAC, lighting, and control strategies.

SUMMARY AND FUTURE DIRECTIONS

The auto-CPP tests in 2005 have demonstrated the technical feasibility of fully automated DR. While there are considerable challenges in auto-DR in general and auto-CPP specifically, the research demonstrates that this can be done with reasonable levels of effort with today's technology. New knowledge on what strategies are available for different types of buildings has been obtained and is the subject of another ACEEE paper (Watson et al, 2006).

During 2006 we will be pursuing a larger number of tests throughout California. The research may also move beyond CPP into other DR programs such as demand bidding. The primary objective of this new research will be to better understand the economics of installing and configuring automated systems, exploring connectivity and control strategies in more building types, including industrial facilities, and evaluating the peak demand reduction levels for different weather. We are also interested in "heat storm" performance that moves beyond single day DR participation, to several hot days in a row.

In the long term this research aims at transform communications in commercial and industrial facilities to explore literally "connecting" the demand and supply side systems with the technologies and approaches explored in this project. Our goal is to understand how to configure buildings to be "DR ready" in a low cost way, developing requirements for new buildings through future codes and embedding such communications directly into future EMCS. Additional research is also needed

Table 29-4. Maximum Demand Response for all Seven Event Days

		Aug-08	Sep-22	Sep-29	Oct-06	Oct-13	Oct-25	Nov-10	2004	Max
ACWD	Baseline Peak kW			330	253	290	238			330
	Max Shed kW			101	74	83	77			101
B of A	Baseline Peak kW			5311		5163	5053			5053
	Max Shed kW			291		219	552			552
Chabot	Baseline Peak kW		225	308	244	270				308
	Max Shed kW		19	88	36	42				88
2530 Arnold	Baseline Peak kW	505	419	431	404	406	345			505
	Max Shed kW	176	119	90	63	89	40			176
50 Douglas	Baseline Peak kW	381					259			381
	Max Shed kW	95					78			95
Echelon	Baseline Peak kW		334	403	363	359	304			403
	Max Shed kW		115	143	132	117	84			143
Gilead 342	Baseline Peak kW		288	384	289	340	278			288
	Max Shed kW		94	75	45	55	80			94
Gilead 357	Baseline Peak kW			607		455	443			607
	Max Shed kW			150		119	145			150
IKEA	Baseline Peak kW					1982	1803			1982
	Max Shed kW					321	223			321
Oracle	Baseline Peak kW							507		507
	Max Shed kW							65		65
Target	Baseline Peak kW		314	364	328	341	296			341
	Max Shed kW		52	53	60	64	49			64
USPS*	Baseline Peak kW								1483	1483
	Max Shed kW								333	333
Total	Baseline Peak kW	886	1579	8138	1881	9608	9020	507	1483	12189
	Max Shed kW	272	399	992	410	1108	1329	65	333	2182

* 2004 data (Oct-13) is used for USPS because USPS failed to conduct demand shed in 2005.

to integrate price and reliability DR signals, which we believe can co-exist on similar communications systems. Finally, there is a need to better understand advanced controls for simultaneous use applications of energy efficiency and demand response. We need to define explicit "low power" building operating modes for DR events. Daily advanced energy efficient operations with granular controls provide the best starting point for DR capability. New technologies such as dimmable ballasts and wireless HVAC control are likely to provide such new levels of granularity that can be optimized to provide both daily and enable advanced DR strategies. Along with such new technology is the need for improved energy management and financial feedback systems. As the DR economics mature, better real-time economic feedback is needed if energy managers and facility operators are going to understand the value of participating in DR events.

Acknowledgements

The authors are grateful for the extensive support from numerous individuals who assisted in this project. Many thanks to the engineers and staff at each building site. Special thanks to Ron Hofmann for his conceptualization of this project and ongoing technical support. Thanks also to Laurie ten Hope, Mark Rawson, and Dave Michel at the California Energy Commission. Thanks to the Pacific Gas and Electric Company who funded the automated CPP research. This work described in this report was coordinated by the Demand Response Research Center and funded by the California Energy Commission, Public Interest Energy Research Program, under Work for Others Contract No.150-99-003, Am #1 and by the U.S. Department of Energy under Contract No. DE-AC03-76SF00098.

References

Piette, Mary Ann, David S. Watson, Naoya Motegi, Norman Bourassa and Christine Shockman. 2005a. "Findings from the 2004 Fully Automated Demand Response Tests in Large Facilities" September. CEC-500-03-026. LBNL-58178. Available at http://drrc.lbl.gov/drrc-pubs1.html

Piette, Mary Ann, Osman Sezgen, David S. Watson, Naoya Motegi, and Christine Shockman. 2005b. "Development and Evaluation of Fully Automated Demand Response in Large Facilities," January. CEC-500-2005-013. LBNL-55085. Available at http://drrc.lbl.gov/drrc-pubs1.html

Piette, Mary Ann., David S. Watson, Naoya Motegi, Sila Kiliccote. 2006. Automated Critical Peak Pricing Field Tests: Program Description and Results, LBNL Report 59351. March.

Piette, Mary Ann., David S. Watson, Naoya Motegi, Sila Kiliccote, and Eric Linkugel. 2006. "Automated Demand Response Strategies and Commissioning Commercial Building Controls, 2006 National Conference on Building Commissioning. April.

Quantum Consulting Inc., and Summit Blue LCC. 2004. Working Group 2 Demand Response Program Evaluation—Program Year 2004, Prepared for the Working Group 2 Measurement and Evaluation Committee, December.

Watson, David S., Naoya Motegi, Mary Ann Piette, Sila Kiliccote. 2006. Automated Demand Response Control Strategies in Commercial Buildings, Forthcoming Proceedings of 2006 ACEEE Summer Study on Energy Efficiency in Buildings. Pacific Grove, CA. Forthcoming.

Xu, Peng, Philip Haves. 2005. Case Study of Demand Shifting With Thermal Mass in Two Large Commercial Buildings. ASHRAE Transactions. LBNL-58649.

Xu, Peng, Philip Haves, and Mary Ann Piette, and James Braun. 2004. Peak Demand Reduction from Pre-cooling with Zone Temperature Reset of HVAC in an Office. Proceedings of 2004 ACEEE Summer Study on Energy Efficiency in Buildings. Pacific Grove, CA. LBNL-55800.

Chapter 30

Industrial Customers Participating in the Deregulated Texas Electric Market's Electric Reliability Council of Texas (ERCOT) Utilizing Dashboards to Integrate Revolutionary Web-based Software Solutions and Enterprise Systems

Joseph Rosenberger, P.E SophNet Inc.,
Christopher Greenwell, P.E. Tridium Inc.,
Michael Cozzi, Cirro Energy Services, Inc., and
Noshad Chaudry, Tridium Inc.

INTRODUCTION

TODAY A SELECT and innovative group of manufacturing plants utilize visualization technologies to display "Electronic Dashboards" enabling them to participate in supply and demand response opportunities. These dashboards are where innovative plant energy personnel receive web-based signals such as hourly plant utility costs (electric, natural gas, etc), and display of real time metrics, including demand-response measures, forecasting energy consumption per pound and measurement & verification (M&V) of integrated supply/demand strategies.

Manufacturers competing in global markets require visualization technologies, such as dashboards, to measure energy and operations performance of a manufacturing plant enabling tight control of its energy budget in their pursuit of the prize as "low cost producer." This requires visualization of both the supply-side and demand-side condition in deregulated markets. The integration of web-based visualization software and wireless technologies combined with virtual real-time price signals from Electric Reliability Council of Texas (ERCOT) and demand-response programs market enables sophisticated large industrial energy users to achieve lower costs. From the manufacturing plant's "device" (Level 1) conductivity whether hard wire or cost effective wireless mesh networks can move two-way TCP/IP information up to the "enterprise dashboard" (Level 2) and out to the web where the distributed "energy team users" (Level 3) can provide the strategy and solutions

to achieve the plants objectives. More detail on how data flow between Level 1 and 2 comes later in the solution part of this chapter. In the deregulated electricity ERCOT market in Texas, an industrial end-user taking advantage of demand-response programs offered by the retail electric providers (REP) can save significant costs on electricity ($10/MW to $20/MW). Using the revolutionary software solutions available supporting "fully open architectures," the plant's enterprise system can maintain the visualization of the supply/demand integration vital to the manufacturing plant's operation.

THE VALUE

In Texas, these demand-response programs include ERCOT's Load Acting As Resources (LaaR), Cirro Energy Services 4CP (four coincidental peaks in kW); and local utilities emergency load management (ELM) standard offers programs mandated by the Public Utility Commission of Texas (PUC). The LaaR program is administered by ERCOT. ERCOT's successful Load Acting as a Resource (LaaR) initiative constitutes the largest demand-side resource program in North America in terms of the quantity of demand that can be interrupted by the independent system operator (ISO) in response to low frequency or system emergencies.

Figure 30-1 depicts an enterprise system managing integrated supply and demand-response operations of a world class manufacturing firm with four plants located in Texas, Illinois, California, and New York (all open access markets).

Below is an example of a world-class plastics manufacturing firm employing an enterprise system utilizing the dashboard technology (See Figure 30-2) to integrate both demand response (LaaR, 4CP, and TXU-ELM) and portfolio-based procurement approach into the end-user's energy cost picture. The end-user has two manufacturing plants with a combined peak demand of 20 MW and load factor of 93% (see electric load summary below).

Manufacturing Plant 1
Electric Load Profile
- 16.0 MW Peak Demand @93% Load Factor served at transmission service voltage by TXU
- 130,000 MWh annual electric consumption

Demand-Response Profile
- 12.0 MW LaaR (50% Fixed + 50% Day Ahead Pricing)
- 1.5 MW TXU-ELM
- 1.5 MW 4CP

Electric Commodity Product
- 12.0 MW Block + 4 MW Marginal Clearing Price of Energy (MCPE)

Projected savings (demand response and supply integration strategies) $1,400,000

Manufacturing Plant 2
Electric Load Profile
- 4.0 MW Peak Demand @93% Load Factor served at transmission service voltage by TXU
- 35,000 MWh annual electric consumption

Demand-response Profile
- 4.0 MW LaaR (50% Fixed + 50% Day Ahead Pricing)
- 1.0 MW TXU-ELM
- 1.0 MW 4CP

Electric Commodity Product
- 2.0 MW Block + 2.0 MW Marginal Clearing Price of Energy (MCPE)

Projected annual savings (demand response and supply integration strategies)—$ 600,000

Resource and Qualified Scheduling Entities
In the ERCOT operating scheme, all LaaRs are represented by a resource entity (RE). These REs participate

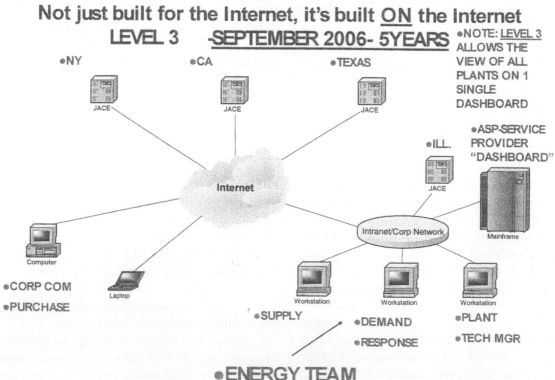

Figure 30-1. Enterprise Dashboard Technology for Manufacturing End-User With Multiple Locations

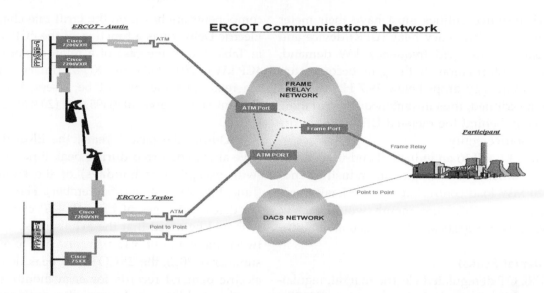

Figure 30-2. LaaR's telecommunications network that the enterprise will integrate with.

in the ERCOT markets through bilateral contracts with qualified scheduling entities (QSEs) and load serving entities (LSEs) or by bidding their resources directly into the ERCOT markets thru their QSE. There are 34 different resource entities currently registered with a LaaR portfolio. All resource entities at ERCOT must be scheduled through QSEs. There are currently 10 QSEs which schedule LaaRs.

Load Acting as a Resource (LaaR)

With the LaaR program, the industrial end-user or resource entity typically enters into a "bilateral" contract

agreement with a retail electric provider or LSE and/or directly with a qualified scheduling entity (QSE). There are typically two types of LaaR programs available. One type is to bid prices in on a day-ahead basis for responsive reserve services (RRS). If the end-user's offer is accepted, then the end-user is provided a payment whether curtailed or not. The other type of LaaR is a fixed-price for a set term. In this case, the end-user is provided with a monthly check based on contracted load (MW) and fixed-price ($/MW). In Figure 30-3 is the historical pricing for LaaR responsive reserve services in ERCOT.

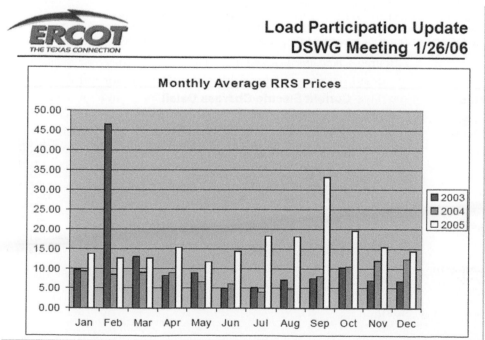

Figure 30-3. ERCOT Monthly Average RRS Prices ($/MW).

All LaaR resource entities must have their meter installed to an under-frequency relay (UFR) device and monitored real-time for grid frequency, kW demand, kWh, and other information. If the grid becomes unstable and the frequency drops below 59.7 Hz, then the UFR devices are tripped, thus instantaneously curtailing all power directly behind the metered UFR point of delivery to the resource entity.

The value of LaaR to an industrial end-user can be relatively substantial. For example, an industrial end-user with a 20 MW load contracted under a LaaR fixed-price contract of $10/MW per month would receive monthly payments of $200,000 or $2,400,000 annually.

4CP (Coincidental Peaks)

In the ERCOT deregulated electric market, regulated transmission distribution service providers (TDSPs) or utility companies assess end-users with regulated transmission and delivery charges. The transmission components of the regulated delivery charges have tariffs that utilize a component measured by ERCOT called the four coincidental peaks or 4CP kW. The 4CP is determined as described below.

The two electric billing components that are impacted by the 4CP kW are the transmission cost recovery factor (TCRF) and the transmission charge. A sample electric bill is shown in Figure 30-4 identifying the two transmission billing components. The two transmission components are based on the tariff rate charged by TXU Electric Delivery as a function of the 4CP kW as shown in Table XX. In the case of a retail customer that has a 4CP kW of 20,000 kW on TXU's primary voltage service, the retail customer would be assessed an annualized transmission charge of $395,600 (20,000 kW x $19.78/kW).

During the past 5 years, the ERCOT 4CP events have always occurred during peak times and peak days (with exception of holidays) of the summer months (June, July, August, and September). Figure 30-6 reflects the dates and times of ERCOT 4CP events since 2001. It should be noted that the events most often occur between the hours of 3:45 p.m. and 5:15 p.m. During the summer of 2005, the ERCOT grid peak demand set new electric demand records for each month as a result of weather and improved economic conditions.

An innovative energy services company, Cirro Energy Services Inc. (CES), located in Texas, created a warning notification system that warns large end-users when to curtail electric demand in order to avoid or mitigate the costs associated with 4CP. CES has developed a proprietary algorithm model that ultimately predicts the level of probability of a 4CP event occurring. The forecast model then produces a color-coded warning system as follows:

- Green (very low probability)

DETERMINATION OF 4 CP kW

The 4 CP kW applicable under the Monthly Rate section shall be the average of the Retail Customer's integrated 15 minute demands at the time of the monthly ERCOT system 15 minute peak demand for the months of June, July, August and September of the previous calendar year. The Retail Customer's average 4CP demand will be updated effective on January 1 of each calendar year and remain fixed throughout the calendar year. Retail Customers without previous history on which to determine their 4 CP kW will be billed at the applicable NCP rate under the "Transmission System Charge" using the Retail Customer's NCP kW.

Figure 30-4. Billing Components on Typical Electrical Bill in TXU Territory (North Zone)

Service Address

4401 BLUE MOUND RD
FORT WORTH TX 76106

For outages or emergencies
1-888-313-4747

ESI ID:
10443720002476017
IDR Meter

Electric Usage Detail

| Demand | 1,482 KW |
| Load Factor | 49.1 % |

Meter Number: 034219001TD

Current Read 04/12/2005	2863
Previous Read 03/13/2005	1407
kWh Multiplier	360
kWh Usage	523,837
Power Factor	75.5 %

Current Electric Charges Detail

30 Day Billing Period From 03/13/2005 To 04/12/2005

Fixed Price

Actual Consumption * Price	523,837 KWH @ $0.064580/KWH	33,829.42
GRT reimbursement charge		860.02
PUCA reimbursement charge		71.78
Transmission Cost Recovery Factor		362.78
Transmission Charge (TUOS)		2,376.66
Transition Charge		1,071.08
Nuclear Decommissioning (NDF)		83.97
System Benefit Fund (SBF)		333.68
Distribution Charge (DUOS)		5,523.36
Delivery Point Charge		41.56
Total Current Charges		**$44,554.31**

	TXU		
	TXU Sec	TXU Pri	TXU Trans
Transmission	$1.47	$1.43	$1.06
TCRF (effective 3/1/05)	$0.232808	$0.218281	$0.284134
Monthly Cost (per 4CP kW)	$1.70	$1.65	$1.34
Annual Cost (per 4CP kW)	$20.43	$19.78	$16.13

Figure 30-5. Transmission cost elements of regulated delivery tariff

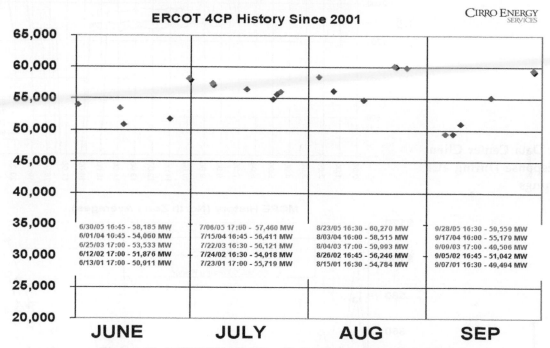

Figure 30-6. ERCOT 4CP Events Each Summer Since 2001

- Yellow (caution low probability, but possibility of increase)
- Orange (moderate probability)
- Red (high probability)

Both the orange and red color-coded warnings are considered "actionable events." This means that clients should take action to curtail or shed load during the specified time period covered by the warning notification. During the summer of 2005, CES successfully predicted the 4CP events 75% of the time. The only month missed was July and CES originally had predicted an orange level for July 6th, but at the last moment downgraded to non actionable event. Figure 30-7 shows an example of a data center client with over 3,000 kW of demand that successfully benefited from cost savings utilizing Cirro's 4CP warning notification system. This particular client has an uninterruptible power source (UPS) system, back-up generators (12,000 kW of generation), thermal energy storage, and was able to curtail load within less than one hour's notice. During the summer of 2005, this client obtained economic value from its thermal energy storage system (~1,250 kW) and an

additional 1,750 kW of curtailed demand as a result of 4CP warnings. The total annualized cost savings on this client's two transmission tariff rate components is approximately $48,000. There were no capital cost requirements incurred. Incidentally, this client was contracted under a market-indexed electric product referred to as the ERCOT marginal clearing price of energy (MCPE) in which the price of electricity changes every 15 minutes and was extremely volatile during the summer of 2005 as evidenced by the chart.

The avoided cost of electricity (MCPE) in the ERCOT north congestion zone during the 4CP events over the summer of 2005 are shown in Figure 30-9. The MCPE prices in some cases were as high as $400/MWh. End-users that have the ability to curtail load and have a price signaling mechanism might be able to avoid high electric costs.

In addition to avoiding the cost of electricity of MCPE during the 4CP events in the summer of 2005, end-users in ERCOT contracted under a portfolio-based electric product with the ability to either buy or sell blocks of power into the ERCOT balancing market. The benefit of managing one's electric load in this fashion enables an

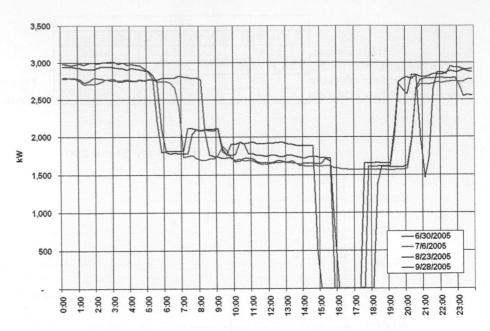

Figure 30-7. Data Center Client Demand Response During 4CP Events/Warnings

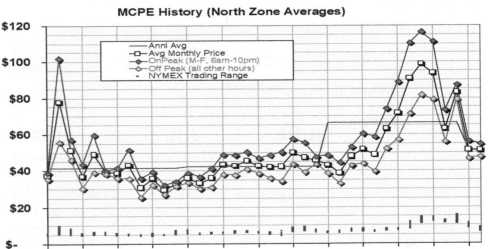

Figure 30-8. ERCOT MCPE Price History (North Congestion Zone)

Figure 30-9. Avoided Cost of Electricity (MCPE ERCOT North Zone) during 2005 4CP Events

15-min Interval	63	64	65	66	67	68	69
4CP Date/time	3:45pm	4:00pm	4:15pm	4:30pm	4:45pm	5:00pm	5:15pm
6/30/2005	$117.75	$126.03	$110.01	$110.01	$110.01	$112.30	$121.15
7/6/2005	$435.23	$435.23	$411.59	$411.59	$411.59	$411.59	$417.05
8/23/2005	$233.08	$250.00	$250.00	$295.00	$276.62	$250.00	$250.11
9/28/2005	$233.52	$234.92	$225.77	$225.00	$229.89	$242.14	$224.26
Avg MCPE Price	$254.90	$261.55	$249.34	$260.40	$257.03	$254.01	$253.14

end-user the flexibility to avoid buying high spot prices and/or to sell a block of power at high price thus mitigating energy costs. Attempting to manage one's electric load in such a manner requires visualization of real time information (e.g. prices, current supply position, current demand position, and status of counter-measures to activate load shedding and/or self-generation).

Figure 30-10 is an example of the potential value of managing one's load in this manner. Let us refer to the manufacturing end-user described earlier using the dash-board technology and assistance from an energy advisor, assuming the Corsicana plant and the Temple plant can shed 1 MW and 4 MW blocks of power, respectively. Further, let us assume that the end-user set an alarm so that whenever the MCPE price hit $200/MWh or higher, then load shedding would commence resulting in the end-user avoiding the high price of electricity. During the entire year of 2005, the end-user would have avoided 212 hours of MCPE prices at or above $200/MWh resulting in over $276,300 in avoided electricity costs.

						ERCOT NORTH CONGESTION ZONE MCPE PRICE ENTIRE YEAR - 2005								
MCPE Price Threshold ($/MWh)		$100	$125	$150	$175	$200	$250	$300	$400	$500	$600	$700	$800	$900
% Intervals above Threshold		14.04%	7.80%	4.95%	3.35%	2.42%	1.34%	0.59%	0.42%	0.31%	0.05%	0.05%	0.05%	0.05%
Hours Above Threshold		1230	683	434	293	212	117.5	52	36.5	26.75	4.75	4.5	4.5	4.25
# of Hours Between Next Threshold		547	250	140	81	95	66	16	10	22	0.25	0.00	0.25	4

Avoided Cost of MCPE	Block Size MW	$100	$125	$150	$175	$200	$250	$300	$400	$500	$600	$700	$800	$900
Corsicana	1	$54,650	$31,219	$21,038	$14,175	$18,950	$16,375	$4,650	$3,900	$11,000	$150	$0	$200	$3,825
Temple	4	$218,600	$124,875	$84,150	$56,700	$75,800	$65,500	$18,600	$15,600	$44,000	$600	$0	$800	$15,300
Total	5	$273,250	$156,094	$105,188	$70,875	$94,750	$81,875	$23,250	$19,500	$55,000	$750	$0	$1,000	$19,125

Cumulative Avoided Cost of MCPE	Block Size MW	$200	$250	$300	$400	$500	$600	$700	$800	$900
Corsicana	1	$59,050	$40,100	$23,725	$19,075	$15,175	$4,175	$4,025	$3,825	
Temple	4	$236,200	$160,400	$94,900	$76,300	$60,700	$16,700	$16,100	$15,300	
Total	5	$276,300	$184,125	$113,975	$91,475	$64,875	$20,725	$20,125	$19,925	$15,300

Figure 30-10. Avoidance of High ERCOT MCPE Prices (2005)

THE SOLUTION

The enterprise system integrates the plant's level 1 sensors and sub-metering systems with the web-based level 2 dashboards for visualization of daily energy and operational performance conditions:

Level 1

At the device level both hard wire and wireless mesh sensors (See Device Level, Figure 30-11) bring all energy related information (kW, kWh, temperature, flow, pressure) to the JACE 2 controller. On its route to ERCOT it is converted from machine language to common object model language suitable for direct internet communications.

Level 2

At the PC level the Niagara framework database stores the formatted web-based data and uses the utility rate tool VES to provide the graphic display which appears on the local dashboard, ERCOT's display and the energy team's dashboards respectively. VES suites converts the data into reports displayed on the dashboard for daily energy cost, sub-metered areas, kWh/lb forecasting and ERCOT demand-response countermeasures (LaaR, 4CP, TXU ELM).

Level 3

To display all plants and provide a toolbox of dashboards for multiple plants the energy team depends upon the data floor of an ASP. Here centralized programming and analysts can review the plants on a cooperative basis to review the opportunities of supply, demand response and energy metrics kWh/lb produced to provide future strategy in effort to hold energy costs to more predictive levels.

Niagara Framework®

The underlying technology that allows the visualization of the "electronic dashboards" and transmitting the information from device (Level 1) out to the web (Level 2) is the Niagara framework developed by Tridium.

Niagara is a software framework specifically designed to address the challenges of building device-to-enterprise applications, internet-enabled products and internet-based automation systems. Niagara provides a unified, feature rich platform which streamlines the development process significantly reducing implementation costs and time to market. The framework creates a common environment that connects to almost any embedded

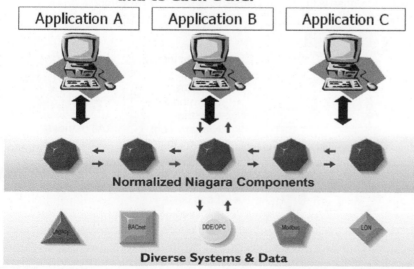

N–1
Niagara Normalizes All Systems to the Enterprise and to each Other

Figure 30-11. The Common Object Model within the Niagara Framework

device imaginable, regardless of manufacturer or communication protocol. It models the data and behavior of the devices into normalized software components, providing a seamless, uniform view of device data to the enterprise via a wide variety of IP-based protocols, XML-based connectivity options, and open APIs[1].

This is done by modeling the data and the behavior of the devices into normalized software components, the common object model, providing a uniform view of the device data to the enterprise over standard IP based protocols. This allows data from the field devices to be converted into a single universal protocol which is then passed on in real-time to the external user(s) and applications over the internet or any intranet which in turn can communicate back down to the Niagara framework relaying scheduling directions and commands from any user.

By transforming the data from diverse external systems into normalized components, Niagara creates a development architecture that provides substantial benefits over gateway-based integration methods that suffer the complexity of an N to N architecture. The benefit—any device or system normalized by Niagara immediately becomes compatible with any other system connected to the framework, providing true inter-system interoperability and uniform data presentation to enterprise applications. As a result, developers don't have to spend any time creating, testing and revalidating multiple gateways. Figure 30-12 further clarify the framework's concept.

JACE

Niagara framework is portable and can be embedded at the controller level. These controllers are called JAVA application control engines or simply known as JACEs. Energy management systems require a link between the meter and the technologies that control electrical consuming equipment. The Java application control engine (JACE) is the mechanism that provides connectivity between systems within or between buildings. These controllers enable large system scalability and a distributed processing architecture over the internet.

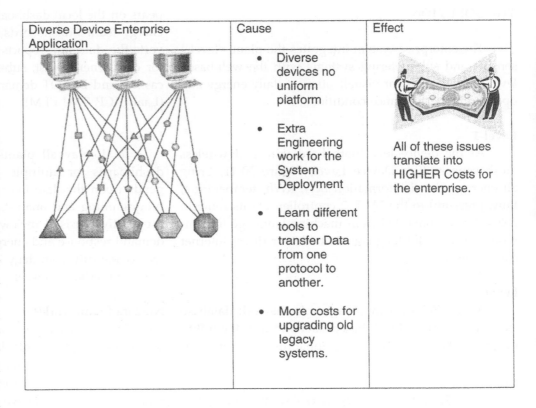

Figure 30-12. Comparison of Diverse Device Architecture

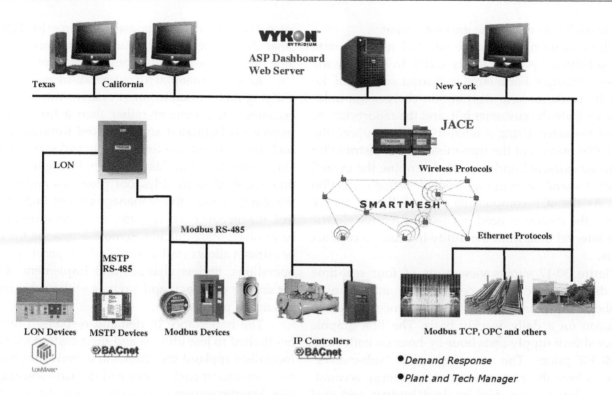

Figure 30-13. Typical Tridium Architecture

JACEs have the ability to integrate with a wide variety of devices via input/output (I/O) modules, serial and ethernet ports. The data from the metered device are modeled in the JACE, data are brought to a single universal protocol, and forwarded to the enterprise level via I/P based protocol.

In Figure 30-13 the data from field devices that communicate commonly used protocols of LonWorks, BACnet and legacy systems are modeled at the JACE level and then communicated upwards to the enterprise level.

Vykon Energy Suite-VES

Vykon Energy Suite is an application that runs on top of the Niagara framework (Figure 30-14). The common object model of the Niagara framework would provide little help if there were no way to validate or view the data it aggregates.

This can be achieved by an aspect of the Vykon Energy® Suite (VES); the E2 profiler. It allows the monitoring and measurement of energy.

Figure 30-15 shows how the data are represented and displayed graphically to the user. The E2 profiler is a measurement and verification application that ties the

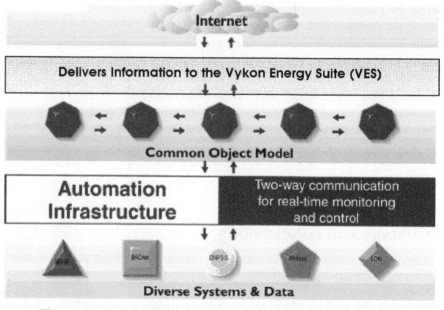

Figure 30-14. VES Runs on Top of the Niagara Framework.

various elements of demand response together. These reports display energy consumption before and after the demand response programs have been implemented, and also allow the user to define the period, baselines, and data normalization parameters. These reports also allow the energy manager or end user to rank specific sites or meters in terms of usage and efficiency so they can more easily see where demand response and load shedding can be implemented to obtain the greatest benefit.

VES cost profiler takes the demand response data one step further by using utility tariffs to calculate the end user's charges. For a large industrial user, all of the typical transmission and distribution charges are modeled to replicate the customer bill, and the reports can be accessed instantly. Using a what-if analyzer report, the controllable portion of the transmission and distribution cost can be extracted and used to determine the overall value of demand response and power factor correction efforts. As the 4CP components of the bills are readily obtained, the customer need not wait to receive electric bills or interval data from the utility in order to conduct analysis.

Figure 30-17 shows view-ports of four real-time views displaying economic performance and technical data about the plant's energy performance related to production on a daily, weekly basis. The first graphic interface shows supply costs hour-by-hour on both block and MCPE prices. The second graphic "sub-meters" displays where the energy went as an energy accounting tool allowing for dept-by-dept budget and cost allocation to the product. The third graphic "Forecast" displays the kWh/lb for production "real time." This allows the plants to complete and measure metrics on good, better, and best plant performances with production yields. The final graphic, demand-side programs, displays the performance of the ERCOT-based demand response programs at the plant (LaaR, 4CP, TXU ELM) and conservation projects. Ultimately the decision team can use these MIS reporting dashboards to shape the short, medium-range strategies for integrating supply and demand measures with the production of product.

SUMMARY AND CONCLUSIONS

Traditionally the highest operating expense for organizations has been "energy" and now with the cost of crude oil increasing more emphasis needs to put on energy costs. More industrial customers will seek to participate in real time energy markets seeking and use their discretionary load as a countermeasure to offset spot electricity at times of high price. In the past, integrating information from real time metered devices, automated controls and real time markets was difficult at best and often impossible. Visualization technologies referred to as dashboards provide the tool to view the

integration demand response programs by ERCOT or other deregulated electric markets and incorporate real time decisions using data from a variety of sources.

As smart manufactures prevail in modern society, unifying many discrete smart systems into one network becomes a requirement rather than a luxury. Demand response is facilitated and enhanced through the use of such networks, saving time, money and scarce resources. Other important benefits include reducing the amount of time required to troubleshoot problems and enhancing the ability to proactively manage energy. Important control, maintenance, energy use, and operating issues can be resolved more quickly. Remote access via the internet or intranet allows staff to troubleshoot problems, modify operations, review changes and implement strategies for energy efficiency and savings without having to be physically present.

The importance of demand response programs is not limited to just utilities and their customers. Environmentalists applaud the reduction in emissions that result from intelligent curtailment and demand-shifting strategies. Smarter energy use also decreases our country's dependence on foreign energy sources. The time to install energy-smart systems is now. You can't afford not to.

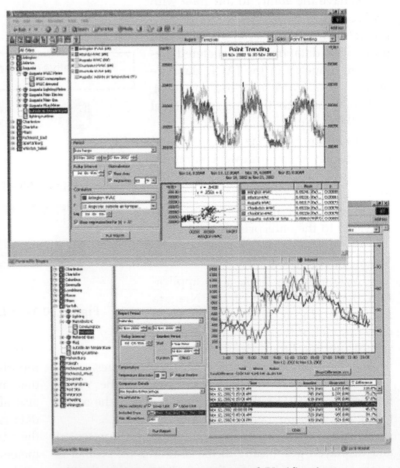

Figure 30-15. Measurement and Verification

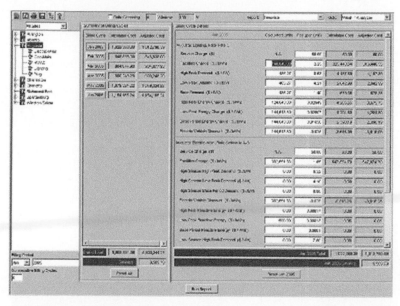

Figure 30-16. The "What-if Analyzer" a Forecasting Tool in the Cost Profiler Suite of Reports.

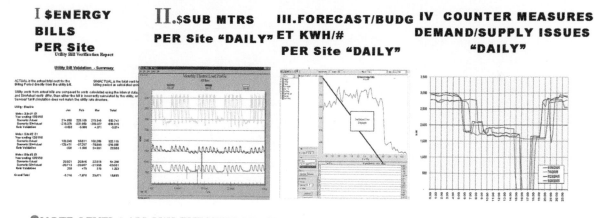

●NOTE: LEVEL 3 ALLOWS THE VIEW OF ALL PLANTS From anv browser

Figure 30-17. Real-time Dashboards

Reference

[1]   Product Model—Niagara AX Product Overview" A White Paper. Tridium Inc.

Chapter 31

Web Enabled Metering and Controls for Demand Response

Rahul S. Walawalkar[a,b], Dr. Rahul Tongia[a], and Dr. Bruce K. Colburn[c]
a: Carnegie Mellon University, Pittsburgh, PA USA
b: Customized Energy Solutions Ltd., Philadelphia, PA USA
c: EPS Capital Corp., Inc., Villanova, PA USA

ABSTRACT

THE USE OF THE INTERNET as a communications medium for energy management has increased substantially over the past decade. With improved reliability and availability, the internet has expanded into mainstream business and residential life, and data transmission speeds have dramatically increased while the costs have come down. By harnessing this information revolution, the electric power sector can be made more efficient through web-enabled metering and control for demand response (DR) on the grid. This can help keep electricity costs low, improve grid security and power quality, and offer new services not previously available through the power utility. Historically, electric utilities were motivated by regulatory commissions to construct more systems, as they could recoup all such costs through (almost automatic) rate increases. Today, the demand for electricity has outstripped the ability to economically generate and distribute power, and the public no longer wants to see new transmission lines and generators next door. In competitive electricity markets, new independent system operators (ISOs) have developed DR programs to influence users to better utilize those same assets in competitive electricity markets. To make this happen, an integration of demand response technologies with more traditional energy efficiency techniques has occurred, with the web being a natural means of allowing high speed communication and control. Advanced metering will become the norm in the coming decades, due to improvements in price-performance, regulatory push, and a cyclic investment in metering (especially given half the world lacks electricity today). A convergence is emerging of past energy management experience, new metering and hardware, and new DR grid programs aimed at encouraging avoidance of peak demand during critical or expensive periods. In this chapter, we provide some of the methods and procedures for control, and future projections of the direction of DR in fine tuning asset usage and operating costs.

INTRODUCTION

The use of the internet as a communication media for energy management has increased substantially over the last 10 years, as the web has been improved, expanded into the mainstream of business and residential life, and data speeds have dramatically increased while costs have come down, just as with computers through "Moore's Law." The electricity industry, sadly, has a much flatter learning curve, and electric generation is getting more expensive, especially with rising fuel costs. However, by harnessing the information revolution, the power sector can be made more efficient. One intersection of these industries is web enabled metering and control for demand response (DR) in order to reduce peak MW usage on the grid and thereby keep electricity costs lower than would otherwise be the case. In addition to lowered costs, this can offer greater grid security, power quality, and new services previously unavailable.

Historically, electric utilities in North America and western Europe have been authorized by their regulatory bodies to recoup their costs of generation, transmission, and distribution on an average-cost basis. In addition, utilities were generally allowed a regulated rate of return based on installed assets. Thus, utilities traditionally had been motivated to obtain more equipment and systems, within reason. Most customers are served under rates based on average embedded costs. As a result, consumers have had little incentive to control their overall electric costs by matching their preferences regarding the cost, timing, and reliability of service to the price and character of the electric services purchased.

Except for larger consumers who were billed for time of use (TOU), most consumers just paid for the energy they consumed (kWh), regardless of when consumed. At the same time, the ever-increasing demand for electricity has finally outstripped the ability to economically generate and distribute electricity on an uncontrolled basis. As a result, new independent system operators (ISOs) such as NYISO, and PJM are working on developing effective DR programs to influence the grid kW load shapes and help provide efficient operations in competitive electricity markets. In some areas, DR including active load management has become mandatory to avoid blackouts and service interruptions due to lack of peak carrying capacity of the grid.

Fundamentally, DR programs require cooperation (either explicit or implicit) between the consumers and the service providers (utilities, ISOs, etc.). Consumers need to know the state of the system (such as system load and pricing) and they can then respond as they see fit. There are many small consumers in a system, both residential and commercial, but their aggregate load is substantial, and any demand side management requires significant information flow across this distributed network of consumers. High-speed if not real-time communications are a vital requirement for energy management systems (EMS). One critical requirement for successful participation in DR programs is availability of cost effective metering data, at least interval if not real-time. Web enabled communication procedures have vastly increased the potential for integrating DR programs with traditional EMS functions.

The good news is that most energy management systems and techniques from the past are fully compatible with the new DR program requirements, but simply not yet fully implemented. In the future, the problem for end users is to judge which technologies will be the best, and be in operation five and ten years hence.

To cost-effectively change part of the current paradigm of thinking, an integration of demand response technologies with more traditional energy efficiency techniques is in order. The web is a natural means of transmitting the vast amounts of information required in order to meaningfully shed electric load at high cost times. Without the use of advanced communications, large-scale DR programs are likely not to work,

especially as the variability of the times "peaks occur" is changing, unlike in the past. In addition, without greater automation and computerization of the process, significant DR achievements are likely to be low—relying on people to manually take actions has typically had modest impact when we consider the broader population instead of early adopters or motivated individuals.

ISO AND DEMAND RESPONSE PROGRAMS

Demand response programs are utilized by electric utilities and ISOs to encourage consumers to modify their electric demand level and pattern of electricity usage. DR refers only to energy and load-shape modifying activities undertaken in response to economic or reliability signals provided by utilities or ISOs and not to load-shape changes arising from the normal operation. Based on the type of signal used to activate the DR program, these programs can be categorized as either emergency (or reliability based) DR programs or economic (price based) DR programs. Figure 31-1 shows the classification of various DR programs and criteria for participation.

Most emergency DR programs aim to provide cost-effective capacity resources to help avoid system outages in case of severe grid stress. On the other hand economic DR programs are developed to exert a downward pressure on electricity prices, by allowing demand side participation in electricity markets through voluntary reductions in what otherwise would have been peak electric use. In recent years ISOs have started to explore ways to utilize DR resources for also providing ancillary services (frequency regulation or spinning reserves). PJM has started allowing DR resources to participate in ancillary services markets in 2006. These programs can help defer

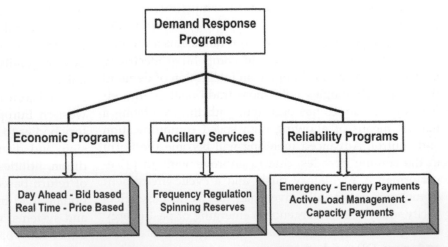

Figure 31-1. Classification of Common Demand Response Programs in US

the need for new sources of power, including generating facilities, power purchases, and transmission and distribution capacity additions. All DR programs rely on end users deliberately altering use of equipment and systems, which generally means lifestyle or comfort changes, or changes in operating procedures. Such changes would be acceptable to end users only if the consumer has a stake in the process either through financial compensation or through improved reliability of power supply.

When electric demand is at or near its peak level, less efficient or higher cost generating units* must be utilized to meet the higher peak demand. In some cases, electricity prices in wholesale markets could fluctuate from less than 3 cents per kWh to as much as 30 cents per kWh on a significant number of days per year. During capacity shortages, prices could increase to 50 cents per kWh or higher, reflecting the cost of building new generation to serve peak loads and the price signals that might be required to match demand to available supply. Under these circumstances, even a small reduction in demand can result in an appreciable reduction in system marginal costs of production. These marginal costs, although high in cost and short in duration, add to the average cost per kWh to the consumer.

The system-wide benefits from DR can also include reduced volatility in wholesale electricity prices, market power mitigation and enhanced reliability [1,2]. Today it seems that DR programs are viewed not as a means of saving costs, but avoiding unreasonable cost increases. An interesting clarification is also that in the past, demand side management (DSM) initiatives usually meant "peak shaving" kW (since the main incentive was to avoid demand charges), but now it also means reducing the cost of a kWh during a peak use period (as the consumers can receive DR participation payments for energy curtailment), which is a very different meaning than in the past.

In recent years, various research groups have tried to quantify the social benefits of DR in US markets. A 2001 study by McKinsey & Company estimates that, at the US level, $10-15 billion per year in benefits can be achieved from participation of all customers in DR programs on a wide scale, with the majority of the potential, contrary to conventional wisdom, from residential sector DR efforts [3]. Another study by Rocky Mountain Institute suggests that lowering demand by 5 percent of the system's maximum can reduce peak wholesale power market prices by 90 percent, as utilities and independent

system operators reduce their need to purchase on-peak power [4]. Such a drop in wholesale peak prices also means that non-participants in demand response also share in the benefits, as prices for everyone are held in check. Based on a review of current utility programs, EPRI estimates that DR has the potential to reduce current U.S. peak demand by 45,000 MW [5]. FERC released a cost-benefit analysis that showed a $60 billion savings over the next 20 years if DR is incorporated into RTO market design and operations [6].

Prior to electric deregulation, timely information on detailed electricity pricing breakdowns and consumption patterns was generally not readily accessible to end users. In fact, customers were typically not able to determine accurately how much they were paying for energy until they received the bill the following month after the usage and load profile (i.e., peak) had been set, unless they had separately installed an electrical submetering system and were tracking it [7]. With electricity market transparency, the demand side of the market can now take part on more of a real-time basis. The customer with demand response capabilities now has the opportunity to get paid by the ISO or local power company as much as an energy generator would get paid for selling electrical generation into the market, helping make it a more level playing field. Most of the current DR programs can be traced back to the DSM initiatives taken by utilities in 80s and 90s. Table 31-1 shows the evolution of DR since 1900.

Independent of DR programs, merely having the load profile and electricity cost data in real time can allow users to decide if they are willing to accept operating or lifestyle disruptions, in order to avoid large prices. There is a perception that if full grid price deregulation occurred, that end user electric costs would skyrocket for consumers whose load profile matches the system load profile, if no intervention occurred by the customer. In theory, over a year, the prices should be neutral (with some winners and some losers), but in fact the vast majority of end users who can do something about their usage on a large scale have loads coincident with the system peak, and have little need for power when prices could be very low; hence their need for positive interaction in order to avoid large price increases, such as have occurred in the past.

While DR is not the answer to all the difficulties in electricity restructuring and wholesale market design, it is certainly one of the missing links. Today, various ISOs have developed demand response programs designed to advance price responsive demand from electricity customers, to "entice" end users to voluntarily curtail for their own financial benefit (as well as for the

---

*The cost of electricity generation varies by not only fuel technology but also how long the plant operates per year. Peaker units might operate only 500-1000 hours per year, or less, and so their electricity is inherently expensive.

Table 31-1. Evolution of Demand Response (US)

	Phase I Options Define Service 1800s to Early 1900s	**Phase II** Option Consolidation 1920s to 1960s	**Phase III** Separate Options 1970s to 2000	**Phase IV** Integrated Options After 2000
Pricing	End-Use Rates	Usage Based Rates	TOU Rates	Real Time Pricing
Metering	*None*	*Total kWh Usage*	*Time Period Loads*	*Hourly Loads*
Load Shape Objectives	Load Growth	Load Growth, Valley Filling	Peak Shaving, Shifting, Conservation	Preserve Electric Reliability, Customer Cost Management
Customer Involvement	Active, Fuel Switching	Passive, few options	Utility Command and Control	Interactive Participation
Demand Response	Contracts for Service	Water Heater Time Clocks	Curtailable, Interruptible, Direct Control	Demand Bidding, Risk Management
	⇩	⇩	⇩	⇩
	Increased choice, Service tailored to customer needs	Reduced choice, Increasing value to customers, Declining cost	Reduced choice, Increasing costs, Loss of control, Declining value to customers	Increasing choice, Cost Volatility, Value of Information

Source: EPRI 2002 [8]

utilities), or for grid reliability. There are a variety of mechanisms for customers to respond to price signals, but the economic DR programs advanced by ISOs offer incentives for end users to change their ways, and be paid for the inconvenience, along with additional pure energy cost savings anyway. These new programs are considered an essential component for "competitive markets" being developed across the country by allowing interaction of supply and demand curves. Table 31-2 shows the types of typical demand response programs offered by ISOs in US.

As can be seen from Table 31-2, today some form of demand response programs are offered to customers in almost all deregulated electricity markets. These price responsive load programs allow approved curtailment service providers to aggregate end users that are capable of load reduction and with whom they have a contractual arrangement. PJM and NYISO offer bid-based wholesale markets to get participation in both economic and reliability DR programs. In addition to consumer control solutions, direct load control programs remotely cycle off or throttle customer appliances, such as air conditioners, water heaters, and pool pumps during times of high peak demand [11,12]. At this time the majority of DR program participation is through use of backup generators or use of curtailable loads. The vast potential of existing building EMS systems and energy management strategies is still untapped, not to mention future possibilities as home automation and smart appliances become standardized. Although the curtailment for each customer may be only 10-500 kW, or lower,

overall DR can occur on a vast scale if adopted widely and aggregated. Table 31-3 shows a sample of some of the control strategies that can be used by different types of customers for participating in various DR programs. In the long term, use of standby or backup generators alone for grid load shedding will not be viable for tight control because emissions regulations, fuel prices, and lack of long term DR programs might discourage their use, and this leaves only fundamental energy efficiency techniques, i.e. shutting down loads, as feasible.

In a competitive market, new generating capacity will not be added until prices have risen sufficiently above the cost of new facilities to ensure generation suppliers a reasonable return at variable and uncertain market prices. The existing EMS infrastructure, however, can be a great tool in implementing new DR programs. As shown in Table 31-3, by using direct load control or pre-programmed optimization strategies in an EMS or even by scheduling use of household appliances, curtailment service providers (CSP) can respond to load reduction requests by the ISOs and then distribute the incentives to program participants. These data can be best made available through the internet. The programs offer incentives that are directly linked to the day ahead or real-time locational marginal prices (LMPs). The financial benefits to end users are both a payment from the ISO for curtailing, as well as the actual kWh cost reductions by not consuming that energy during that period.

New communication technologies are making it practical to provide consumers variable price signals and a range of other demand-side services [9,13]. Cur-

Table 31-2. Comparison of Demand Response Programs
Available to Customers in Different Regions in US

	NYISO	ISO NE	PJM	MISO	ERCOT	CALISO
Population Served	19 Million	14 Million	51 Million	15 Million	20 Million	31 Million
States	NY	CT, MA, ME, NH, RI, VT	DE, IL, IN, MD, MI, NJ, OH, PA, VA, WV,	IA, IL, IN, KY, MI, MN, MO, MT, ND, NE, OH, PA, SD, WI, WV	TX	CA
Types of DR Programs Offered	Day Ahead, Emergency (Energy and Capacity)	Day Ahead, Real Time, Emergency (Energy and Capacity), Ancillary Services	Day Ahead, Real Time, Emergency (Energy and Capacity), Ancillary Services	Day Ahead, Real Time, Emergency (Energy)	Real Time, Ancillary Services	Emergency, Direct Load Control
Peak Load (MW)	30,983	25,158	134,017	112,197	60,157	45,900
Registered DR (MW)	1,754	368	3,419	382	1,718	0 Voluntary
DR Share of Peak Load	5.7%	1.5%	2.6%	0.3%	2.9%	0.0%

Adapted from Heffner & Sullivan 2005 [9], FERC 2004 [10], and ISO websites.

Table 31-3. Control Strategies and Applicable DR
Programs for Different Customer Types

Customer Type	Equipment / Building Component	Control Strategy	Applicable DR Program
Residential	Air Conditioners	Cycling/Forced Demand Shedding	DLC/ALM/Real Time
	Water Heaters	Cycling	DLC/ALM/Real Time
	Pool Pumps	Cycling	DLC/ALM/Real Time
	Electric Stove	Scheduling	Day Ahead
Commercial	Chillers	Demand limiting during on peak period	DLC/ALM/Real Time
	Chillers HVAC	Pre-cool bldg over night- storage	Day Ahead
		DX Forced Demand Scheduling	DLC/ALM/Real Time
	Refrigerator/Freezer	Prioritized Demand Shedding	Day Ahead
	Lighting	Scheduled on/off	Day Ahead
	Lighting	Scheduled dimming of selected circuits	Day Ahead
Industrial	Chillers	Demand Limiting on time Schedule	Day Ahead
	Electric Furnace	Demand Limiting through Heat Stages	Day Ahead
	Electric Furnace VSDs	Curtail (during peak period)	DLC/ALM/Real Time
		Limit Output on Scheduled basis	Day Ahead
	Well pumps	Defer during peak	DLC/ALM/Real Time
	Production Eqpt	Prioritized demand on selected units	Day Ahead
Restaurants / Shopping Malls	HVAC	Chillers- Demand limiting during peak	Day Ahead
	DX Compressor	Forced demand shedding of multiple units	DLC/ALM/Real Time
	Refrigerator/Freezer	Prioritized demand shedding	Day Ahead
	Electric Stoves	Scheduled pre cooking	Day Ahead

DLC: Direct Load Control; ALM: Active Load Management;
VSD: Variable Speed Drives; DX: Direct Expansion

rently all ISOs use web-based systems to provide users access to day ahead and real time energy market prices, and also inform customers about system emergencies [2,6]. One such web based tool utilized by PJM is called eData, which provides customers not only real time price information, but also information on system load, transmission congestion occurring at various parts of the grid and weather data. The future of DR will be determined by the choices that consumers, utilities, other service providers, regulators, and legislators make during the transition to competitive electric power markets. Various recent studies conducted all over the world indicate that the short-term retail electricity demand does not have to be static with zero elasticity. With right information and incentives, even retail customers can respond to prices and other incentives in the wholesale power markets. Such price responsive load can help electricity markets in lowering prices in tight capacity situations and mitigate reliability and market power concerns.

EVOLUTION OF METERING

An important step for all DR programs is information on the state of the system and compliance data for each DR program participant. Utilities also make metering a priority as payments depend on it. From its early days, electricity was recognized as a product to be charged for, and the early history of electricity services around the world has often been through private companies, even in countries that today have public provision of power [14]. The structure of the electricity industry might have evolved over time, but the fact remains that it is generated, transmitted, distributed, and retailed (the latter being the interface with the customer). Electricity meters were important at all segments of the supply chain, not merely to bill the end-user but also to recognize where electricity was flowing, more important if it went over infrastructure owned by third parties (such as transmission wires). Not every net unit of power generated is consumed in end-user devices. A small fraction is lost due to resistive (I^2R) and other losses, typically less than 10%, while a few percent may be deemed "missing" due to theft, poor accounting, etc. In reality, in the developing world, total losses can be as high as 30%, and the distinction between technical and "commercial" losses is murky.

While the technology underlying end-user metering technology has evolved over time, there have been parallel changes in the functionality required of meters. Of course, a residential meter is not the same as a com-

mercial meter, which is not the same as an industrial meter. Larger users may connect at much higher (even transmission) voltages, be three-phase connections, and be subject to time-of-use (if not real-time) rates. They may also pay for peak power consumption (capacity charges), power factor, and other ancillary services. In contrast, the vast majority of residential consumers today pay a flat rate for their electricity, simply per kWh.* To that end, a "dumb" meter that simply counted the kWh as they went past was sufficient. But the advent of digital metering allows recording time of day use, provides higher accuracy, and can be electronically reprogrammed, also allowing for remote meter reading via a communications channel, which has been termed automated meter reading (AMR). Table 31-4 summarizes the newer metering options for service providers, and new utility tariff procedures.

Figure 31-2 shows the daily variations in system load during various seasons. The load curves were obtained from 2005-06 PJM ISO load data for four separate days across each season. As one can see during a hot summer afternoon or cold winter evening, the system load is very high, and these are times which typically correspond to the price spikes in the wholesale electricity markets.

Metering technology has evolved not only from a supply (technology push) direction, but also due to different functionality requirements (Table 31-1). We can see that offering enhanced demand response services or capabilities requires enhanced communications, perhaps even bi-directional communications. In contrast, simple AMR requires only upstream data flows. Although most of the ISO programs today need hourly integrated meter data, there is some evidence that successful demand response may require 15-minute intervals for measurement, not simply hourly reads [16]. If we project out to the future, one can envision competitive markets where real-time metering is an integrated feature of retail competition. In addition, if one expects increased use of renewable or distributed generation, these also require upgrades to bidirectional meters, ones where power can be measured flowing in both directions (often implemented as "net metering") [17].

DESIGN OF WEB-BASED, CONTROLLABLE METERING

When considering advanced metering and control, fundamental design questions include who is respon-

---

*Utilities, who faced varying costs of generation over time, would average out the costs to their consumers.

Table 31-4. Capabilities of Different Metering Systems

	Accuracy	Theft Detection	Communications	Control	Capabilities
Electro-mechanical Meter	Low (has threshold issues for low usage)	Poor	Expensive add-on	Nil	Total kWh usage
Digital (solid state)	High	Node only	External or packaged into meter (but not chip)	Limited	Historical usage reads (time of use, peak, etc. possible)
Next Gen. Meter (upcoming)	Arbitrarily high	High (allows network level accounting)	Built-in (incl. on-chip)* *often bi-directional	Full (connect / disconnect); Extending signaling to appliances possible	Real-Time control; DSM/Demand Response

Adapted from Tongia 2003 [15]

sible for the information flows, who has access to them, and over what medium(s)? Is the power utility distributing the information, or some third party? Is it sent via a public webpage, using the public internet, or is there a closed/proprietary/private network established for such purposes? Internet protocol (IP) based communication can be used in a private network (an intranet), which may even use standard web protocols for creating web interfaces—but this would not be a public web page.

At a high level, a DR system for any end user (residential or commercial or industrial) has several components. The system operator monitors the entire grid by controlling and exchanging information with generators, transmission owners and distribution networks. The sys-

tem operator simultaneously shares the aggregate system information such as load and price signal with end users. This can be achieved either through the public internet using websites for information display or such information can also be transmitted over a radio network such as FM, or via other means. At the consumer end, there is the advanced meter, which can not only store and perhaps process data, but also communicate and effect control if so designed. It may have a (remote) connect/disconnect capability, and the communications would likely be every 5, 15, 30, or 60 minutes "upstream" to the utility, or maybe just once per day. The utility could also signal "downstream" information such as pricing, critical conditions, etc. Figure 31-3 shows one system for achieving such a design. The metering data

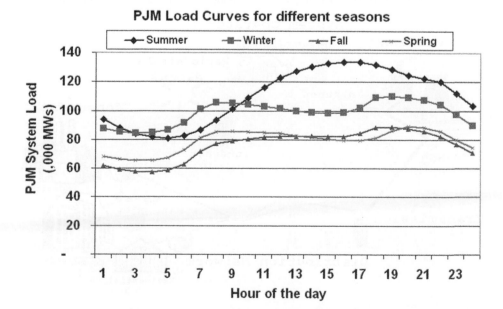

Figure 31-2. Typical Electric Load Swing on PJM Grid During A Day

are crucial for assessing the compliance with the DR signal and determining the demand response payment to end user.

Once a signal reaches the meter, another design challenge is how to send that information further downstream, to the end-user and/or their appliances and devices. In-home signaling and communications can be achieved through several technologies, but the market has not yet chosen a winning standard. It remains to be seen what designs will ultimately be deployed by different utilities. One suggestion has been for the utility to "make available" real-time pricing, criticality, and other information through a multitude of means, and the consumer would then determine how they want to utilize such information for demand response or other uses. How the information is disseminated, via high speed internet, or digital radio, is yet to be determined.

All of this still requires some form of electronic metering. What functionality would such a meter need to have? Given a target minimum lifespan of 15 years, it is very difficult to predict what feature set will be required, either from a regulatory perspective, or from a business case perspective. Even relying on a third-party communications network may be difficult, as they migrate to new technologies at higher speeds (e.g., paging networks have been usurped by cellular and SMS networks, and even GPRS is making way for 3G and other solutions). Regulators may demand different capabilities, in synch with different levels of deregulation, decentralized generation, and system reliability requirements.

Data and system security would be another major issue for any system, one that would be especially important if a public communications system or public internet addressing schemes were chosen. While there are some usability benefits to having a meter as a node on the internet,* hackers would be of major concern, not merely from a financial perspective, but also from a reliability and safety perspective. This is one reason utilities are often hesitant to committing to full signaling to within consumer homes, for demand response or home automation programs. The worst-case scenario is if a hacker or accident turns on somebody's programmable controllable stove! Beyond safety issues, the economic implications of failing to meet contractual obligations may be enormous (which escalates with increases in volatility in competitive markets). This is one reason that many designs separate the wide area network (WAN) from the in-home portion, perhaps through a gateway (which may or may not be the meter itself). In the absence of direct utility-based signaling and/or control, end-users need to themselves prepare for DR programs, perhaps using the web.

Costs and Implications

Estimates for the US indicate that metering, billing, and customer relations management (CRM) might cost as much $2-5/month per consumer [18], if not more. In many locations, meters are not ready every month, and in some sparse areas, consumers must phone in their meter readings. However, most public estimates for savings from avoiding manual meter reads are not enough to justify investing in advanced metering technologies,

---

*Given IP address limitations, one suggestion has been to use the next generation of internet protocols, IPv6. Additionally, there are designs that isolate the power communications network behind a gateway, through solutions such as Network Address Translation (NAT).

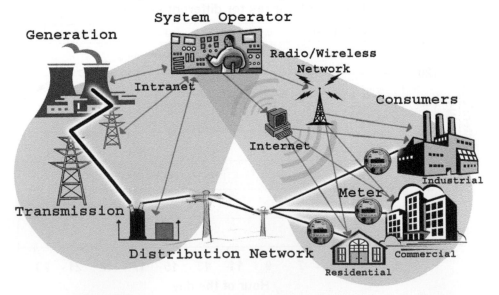

Figure 31-3. Components of a Demand Response Program Implementation

but the additional features (including DR potential) and benefits to the consumer can make up the difference. The costs for such systems do not have an easy answer because technologies and capabilities vary dramatically, and published figures for US deployments range from $60+ for simple meter reading technologies to hundreds of dollars for daily meter reads. US data for 2002 for all AMR—spanning simple kWh meters to advanced meters capable of time-of-day pricing—show 16.9% penetration by meter, and almost one-third of utilities have some level of AMR in place [19]* As of now, there are few standards today for such technologies (though standards bodies and consortia like ANSI, IEC, Grid-Wise, Intelligrid, Consortium for Electric Infrastructure to Support a Digital Society [CEIDS] etc. are working in this space), and many deployments thus far have been one-off or based on proprietary technologies.

Consumers are beginning to anticipate changes in their power delivery and metering, in part due to the evolution of the market and competition. Regulators are experimenting with time of use and even real-time metering, but this has usually been for larger (non-residential) users or pockets of population as part of trials—the bottleneck has been the infrastructure to allow such schemes.

One additional intersection of electricity and IT may be in home automation, the so-termed "smart home." With decreasing costs and emerging standards, many appliances (especially heating, ventilation, and air conditioning—HVAC) may be able to respond to electricity pricing signals. Leading the charge in smart home networking has been entertainment, where distributed audio, video, and computer signals are already a part of many households.

COMMUNICATIONS AND CONTROL SOLUTIONS

Controls

Most of the larger commercial and industrial facilities today have large energy management systems in place. In order to take advantage of real time DR programs on a large scale, some form of EMS is almost essential, including even for residential users. This is because DR is designed to reduce the electric demand, for whatever rationale. Some method of selecting the equipment to be curtailed, and time period, has to be

determined and rapidly implemented. This means that high-speed communications is essential. If one is also trying, independent of ISO DR programs, to control the demand of a "system of buildings," scattered throughout a utility service territory, data must be available at a central location in real time, and decisions made and communicated to a wide variety of locations, so as to control the coincident peak of all these scattered units; this has not been done in the past, and can only occur today due to high speed communications, in the hundreds of kbps or even Mbps data rate. Manual systems can still be utilized for some of the DR programs where customer can receive a larger notice period, as is the case with day ahead DR programs. In such programs the customer is usually notified at end of the day for the schedule and amount of load curtailment required for the next day.

EMS has increasingly become important as the need for more electricity demand control occurs. However, what was really sought was a simple means of tying end user electric equipment together cost-effectively, and allowing large amounts of information to be exchanged for energy control in real time, and not just information gathering for off line evaluation [13,20]. Technology has created a large variety of communications media for both getting information to the consumer or EMS, and then distributing the information and/or control signals across the end-user facilities. These include traditional broadband solutions like cable modem, DSL, and wireless (WiFi, Cellular, etc.), as well as broadband over power lines (BPL), optical fibers, and emerging solutions such as based on the ZigBee standard (wireless sensor networks). Except for those DR programs using generators, widespread adoption of load shedding would not be possible without use of the web and high-speed communications.

There is much talk about web-based control systems, but the full promise is yet to come. The approaches taken vary considerably, but have three major element directions: (a) how much control is being provided through the particular interface from the browser, (b) to what extent software products support the various open protocols available today on the market, and (c) the manner and means by which web pages are generated, or the manner in which the software takes and uses the data. One of the problems today is that manufacturers of controls often view web access as an add-on to their workstation software, and the concepts are not integrated into the design. For example, sensor readings can be displayed on a screen web page, and there are usually methods for the operator to remotely lock in set points, but otherwise proprietary software is needed at each workstation. Today, some of the most important

---

*AMR deployments were found in 65.8% of Cooperative Utilities, 63.7% of Investor Owned Utilities and 16.8% of Public Utilities in the US (2002). This might be indicative of public utilities being less aggressive to use new technologies to lower costs—but generalizations across AMR technologies and capabilities are also misleading.

factors in controls, such as configuration, trend data, alarm reporting, equipment operating schedules, data summary reports, benchmark comparisons, as well as other standard control tracking procedures, are not yet available through the web interface. Because these data requirements tend to add significant workload on software developers, many systems in the past simply bypassed the web, or the web system feature sets generally were limited in size and scope, partially defeating the whole concept of web interfaced control. The future, however, lies clearly in development of new toolkits specifically designed for web-based interface, even to the extent that features are completely redesigned to make the web-based aspect a fundamental tool. Then a standard browser, a convenient platform for consumers and energy managers, can enable most of the practical control, monitoring, and reprogramming features originally envisioned. This also leads to whole new areas currently under development, such as use of cell phones for data interface, or new technologies yet to be developed, all designed to provide easy end user interface quickly, easily, and efficiently [21].

BACnet is an example of a developed standard which purportedly offers standard protocols, and thus provides a logical way to interface control systems from various manufacturers or suppliers of a control system. This has become a key element in that "god, mother, and country" leads to interoperability of all different systems, just as the PC has evolved as far as automatic driver installation; how many remember the days in the early 80s when it was a nightmare to get any device to interface to a new computer. However, the definition of "interoperability" varies greatly from one manufacturer to another. BACnet, LonWorks, MODBUS, and other protocols are often touted as the "one true path" to interoperability. At the controller level, there is not a clear winner today. The most interoperable system may be the one that supports all of the popular protocols. This is because BACnet is one of the few open protocols that defines a standard way to handle all of the higher workstation level functions, such as trending, alarming, scheduling, and file transfers. A workstation that uses BACnet to implement these functions will be able to interoperate with other systems that support these BACnet services. This is true whether the workstation is web-based or not. A web-based system may have a slight advantage in this regard because it will of necessity support internet standards such as TCP/IP, HTML, and XML, but it is important to realize that these standards do not in themselves guarantee interoperability. TCP/IP, for example, is the standard used by the internet for packaging, addressing, and delivering information

"packets." TCP/IP imposes no standard upon the information contained in these packets. The content makes no difference to TCP/IP, but it definitely makes a difference to the control system that receives this packet! Similarly, HTML simply defines the way information will be presented on your computer screen. While this may provide a degree of interoperability if you are simply viewing a graphic page from another system, it doesn't help much if your system is receiving an alarm from another system and needs to decide what actions to take in response. XML is a system for creating and documenting data structures. The ASHRAE BACnet committee is developing an XML-based structure called "BML" for better markup language [22].

Communication Methods in Facilities and End-User Configurations

The idea of the web as merely a "wire conduit" fails to convey the vast interactive communications possible, and that it can be done inexpensively, across continents, using a variety of otherwise incompatible equipment and systems through plug and play. It has created opportunities not imagined even 15 years ago, and there is no reason to believe that the opportunities will not increase, especially deeper within the home or business.

The global internet and the web (a special system overlaid on the internet), are widespread and growing, and over 40% of US households now have always-on broadband connectivity. Unfortunately, the reliability of such connections is not as high as that of the power (or even traditional telephone) systems, and many web applications are subject to security risks from viruses, denial of service attacks, spoofing, etc. These can certainly be mitigated through firewalls, encryption and digital signatures and other means to a large extent, but this imposes costs and operational burdens.

To talk about using the web as a tool in energy management systems monitoring and control, one should understand what the procedures are for this process. Web-based systems are those that allow two way, simultaneous communication with energy management equipment and systems, via a web-based interface, to another place with equipment and software which can retrieve and process information. This may or may not be over the public internet, instead utilizing private connections (an intranet). One of the challenges is that there is a great variety of legacy, wired EMS units, with no connection to the internet, and no easy means at present of connecting them. Yet the building automation industry and SCADA (supervisory control and data acquisition) systems have seen a dramatic increase in high powered, inexpensive processors, wireless technology, multiple means of achieving high speed connection cost effectively to the internet, and much greater control capabilities through improved adaptive and learning control techniques, which 10 years ago were more in the realm of university researchers rather than in hardware in the field. Typically EMS vendors only made web access available to their new products, not their legacy systems.

There are a number of common elements that can be addressed through a web-based interface for an existing legacy EMS unit; these are:

- Easy access (including through PDAs as well as PCs) and graphical and text interfaces,
- Reduced total installed cost, including software, hardware, wiring, maintenance and debugging as well as training,
- Commonality of the front end for dissimilar EMS systems at different sites (this is often called the "front end transparency" issue, and is aided by making a standard web browser the graphical user interface),
- Reducing operating costs through near real time data tracking and benchmarking, such as chiller kW/ton or total lighting circuit kW,
- Extending the useful life of the installed EMS and its sunk cost, by providing new software front end capabilities and features, without need to go to the original vendor to completely upgrade or replace the entire system,
- Support for different operating systems and offering a scalable solution
- Drivers to convert proprietary protocols, including native drivers, DDE drivers and OPC,
- Reliable and Secure operations,
- Easy networking of the EMS to a LAN or WAN,
- Plug and play features (allowing rapid installation and commissioning, much as is done now with new printers, Bluetooth devices, etc.),
- Alarm notification through email or pager,
- Multi-user capability (connect to multiple sites from the same remote setup, without need for separate software licenses),
- Ability to use an SQL compliant database, allowing use of enterprise solutions software for large scale analysis, summarizing, and reporting for multiple sites and multiple users, and
- Expandability (easily adding graphics, other plug and play features, etc.).

There are a variety of communications media for getting data from various end users to common points of control and monitoring. If these are on the same premises, wired or wireless solutions are easier. Overall, the speed of data transfers has been increasing rapidly, from about 300 bits/s in 1971 to 10 Mbps (megabits/second) or higher today. These cost effective communication media for wide area connectivity, available to the majority of business and residential users, include:

- Cable Modems: These are the dominant form of broadband connectivity in the US, and offer speeds in the megabit range (per user).
- DSL: These are also major players in broadband, and growing rapidly. These have distance limitations, and so are only available in certain (populated) areas. Typical speeds are also Mbps class.
- Optical Fibers: These are the ultimate connectivity solution, offering virtually limitless (sometimes 1,000 Mbps or higher) connections. The electronics and installation are both expensive, and this is thus far limited in the US.
- Broadband over Powerline: This emerging solution has technological potential, but the business case is not yet proven. This relies on the participation of the power utility, who may use the system for SCADA, metering, and other needs as well.
- Wireless: There are several wireless technologies deployed, ranging from mobile telephony (cellular) systems to unlicensed (WiFi) connections that are growing in popularity.

All of these *could* offer sufficient bandwidth for EMS needs, but wireless is slightly more at risk of congestion and failure because of its inherently shared design, and limits to spectrum availability.

Some parallel innovations in connectivity include:

- **Wireless EMS and sensors**: Another growth area over the last few years is the use of wireless technology for installing and communicating with control devices and sensors. This has been speeded up by adoption of new standards including IEEE 802.11 (WiFi) standard, and the upcoming IEEE 802.15.3 standard (which is implemented as the ZigBee Alliance open standard) for sensor and control networks. These are especially attractive as they are unlicensed and inexpensive (because of large volumes and by design). Sensors are growing in capabilities (including auto-configuration and self-meshing) while shrinking in size, cost, and power requirements. There are prototypes of tiny sensor motes that one could sprinkle around the home or business to monitor and even control temperature, humidity, etc. There are a number of other wireless EMS systems already on the market, including Profile Wireless EMS Control, Sensicast, Inncom International, WebGen Systems, Senercomm (Tesa Entry Systems), WisNet (a British company using a Bluetooth network system), eLutions, Profile Systems LLC's Wireless Energy Management, Advanced AMR, Energy Eye, Itron, GE Wireless Energy Management, AmerenUE and TRS Systems (with their new RC2100 Wireless Hydronic Controller with wireless space/outdoor temperature sensors, a wireless energy management interface and an intelligent sensor reading control algorithm in one product).

- **Power over ethernet (PoE)**: Ethernet is a ubiquitous wired standard for connectivity in a local area. Power over ethernet (PoE) is a new standard (IEEE 802.3af) that allows small amounts of electricity (13W, soon to be raised to 30W) to be transmitted alongside data signals on the same wire. This is enough for remote sensors and embedded controllers and even specialized PCs. This makes installation, management, and power back-up of critical EMS elements significantly less expensive.

Typically, an EMS or SCADA system has been implemented to reduce costs based on the rate tariffs in place with the local power company. However, over time various other problems have arisen, manifesting themselves with a vengeance in California during the 2001 energy crisis, when major state blackouts occurred on a continuing basis. This same problem has been recurring in 2006, and some power companies are actively working with larger industrial and commercial customers on curtailment programs to cut peak demand voluntarily. If not, as in 2001, brownouts and blackouts may ensue, with negative impacts on productivity estimated at ~100 times the cost of the electricity denied. Therefore, due to the need to cut demand, and a growing emphasis on "green" options, many large companies are implementing further DR and energy efficiency measures not solely based on normal paybacks, say 3-4 years, but also on the basis of their kW demand reduction, and the "bragging value" from energy efficiency in their annual reports and advertising to the community.

A standard option, which has been available for decades, is a priority based kW demand shedding, to limit the kW demand on the building/factory to a preset maximum, decided by management. In practice, however, these procedures have rarely been used because in buildings most of the equipment needs to run to provide

for both productivity and comfort. However, experience has shown that demand limiting chillers can allow for controlled kW cutback without major comfort problems, or using variable air volume overrides on fans, or raising chilled water temperatures through override of chilled water cooling coils. If the time-of-use price signal exists, some larger consumers attempt to store "cool" during off-peak hours, e.g., through refrigerated water overnight. Web-based tools will allow for more accuracy in deploying such systems than simple time-of-use rates, which might help smooth statistical load variations somewhat, but are of little value during unexpected or emergency conditions.

CASE STUDIES AND REAL-WORLD EXPERIENCE

Utilities

Most utilities (load serving entities) around the world are highly regulated, and they face a paradox in terms of investing in new technologies. As long as the regulator allows a capital expenditure, they can recover their rate base. If they deploy a solution that enables operating savings, they may have to be passed on entirely to the consumer. So, why would a utility invest?

Regulatory requirements often drive AMR and automated meter management (AMM) development, and the goals often span to systems capable of multi-meter reading (water, gas, etc.) Sweden recently passed legislation mandating electricity AMR for all users by 2009 as part of the EU push towards deregulation [23], and California is holding hearings to consider peak pricing and advanced metering statewide. This includes a move to a standard for programmable, controllable thermostats for all new deployments after 2008, which would interface with external communications, perhaps over radio signals.

Utilities often push such technologies as part of their cyclic investment in metering. They may also envisage new businesses and services, or competitive differentiation, especially in a deregulated market. An extreme form of the former includes utilities offering consumers internet access, either by entering a new market on their own, or through a partnership with an internet service provider. Given the communications requirements for meters, there is a potential synergy for broadband provision for consumers. The power utility TXU recently announced the use of broadband over powerline (BPL) for both smart metering as well as competitive internet provision for consumers working with a BPL specialist, Current Technologies.

One of the most widely cited examples of smart

metering systems, with two-way communications, was the pioneering deployment by the Italian utility Enel, which begin in the late 1990s. As of this writing, they were close to completing full deployment across their entire service territory, which covers over 27 million users. This system was designed by Enel, and was based on a hybrid of powerline carrier and GSM/GPRS communications. The meters were not "off-the-shelf" but rather custom designed and manufactured on contract. The Enel deployment was based on a store-and-forward architecture, and there is no short-term plan to signal real-time or near real-time pricing to consumers.

The costs for the system were initially quoted at ~$80/node installed, inclusive of all costs, but since that time, the euro appreciated in value, and Enel now factors in some implicit R&D costs into per node costs, bringing the total closer to $100/node, if not higher. They announced a partnership with IBM to extend the functionality of the devices, and make it available to utilities worldwide. Enel cites a payback of only about 4 years for their system, driven by reduced labor costs (some of which may not be applicable to other utilities—their earlier meters often needed upgrading), better load profiling and management, and improved supply chain management (especially transformer operations and distribution equipment spares).

There are a number of global initiatives on advanced metering, especially in Asia and Europe. East Asia has a number of vendors pushing the technologies, and the low-cost manufacturing base (along with homegrown standards or proprietary solutions) helps bring the price-point down.

The Ontario Energy Board (Canada), as part of the 1998 Ontario Energy Board Act, is pushing utilities to deploy smart metering infrastructure. This mandates utilities to examine time of use rates, and submit proposals for the metering and communications architectures. To this end, Toronto Hydro Corporation has announced building a city-wide WiFi (802.11) based wireless across Toronto, which would serve as communications for their smart metering initiative, and also enable retail internet connectivity across the city. While there are a number of high-profile city level wireless (WiFi) initiatives, especially Philadelphia, this one is unique in harnessing smart metering and consumer connectivity synergy. Of course, such deployments raise a number of regulatory issues, especially cost allocations and affiliate transactions.

Until now, the US has not been a pioneer in widescale deployment of AMM and smart IT systems. The US led the way with early adoption of some AMR technologies, especially based on wireless, but solutions suffered from the "first-mover's curse"; the technology was much more expensive in the mid-nineties, and less capable.

This may change with regulatory impetus. The U.S. Energy Policy Act of 2005 directs state regulators to determine whether utilities should deploy advanced metering infrastructure. A number of states were ahead of the national average, in particular California (which has persistently faced peak shortages of power, and implemented critical peak pricing programs). Different utilities have proposed different advanced metering solutions, but some have been simpler, closer to AMR solutions. Southern California Edison (SCE) has proposed a more advanced solution,* which requires new technology integration instead of combining existing products. This raises the costs somewhat in the short term, but also allows much greater future flexibility, standardization, and additional savings beyond metering *per se*. In particular, increases in fuel prices can be better signaled to consumers during peak and critical peak periods, and consumers can have more choices regarding supplies and pricing.

EMS Deployments by Consumers

EMS is a rapidly growing field, driven by economic considerations and new market opportunities. There are already automated products for the residential environment for energy management. One such system is the Automated Home "Comfort Web Interface" for controlling lighting and HVAC functions. This can control the timing of the DX cooling, and other devices, such as suitably designed electric stoves (limit elements on cook top and oven, without shutting down all cooking). A UK residential EMS retailer example is UK Automation which has a line of home automation gear, controlling appliance and lighting, gate and door automation, and home security; users can program a computer interface.

Small convenience stores, such as Hucks Convenience Stores, utilized a Profile Systems LLC wireless EMS tool for centrally controlling lighting, HVAC, and monitoring freezer/cooler functions in their stores. They found that the results were quite good, with less than a 2-year payback, saving 21 % on lighting, 14% on HVAC, and reducing maintenance costs by extending equipment life. This is similar to findings for other demand controllers and scheduling systems for restaurants.

Harwood Management Services is employing an internet-based energy management system in its 24/7 command center to monitor widespread properties in California, Pennsylvania, Dallas and London, using commodity internet connections as opposed to expen-

---

*Disclosure: one of the authors, Dr. Rahul Tongia, is on the Technology Advisory Board for SCE's Advanced Metering Infrastructure project.

sive dedicated ISDN or T1 lines. Uses include not only typical lighting and HVAC functions, but also load metering of utility meters for control of peak usage via load shedding. An important element of this is ensuring password-protected sites, especially since they are interwoven with hyperlinks. It is important to ensure data integrity because with today's legal climate, sensitive internal data could be used against one in a lawsuit.

As an example of where useful monitoring data could also plant the seeds of future problems, consider a cooling tower water treatment supplier using wireless technology to monitor the building's cooling tower water chemistry. The information could be gathered on the east coast and travel through communication lines to a display in California. If there is a chemical deviation that causes the cooling tower to stray from accepted benchmarks, it will be caught and corrected. In the wrong hands, of course, that data could give the competition an edge or even open a company up to legal problems should the cooling tower chemistry ever stray out of compliance for even a very short period of time. Similarly, there is the danger of pure hackers, who break into company data files merely to wreak havoc; this is especially important in EMS, where in the past the potential for that type of mayhem was virtually nonexistent. As EMS systems integrate with business process systems, the opportunities for savings and optimization grow, but so do the risks. Because of this, only those methods and procedures which are quite secure will be successful; the most likely scenario is to receive signals to shed, and then only the local controllers, in their local, protected world will respond, as they are locally programmed.

During the 2001 energy crisis, the California State Department of General Services (DGS) installed the WebGen Systems "intelligent use of energy" (IUE) system to 65 buildings statewide, using the internet to link buildings into an efficient energy management network. The system then began forecasting and assessing DGS' energy-consumption data in conjunction with other variables that drive energy use and cost: building use, time of day, space occupancy and temperature, market demand, and weather forecasts. Since it began its energy conservation program, DGS energy savings have consistently been well over 20 percent as compared to 2000.

Another web-based energy management software system is "Power Measurement," utilizing the ION Enterprise 4.0 software WebReach. Data are made available through a web browser. One can access meter events, waveforms, data logs, and real-time system information from any computer with internet access, displaying information through a standard web browser. In doing so, it leverages standard communications infrastructures—internet, ethernet, telephone, and wireless.

Another system is RSEnergyMetrix Manager, a core data logging, reporting, charting and billing package. This is a server-based, web-enabled application that runs on a workstation. Manager is available with licenses of 8, 64 or 10,000 meters, and comes with Microsoft Internet Explorer which is used for access and configuration. Along with that is RSEnergyMetrix RT, the real-time communications, configuration and data display package. Another approach is ADI Integrated Technologies' web/i, a web based interface solution for monitoring and control of EMS.

FUTURE TRENDS

The trend towards more control of the grids is forthcoming, everywhere in the world. There is a lack of interest on the part of the public in easily accepting new transmission tower right of ways, more electric generating plants, and the associated environmental pollution from these plants. Additionally, there is more stress on energy efficiency, which in conjunction with demand side management, can allow existing generation and transmission capacity to cover more end users, and in a cost effective manner. Also, the grid will be more loaded, because there will be flatter load profiles, due to a combination of controls for peak periods, and investments in technologies which in and of themselves could permanently shift former peak loads to off-peak periods.

IT and communications continue to improve in price-performance. The voracious appetite of computers and software systems for handling vast amounts of data in near real time clearly indicates the demands for higher data transfer rates. Today the upper limit for low cost data transfer is about 6-10 Mbps, but in the next 10 years it would not be surprising to see this go up by a factor of 10. Simultaneously, there is a continuing trend towards more sophisticated software analysis of data through data mining techniques, pattern recognition, and predictive control, which will continue to expand potential features for web-based energy management system operation [24]. Such features can allow a user to interface to multiple existing sites, as well as add new EMS features with systems developed from scratch for the EMS field with ideas of wireless access, fast installation, wide compatibility, ease of use, and related factors.

Due to the plethora of equipment, systems, software, and continuing concern and awareness for utility costs, it can be anticipated that further development and expansion of web-based options will occur. It can only be assumed that expansion of the web will continue at the breakneck pace it has; there is already even more advanced broadband available to some end users, but at greater cost. Some years ago, the fear was that the web, although conceptually a good communication media, was not reliable enough for online control. Now, with the advent of all the distributed controls and security features, local control, with global monitoring and oversight, can occur, and if and when the web fails, then local controls continue to function. No system is ever foolproof, and when communications are restored, improvements in control and information are available in data bursts, so even the interruptions in communications do not disturb the robust nature of the media and the system.

Advanced metering will certainly become the norm in the coming decades, due to various reasons including improvements in price-performance, regulatory push, and continued investment in metering (especially given half the world lacks electricity today). The question remains what technologies will play out in the short term, and would early adopters hurt themselves through proprietary, non-scalable, or interim solutions? The use of modular, upgradable, and standards-based technologies can mitigate many of these concerns, and web-based systems fit this bill.

The landscape for advanced metering may be shaped not only by metering requirements directly, but also shifts in power systems, e.g., towards distributed generation. The Netherlands already has roughly a third of its power from DG [25], and as combined heat and power solutions become more robust at smaller scale, their deployment may spread. One particular trend may be the advent of plug-in hybrid vehicles, which could dramatically impact time-of-use consumption of electricity. Hybrid vehicles are already a growing niche of cars, but today's models do not plug-in, instead utilizing batteries for transient or limited propulsion energy. With advanced batteries, cars should be able to mange the median commute only on electric power, which would be meant to be plugged in overnight to take advantage of inexpensive, off-peak power. Given that on a household level, the typical load is only a few kilowatts, and even modest-size hybrids today have electric motors of an order of magnitude greater in size, and with larger electric-mode capacities, even a 5-10% shift towards plug-in cars over a decade* may dramatically change the electricity system, with attendant changes required in metering and information flows.

The pressing needs of the electric utility industry, and the continuing demands for more load control in real time, on a much larger scale than in the past, strongly suggests that more individual load control will occur, across all types of buildings, including industrial plants, residences, restaurants, and shops. EMS units, and low cost means for distributing the control and sensors, only add to the ability to cost-effectively control loads unheard of in the past in residences, like electric ranges, small DX cooling units, lighting circuits for dimming or on/off, pool pumps, variable speed air handlers (where we can further modulate the air and the cooling coil delivery, but in ways which still provides "control" or at least "controlled discomfort"). Smarter appliances can play a role, e.g., freezers that delay their defrost cycle until off-peak periods. A convergence is emerging of past energy management experience, hardware, and procedures with the "threats" and "potentials" of new DR grid programs which are aimed at encouraging avoidance of peak demand during certain critical periods, but in a way which does not add discomfort or business interruption beyond that which is desirable or acceptable by a building or industrial plant owner. This approach will also become more acceptable to the residential market as end users realize that failure to adjust their electric demand could lead to skyrocketing electric bills which otherwise need not occur.

To accommodate this communication ability, new electronic metering techniques, already available in some areas, are demonstrating the ability to also capture substantial energy use data, electrical as well as for other utilities like gas and water, so that various time of day factors can be utilized, and that rate tariffs as currently existing may be transitioning to hourly published data, which will determine the real cost of all utilities in the future (not just electricity). There is no reason that peak water usage can not be controlled the same way as electrical consumption, by putting a premium on the cost for use during heavy use periods; such is already performed for natural gas usage in peak winter periods.

As consumers begin to transact more and more business online, they may also expect their interactions with their power company to be online. This represents not only billing and payments, but also the ability to monitor and control their usage of power. Increasing home automation (and the multitude of companies in this space) may accelerate the trend towards advanced metering infrastructure using web-based controls.

Real time pricing (RTP) is likely to have a dramatic effect on the use of embedded EMS systems to better

---

*While capital costs today are high, these are expected to fall over time. The operating costs are demonstrably lower for plug-in electrics harnessing cheap off-peak power, not the least because of high oil prices. Estimates in the EPRI Journal show the equivalent for a plug-in-hybrid to be ~$0.75/gallon, many times cheaper than conventional gasoline. [26]

control energy usage in buildings and industrial plants. The risk and fear that peak kWh prices will skyrocket will force end users to pay more attention, and perhaps be more amenable, to modest inconveniences in comfort or production, in return for much less increases in electric kWh costs or unexpected outages. The theory of having the overall kWh price, on average, not change is purely academic if the vast majority of the usage is during the day as in the case of commercial and many residential customers; in such a case an "average" cost is irrelevant, since they would rarely be able to take advantage of "cheap" off-peak rates, but would be disproportionately saddled with potentially very high peak kWh rates at times. Therefore, more proactive controls for electric power can be expected to become the norm in the coming years.

References

[1] "Electricity Markets: Consumers could benefit from Demand Programs, but Challenges Remain," Report to Chairman, Committee on Governmental Affairs, U.S. Senate, Aug 2004

[2] "DRR Valuation and Market Analysis (Volume I and II)," A report for IEA Demand Side Program, Violette, Freeman, Neil, Jan 2006

[3] "White paper: The benefits of Demand Side Management and Dynamic Pricing Programs," McKinsey and Company, May 2001

[4] "Small Is Profitable: The Hidden Economic Benefits of Making Electric Resources the Right Size"; Rocky Mountain Institute (RMI). 2002.

[5] "The Western States Power Crisis: Imperatives and Opportunities." EPRI White Paper, 2001

[6] "Economic Assessment of RTO Policy," by ICF Consulting for FERC, Feb 2002.

[7] "Advanced metering & billing strategies for DSM"; R. Walawalkar, Dr. B. Colburn, J. Jakubiak and R. Modak; Metering, Billing and CRM/CIS Americas 2004, USA; 2004.

[8] "EPRI" New Principles for Demand Response Planning"; 2002, EPRI: Palo Alto.

[9] "Do enabling technologies affect customer performance in Price Responsive Load Programs?," Goldman, Kintner-Meyer, Heffner, Lawrence Berkeley National Laboratory

[10] "2004: State of the Markets Report"; Federal Energy Regulatory Commission, Office of Market Oversight and Investigations, June 2005

[11] "A study of NYISO 2003 Price Responsive Load Program Performance"; NYISO; 2004

[12] "Reduce Energy and Get Paid, NYISO DSR Programs"; Aaron Breidenbaugh; 2005

[13] "The Power of Energy Information: Web-based Monitoring, Control and Benchmarking," by R. Walawalkar, B. Colburn, and R. Divekar, chapter in *Information Technology for Energy Managers Volume II– Web Based Energy Information and Control Systems—Case Studies and Applications*, Fairmont Press, edited by B. Capehart and L. Capehart, 2005.

[14] *The Political Economy of Power Sector Reform: The Experiences of Five Major Developing Countries*; Victor, D. and T. Heller, eds. 2006 (in press), Cambridge University Press: Cambridge.

[15] "Power Sector Development—The role of IT and new technologies"; Tongia, R.; July 21, 2003. in Portland International Conference on Management of Engineering & Technology (PICMET). 2003. Portland.

[16] "Demand Trading: Building Liquidity," 2002, EPRI.

[17] "Electric Power Reform: Equity and Environment"; Dubash, N.K.; in Encyclopedia of Energy, C.J. Cleveland, Editor. 2004, Elsevier.

[18] "Business Case for Energy Service Portal"; CEIDS; 2004, Electricity Innovation Institute.

[19] "The Scott Report: Insights on AMR Deployments in the United States. 2003"; Scott, H.; Cognyst Consulting.

[20] "The Power of Energy Information: Using the Web for Monitoring, Control and Benchmarking," by Bruce K. Colburn, Proceedings of the 2005 GLOBALCON Energy Conference, Atlantic City, NJ, March 2005

[21] "Web Based Monitoring & Control Technology Update," by B. Colburn, Proceedings of the 2006 GLOBALCON Conference, Philadelphia, PA, March 29-30, 2006.

[22] "Web Based Control system—The Devil is in the Details"; by Steve Tom; AutomatedBuildings.com Article—July 2001,

[23] "The future of AMR or why you better brush up on your Mandarin"; Price, E.; Metering International, 2003(3): p. 52.

[24] "Intelligent Tools for Reducing Energy, Facility-Management Costs," D.E. Mahling (WebGen Systems), Networked Controls, October 2005

[25] "IEA: Distributed Generation in Liberalised Electricity Markets." 2002, International Energy Agency: Paris.

[26] "Driving the Solution: The Plug-in Hybrid Vehicle"; Sanna, L.; *EPRI Journal*, 2005(Fall).

Acknowledgement

The authors gratefully acknowledge discussions and feedback from Dr. Jay Apt, Dr. Lester Lave, Mr. Stephen Fernands, Mr. Rick Mancini and Mr. Constantine Samaras. The authors would also like to thank Mr. Amit Gadkari for help in developing graphic images required for the chapter. Portions of this chapter build off analysis supported by the Alfred P. Sloan Foundation and the Electric Power Research Institute under grants to the Carnegie Mellon Electricity Industry Center (CEIC) at Carnegie Mellon University, Program on Energy and Sustainable Development (PESD) at Stanford University, American Public Power Association through DEED program, Customized Energy Solutions, and EPS Capital Corp.. Any errors and omissions are those of the authors.

Section VI

Hardware and Software Tools and Systems for Data Input, Data Processing and Display in Enterprise Systems

Chapter 32

Wireless Metering: The Untold Story

Jim Lewis, President
Obvius LLC

THE GENERAL CONSENSUS is that the pace of change in the metering/submetering world could be described as glacial, at best. Much of this perception stems from the fact that the essential elements being measured by the meters (volts and amps) haven't changed much since the dramatic shift from DC to AC many long years ago. Try as they might, makers of meters at the end of the day have to focus on using the more powerful tools (microprocessors, A/D converters, etc.) available today to produce essentially the same data that they did 30 years ago. The size of the meters has been reduced and the meters can detect and communicate more high speed events (just what do you do with the 51st harmonic data, anyway?), but beyond that, the changes have been minimal (volts x amps = watts just like always). For building owners and managers simply looking for metering data to help in managing their facilities, there is limited benefit to the advances in technology that have provided such radical change in other areas of business.

Happily, there are some areas of metering where improvements in technology are providing benefits for the owners of commercial, industrial and campus facilities. While the meter technology produces essentially the same outputs, the costs of adding submeters to existing facilities has been dramatically reduced through the use of wireless communications.

A SUBMETERING PRIMER

For most end users, when we use the term "submetering," the immediate image is of the hardware installed on the electrical system to measure kW and kWh. While these devices are certainly an indispensable part of a submetering system, they are only one of the components. A successful submetering system takes the raw data from one or more meters and converts them into useful and timely information to be used for:

• Cost allocation to tenants

• Operational analysis and improvement
• Measurement and verification of energy savings
• Benchmarking and accountability

In order to convert the raw data at the meter level to actionable information requires several steps. Within the metering industry, this process can best be illustrated by looking at five key components:

1. **Meters and sensors**—this is the hardware level where the devices are actually installed in the electrical (or gas or water) system to capture energy consumption for one or more different areas of the building. In the electrical system, these meters might be the traditional round glass meters or they might be meters designed especially for submetering.

2. **Internal communications**—this is the mechanism used to communicate data from the meters and sensors to the data acquisition server. In traditional applications, this is usually a simple twisted pair of wires connected on an RS 485 serial daisy chain.

3. **Data acquisition**—the meters and sensors from level 1 produce industry standard outputs (e.g., pulse or Modbus) that correspond to the real time outputs being monitored. For example, each pulse from an electrical or flow meter has an assigned value (e.g., 1 pulse = 10 kWh). In order for this raw data to be utilized, they must be captured in a timely manner, time stamped and made available for presentation. The data acquisition server (DAS) takes this raw data, time stamps them and then sends them to a local or remote server for storage and report generation

4. **External communications**—once the data are collected by the DAS, they are sent to a local or remote server for reporting. In traditional metering applications, this communication may use a local area network (LAN) connection or a modem.

5. **Storage and reports**—once the data from the meters are gathered at the building level, they still requires further processing to produce user friendly reports and information. This processing occurs at the user interface level where information from one or more buildings is collected into a traditional database for reporting and storage. This process may be accomplished on a local server using software installed on a PC or may be pushed to a remote server providing data management and reporting services.

A detailed discussion of each of these components is beyond the scope of this chapter, but most successful submetering systems have the following characteristics:

*   "Open" meter protocols that provide the capability for any DAS to connect to any meter or sensor, regardless of what is being measured (e.g., electricity, gas, water, flow, Btus). Most meter manufacturers use Modbus RTU as the primary protocol for communications with submeters, but the DAS should also accept pulse or analog values that meet industry standards.

*   Non-volatile storage of data—the DAS should be able to store at least 30 days of data from multiple meters without risk of data loss in the event of power failure,

*   Options for user-selected data intervals—the user should be able to select data intervals to match the utility interval (e.g., 15 minutes)

*   "Open" protocol support at the DAS level that allows the data collected to be sent to any web server (i.e., non-proprietary format). In most cases, this can be accomplished using standard internet protocols such as HTTP, FTP, and XML. This allows the end user to select virtually any software program (including custom programs) regardless of the DAS or meter brand.

Figure 32-1 shows a graphical representation of a traditional wired submetering system, starting with the meters and sensors at the bottom and moving up to the storage and reports level at the top of the figure.

WHY DO WE NEED WIRELESS?

As outlined in the previous section, there are readily available wired solutions that provide reliable submetering data using commonly available tools such as serial ports and local area networks, so why do we need wireless? The simplest answer is that in many applications (particularly retrofits), the cost of running wire between meters and the DAS is prohibitively expensive and using wireless components can dramatically reduce both the cost and the installation time. In addition, in many submetering projects, the cost and time involved in securing a network connection or phone line for communicating from the DAS to the internet becomes a major headache.

Using wireless solutions in submetering not only minimizes the costs of many projects, but also greatly reduces the disruption to day-to-day operations caused by running wire several hundred feet or more through an existing facility. Wireless also provides a very attractive alternative to trenching between buildings in a campus environment and eliminates the need for coordinating with the IT department for use of an existing LAN connection.

WHICH WIRELESS ARE WE TALKING ABOUT?

Ask anyone what the term "wireless" means and you'll probably find yourself on the receiving end of one of those looks that implies you must be one of the dumbest people on earth. Wireless, as any fool knows, means no wires and therein lies the problem with understanding wireless submetering. A short list of wireless terms includes the following:

*   Satellite
*   Cell phone
*   Wi-fi
*   Pager
*   Proprietary radio
*   Bluetooth
*   Zigbee

Clearly, each of these terms does refer to a some aspect of wireless communications, but there are vast differences between them and their applications. This overlap has contributed significantly to the confusion surrounding wireless metering and therefore one of first points to be addressed in this chapter is to explain just which wireless we are talking about.

In essence, wireless (or for that matter, wired) communications form the bridge between each of the levels of metering discussed above. We will be examining two different wireless applications in this chapter, the first being communications between meters and the DAS and

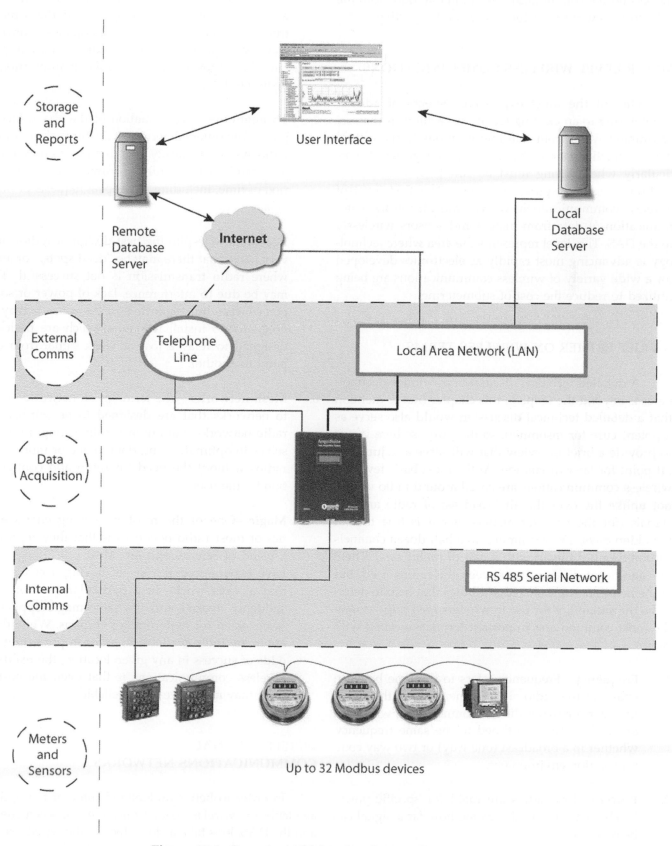

Figure 32-1. Overview of Metering System Components

the second the communications of interval data from the DAS to a remote server for storage and reporting.

METER LEVEL WIRELESS COMMUNICATIONS

One of the most expensive aspects of adding submetering to an existing facility or facilities is the cost of running wires from multiple locations to the central data acquisition point. This cost can be significant, particularly when wiring must be run between multiple buildings (i.e., campuses or bases). The first level of wireless communications we will consider is the communication of data from meters and sensors wirelessly to the DAS. This level represents the area where technology is advancing most rapidly as electronics developed for a wide variety of wireless communications are being utilized to reduce the cost of submetering.

A BRIEF PRIMER ON WIRELESS TERMS

A detailed technical discussion of wireless technologies is beyond the scope of this chapter (not to mention that a detailed technical discussion would also serve as a potent cure for insomnia), so the purpose here is just to provide a brief overview that will serve as a jumping off point for later discussion. At the most basic level, all wireless communications are based around radio waves, not unlike the over the air broadcast of radio and TV signals (for the younger readers, yes it is true that in the olden days, TV was limited to a half dozen channels captured out of the air with a pair of rabbit ears). There are significant differences between the radios used, but at the heart are some shared attributes that help to determine the suitability for use in wireless metering. Among the most common and important terms associated with radios:

1. **Frequency**—Frequency refers to the time between peaks of the radio signal generated by the radio. In order for two radios to communicate with each other, they must be tuned to the same frequency, whether in a broadcast (one way) or two way communication environment.

2. **Power**—Most radios are rated for specific power levels that serve to determine how far a signal can be detected.

3. **Interference**—As anyone who has ever driven under a power line or tried to use a cell phone knows, there are a number of things that can cause interference with a radio signal and the same is true for wireless metering technologies. Concrete walls, steel panels and other radio sources all provide challenges to the successful communication of metering data.

4. **Throughput**—A combination of the above factors plus a few others determines the throughput of the radio system. Basically, this just means how much data can be successfully transmitted in a give period of time, including any repeat requests or other delays.

5. **Repeaters**—Despite the best design of radios, it is very likely that there may be "dead spots," or areas where radio transmission is not successful. This may be due to interference, lack of power or some other factor, but regardless of the cause it may be necessary to install repeaters which are basically designed to relay a lost or weak signal from one point to another.

6. **Mesh networks**—The term "mesh network" refers to networks that are designed to be self-healing radio networks that automatically configure themselves to optimally send data to one or more other radios without the need for setup and configuration by the user.

7. **Magic**—One of the most interesting characteristics of most radio networks is that they represent a blend of science and art. Most cell phone users have experienced the joy of having a cell phone that always works in a particular location, but suddenly doesn't and the same magic applies to radios used for metering applications. While there are many tools available to assist in predicting the odds of success in any given location, the reality of wireless communications is that even the best of tools may not prove 100% reliable.

WIRELESS SERIAL COMMUNICATIONS NETWORKS

In order to better understand some of the options available for wireless communications between meters and the DAS, let's take a closer look at the wired version of this same communication. Figure 32-2 shows a section of the previous figure that focuses on the internal serial communications networks for submeters.

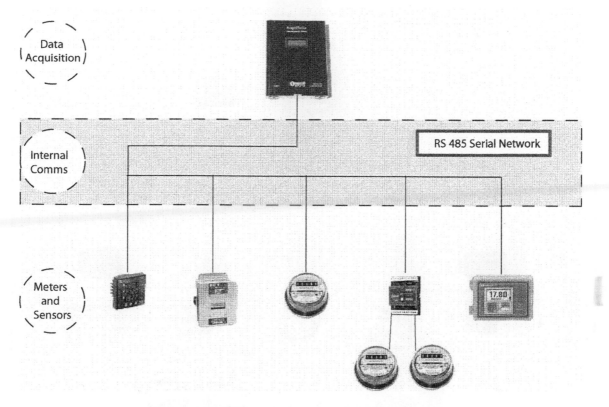

Figure 32-2. Internal Metering Components View

As Figure 32-2 shows, in a wired sensor level communications network, a twisted pair of wires is run from a serial port (RS485) on the data acquisition server (DAS) and daisy-chained to each of the Modbus meters or devices. The DAS uses plug and play technology to detect the type of meter and load the appropriate drivers to interrogate the meter(s) to obtain the desired data. On pre-selected intervals (typically 15 minutes or one hour), the DAS will gather data from the meters that may be limited to consumption (kWh) or may include other parameters such as power factor, THD, etc. if the meter supports those additional functions. The readings are then stored by the DAS until they are "pushed" or "pulled" to a remote server for storage in a standard database (more on that later).

This system is quite reliable and provides an excellent method of communication if the meters and the DAS are all in close proximity. This type of serial network is typically capable of reliable communications up to 4000 ft. If, on the other hand, there are multiple locations throughout the facility or campus that make wiring difficult and/or expensive, wireless nodes near the meters can provide a very cost effective alternative to wiring.

HOW DOES WIRELESS
SERIAL COMMUNICATIONS WORK?

There are at least a couple of different ways of getting data from meters wirelessly. One method involves using low power battery devices to capture and transmit pulses from meters to the DAS, which represents a very low cost means of communications, but is not a replacement for the ideal serial network shown above. The wireless nodes function in one way communication only and cannot provide the reliability or the additional data available from reading and sending Modbus serial data to the network.

The second (and preferred) method is to simply use wireless nodes on either end of the transmission to seamlessly replace the twisted pair of wires in Figure 32-2. The newest technologies on the market accomplish this in a mode that is completely transparent to the user (i.e., the data transmitted wirelessly is indistinguishable from the data gathered on a wired network.

Figure 32-3 shows a typical system that substitutes wireless Modbus communications for wired networks. In this case, we will assume that the two meters on the left of the graphic are located in one electrical room, the

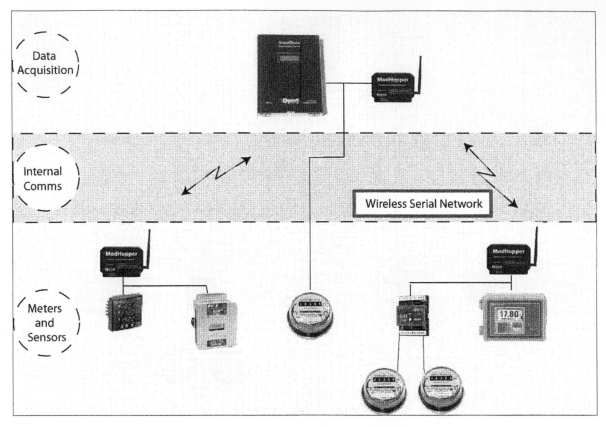

Figure 32-3. Metering Components with Wireless Communications

three on the right are located in a second electrical room, and the single meter in the center is the primary meter for the facility and is located in close proximity to the DAS and thus can be hard-wired.

In this example, all of the meters can communicate all of their data using the standard Modbus protocol, regardless of whether they are connected via wires or wirelessly. In the case of the remote meters, we simply have wired the RS485 outputs from the meters into a wireless node that puts the Modbus data into a wireless form and sends the data via radio to another node connected directly to the DAS where the data are once again converted to RS485.

SO HOW DOES THIS
MESH NETWORK FUNCTION?

The two keys to making the wireless communications network shown in Figure 32-3 actually work are: 1) that the individual radio transceivers (or nodes) are specifically designed for transmitting Modbus data; and 2) that the radios contained within the transceivers can function as part of a "mesh" network. A brief overview of the functioning of a mesh network will

provide a better understanding of the benefits to using wireless mesh networks to replace wired serial communications.

All mesh networks (regardless of the application) share certain common characteristics that make them valuable for replacing or augmenting wired networks. Among the most important:

• **Self-configuring nodes**—The whole concept of mesh networks is that the radio nodes will be aware of other nodes and will automatically configure the network to optimize throughput of data, without the need of programming or configuration on the site

• **Multiple routing paths**—Since each node is "aware" of all the other nodes it can communicate with, there are multiple options available for routing the data. As new nodes are added to the network, the network should recalculate the optimal routing paths to take advantage of more efficient routing options

• **Spread spectrum radios**—This feature allows the radios to work on a variety of predetermined frequencies, minimizing the likelihood of interference from other radio sources. Essentially, the radios are designed to "hop" to different frequencies if interference is encountered until the two radios find a frequency that minimized interference.

• **ISM band radios**—Transceivers that use the ISM (instrumentation, scientific and medical) frequencies set aside by the FCC do not need to have a local license for each site. These radios are designed to share the frequency bands with other systems and devices with a minimum of interference.

Figure 32-4 shows a diagram of a typical wireless mesh network. Each of the transceivers can accept up to 32 Modbus devices and 2 pulses.

This figure shows a mesh network with eight transceivers and nine meters, all connected to a DAS for data collection. Two of the transceivers are functioning as repeaters since they have no meters connected and serve only to route traffic to other nodes. The

dashed lines between the transceivers shows the other nodes that each can communicate with and most of the nodes can see multiple other nodes, providing multiple routing paths for requests for data from the DAS. The network will automatically figure the optimal routing to reach each of the meters and will adjust these routing paths based on the success rate of transmissions. If other Modbus devices are added to any of the nodes, their presence is detected by the network and optimal routing paths are determined. If other transceivers are added, they will similarly be detected by the network and added to the routing paths.

SUMMARY

The technology available today makes wireless submetering a viable alternative to hard-wired systems. The installation of these products, whether as the total solution or in conjunction with wired Modbus meters, provides a transparent and cost effective option where wiring costs are prohibitively expensive.

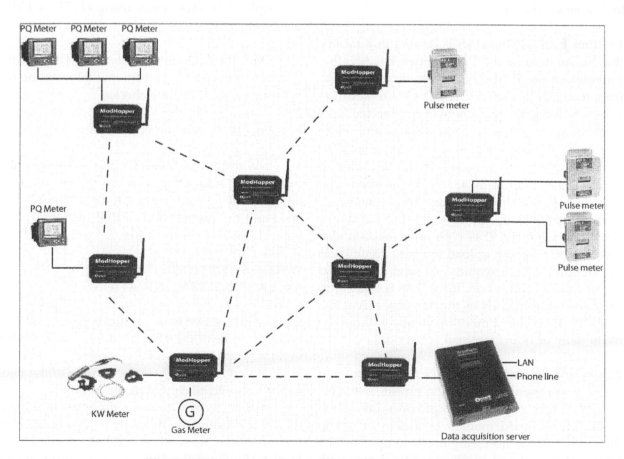

Figure 32-4. Typical Mesh Network Metering System

PART 2—
EXTERNAL WIRELESS COMMUNICATIONS

Summary

Just as mesh networking provides an excellent alternative to a wired solution at the sensor communication network level, advances in wireless technology allow the user to send data from the DAS to a remote server without the need for a hard-wired connection. The use of GSM/GPRS (cell) modems in the DAS provide a very cost effective means of communications that piggybacks on the structure built out for cell phone communications.

WIRED EXTERNAL SOLUTIONS

Figure 32-5 shows a closer look at the upper part of Figure 32-1 and provides some details about how data are typically sent from the DAS to a remote server for logging and reports. The DAS provides several options for communications, depending on the needs of the user and available communications options. There are usually two alternatives for connecting to either a remote or a local server:

- **Existing LAN**—If the DAS is located in a facility that has an existing local area network (LAN) this provides an excellent mechanism for getting data from the DAS to a server. The DAS is IP-addressable, which means that a network-connected DAS can provide two-way communications and allow the user to see near real time data in addition to interval information. This is also the easiest solution to implement technically as the DAS simply functions like any other workstation on the network, opening connections to the internet as needed to send data remotely. This type of connection also provides the fastest upload speeds, minimizing the amount of time required to send data out to the remote or local server. If the DAS is located on the same LAN or WAN as the database server, the data can be uploaded more frequently without the usual daily or hourly upload schedule.

- **Phone line**—If there is no LAN connection (or if the IT department can't or won't make a connection available), the DAS also provides the option to use an on-board modem to dial in or out to the system. If the user only needs to upload data (dial-out) then the phone line can be shared with other devices such as fax machines. In this case,

the DAS is programmed to call out (usually once a day) to a local ISP to upload data via the internet. If the preferred method is to have the remote server call in to the DAS (dial-in), then the DAS will require a dedicated phone line and number. Uploads are generally limited to daily transmissions as the speed at which data transfer occurs is substantially slower than using a LAN connection.

Details of the file upload in a wired system are beyond the scope of this chapter, but s typical upload session on an existing LAN would have the following elements:

1.   The DAS initiates a connection to the remote server via the LAN by accessing a URL (e.g., http://www,obvius.com) where the server is located. The DAS provides a user name and password to the server, which uses this information to access the existing database records;

2.   Once the connection is established and verified, the DAS will upload data collected since the last upload, in most cases using HTTP or FTP protocols;

3.   Once the data are uploaded and verified, the DAS can optionally get additional information from the server, such as time checks;

4.   Finally, the session is terminated.

The process for using a phone line connection is very similar, with the primary difference being that the upload speeds are slower (think dial-modem versus DSL) and the connection is a PPP session.

WIRELESS OPTIONS FOR
EXTERNAL COMMUNICATIONS

There are two basic applications for wireless communication from the DAS, one for inside the building to a LAN and one for totally wireless communications. We will first look briefly at the simpler of the two, wireless communications within the building.

In this scenario, we are simply substituting a wireless access point for the wired LAN connection that them connects to another wireless access point on the local LAN. Since the DAS is an IP addressable device like any workstation, there is no special setup required

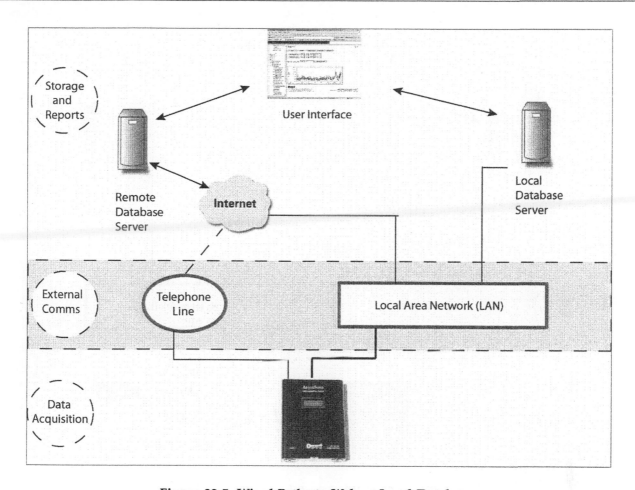

Figure 32-5. Wired Paths to Web or Local Database

and the functionality is exactly the same as that shown for the wired LAN connection in Figure 32-5.

The second option for wireless external connectivity uses the existing cell phone networks built out around the country to bypass any local connection. The primary reasons for doing this are:

- Lack of available LAN or phone lines;
- Security concerns from IT personnel to using existing LAN's;
- High cost of acquiring and maintaining either phone of LAN connections;
- Delays in getting phone or LAN connections installed
- Ease of installation

WHAT ADDITIONAL HARDWARE IS REQUIRED?

In the simplest terms, the only real change from the LAN or phone connected version is that a cell phone modem is used in place of the typical RJ11 modem. This modem is connected to an antenna mounted

on the DAS, but these are the only functional differences between the cell modem version and any other DAS.

HOW DOES IT WORK?

In the olden days (think 1990's) cell phone communication was based on taking analog voice signals from one user and transmitting them to an analog receiver held by another user on the other end. Today, analog transmission is virtually non-existent and has been replaced by digital cell technology (basically a process that converts the analog voice signal to digital for transmission that is decoded on the other end). The evolution from analog to digital is beneficial for data communication, because data transmission on the internet is inherently digital and thus ideally suited for the digital networks.

The DAS is functionally the same device used in the hardwired (LAN or phone modem) connection shown earlier in this chapter. To use the cell phone

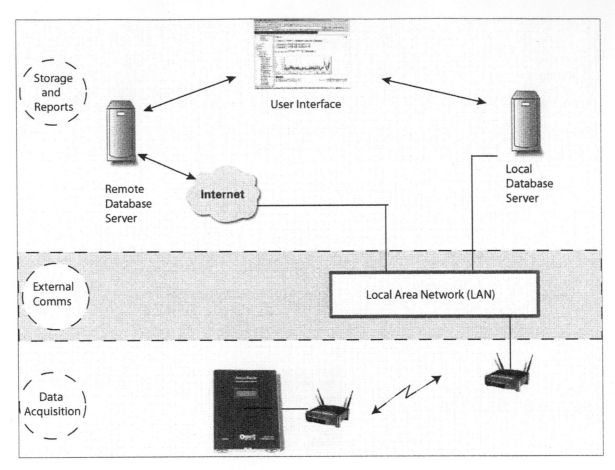

Figure 32-6. Wireless Ethernet Connection

system for data transmission, the phone modem module is replaced with a cell modem that connects to the cellular network just like any cell phone. The system relies on the use of the GSM network using GPRS (general packet radio service) to transmit data packets wirelessly over the cell network. This is the same system and service used by cell providers to allow users to view email and browse the internet using cell phones and PDA's.

To implement this method of communication requires a SIM card (available from the wireless carrier) and a monthly data plan from the same carrier (e.g., Cingular) that typically runs $20 to $30 per month for most DAS installations. When an upload is due (e.g., daily or hourly) the DAS uses the cell modem to connect to the cell carrier network and then uses the carrier's ISP service to establish a connection with the remote server. The data are then transferred to the remote server for storage and reporting.

WHY USE THE CELL NETWORK INSTEAD OF A LAN OR PHONE LINE?

In addition to the ongoing monthly costs of cell service, the GSM version of the DAS is also typically 20 to 30 % more expensive than an identical LAN version and the upload times are much slower, so the obvious question is why would anyone use the cell system. There are a number of reasons why the GSM version may be preferable in some cases, despite the higher costs, principal among them:

- **Costs of LAN or phone connections**—in many cases, it may be prohibitively expensive to add a network drop or phone line, either because of the end user's policies or physical limitations (long wire runs, firewalls, etc.). In cases like this, the incremental costs of installing GSM may be less than the hard-wired costs;

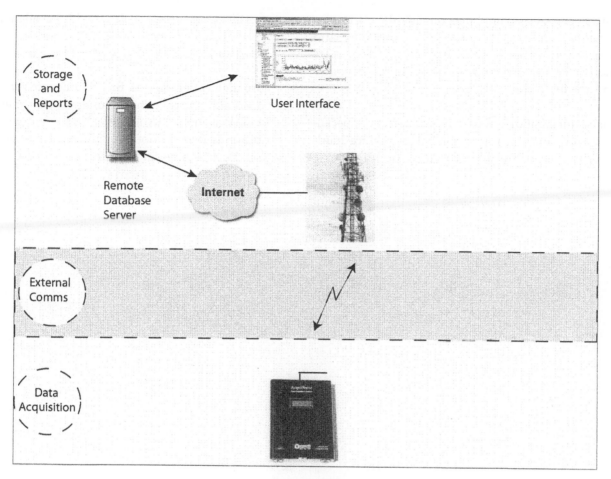

Figure 32-7. Cell Based Web Connection

- **IT security concerns**—the LAN-based DAS is designed to function using the existing network without creating any security concerns. If the system is installed and connected behind the firewall, it functions in an identical manner to any connected workstation. Despite this design, many IT managers are reluctant to allow access to the network which can at best delay the installation and at worst prevent it entirely. Because the GSM DAS has no connection to the existing network or phone system, any concerns are alleviated.

- **Installation delays**—adding a network drop or phone line frequently involves coordinating with either other departments or contractors. In many installations, the lead times for getting these lines run and activated is the longest part of the installation whereas the GSM system will be operating without the need for any outside resources the same day it is installed.

The most appealing aspects of the cell based DAS

are that it functions as a self-contained system without relying on existing networks or departments for installation or operation and can be up and sending data within minutes of the installation.

WHAT DOES THE FUTURE HOLD?

There are two primary forces driving improvements in the market for wireless submetering systems. First, advances in radio technology (both at the RS485 meter level and at the cellular system level) that are focused on broader markets such as consumer cell phones will be adopted by the energy information industry. Second, as the market for energy information grows, companies (such as Obvius) will emerge that specialize in the development of hardware and software specifically for gathering energy information, which will improve the functionality and ease of installation.

In general, it is reasonable to expect the following changes in the market:

- **Lower costs**—as companies outside the energy information industry (such as cell carriers) expand their offerings and increase the volume of sales there will be lower costs for radio technologies and for the cellular service

- **Easier installation**—the increased focus of specialized companies in this market will produce hardware and firmware that is focused on minimizing the time and costs associated with installation. Next generation DAS hardware will have much higher levels of integration and support for wireless

- **Broader coverage**—as the cellular companies expand and improve the coverage of their networks, it is reasonable to assume that some areas and locations that do not provide cell coverage today will be accessible in the future.

Chapter 33

An Energy Manager's Introduction to Weather Normalization of Utility Bills

John Avina, Director
Abraxas Energy Consulting

UTILITY BILL TRACKING: THE REPORT CARD FOR FACILITIES AND ENERGY MANAGERS

ENERGY MANAGERS all too often have to justify their existence to management. They may be asked: "How much did we save last year?"; "Did your recommendations give reasonable paybacks?"; "Since the last project didn't save any money, why would we expect the next one to?"

Since over the reign of an energy manager, many energy conservation projects, control strategies, and operation and maintenance procedures may be employed, the simplest method to report on the energy manager's complete performance is to look at the utility bills. Management often sees it quite simply—it is all about the utility bills, since the bills reflect how much you are paying. Did the energy manager save us money or not?[1]

Since most energy managers are already tracking their utility bills, it should only take an additional step to see whether you have saved any energy and costs from your energy management program. In theory, you could just compare the prior year's bills to the current year's bills and see if you have saved.

But if it is so easy, why write a chapter on this? Well, it isn't so easy. Let's find out why.

Suppose an energy manager replaced the existing chilled water system in a building with a more efficient system. He likely would expect to see energy and cost savings from this retrofit. Figure 33-1 presents results the energy manager might expect.

But what if, instead, the bills presented the disaster shown in Figure 33-2?

Imagine showing management these results after you have invested a quarter-million dollars. It is hard to inspire confidence in your abilities with results like this.

How should the energy manager present these data to management? Do you think the energy manager is feeling confident about his decisions and about getting

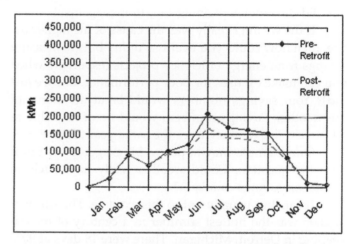

Figure 33-1. Expected Pre- and Post-Retrofit Usage for Chilled Water System Retrofit.

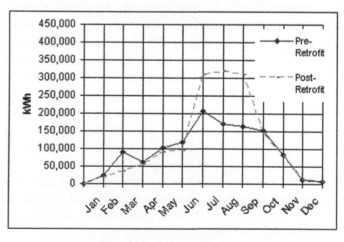

Figure 33-2. A Disaster of a Project? Comparison of Pre-Retrofit and Post-Retrofit Data

funding for future energy savings projects? Probably not. Management may simply look at the figures and, since figures don't lie, conclude they have hired the wrong energy manager!

There are many reasons the retrofit may not have delivered the expected savings. One possibility is that the project is delivering savings, but the summer after the retrofit was much hotter than the summer before the retrofit. Hotter summers translate into higher air conditioning loads, which typically result in higher utility bills.

Hotter Summer ⟹ Higher Air Conditioning Load ⟹
Higher Summer Utility Bills

In our example, we are claiming that because the post-retrofit weather was hotter, the chiller project looked like it didn't save any energy, even though it really did. Imagine explaining that to management!

If the weather really was the cause of the higher usage, then how could you ever use utility bills to measure savings from energy efficiency projects (especially when you can make excuses for poor performance, like we just did)? Your savings numbers would be at the mercy of the weather. Savings numbers would be of no value at all (unless the weather were the same year after year).

Our example may appear a bit exaggerated. But it begs the question: Could weather really have such an impact on savings numbers?

It can, but usually not to this extreme. The summer of 2005 was the hottest summer in a century of record-keeping in Detroit, Michigan. There were 18 days at 90°F or above compared to the usual 12 days. In addition, the average temperature in Detroit was 74.8°F compared to the normal 71.4°F. At first thought, 3 degrees doesn't seem like all that much, however, if you convert the temperatures to cooling degree days[2], as shown in Figure 33-3, the results look dramatic. Just comparing the June through August period, there were 909 cooling degree days in 2005 as compared to 442 cooling degree days in 2004. That is more than double! Cooling degree days are roughly proportional to relative building cooling requirements. For Detroit then, one can infer that an average building required (and possibly consumed) more than twice the amount of energy for cooling in the summer of 2005 than the summer of 2004. It is likely that in the upper Midwestern United States there were several energy managers who faced exactly this problem!

How is an energy manager going to show savings from a chilled water system retrofit under these circumstances? A simple comparison of utility bills will not work, as the expected savings will get buried beneath the increased cooling load. The solution would be to somehow apply the same weather data to the pre- and post-retrofit bills, and then there would be no penalty for extreme weather. This is exactly what weather normalization does. To show savings from a retrofit (or other energy management practice), and to avoid our disas-

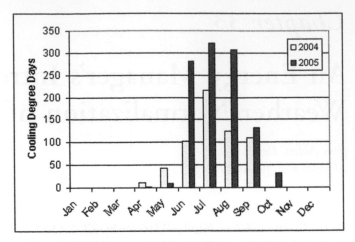

Figure 33-3. Cooling Degree Days in Detroit, Michigan for 2004 and 2005

trous example, an energy manager should normalize the utility bills for weather so that changes in weather conditions will not compromise the savings numbers.

More and more energy managers are now normalizing their utility bills for weather because they want to be able to prove that they are actually saving energy from their energy management efforts. This process has many names: weather correction, weather normalization, tuning to weather, tuning or weather regression.

HOW WEATHER NORMALIZATION WORKS

Rather than compare last year's usage to this year's usage, when we use weather normalization, we compare *how much energy we would have used* this year to *how much energy we did use* this year. Many in our industry do not call the result of this comparison, "savings," but rather "usage avoidance" or "cost avoidance" (if comparing costs). Since we are trying to keep this treatment at an introductory level, we will simply use the word *savings*.

When we tried to compare last year's usage to this year's usage, we saw Figure 33-2, and a disastrous project. We used the equation:

Savings = Last year's usage – This year's usage

When we normalize for weather, the same data results in Figure 33-4 and uses the equation:

Savings = How much energy we would have used this year – This year's usage

The next question is how to figure out *how much energy we would have used this year*? This is where weather

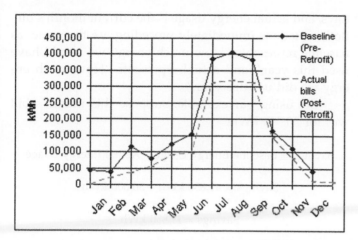

Figure 33-4. Comparison of Baseline and Actual (Post-Retrofit) Data with Weather Correction

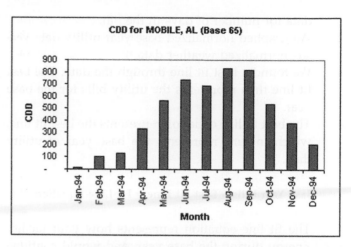

Figure 33-5. Cooling Degree Days

normalization comes in.

First, we select a year of utility bills[3] to which we want to compare future usage. This would typically be the year before you started your energy efficiency program, the year before you installed a retrofit, the year before you, the new energy manager, were hired, or just some year in the past that you want to compare current usage to. In this example, we would select the year of utility data before the installation of the chilled water system. We will call this year the **base year**[4].

Next, we calculate degree days for the base year billing periods. Because this example is only concerned with cooling, we need only gather cooling degree days (not heating degree days). A section on calculating degree days follows later in the chapter. For now, recognize that only cooling degree days need to be gathered at this step.[5] Figure 33-5 presents cooling degree days over two years.

Base year bills and cooling degree days are then normalized by number of days, as shown in Figure 33-6. Normalizing by number of days (in this case, merely, dividing by number of days) removes any noise associated with different bill period lengths. This is done automatically by canned software and would need to be performed by hand if other means were employed.

To establish the relationship between usage and weather, we find the line that comes closest to all the bills. This line, the **best fit line**, is found using statistical regression techniques available in canned utility bill tracking software and in spreadsheets.

The next step is to ensure that the best fit line is good enough to use. The quality of the best fit line is represented by statistical indicators, the most common of which, is the R^2 value. The R^2 value represents the goodness of fit, and in energy engineering circles, an R^2

> 0.75 is considered an acceptable fit. Some meters have little or no sensitivity to weather or may have other unknown variables that have a greater influence on usage than weather. These meters may have a low R^2 value. You can generate R^2 values for the fit line in Excel or other canned utility bill tracking software.[6]

This best fit line has an equation, which we call the **fit line equation**, or in this case the **baseline equation**.[7] The fit line equation from Figure 33-6 might be:

$$\text{Baseline kWh} = (5 \text{ kWh/Day} * \#\text{Days}) + (417 \text{ kWh/CDD} * \#\text{CDD})[8]$$

Once we have this equation, we are done with this regression process.

Let's recap what we have done:

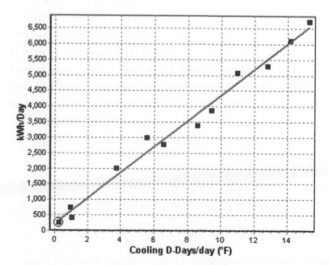

Figure 33-6. Finding the relationship between usage and weather data. The dots represent the utility bills. The line is the best fit line.

data for number of days in the bill.

2. We graphed normalized base year utility data versus normalized weather data.

3. We found a best fit line through the data. The best fit line then represents the utility bills for the base year.

4. The best fit line equation represents the best fit line, which in turn represents the base year of utility data.

Base Year bills ≈ Best Fit Line = Fit Line Equation

The fit line equation represents how your facility used energy during the base year, and would continue to use energy in the future (in response to changing weather conditions) assuming no significant changes occurred in building consumption patterns.

Once you have the baseline equation, you can determine if you saved any energy.

How? You take a bill from some billing period after the base year. You (or your software) plug in the number of days from your bill and the number of cooling degree days from the billing period into your baseline equation.

Suppose for a current month's bill, there were 30 days and 100 CDD associated with the billing period.

$$\text{Baseline kWh} = (5 \text{ kWh/Day} * \text{\#Days}) + (417 \text{ kWh/CDD} * \text{\#CDD})$$

$$\text{Baseline kWh} = (5 \text{ kWh/Day} * 30) + (417 \text{ kWh/CDD} * 100)$$

$$\text{Baseline kWh} = 41{,}850 \text{ kWh}$$

Remember, the baseline equation represents how your building used energy in the base year. So, with the new inputs of number of days and number of degree days, the baseline equation will tell you how much energy the building would have used this year based upon base year usage patterns and this year's conditions (weather and number of days). We call this usage that is determined by the baseline equation, **baseline usage**.

Now, to get a fair estimate of energy savings, we compare:

Savings = How much energy we would have used this year –
How much energy we did use this year

Or if we change the terminology a bit:

Savings = Baseline Energy Usage – Actual Energy Usage

where baseline energy usage is calculated by the baseline equation, using current month's weather and number of

days, and actual energy usage is the current month's bill. Both equations immediately preceding are the same, as baseline represents "How much energy we would have used this year," and actual represents "How much energy we did use this year."

So, using our example, suppose this month's bill was for 30,000 kWh:

Savings = Baseline Energy Usage – Actual Energy Usage

Savings = 41,850 kWh – 30,000 kWh

Savings = 11,850 kWh

CALCULATING DEGREE DAYS AND FINDING THE BALANCE POINT

Cooling degree days (CDD) are roughly proportional to the energy used for cooling a building, while heating degree days (HDD) are roughly proportional to the energy used for heating a building. Degree days, although simply calculated, are quite useful in energy calculations. They are calculated for each day, and are then summed over some period of time (months, a year, etc.).[9]

In general, daily degree days are the difference between the building's balance point and the average outside temperature. To understand degree days then, we first need to understand the concept of balance points.

Buildings have their own set of **balance points** for heating and for cooling – and they may not be the

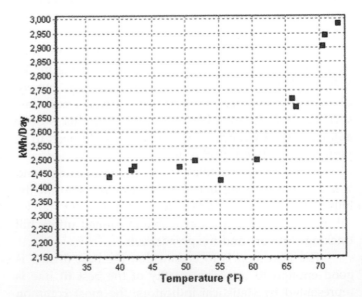

Figure 33-7. Determining the Balance Point using a kWh/day vs. Outdoor Temperature Graph

same. The **heating balance point** can be defined as the outdoor temperature at which the building starts to heat. In other words, when the outdoor temperature drops below the heating balance point, the building's heating system kicks in. Conversely, when the outdoor temperature rises above the **cooling balance point**, the building's cooling system starts to cool.[10] A building's balance point is determined by nearly everything associated with it, since nearly every component associated with a building has some effect on the heating of the building: building envelope construction (insulation values, shading, windows, etc.), temperature set points, thermostat set back schedules if any, the amount of heat producing equipment (and people) in the building, lighting intensity, ventilation, HVAC system type, HVAC system schedule, lighting and miscellaneous equipment schedules among other factors.

In the past, before energy professionals used computers in their everyday tasks, degree day analysis was simplified by assuming balance points of 65°F for both heating and cooling. As a result, it was easy to publish and distribute degree days, since everyone calculated them using that same standard. It is more accurate, however, to recognize that every building has its own balance points and to calculate degree days accordingly. Consequently, you are less likely to see degree days available, as more sophisticated analysis requires you to calculate your own degree days based upon your own building's balance points.[11]

A way to find the balance point temperature of a building is to graph the usage/day against average outdoor temperature (of the billing period) as shown in Figure 33-7. Notice that Figure 33-7 presents two trends. One trend is flat, and the other trend slopes up and to the right. We have drawn lines signifying the two trends in Figure 33-8. (Ignore the vertical line for now.) The flat trend represents **non-temperature sensitive consumption**, which is electrical consumption that is not related to weather. In Figure 33-7, non-temperature sensitive consumption is roughly the same every month, about 2450 kWh per day. Examples of non-temperature sensitive consumption include lighting, computers, miscellaneous plug load, industrial equipment and well pumps. Any usage above the horizontal line is called **temperature sensitive consumption**, which represents electrical usage associated with the building's cooling system. Notice in Figure 33-8, the temperature sensitive consumption only occurs at temperatures greater than 61°F. The intersection of the two trends is called the **balance point** or balance point temperature, which is 61°F in this example.

Notice also that, in Figure 33-8, as the outdoor temperature increases, consumption increases. As it gets hotter outside, the building uses more energy, thus the meter is used for cooling, but not heating. The balance point temperature we found is the cooling balance point temperature (not the heating balance point temperature).

We can view the same type of graph for natural gas usage in Figure 33-9. Notice that the major differ-

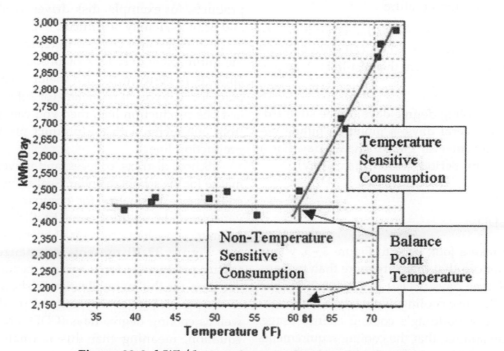

Figure 33-8. kWh/day vs. Average Outdoor Temperature

ence between the two graphs (electric and gas), is that the temperature sensitive trend slopes up and to the left (rather than up and the right). As it gets cooler outside, they use more gas, therefore, they use gas to heat the building.

Now that we have established our balance point temperature, we have all the information required to calculate degree days. If your graph resembles Figures 9, you will be using heating degree days. If your graph resembles Figure 33-8, you will be using cooling degree days. If you calculate degree days by hand, or using a spreadsheet, you would use the following formulae for your calculations. Of course, commercially available software that performs weather normalization handles this automatically.

For each day,

$$HDD_i = [\ T_{BP} - (T_{hi} + T_{lo})/2\ ] \times 1\ Day^+$$

$$CDD_i = [\ (T_{hi} + T_{lo})/2 - T_{BP}\ ] \times 1\ Day^+$$

Where:

HDD_i = Heating Degree Days for one day

CDD_i = Heating Degree Days for one day

T_{BP} = Balance Point Temperature,

T_{hi} = Daily High Temperature

T_{lo} = Daily Low Temperature

$^+$ signifies that you can never have negative degree days. If the HDD_i or CDD_i calculation yields a negative number, then the result is 0 degree days for that day.

Heating and cooling degree days can be summed, respectively, over several days, a month, a billing period, a year, or any interval greater than a day. For a billing period (or any period greater than a day),

$$HDD = \Sigma HDD_i$$

$$CDD = \Sigma CDD_i$$

Now, let's take a look back to Figure 33-3, where you may have noticed that there are more than twice as many cooling degree days (CDD) in August 2005 than in August 2004. Because cooling degree days are roughly proportional to a building's cooling energy usage, one could rightly assume that the cooling requirements of the building would be roughly double as well.

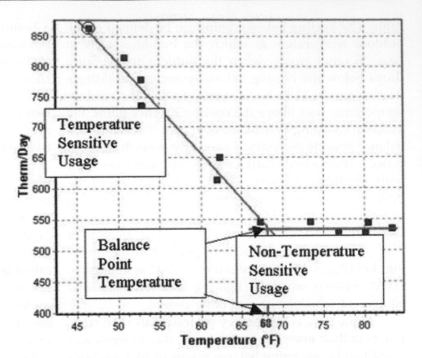

Figure 33-9. Therm/day vs. Average Outdoor Temperature

NORMALIZING FOR OTHER VARIABLES

More and more manufacturing energy managers are coming to understand the value of normalizing utility data for production in addition to (or instead of) weather. This works if you have a simple variable that quantifies your production. For example, a computer assembly plant can track number of computers produced. If your factory manufactures several different products, for example, disk drives, desktop computers and printers, it may be difficult to come up with a single variable that could be used to represent production for the entire plant (i.e. tons of product). However, since analysis is performed on a meter level, rather than a plant level, if you have meters (or submeters) that serve just one production line, then you can normalize usage from one meter with the product produced from that production line.

School districts, colleges, and universities often normalize for the school calendar. Real estate concerns, hotels and prisons normalize for occupancy. Essentially any variable can be used for normalization, as long as it is an accurate, consistent predictor of energy usage patterns.

Figure 33-10 presents normalized daily usage versus production for a widget factory. The baseline equation for this normalization is shown at the bottom of the figure. Notice that units of production (UPr) as well as cooling degree days (CDD) are included in the equation, meaning that this normalization included weather data and production data.

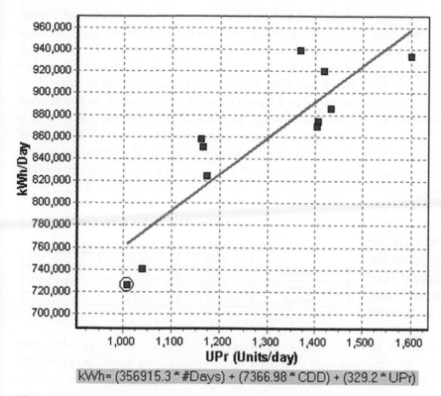

$$kWh = (356915.3 * \#Days) + (7366.98 * CDD) + (329.2 * UPr)$$

Figure 33-10. Daily Usage Normalized to Production and Weather. The Baseline Equation is Shown at the Bottom of the Figure.[12]

MANAGING UNEXPECTED CHANGES IN ENERGY USAGE PATTERNS

The greatest difficulty involved in using utility bills to track savings occurs when there are large, unexpected and unrelated changes to a facility. For example, suppose an energy manager was normalizing usage to weather for a building in order to successfully determine energy savings, and then the building was enlarged by several thousand square feet. The comparison of baseline and actual usage would no longer make any sense, as the baseline number would continue to be determined based upon usage patterns before the building addition, whereas the actual bills would include the addition. Now we would be comparing two different facilities, one with the addition and one without. If there were any energy savings, they might be buried in the additional usage from the new addition. Figure 33-11 presents our hypothetical case in which the new addition came online in August.

Notice that in Figure 33-11 the actual usage has increased while the baseline did not. As a result, savings are hidden by the increase in usage from the building addition. Here, the energy manager would need to make a baseline adjustment (also known as baseline modification) to handle the increase in usage due to the building addition (since the actual bills already include it). The energy manager would make a reasonable estimate of the additional usage and add that onto the baseline. Our earlier equation, then, becomes:

$$\text{Baseline kWh} = (5 \text{ kWh/Day} * \#Days) + (417 \text{ kWh/CDD} * \#CDD) + \text{Adjustment}$$

where the adjustment represents the additional usage due to the building addition. Figure 33-12 presents an example with the addition of the baseline adjustment.

Baseline adjustments are the most troublesome part

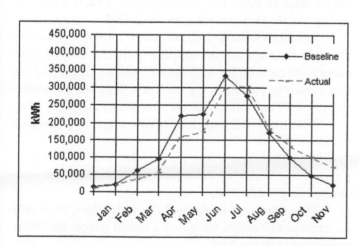

Figure 33-11. Example of increase in energy usage due to increase in square footage starting in August.

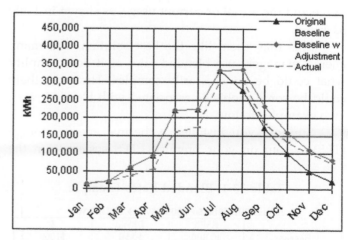

Figure 33-12. Baseline now Adjusted to Account for Increase in Usage Due to Building Addition

of using utility bills to analyze building usage. Buildings continue to change their usage patterns regardless of the energy managers' efforts. To maintain usefulness, baseline adjustments must be added to the analysis.

APPLYING COSTS TO THE SAVINGS EQUATION

Energy managers often need to present their savings numbers to management in a form that managers can comprehend, which means showing savings in cost, rather than energy or demand units. Transforming energy savings into cost savings can be done quite simply, and there are several methods by which this can be done. As in most things, the simplest methods yield the most inaccurate results. The methods investigated here are blended rates and modeled rates. There are some variations on these themes, but they will not be covered here.

In many areas, utility rates may be difficult to understand and model. Once the energy manager understands the rate, he might have to explain it to management, which can be even more difficult. Simplicity is always worth striving for, as many energy managers don't have the time to learn their rates and model them explicitly.

Blended rates (also called average costs) are the simplest way to apply costs to energy units. Suppose for a billing period that baseline usage was 10,000 kWh and the current usage was 8,000 kWh, and current total cost was $800. It doesn't matter how complex the rate is, to apply blended rates, we just consider total cost. The simplest application of blended rates would be to determine the average $/kWh of the *current bill*. In this case, we have $800/8000 kWh = $0.10/kWh. So, the blended rate ($0.10/kWh) would be applied to both the baseline usage and the actual or current usage, as shown in Table 33-1.

This may seem like the best solution, and many energy managers use blended rates as it does simplify what could be unnecessarily complex. However, there could be some problems associated with blended rates. Two examples follow.

Suppose you installed a thermal energy storage (TES) system on your premises. TES systems run the

Table 33-1. Blended Rates Example

Actual Bill: $800 for 8000 kWh → $0.10 /kWh

	KWh	$/kWh	$
Baseline	10,000	$0.10	$1000
Actual	8,000	$0.10	$800
Savings	2,000	$0.10	$200

chillers at night when electricity is inexpensive and stores the cooling energy as either ice or chilled water in large storage containers. Then during the day when electricity is expensive, the chillers either don't run at all, or run much less than they normally would. This strategy saves money, but it doesn't usually save energy. In fact it often uses more energy, as some of that cooling energy that is stored in the storage container is lost through the walls of the container, and the extra pump that runs the system consumes energy. If you applied a blended rate to the TES system you might see the numbers in Table 33-2.

Table 33-2. Where Blended Rates Can Go Wrong

Actual Bill: $10,000 for 1000 kWh → $0.10 /kWh

	KWh	$/kWh	Cost
Baseline	9,500	$0.10	$950
Actual	10,000	$0.10	$1000
Savings	-500	$0.10	-$50

If you modeled the rates, you would see that even though you used more energy, you saved on electricity costs. On the other hand, if you used blended rates, you might see a net increase in energy costs. Blended rates would deliver a dramatically incorrect representation of cost savings.

Most energy managers don't employ thermal energy storage, but they may shift demand to the evening. Suppose a facility is on a time-of-use rate and there is a small net increase in energy usag, combined with a significant shift in energy usage to off peak (less expensive) periods. What happens then? Since less energy is consumed during the more expensive on peak period and more is consumed during the less expensive off peak period, the total costs might decline (in real life). But if the usage increases, your blended rate strategy will show a net increase in costs (in your analysis), which is exactly wrong. Again you can refer to Table 33-2.

Another example demonstrates a weakness in the blended rate approach. Suppose you installed a new energy efficient boiler and boiler controls in a building that is mostly vacant in the summer. Suppose the utility charges a $25 monthly charge plus $1.30/therm.

A January bill, with 100 therms usage, is presented in Table 33-3.

If our baseline usage for January was 120 therms, then savings would be calculated using the blended rate, as shown in Table 33-4.

That seems to work well. Now try July, in which the current bill might have had 1 therm usage, the bill is presented in Table 33-5.

And suppose baseline usage in July was 4 therms, then savings would be calculated as in Table 33-6.

Table 33-3. A Hypothetical Winter Gas Heating Bill

Charge	Usage	Rate	Cost
Monthly Charge	N/A	$25	$25
Usage Charge	100 therms	$1.30/therm	$130
Total Bill			$155
Blended Rate $/therm = $155 / 100 therms = $1.55/therm			

Table 33-4. Savings Calculations Using a Blended Gas Rate

	Therms	$/therm	Cost
Baseline	120	$1.55	$186
Actual	100	$1.55	$155
Savings	20	$1.55	$31

Table 33-5. Problematic Hypothetical Summer Gas Heating Bill

Charge	Usage	Rate	Cost
Monthly Charge	1	$25	$25
Usage Charge	1	$1.30/therm	$1.30
Total Bill			$26.30
Blended Rate $/therm = $26.30 / 1 therms = $26.30/therm			

Table 33-6. Hypothetical Gas Heating Savings Problems

	Therms	$/therm	Cost
Baseline	4	$26.30	$105.2
Actual	1	$26.30	$26.30
Savings	3	$26.30	$78.90

The blended rate calculation told us that the customer saved $78.90, whereas the actual rate calculation would have told us that the customer saved 3 therms * $1.30/therm = $3.90. This problem is not unusual. Often, this type of overstatement of savings occurs without anyone noticing. Blended rates can simplify the calculations and on the surface may return seemingly correct savings numbers. However, upon further analysis, it can usually be found that using blended rates introduces inaccuracies that can, at times, prove embarrassing. The whole point of weather normalization was to reduce the error (due to weather and other factors) in the savings calculations. What is the point of going through the weather normalization procedure if you are only going to reintroduce a potentially even greater error when you apply costs to the savings equation?

If you want to get more accurate cost savings numbers then you would elect to model the rates, which unfortunately means that you will have to understand them.

This would involve retrieving the rate tariff from the utility (usually, they are on the utility's website), and then entering all the different charges into your software

or spreadsheet. There are a few difficulties associated with this approach:

1. Many rates are very difficult to understand
2. Some tariff sheets do not explain all the charges associated with a rate.
3. Some software packages have limitations and can model most but not all of the different charges, or even worse, some packages don't model rates at all.
4. Rates change often, which means you will have to continually keep updating the rates. The good news on this front is that once the rate is modeled, the changes are usually very minor.

As mentioned before, if you are modeling rates, then usually the same rate is applied to both baseline and actual usage and demand. There are exceptions of course. If you changed your facility's rate or changed utility providers, then you should apply your old rate to the baseline, and your new rate to your actual usage. To understand which rate should be used for the baseline, answer the same question: "How much would we have spent if I had not run the energy management program?" The answer is, you would still be on the old rate, therefore, baseline gets the old rate, and actual gets the new rate.

Regardless of how you apply costs to your savings equation, good utility bill tracking software can handle all of these situations.

WEATHER NORMALIZATION IN EXCEL VS. SPECIALIZED UTILITY BILL TRACKING SOFTWARE

Weather normalization can be done in Excel; however, it can be laborious and oftentimes may not be as rigorous as when done using specialized software. Excel will give regressions, fit line equations and statistical indicators which show how well your usage is represented by the fit line.

However, it is difficult to find the best balance point in Excel, as you can in specialized software.[13] If you use Excel, the steps we outlined in this paper will have to be done manually, whereas with canned software, most of it is done for you automatically. In addition, in Excel, if you want to achieve a good fit to your data, you may have to iterate these manual steps for different balance points. The most tedious process in Excel is matching up the daily weather to the billing periods. Try it and you will see. Assuming the weather and bill data are already present, it should take less than two minutes in canned software to perform weather normalization, versus at least 30 minutes in Excel.

AVAILABLE WEATHER NORMALIZATION DESKTOP SOFTWARE

All of the major desktop utility bill tracking software packages will now normalize for weather data. All of them will correct for your own variables as well; however, only some of them will normalize for weather in addition to your own variables. The major desktop programs are *Energy CAP®*, *Metrix™* and *Utility Manager™ Pro*. You can find information on all of them online.

AVAILABLE WEATHER NORMALIZATION IN WEB SOFTWARE

At the time of this writing, only one of the above desktop programs is also offered on a web platform, though a web front end is available from some of the other providers which allow users to enter bill data, perform diagnostic tests and make reports online.

WEATHER NORMALIZATION IN INTERVAL DATA WEB SOFTWARE

There are some interval data programs that perform weather normalization as well, but for these packages, weather normalization is done primarily for forecasting applications, not for verifying energy savings. The method is more complex as the data are in finer increments. Weather forecasts are downloaded and then projected usage is then calculated. At least one of the programs uses weather normalization, or any of a handful of other techniques to forecast energy usage. Energy managers can use these forecasts to adjust their energy consuming activities to prevent high peak demands.

CONCLUSION

Weather varies from year to year. As a result, it becomes difficult to know whether the change in your utility bills is due to fluctuations in weather, your energy management program, or both. If you wish to use utility bills to determine energy savings from your energy management efforts with any degree of accuracy, it is vital that you remove the variability of weather from your energy savings equation. This is done using the weather normalization techniques described in this chapter. You may adjust your usage for other variables as well, such as occupancy or production. You may have to make baseline adjustments to further "correct" the energy savings equation for unexpected changes in energy usage patterns such as new additions. Finally, the method in which you apply costs to your energy savings calculations is very important. Blended rates, although simple, can result in inaccurate cost savings numbers, while more difficult modeling rates, are always right.

Footnotes

1. What are the alternatives? The most common might involve determining savings for each of the energy conservation activities using a spreadsheet or perhaps a building model. Both of these alternative strategies could require much additional work, as the energy manager likely has employed several strategies over his tenure. One other drawback of spreadsheets is that energy conservation strategies may interact with each other, so that total savings may not be the sum of the different strategies. Finally, spreadsheets are often projections of energy savings, not measurements.
2. Cooling degree days are defined in detail later in the chapter; however, a simplified meaning is given here. Warmer weather will result in more cooling degree days; whereas a colder

day may have no cooling degree days. Double the amount of cooling degrees should result in roughly double the cooling requirements for a building. Cooling degree days are calculated individually for each day. Cooling degree days over a month or billing period are merely a summation of the cooling degree days of the individual days. The inverse is true for heating degree days.

3. Some energy professionals select 2 years of bills rather than one. Good reasons can be argued for either case. Do not choose periods of time that are not in intervals of 12 months (for example, 15 months, or 8 months could lead to inaccuracy).

4. Please do not confuse base year with baseline. Base year is a time period, from which bills were used to determine the building's energy usage patterns with respect to weather data, whereas baseline, as will be described later, represents how much energy we would have used this month, based upon base year energy usage patterns and current month conditions (i.e. weather and number of days in the bill).

5. Canned software does this automatically for you, while in spreadsheets, this step can be tedious.

6. The statistical calculations behind the R2 value and a treatment of three other useful indicators, T-Statistic, Mean Bias Error, and CVRMSE are not treated in this chapter. For more information on these statistical concepts, consult any college statistics textbook. (For energy managers, a combination of R2 values and T-Statistics is usually enough.)

7. Baseline equation = fit line equation +/- baseline modifications. We cover baseline modifications later in this chapter.

8. The generic form of the equation is:
Baseline kWh = (constant * #days) + (coefficient * #CDD)
where the constant and coefficient (in our example)

are 5 and 417.

9. Summing or averaging high or low temperatures for a period of time is not very useful. (Remember the Detroit example mentioned earlier.) However, you can sum degree days, and the result remains useful, as it is proportional to the heating or cooling requirements of a building.

10. If you think about it, you don't have to treat this at the building level, but rather can view it at a meter level. (To simplify the presentation, we are speaking in terms of a building, as it is less abstract.) Some buildings have many meters, some of which may be associated with different central plants. In such a building, it is likely that the disparate central plants would have different balance points, as conditions associated with the different parts of the building may be different.

11. Some analysts had separate tables of degree days based upon a range of balance points (65, 60, 55, etc.), and analyzed their data painstakingly with several balance points until they found the best balance point temperature for their building. On the other hand, other analysts believe that all degree days are calculated assuming the standard balance point of 65 °F.

12. A better presentation of the data would be in 3 dimensions (with Units Produced in X axis, Weather in Y axis, and kWh in Z axis), however due to limitations of a printed page, a 2 dimensional image is shown here, which is just one slice (or plane) in the actual relationship. This explains why the line may not look so close to the bills.

13. It is not necessary to find the best balance point, and you might choose instead to use published tables of degree days, which are often based on a 65-degree balance point. Using these standard degree days will in most cases lead to decreased accuracy and poorer fits. Using the base 65-degree balance point, many meters will not have an acceptable fit (R2 > 75%) at all.

Chapter 34

The Energy and Cost Savings Calculation System [ECSCS]

Jim McNally PE
Siemens Building Technologies, Inc.

INTRODUCTION

AN ENERGY AND COST SAVINGS calculation system (ECSCS) based on highly accurate MVNL load forecasting technology has several practical uses. These include: calculating the energy and cost savings of conservation measures, forecasting future energy loads and peaks, calculating energy alerts of *future* conditions (i.e. a new peak in power use), and calculating on-going energy use comparisons. This chapter describes ECSCS applications, gives example uses of the technology, provides a brief overview of the ECSC system, and identifies system problems commonly encountered.

Note: The ECSC system embodies the MVNL load forecasting technology described in the *IT for Energy Managers*, volume II, chapter: "Load Forecasting."

ECSCS APPLICATIONS

Using the ECSCS basic capabilities of load forecasting, flexible utility rates, data quality control, data repair, and data source flexibility, a number of applications are possible. What began as an effort to automate performance assurance computations (M&V) has developed into a series of applications to identify conservation opportunities, help operations become more efficient, and quantify the resulting savings.

The basic applications of the ECSC system may be divided into two applications dealing with the *past* and four applications dealing with the *present* and *future*. They are as follows:

Past Applications

A. Quantify Energy Savings of conservation measures.
B. Calculate Cost Savings of conservation measures.

Present and Future Applications

C. Statistically Identify Additional Energy Cost Savings Potential.
D. Forecast Future Loads. [Rolling 5-day look-ahead, updated daily.]
E. Smart Alerts. [Notify operators of future increases in billing demand.]
F. Energy Cost Budgets.

Let's use the Siemens Building Technologies, Inc., Buffalo Grove, IL, campus to illustrate these applications. The plan of the campus is shown in Figure 34-1.

Actual Energy Savings of Conservation Measures
Reporting Periods

Many times data are tied to utility bills. The user has the option of defining utility billing periods as the reporting periods or of using calendar months. In our example, we are using calendar months.

Comparative Energy Report

The comparative energy report shows gross energy consumption differences between the reporting periods

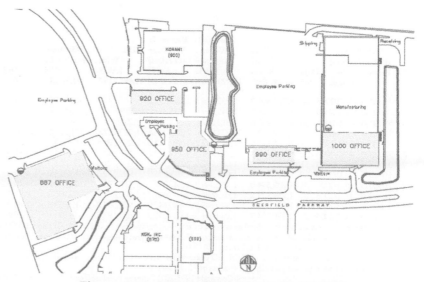

Figure 34-1. SBT Buffalo Grove, IL Campus

and the reference periods after having been adjusted by the MVNL process. The comparative energy report is shown in Figure 34-2. Each reporting period has three bars. The bar on the left represents the unadjusted energy use of the reference period. The bar in the center represents the reference period energy use, adjusted by the MVNL process to the weather and operations conditions of the reporting period. The bar on the right shows the actual energy consumption of the reporting period. Because the MVNL forecast process is used, it is not necessary to have a complete year of reference data; as little as 6 to 7 months may be sufficient.

The uses of the report are many. It shows:

* Annual overall energy use,
* Energy savings,
* Percent savings,
* Heating and cooling degree-days,
* Energy consumption per heating and cooling degree-day.

In monitoring a performance contract, the comparative energy report can serve as a first view of the savings attained. It provides a high level overview of energy consumption differences from one year to another. It breaks out performance by month or billing period with an annual summary. Because a report such as this may be automatically calibrated, built, and distributed. It is the first responder so-to-speak in assessing savings. In the example above, the annual electricity consumption savings was 837,823 KWH—a 6.5% reduction (the reference or baseline period was from 12/17/01 to 1/29/03).

Cost Savings of Conservation Measures

The next logical question arising is: "How much money was saved due to the conservation measure?" With the ECSC system, once the utility rate is determined, the answer is one step away. Identifying the exact utility rate, however, can sometimes be time-consuming. Hence, the comparative energy report gives a good idea of the savings by reporting on the energy reduction alone in the comparative energy report above.

The comparative energy report conceivably could be run and distributed weekly or monthly to keep tight

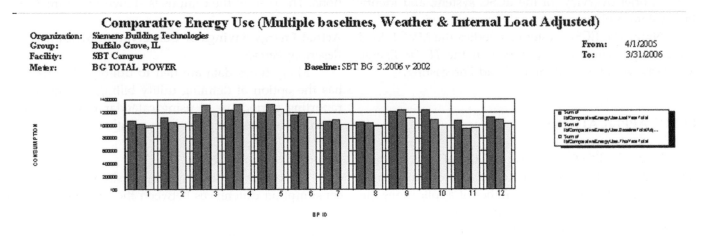

Comparative Energy Use (Multiple baselines, Weather & Internal Load Adjusted)

Organization: Siemens Building Technologies
Group: Buffalo Grove, IL
Facility: SBT Campus
Meter: BG TOTAL POWER   Baseline: SBT BG 3.2006 v 2002

From: 4/1/2005
To: 3/31/2006

BP ID	Date Ranges [Billing Periods] From	To	Days	Last Year Total KWH	Baseline1: Weather Adjusted KWH	Baseline2: Weather+ Int'l Adjust KWH	This Year Total KWH	Difference ThisYr-BL2 KWH	% Diff	This Year HDD	This Year CDD	Baseline KWH /HDD	Baseline KWH /CDD	This Year KWH /HDD	This Year KWH /CDD
		Totals:	365	13,747,282.	13,723,554.4	13,723,554.4	12,885,730.7	-837,823.8	-6.50%	6,094	1,050				
1	4/1/2005	4/30/2005	30	1,076,684.2	1,022,326.3	1,022,326.3	967,493.0	-54,833.3	-5.70%	420	3	2,434	340,775	2,303	322,497
2	5/1/2005	5/31/2005	31	1,125,261.2	1,045,289.0	1,045,289.0	1,019,633.2	-25,655.8	-2.50%	303	16	3,449	65,330	3,365	63,727
3	6/1/2005	6/30/2005	30	1,174,938.6	1,313,232.0	1,313,232.0	1,208,709.7	-104,522.4	-8.60%	14	262	93,802	5,012	86,336	4,613
4	7/1/2005	7/31/2005	31	1,239,187.4	1,332,008.4	1,332,008.4	1,200,429.4	-131,579.0	-11.00%	0	311	0	4,282	0	3,859
5	8/1/2005	8/31/2005	31	1,196,137.9	1,322,912.3	1,322,912.3	1,253,094.4	-69,817.9	-5.60%	1	271	1,322,912	4,881	1,253,094	4,623
6	9/1/2005	9/30/2005	30	1,163,463.2	1,199,999.5	1,199,999.5	1,122,568.0	-77,431.6	-6.90%	38	154	31,578	7,792	29,541	7,289
7	10/1/2005	10/31/2005	31	1,065,190.0	1,086,742.9	1,086,742.9	1,006,879.6	-79,863.3	-7.90%	355	33	3,061	32,931	2,836	30,511
8	11/1/2005	11/30/2005	30	1,054,049.7	1,041,276.3	1,041,276.3	992,558.3	-48,718.0	-4.90%	713	0	1,460	0	1,392	0
9	12/1/2005	12/31/2005	31	1,209,374.0	1,242,349.1	1,242,349.1	1,118,696.5	-123,652.6	-11.10%	1,292	0	961	0	865	0
10	1/1/2006	1/31/2006	31	1,239,179.2	1,083,684.4	1,083,684.4	1,002,879.2	-80,805.2	-8.10%	998	0	1,085	0	1,004	0
11	2/1/2006	2/28/2006	28	1,074,530.8	949,642.9	949,642.9	963,596.7	13,953.8	1.40%	1,071	0	886	0	899	0
12	3/1/2006	3/31/2006	31	1,129,285.7	1,084,091.2	1,084,091.2	1,029,192.7	-54,898.6	-5.30%	889	0	1,219	0	1,157	0

Notes: 1. Baseline1 is weather-adjusted simulation based of actual energy use data from 12/17/2001 through 1/29/2003. It is adjusted to "This Year's" weather.
2. Baseline2 is weather-adjusted simulation further adjusted by internal changes such as floor area additions, lighting changes, etc.
3. "This Year" is the period from 4/1/2005 to 3/31/2006
4. "HDD" = Heating degree-days   "CDD" = Cooling degree-days.
5. Values in Blue represent partially filled and empty Date Ranges

© Copyright 1998-2006 Siemens Building Technologies, Inc. Utility Information Services

Figure 34-2. Comparative Energy Report

watch on the progress of the conservation measure, while the M&V annual energy and cost report—because of the additional step required to collect and verify utility rate information—could be distributed one-to-four times per year.

M&V Annual Energy/Cost Savings Report

The M&V annual meter report [shown in Figures 34-3a and 34-3b] answers the question: "How much money is being saved?" This is a two-page report. The graphic shows the cost savings achieved during the reporting period. [Savings is positive.] The remainder of the first page shows actual energy use, monthly billing demand values, and costs for each rate component. The second page shows the energy use and cost of the baseline month reference periods (2002 in this case) extended to the reporting period (2005). In the example below, the electric rates are the same, but they don't have to be identical.

Notice in this example, that the annual cost savings was $36,378—a 4.3% cost reduction.

Electric Rate Used

The electric rate used allows for time-of-use consumption blocks—varying the charge of energy throughout the day. It is also possible to vary the charge of energy seasonally within the rate. Accordingly, peak billing demand values may be sensed only during demand charge windows. Weekends and rate holidays may charged for energy consumed at a lower rate and also may not be susceptible to demand charges. The demand time-of-use windows do not have to be the same time frame as the on-peak energy consumption time-of-use window.

Taxes

Two types of taxes may be assessed. The first tax is assessed on the entire pre-tax amount. The second tax type is assessed on the pre-tax amount plus the amount of the first tax.

Fees

Flat fees per billing period also may be entered.

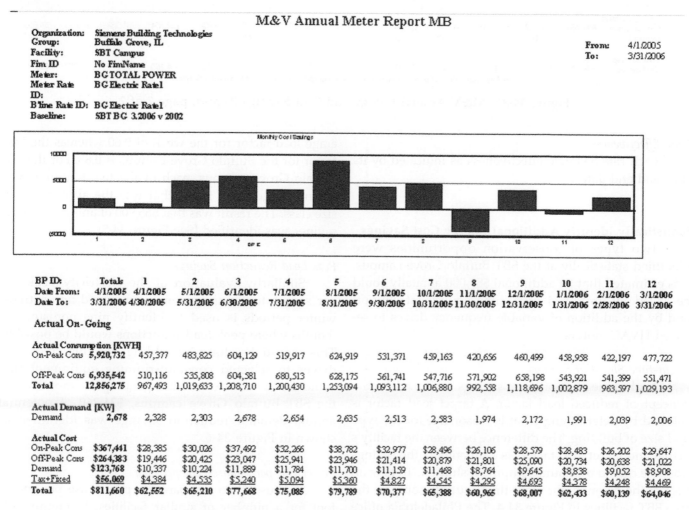

Figure 34-3a. M&V Annual Energy and Cost Savings Report, page 1 of 2.

M&V Annual Meter Report, continued

Organization: Siemens Building Technologies
Group: Buffalo Grove, IL
Facility: SBT Campus
Firm ID No FirmName
Meter: BG TOT PWR
Meter Rate ID: BG Electric Rate1
B'line Rate ID: BG Electric Rate1
Baseline: SBT BG 3.2006 v 2002

BP ID:	Totals	1	2	3	4	5	6	7	8	9	10	11	12
Date From:	4/1/2005	4/1/2005	5/1/2005	6/1/2005	7/1/2005	8/1/2005	9/1/2005	10/1/2005	11/1/2005	12/1/2005	1/1/2006	2/1/2006	3/1/2006
Date To:	3/31/2006	4/30/2005	5/31/2005	6/30/2005	7/31/2005	8/31/2005	9/30/2005	10/31/2005	11/30/2005	12/31/2005	1/31/2006	2/28/2006	3/31/2006

Baseline

AdjustFactor		1	1	1	1	1	1	1	1	1	1	1	1

Baseline Consumption [KWH]

On-Peak Cons	6,143,536	464,217	470,738	628,735	544,342	642,333	574,777	485,073	437,816	502,572	475,948	428,914	488,071
Off-Peak Cons	7,607,592	558,005	574,615	684,753	787,531	681,818	650,861	601,361	603,566	739,778	608,319	521,007	595,976
Total	13,751,129	1,022,222	1,045,353	1,313,489	1,331,873	1,324,151	1,225,638	1,086,434	1,041,383	1,242,350	1,084,267	949,921	1,084,048

Baseline Demand [KW]

Demand	3,037	2,179	2,306	2,726	2,686	2,708	3,037	2,647	2,468	1,815	1,980	1,956	2,026

Baseline Cost

On-Peak Cons	$381,268	$28,809	$29,214	$39,019	$33,782	$39,863	$35,671	$30,104	$27,171	$31,190	$29,537	$26,618	$30,290
Off-Peak Cons	$290,001	$21,271	$21,904	$26,103	$30,021	$25,991	$24,811	$22,924	$23,008	$28,200	$23,189	$19,861	$22,719
Demand	$118,641	$9,676	$10,239	$12,104	$11,928	$12,022	$13,485	$11,754	$10,980	$0	$8,792	$8,683	$8,998
Tax+Fixed	$58,128	$4,480	$4,576	$5,528	$5,438	$5,567	$5,332	$4,781	$4,563	$4,458	$4,586	$4,204	$4,615
Total	$848,038	$64,236	$65,933	$82,755	$81,168	$83,443	$79,299	$69,563	$65,702	$63,848	$66,104	$59,366	$66,621

Savings

Energy Savings

Consumption	894,854	54,729	25,720	104,779	131,443	71,057	132,526	79,555	48,824	123,654	81,388	-13,676	54,855
Demand		-149	3	48	32	72	524	64	495	-358	-10	-83	20

Cost Savings

Net Savings	$36,378	$1,684	$722	$5,086	$6,084	$3,654	$8,922	$4,175	$4,737	$-4,160	$3,671	$-773	$2,574

Figure 34-3b. M&V Annual Energy and Cost Savings Report, page 2 of 2.

Date Effectiveness

Each rate has date effectiveness as indicated by its start and end dates.

Statistically Identify Additional Energy Cost Savings

Two types of conservation opportunities were identified statistically at the SBT Buffalo Grove campus. It is estimated that an additional $87,000 annually could be saved by improved night-time equipment shut-off and by the addition of variable frequency drives to selected HVAC motors.

Night-time Shut-Off Savings

Night-time shut-off type savings are based on the concept of reduced load factor. A target load factor is selected by determining what is reasonable for the type and size of building. The difference between the facility's actual and target load factors is a function of the savings possible due to off-hours equipment shut-off. To give an idea of the potential for night-time shut-off, compare the two SBT facilities in Figure 34-4. The Philadelphia office serves as a reference of what is attainable having an av-

erage load factor for the week of 0.60 whereas the load factor for the Buffalo Grove campus is 0.83. For the SBT Buffalo Grove campus, each of the five buildings was given a target load factor that was the average for its size class. The result was that $65,900 of annual potential savings was identified [See Figure 34-5].

Peak Load Reduction Savings

Similarly, evaluation of the electricity peak use per square foot during non-summer and non-extreme winter periods is used to identify nine non-summer months where peak load reductions would be possible. This would be done by adding variable frequency drives to selected constant-running equipment that could be turned down during non-peak months. For the SBT Buffalo Grove campus, $21,260 of potential annual demand reduction savings was identified as shown in Figure 34-6.

Power Use per Square-Foot per Day

It is useful to compare the power use per square foot for a number of similar facilities. In Figure 34-7, several SBT facilities are compared (watthours/sq.ft./

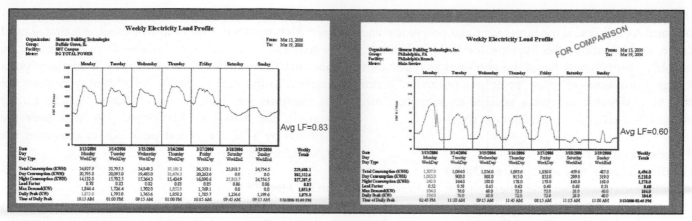

Figure 34-4. Compare Load Factors of Buffalo Grove Campus with that of the Philadelphia Office

		****** Load Factor Savings ******			
	% over avg. for the Class		$/KWH		
	0.00%		$0.038		
Facility	Target LF	Annualized Savings KWH	Annualized Savings $$	Project Areas	Svgs. $$/ Sq.Ft.
TOTALS: ALL	0.489	16,056,276	$612,065		$0.078
SBT BG	0.473	1,728,401	$65,887		0.230
SBT Branches	0.433	274,415	$10,461		0.021
25. 887 Deerfield Pwy.	0.54	126,721	$4,831	Better Nite shut-off.	$0.05
26. 950 Deerfield Pwy.	0.43	284,837	$10,858	Better Nite shut-off.	$0.27
27. 920 Deerfield Pwy.	0.43	230,776	$8,797	Better Nite shut-off.	$0.32
28. 1000 Deerfield Pwy.	0.54	778,758	$29,686	Better Nite shut-off.	$0.13
29. 990 Deerfield Pwy.	0.43	307,308	$11,715	Better Nite shut-off.	$0.39

Figure 34-5. Improved Night-time Shut-off Savings at SBT BG Campus

		****** WSF-Based Savings ******		
	% over avg. for the class		$/KW/Mon	
	0.0%		$4.440	
Facility	Target WSF	KW Savings	Annl Svgs. $$	Project Areas
TOTALS: ALL		4,644	$185,577	
SBT BG		532	$21,261	
SBT Branches		197	$7,873	
25. 887 Deerfield Pwy.	4.13	30	$1,208	Add HVAC Motor VFDs
26. 950 Deerfield Pwy.	3.63	19	$743	Add HVAC Motor VFDs
27. 920 Deerfield Pwy.	3.63	40	$1,594	Add HVAC Motor VFDs
28. 1000 Deerfield Pwy.	4.13	307	$12,257	Add HVAC Motor VFDs
29. 990 Deerfield Pwy.	3.63	137	$5,460	Add HVAC Motor VFDs

Figure 34-6. Peak Load Reduction Savings at SBT BG Campus

day). Interestingly, the Buffalo Grove facilities are at the bottom! As a check, if the target unitized power use were taken at 50% above the average for the population, it would approximately equal the two savings types above ("night-time shut-off" plus "peak load reduction"). Since this view doesn't suggest a specific conservation measure, it is used as a reasonableness check.

Forecast Future Loads
(Rolling 5 Days Ahead, Updated Daily)

Five-day weather and energy peak and consumption forecasts are included in daily emails sent to customers. Each report set emailed to a customer has a summary in the email and a number of reports attached to it. The summary is for the day just completed and the next five days. Weather conditions and temperatures are given as are the expected daily consumption [all energy forms] and the electric daily peak. See Figure 34-8. Forecast and expected loads are calculated by MVNL load forecasting technology.

Use actual vs. expected value to review "yesterday." Two numbers (the ratio of expected:actual should be near 1.0. When "actual" deviates from "expected" by more than 20% it should be investigated. Energy reports attached to the email usually help identify reasons for the variances of "expected" vs. "actual" loads. In our example, the actual facility power use was over the expected by 17%; the actual peak power was over the expected by 18%.

Smart Alerts™

Smart Alerts™ are energy alarms which notify building operators of impending increases in billing demand. The ECSC system generates three types of alerts: future, present, and past.

Future

The Future Smart Alert™ will advise that tomorrow the facility will hit a new demand peak unless evasive action is taken.

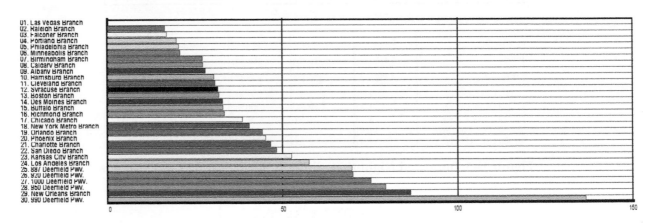

Figure 34-7. Unit Power Consumption Comparison of Selected SBT Facilities

```
*************************************************************************************************
WEATHER AND ENERGY REPORT [1000 Deerfield Pwy.] [4/30/2006] [Buffalo Grove,IL (US)]
*************************************************************************************************
              High  Low  Condition               KW     KWH     Gas     Steam Ch.Water
Actual         57   50   Rain Light Rain         488    10149
Expected                                         412     8675
Percentage                                       118%    117%

FIVE-DAY WEATHER AND ENERGY REPORT [1000 Deerfield Pwy.]  [4/30/2006] [Buffalo Grove,IL (US)]
*************************************************************************************************
Date        Day High Low  Condition              KW     KWH     Gas     Steam Ch.Water
05/01/2006  Mon  59   54  Overcast               667    12477
05/02/2006  Tue  68   53  Few Showers            792    12461
05/03/2006  Wed  75   50  PM T-Storms            740    12515
05/04/2006  Thu  67   45  Partly Cloudy          736    12436
05/05/2006  Fri  55   38  Cloudy                 691    12378


*************************************************************************************************
```

Figure 34-8. Example of Five-Day Forecast.

Present

The Present Smart Alert™ will advise that today the facility will hit a new demand peak unless evasive action is taken.

Past

The Past Smart Alert™ will advise that yesterday, the facility did experience a new demand peak. It will also alert on consumption that is above the weather-adjusted expectation.

Energy alerts look similar to the scheduled emails with attached reports discussed above. They identify the type of alert. A sample of the energy alert is shown in Figure 34-9.

Energy Budgets

Accurate energy budgeting and tracking can be elusive at times. The energy budget application of EC-SCS incorporates actual energy use, forecast energy use, actual rates, and future rates (which may be a function of commodity prices). Energy use impact [a.k.a. "production"] units of the facility being budgeted may also be included. The ECSC system is able to update the energy budget daily. One form of the energy budget report is shown in Figure 34-10.

ECSCS Overview

The ECSC system was developed to support verification of cost savings of performance contracts. The system elements are shown in Figure 34-11. The ECSC system has a few unique aspects as shown below the figure.

- **Improved Accuracy.** In extensive comparisons, the ECSC system has consistently been 25 to 35 times more accurate in calculating energy and cost savings than linear regression techniques based on utility bills.

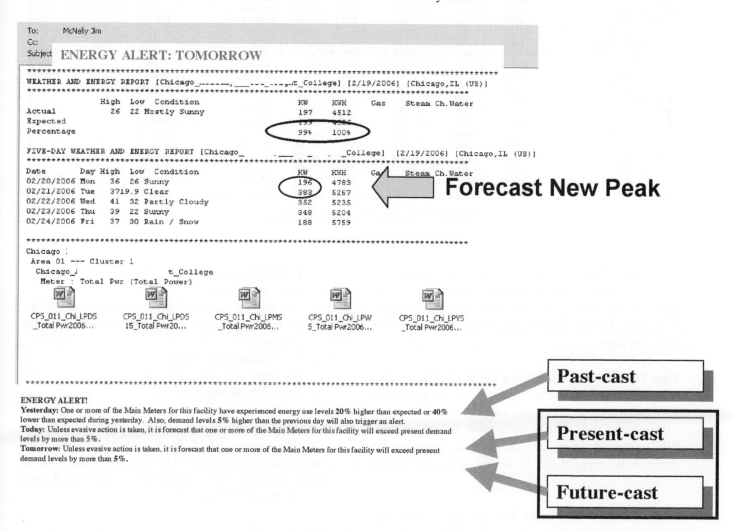

Figure 34-9. Smart Alert™ Example

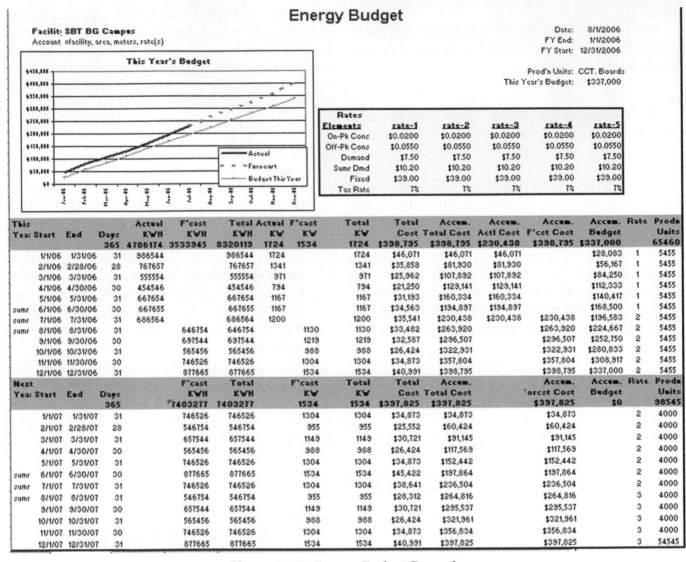

Figure 34-10. Energy Budget Example

- **Complex Utility Rates**. The system was designed from the ground up to accommodate time-of-use utility rates and multiple daily billing demand windows in the load forecasting aspect of the ECSC system.

- **Rate Dynamics**. Recognizing the flexibility needed in rate definition, the utility rates for the reference and reporting data streams can be different. Also, utility rates can change with each billing period. Billing period dates are user-defined. They can be defined to correspond to actual dates on utility bills.

- **Several Data Feed Types**. Recognizing that energy data may come in many forms, the ECSC system is designed to accept data from:

  — Utility meters,
  — Data recorders,
  — Networked meter systems remotely accessed,
  — MeterMail™ from smart power meters via the internet,
  — Purchased data feeds,
  — Building automation systems.
  — Manually input data.

- **Quality Control and Data Recovery Modules**. Not all data sources yield clean and continuous data. Several techniques have been developed to recover from missing and corrupt data which find their way to the ECSC system.

- **Patent**. A U.S. patent covering ECSC system and MVNL technology is pending.

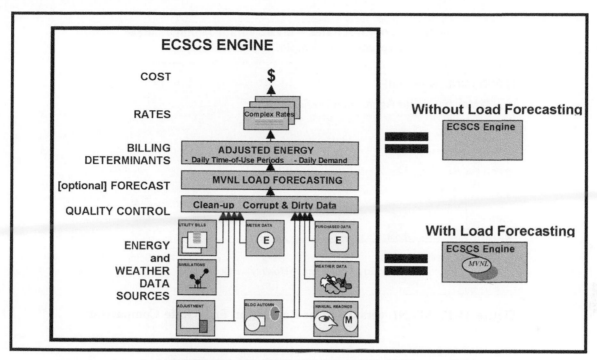

Figure 34-11. ECSCS Functional Overview

MVNL Load Forecasting

The heart of the ECSC system is the MVNL load forecasting technique described in the *IT For Energy Managers*, volume II chapter: "Load Forecasting." Because MVNL load forecasting uses more correlating variables and time-of-use data, its accuracy tests tend to blanket the target data as opposed to drawing a straight line through them. Figure 34-12 illustrates this.

The result is that the MVNL load forecast accuracy surpasses that of linear regression techniques based on utility bills by about 30:1 based on monthly report periods. The graph in Figure 34-13 illustrates this.

Both MVNL and linear regression error rates trend toward zero over long periods of time (such as one year or more). However, since one year is a long time to wait for results, the significantly lower error rates of the MVNL approach usually mean that monthly or quarterly reporting may be done without fear of errors distorting the results.

ECSCS Elements

There are three elements to the ECSC system: 1) Energy data are 2) adjusted and played against a 3) rate to calculate money. This is shown in Figure 34-14.

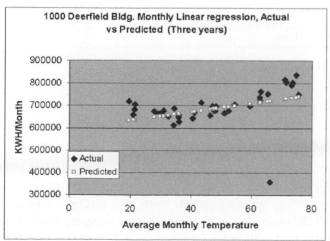

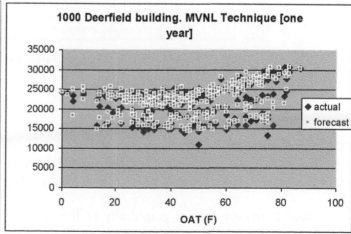

Figure 34-12. Comparison of Linear Regression versus MVNL Load Forecasting

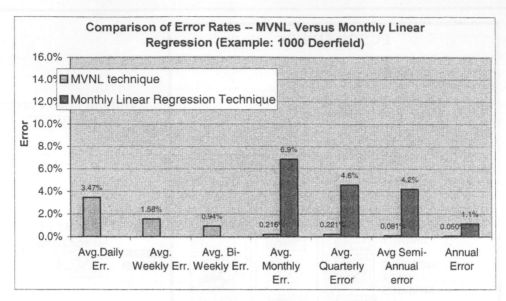

Figure 34-13. MVNL versus Linear Regression Error Rate Comparison

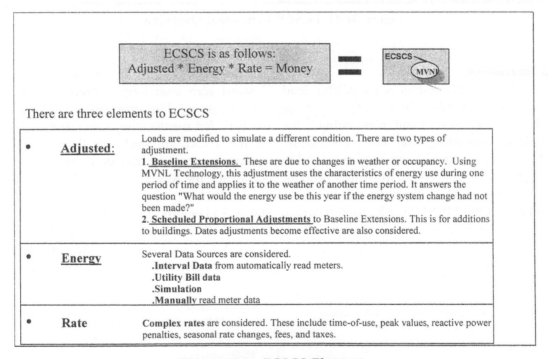

Figure 34-14. ECSCS Elements

ECSCS Building Blocks

The ECSCS building block may be used different ways to obtain different results as shown in Figure 34-15.

These terms are shown graphically in the energy conservation savings determination model in Figure 34-16.

Rates

Rates may change for each billing period (a billing period is roughly one month).

The system forecasts loads over multiple time-of-use (TOU) periods—each period has its separate forecast data stream. Utility rates for baseline and on-going monitoring are independently applied as shown in Figure 34-17.

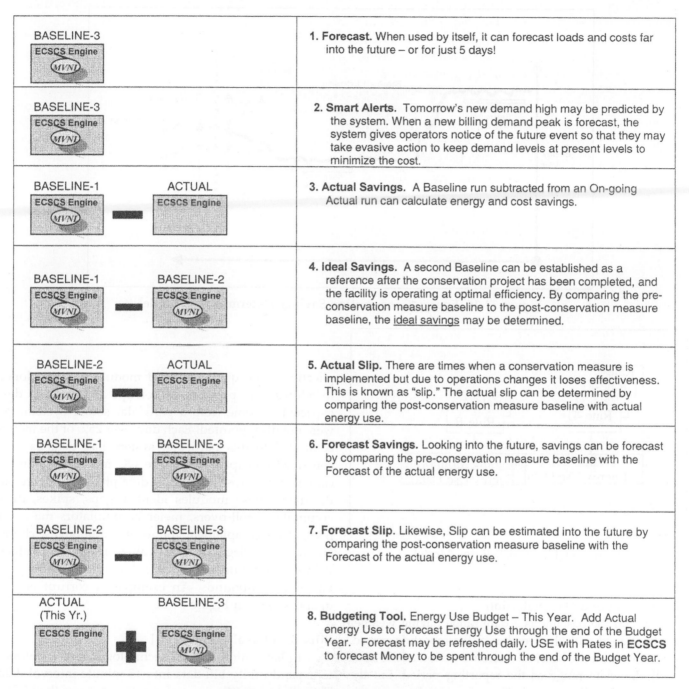

BASELINE-3 ECSCS Engine *MVNI*	**1. Forecast.** When used by itself, it can forecast loads and costs far into the future – or for just 5 days!
BASELINE-3 ECSCS Engine *MVNI*	**2. Smart Alerts.** Tomorrow's new demand high may be predicted by the system. When a new billing demand peak is forecast, the system gives operators notice of the future event so that they may take evasive action to keep demand levels at present levels to minimize the cost.
BASELINE-1 ECSCS Engine *MVNI* — **ACTUAL** ECSCS Engine	**3. Actual Savings.** A Baseline run subtracted from an On-going Actual run can calculate energy and cost savings.
BASELINE-1 ECSCS Engine *MVNI* — **BASELINE-2** ECSCS Engine *MVNI*	**4. Ideal Savings.** A second Baseline can be established as a reference after the conservation project has been completed, and the facility is operating at optimal efficiency. By comparing the pre-conservation measure baseline to the post-conservation measure baseline, the ideal savings may be determined.
BASELINE-2 ECSCS Engine *MVNI* — **ACTUAL** ECSCS Engine	**5. Actual Slip.** There are times when a conservation measure is implemented but due to operations changes it loses effectiveness. This is known as "slip." The actual slip can be determined by comparing the post-conservation measure baseline with actual energy use.
BASELINE-1 ECSCS Engine *MVNI* — **BASELINE-3** ECSCS Engine *MVNI*	**6. Forecast Savings.** Looking into the future, savings can be forecast by comparing the pre-conservation measure baseline with the Forecast of the actual energy use.
BASELINE-2 ECSCS Engine *MVNI* — **BASELINE-3** ECSCS Engine *MVNI*	**7. Forecast Slip.** Likewise, Slip can be estimated into the future by comparing the post-conservation measure baseline with the Forecast of the actual energy use.
ACTUAL (This Yr.) ECSCS Engine + **BASELINE-3** ECSCS Engine *MVNI*	**8. Budgeting Tool.** Energy Use Budget – This Year. Add Actual energy Use to Forecast Energy Use through the end of the Budget Year. Forecast may be refreshed daily. USE with Rates in **ECSCS** to forecast Money to be spent through the end of the Budget Year.

Figure 34-15. ECSCS Building Blocks

The system forecasts billing demand over multiple demand windows per day. Each window has its separate demand forecast data stream.

TOU Groups

To enable accurate forecasting, TOU forecasting periods are fixed for the configuration. TOU periods and demand windows must be the same for "baseline" and "reporting" data streams. This is visualized in Figure 34-18.

Rate Groups

Rate elements may be grouped and given a name. Energy consumed charged to this period is itemized in the M&V annual energy and cost report.

For example, the cost of on-peak and off-peak consumption can be easily broken out—or combined depending on the naming of the rate group for the time-of-use periods. In the example in Figure 34-19, the darker time-of-use rates apply to on-peak time periods, while the lighter time-of-use rates apply to off-peak time periods.

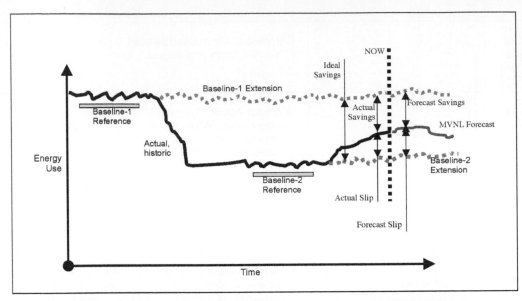

Figure 34-16. Energy Conservation Savings Determination Model

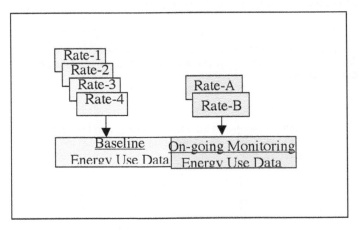

Figure 34-17. ECSCS Rates Model

Reference Period Date Selection

Within the baseline is a date range having weather closely aligned with the weather of the reporting period. Generally, select the baseline reference period as the same month-day date range as the reporting period, but the year of the baseline period. The reference date range is then adjusted so that its high outside air temperature is higher or equal to the reporting period high outside air temperature; and that the low outside air temperature is lower than the reporting period low outside air temperature. This improves accuracy and gives added stability to the MVNL load forecasting. Reference period date selection may be done manually as shown by the in Figure 34-20 or it may be done automatically by the ECSC system.

Quality Control Modules

Not all systems are equal in their ability to collect clean, continuous interval data used in metering ap-

plications. The quality control module was developed to address the problem of corrupt and missing data expected to come from a particular source system. We were not disappointed! Each day, over 25% of the meters connected to this particular system had data gaps or contained corrupted data. The QC modules automatically identify gross and subtle data problems and repair the data. These modules identify gaps, spikes, dips, accumulator roll-overs, meter accumulators manually reset, two's-complement rollovers. Additionally, checks are made for illegal negative values, illegal zero values, daily consumption values out-of-bounds, and daily peak values out-of-bounds. An example of corrupt energy data is shown in Figure 34-21.

Bills-To-Interval-Data [BTID]

Bills-to-interval-data technology was developed to expand utility bill data into 15-minute interval data. This allows a wide range of customers to use the ECSCS applications. Utility bill data automatically finds their way into the ECSC system. To convert utility bill information into interval data, a multi-variant regressive [MVR] approach which adds temperature, dewpoint, a proprietary solar index, and day-type information to utility bill TOU consumption, and peak demand. The result is that weather and day type sensitive consumption is allocated throughout the days of the billing period. The monthly consumption breakout in Figure 34-22 below shows on-peak and off-peak consumption per day. Note that weekend consumption is considered to be "off-peak."

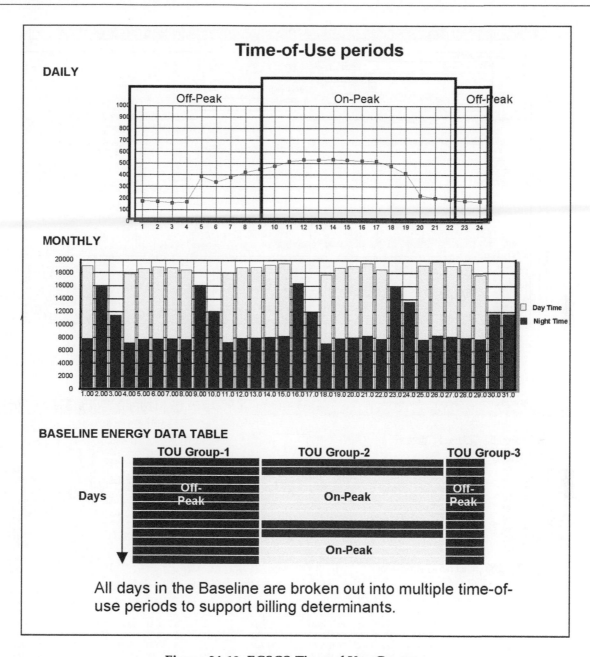

Figure 34-18. ECSCS Time-of-Use Groups

CONCLUSION

ECSCS technology takes metering applications to the next level in managing energy costs. Capabilities such as multiple data feeds, bills-to-interval data capability, advanced quality control features, and MVNL load forecasting technology enable new application concepts with higher reliability and accuracy. Combining energy use, weather, customer production unit, and commodity price feeds facilitate added precision. Savings calculations, load forecasts, smart alerts, and daily updated energy budgets give facility managers new tools to do their job.

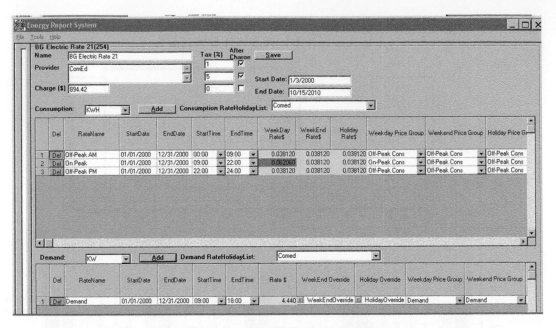

Figure 34-19. ECSCS Utility Rate Form

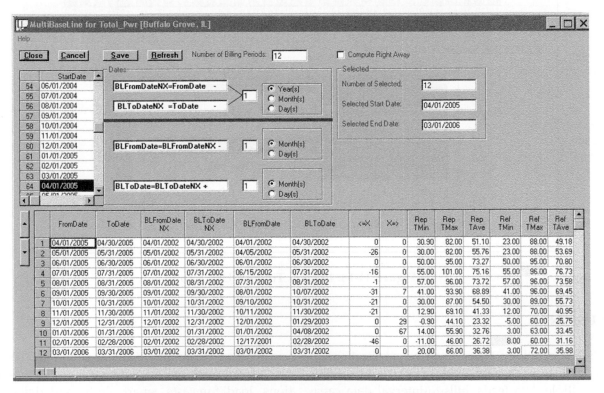

Figure 34-20. Reference Period Date Range Selection

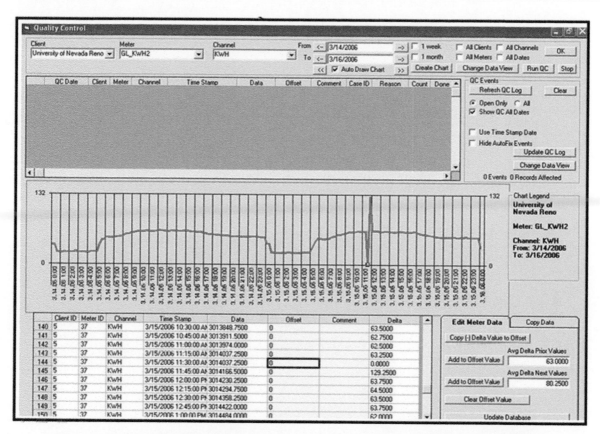

Figure 34-21. Example of Corrupt Data Identified and Repaired by QC Module

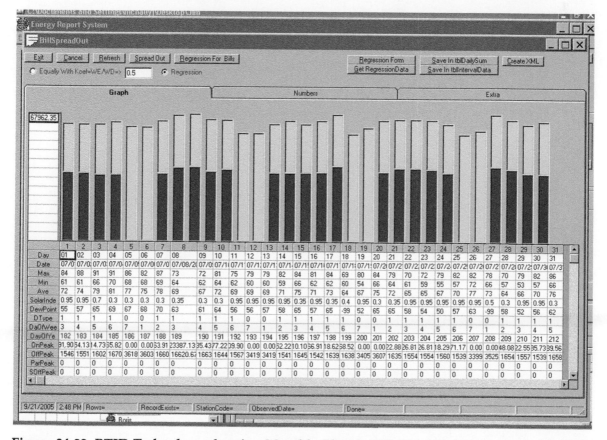

Figure 34-22. BTID Technology showing Monthly Electric Bill "Sculpted" into Daily On-Peak and Off-Peak Loads

Chapter 35

SCADA and DCS Security Vulnerabilities and Counter Measures for Engineers, Technicians, and IT Staff

Ron Brown
Gridlogix, Inc.

ABSTRACT

R EAL-TIME INDUSTRIAL CONTROL and data acquisition systems are typically classified as supervisory control and data acquisition systems (SCADA) or as distributed control systems (DCS). SCADA systems generally operate over a large geographic area such as an electric utility grid, and a DCS system operates in a more discrete setting such as a building or factory. Fundamentally they are the similar when it comes to using TCP/IP and UDP network protocols for controller and device communications. Many organizations allow these systems to communicate over their business networks. Once these systems begin operating on TCP/IP networks they inherit the typical security vulnerabilities associated with all networked systems, and furthermore they add security vulnerabilities that engineers and IT staff might not realize. A compromised SCADA or DCS system can result in financial losses, property and environmental damage, personal injury, or death.

INTRODUCTION

Over the past 30 years, businesses have increasingly added SCADA and DCS systems to their operations to manage every thing from building comfort, energy use, and security to industrial processes, pipelines, and electric grids. Typically these systems have been totally isolated from business networks because they utilized serial communications protocols. Overtime these systems have been adapted to use TCP/IP and UDP network protocols, and as a result SCADA and DCS systems can be managed more effectively and provide real-time data to business decision making systems. Organizations taking advantage of these efficiencies need to understand the security vulnerabilities.

Obviously TCP/IP and UDP enabled SCADA and DCS systems can reside on their own isolated network, but all SCADA systems whether completely isolated, totally serial based or not, have always required some kind of physical security. Typically the field controllers are difficult to access and reside in locked enclosures, and the operator workstations (OWS) are protected in a secure room. Personnel having access to areas where operator workstations reside usually have clearance.

During the nineties as local area networks became more pervasive in business, it became more common place to connect the physically secure OWS to the LAN. The OWS and consequently the SCADA and DCS systems were now vulnerable to being remotely compromised from someone within the organization connected to the LAN. When the internet emerged those same OWSs were allowed internet access, now those systems were more prone to outside vulnerabilities such as hackers, viruses, and worms. Also during that time field devices such as controllers and RTUs emerged with TCP/IP and UDP capabilities. Some even use Linux and Windows operating systems. Consequently, SCADA and DCS systems have become more vulnerable at more points in the SCADA network. If these systems are connected in any way to a business network, then it's very possible that the organization's IT staff and engineers are unaware of these added vulnerabilities.

The newest vulnerability comes from the fact that businesses now recognize SCADA and DCS systems as real-time data sources. Instead of waiting to analyze data dogs from these systems, business software is finding ways to consume that data in real-time—sometimes at the expense of network security. For example, with real-time energy demand data from the grid, electric utilities can implement load curtailment and energy conservation strategies. Facility managers can program their asset management systems to consume real-time DCS information from chillers, air handling units, etc. and auto-generate work orders and notify staff of improper operation or failures.

Third parties can also use real-time information from SCADA and DCS systems. With a secure interface an equipment vendor can remotely monitor their product being managed by their customer's SCADA system and roll a truck before there is a critical failure. A regulatory agency can monitor critical aspects of an electrical grid or pipeline. For example, an electric utility can expose an interface to state and federal agencies so as to monitor real-time loads, voltages, currents, frequencies, and power factors at critical points on a transmission or distribution system.

These examples would not be possible without TCP/IP or UDP enablement. The business advantages are very compelling for allowing business networks to be utilized for SCADA and DCS communication. It's not uncommon that engineers have difficulty convincing IT personnel that SCADA systems should have network access. Sometimes IT management will naively allow SCADA systems to be connected to their networks. They assume it's a collection of embedded equipment resistant to normal network security vulnerabilities. The result is a hidden network within a network that is not only defenseless, but presents additional security problems. Furthermore, this hidden network is as critical to the business as any of its business systems.

Organizations owe it to themselves to educate their IT personnel about SCADA and DCS systems. SCADA and DCS vendors should be expected to point out any security weaknesses their products may have. While many SCADA engineers are network savvy, they

may not be network security savvy. This chapter will provide an overview of security concerns that engineers and IT personnel need to be aware of when connecting their SCADA and DCS systems to their intranet or the internet.

RECENT EXAMPLES OF COMPROMISED SCADA AND DCS SYSTEMS

Vitek Boden sought revenge. After he was turned down for a job with the Maroochy Shire Council in Queensland, Australia, the 48-year-old disgruntled techie unleashed his anger in early 2000 by hacking into the town's wastewater system at least 46 times. On two separate occasions, his electronic attacks (apparently he used a stolen laptop and a radio transmitter) led to pumping station failures that caused as much as 1 million liters of foul-smelling raw sewage to spill into parks, waterways and the grounds of a tourist resort. In the surrounding area on Australia's Sunshine Coast, creeks turned black.

'Paul Blomgren [...] measures control system vulnerabilities. Last year, his company assessed a large southwestern utility that serves about four million customers." Our people drove to a remote substation," he recalled. "Without leaving their vehicle, they noticed a wireless network antenna. They plugged in their wireless LAN cards, fired up their notebook computers, and connected to the system within five minutes because it wasn't using passwords. [...] Within 15 minutes, they mapped every piece of equipment in the operational control network. Within 20 minutes, they were talking to the business

The Japan Times   ⌂ HOME

Power plant security info leaked onto Net

NAGOYA (Kyodo) Security data on a thermal power plant has been leaked onto the Internet from a virus-infected personal computer, the company in charge of the plant's security said Sunday.

The information was passed onto the Internet through a file-sharing program called Share.

The data includes the locations of various facilities in Chubu Electric Power Co.'s thermal power plant in Owase, Mie Prefecture, including the control room, instrument panel room and boilers, officials of the security company, a Chubu affiliate, said.

Also leaked were manuals on how to deal with unconfirmed reports of intruders in the plant, as well as a list of the names and home addresses of the security firm's employees and other personal data on guards, they said.

The data made its way to the Net from a computer belonging to a 40-year-old employee of the security firm, the officials said. He compiled the data on his PC around 2000.

He started to use Share in March, the officials said.

Chubu Power, based in Nagoya, operates five nuclear power reactors in Shizuoka Prefecture.

The Japan Times: Monday, May 15, 2006

network and had pulled off several business reports.'

'In a demonstration at a recent security conference, [Jeff Dagle, a PNNL EE] hacked into his test bed system and tripped an electrical breaker. The breaker then signaled the SCADA software that it had opened. But the SCADA controller did not respond because it had not instructed the breaker to open. It was a classic denial-of-service attack. "We were demonstrating a weakness at the protocol level itself," said Dagle.'

SCADA and DCS Security Vulnerabilities

SCADA and DCS systems are typically compromised by hackers, both internal and external, and by viruses and worms. The easiest form of hacking a SCADA system will come from someone internal to the organization or someone the organization has granted remote access too. The internal breach may be caused by a disgruntled employee, or by a disgruntled individual in the external organization who has been granted remote access to the SCADA and DCS systems.

There is a possibility of new threat coming from disgruntled or politically motivated software engineers that write the source code for SCADA and DCS systems and components. Many SCADA and DCS vendors are international companies with engineering and development teams located around the world. How well are these engineers and developers screened? How well is the source code reviewed for malicious code and backdoors? Reviewing someone else's source code is very difficult and hard to justify when it's working normally. This may be a security threat in the making.

When adding a SCADA or DCS system to a business network new security issues arise that are typically unknown to an organization's IT staff and related to the following:

Operator workstation (OWS) is the console that SCADA and DCS operators use to monitor and control the system. The OWS is also known as a man machine interface (MMI) or human machine interface (HMI). Typically the OWS should be in a physically secure room having an access control system. Physical security perimeters and access control systems are customary in large electric utilities, pipeline companies, and large industrial control facilities, but surprisingly small electric utilities and commercial building control systems have very little if any physical security. The consequences of compromising even a small electric utility's SCADA system can have devastating results beyond their system due to the interconnectivity of the electric grid. Physically securing an OWS should be a requirement for any organization with a SCADA or DCS system.

While physically securing the OWS environment

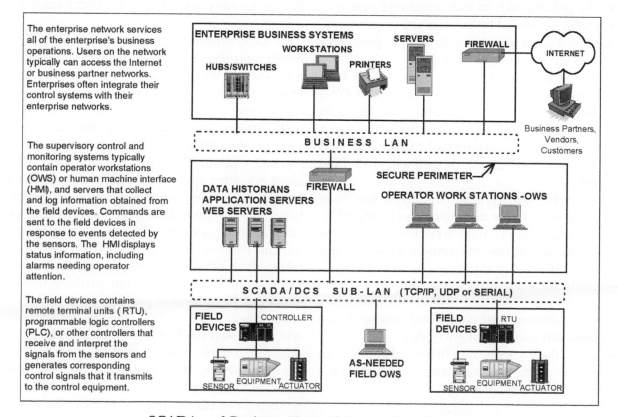

SCADA and Business Network Integration Diagram

should be a requirement for keeping unauthorized people off limits, further security measures must be implemented for an OWS connected to a network. Consider the following:

- Is the OWS connected to internet? Can its user surf the web? If so, the OWS is susceptible to viruses, worms, Trojan horses, etc.

- Is the OWS operating on a computer that is configured to the organization's security standards, or is operating on a computer setup and configured by the SCADA vendor? Often SCADA and DCS vendors will only sell and support an OWS that they turnkey which might conflict an organization's security policies. A vendor supplied computer may be highly suspect. Although the vendor does not intend to provide you a vulnerable computer, consider that the computer might have been configured by a technician in a local field office who is not enterprise IT savvy.

- Typically the OWS computer will or should use the operating system level network security policies assigned and managed by the IT staff such as DHCP or static IP address, user name, password. But the actual OWS software application which has its own security policies such as user name and password are usually controlled by the engineer or technician. Therefore it is not uncommon to discover poor OWS software security policies such as:
  — The SCADA or DCS vendor's default user name and password that is included in OWS software has not been removed. In some cases doing a Google search on a particular vendor's OWS software with "user name" and "password" in the search term will yield those defaults.
  — The OWS software authentication has been assigned a simple user name and password.
  — Several users may use the OWS and consequently use the same user name and password.
  — OWS software user names and passwords never expire.
  — OWS software user names and passwords use plain text and are not encrypted.
  — When there are several OWS installations in an organization, the user names and passwords work at all locations.

- Can any computer running the OWS software

be connected to the network and gain access to the SCADA or DCS system? For example, facility managers responsible for managing several facilities across a campus or across a nation like to run an instance of their OWS software on their laptop. Some laptops are equipped with VPN so the facility manager can remotely access his facilities DCS systems from home or on the road, but consider that manager's OWS enabled laptop is susceptible to being compromised or stolen. It's not physically secure.

- Many OWS software applications are web enabled and can be accessed via a standard browser. Typically these applications are only browser accessible on the organization's intranet, but some organizations allow access from the internet. If internet access is used, then make sure authentication policies are strong, try not to use HTTP port 80, and try to use HTTPs.

Field devices are the SCADA and DCS assets downstream from the OWS. Field devices consist of programmable controllers, remote terminal units (RTU), sensors, and actuators. Historically these components have had very limited computational and communications capabilities. Field devices have used simple microprocessors, limited memory and slow serial communication protocols.

As microprocessor power increased, programmable controllers began running modern operating systems such as Windows and Linux. This has resulted in more robust SCADA and DCS controllers, but inherited the security vulnerabilities associated with those operating systems and added more intrusion opportunity focal points. Often times IT management and staff are stunned and surprised to learn that field devices are using operating systems that they have specific security polices for. It's very common for these controllers not to have any virus protection, recent security patches, or the latest service packs. Patching these components can be difficult because they're often used in situations where taking them offline can cause problems for the system being controlled. However, not patching these components can have serious problems too. This is an area where the SCADA or DCS vendor, the staff engineers, and the IT staff need to work together, preferably at the outset of commissioning a system. If for example Windows enabled SCADA components are discovered later, then everyone needs to get together and form a strategy to keep those components up to date and in adherence to the organizations security policies.

Raw communications protocols are the fundamental protocols used for OWS, controller, RTU, sensor, and actuator communications. The raw communications protocol represents one of the greatest security vulnerabilities in any SCADA or DCS network. These protocols are either closed/proprietary or open. Proprietary protocols are defined by a SCADA or DCS vendor but not published. Open protocols are usually defined by standards bodies and published for all. Most proprietary protocols can easily be reversed engineered with packet sniffing software. All industries using SCADA and DCS are moving to open protocols, but from a security perspective your system and business networks can be vulnerable from either type.

Many protocols are simple, lightweight and use low BAUD rate serial communications. As these lightweight protocols have migrated to TCP/IP networks they've maintained much of their simplicity. In other words, these protocols are simply wrapped inside TCP/IP data packets, but are still simple nonetheless.

There are several freely downloadable client applications that can be used by a hacker or rogue employee to communicate with SCADA and DCS systems at the protocol level. These software applications are typically for non-malicious uses, but in the wrong hands can have devastating results. Client software applications can be downloaded from the web that directly can communicate with protocols such as DNP 3.0, Modbus, BACnet, LonWorks, and OPC.

Here are some additional security considerations about all SCADA and DCS communications protocols:

- **Raw communications protocols are not authenticated**

    Because protocols are simple, access to their data packets flowing on a wire are typically guarded only by an end-user application known as an operator workstation (OWS). The OWS is typically requires a user name and password, but the user name and password have no relationship to the protocol itself.

    Consider a situation where an organization allowes the business network to be used as a SCADA network backbone for an open protocol such as Modbus TCP or BACnet/IP. A disgruntled employee can easily download and install a freely available client software application capable of communicating with one of these protocols. Once he's communicating, he can send malicious and dangerous commands to critical systems. Assuming the business network is not tuned to trap SCADA traffic it's very possible no one would ever know he did it.

- **Protocols are typically not encrypted**

    Raw communication protocol data packets flowing on a network can easily be captured and reverse engineered because they're in plain text. There are many free downloadable protocol scanners designed to packet-sniff unencrypted serial, TCP/IP, UDP, HTTP traffic on a wire.

    The lack of encryption is exacerbated when wireless communication technologies are used. This is discussed in more detail later in the chapter under "Wireless Protocols."

- **TCP and UDP protocol traffic is on one port**

    Protocols capable of using TCP/IP and UDP communicate through specific ports. For example open protocols such as Modus TCP uses port 502 and BACnet/IP uses UDP port 47808. Because these are industry standardized ports, hackers can scan these firewall ports for *internet* device communication, or just exploit these open ports in other malicious ways. If these ports are used for SCADA and DCS protocol communications over the internet, then your systems are open to anyone with a corresponding client application. Surprisingly, this scenario is not uncommon in smaller organizations.

    It's always a good idea to isolate SCADA and DCS networks from the main business network by placing them behind a Firewall on their own sub-LAN. This protects them from not only outside hackers, but rogue employees with malicious intentions. Should client applications need access to the SCADA and DCS system behind the Firewall, then tune the firewall to only accept client connections from a limited number of fixed IP addresses on specific ports such as 502—the Modbus TCP port.

- **Some protocols use simple device addressing**

    Modbus is one the most pervasive raw communications protocols used in SCADA and DCS. Modbus is used on serial and TCP/IP networks, has no built-in security, and can be easily hacked. A hacker who connects to a Modbus network may not need to know a specific device and register address because Modbus device addresses and data registers are based on simple finite ranges of numbers. For example, the hacker could write logical 1's and 0's to a Modbus device's 65535 Boolean coil registers hoping to cause an unexpected equipment operation. Modbus device vendors publish their register descriptions as maps. It would not be hard for a hacker to obtain these maps, build a library, write a program and succeed at a cyber attack.

Other protocols including proprietary ones use simple device address schemes and can be found from internet searches.

- **Some protocols support auto-discovery**

    Some communications protocols support auto-discovery. For example, BACnet/IP is an open protocol widely used in building automation systems. It's used in government and commercial facilities. A BACnet/IP client application connected to a network with Bacnet servers may be able to auto-discover those servers and parse their self-describing device addresses and control critical HVAC, lighting, and security systems.

- **Higher level protocols—OPC, DCOM, SOA, XML Web Services, SOAP and REST**

    As businesses began to recognize that TCP/IP enabled SCADA and DCS systems could use their networks to centralize command and control, they also recognized the opportunity of turning these systems into real-time data sources for enterprise business systems. The primary approach taken so far to maximize the effectiveness of TCP/IP enablement is to use IT friendly methodologies that wrap or encapsulate the protocol communications inside higher level protocols using standardized interfaces.

    Beginning with Ole for Process Control or OPC, SCADA and DCS vendors encapsulated and abstracted raw communications protocols through Microsoft's Distributed Component Object Model otherwise known as DCOM. OPC was designed to *open up* or make accessible SCADA and DCS systems via the DCOM standard. OPC via DCOM makes it easier to integrate disparate SCADA systems with each other and with general business systems.

    DCOM is a Microsoft concept that provides a mechanism for transparently executing function calls across computer systems over a network. In other words, client applications request services over a network from objects on DCOM servers. DCOM has been ported to non Microsoft systems such as Linux, but for the most part clients and servers are entirely Windows based.

    While OPC and DCOM have opened up otherwise closed SCADA and DCS systems, security vulnerabilities abound within the DCOM layer such as:

- *According to Donovan Tindill at Matrikon: DCOM requires many ports for finding other hosts, resolving names, requesting services, authentication, sending data, and more. If these ports are not available, DCOM will automatically search for others. Any port and service used by DCOM is a target to Cyber attacks (viruses and worms). When DCOM security is compromised, all applications are affected including OPC applications. The recent Blaster and Sasser viruses attacked the same components that OPC relies upon. Anyone using OPC may be vulnerable to both of these viruses and more in the future. Added OPC Tunneling technology can also make use of port restriction, user authentication, and data-stream encryption to overcome most of IT's security issues.*

- Typically the objects on a DCOM server are published and self describing. This makes a SCADA or DCS system easy to exploit once a client application connects.

- Creating client applications is very easy for someone with basic programming skills. For example, an Excel power user with VBA skills can connect Excel to an OPC server running on a network and take over the SCADA or DCS system that the OPC server is connected too.

- OPC servers could be setting dormant on SCADA and DCS operator workstations (OWS). It's not uncommon for a vendor to include an OPC server along with its OWS software. When the OWS software is installed the included OPC server and its DCOM settings may have been installed and activated as well. A hacker with knowledge of OPC could probe the network with a general purpose OPC client. Once an OPC server is found it may not require any authentication thereby allowing the hacker to gain control of the SCADA or DCS system.

- There are 3rd party vendors of OPC servers that can directly connect to a multitude of SCADA and DCS communications protocols. In some cases these 3rd party vendors allow you to download fully functional evaluation versions of their OPC servers. These evaluation servers may limit functionality, or only work for a few days, but they may be capable enough for a rogue employee to gain control of a system.

The OPC Foundation at *www.opcfoundation.org* has published documents on their web site that explain how to secure OPC servers. OPC is a very pervasive technology within SCADA and DCS universe. IT management should identify and apply industry standard DCOM security policies when OPC servers are discovered.

The latest move in making SCADA and DCS systems more open is by encapsulating them in a *services oriented architecture* (SOA). In an SOA environment SCADA and DCS systems are turned into services accessed by client applications without the knowledge of their underlying platform, implementation, or protocols. SOA is typically based on XML web services and its underlying protocols known as *SOAP* and *REST*. SOAP and REST are similar to DCOM because they too are a mechanism for transparently executing function calls across computer systems over a network. Where DCOM is primarily Microsoft Windows based, SOAP and REST are not. SOAP and REST are completely platform independent. SOAP and REST client applications running on any operating system can transparently execute SOAP and REST functions running on any other operating system.

REST (representational state transfer) is the simpler of the 2 web services. REST uses XML and HTTP without the extra abstractions and complex XML message payloads used by SOAP. A client application simply makes a web service request to a hosted service via a URI. For example, the following URI call to a Gridlogix SOA web service platform called EnNET® will return the current value from Holding Register 100 on Device 10 on a Modbus TCP gateway:

http://192.168.1.25/EnNET-Modbus/Servers/TCP-IPDevices/StlGateway/Device10/HoldingRegisters/100

This simple REST request can be password protected and encrypted with HTTPS thereby making it secure. REST web services are very useful for read only requests. Client applications needing real-time read/write connectivity to SCADA systems can use the REST method very effectively.

SOAP is based on a set of open specifications that include established internet standards and newer standards managed by the W3C. The basic stack is comprised of HTTP, SMTP, XML, WSDL, UDDI, and WSE which includes WS-Security (WSS).

While REST is much more lightweight, so is its security. SOAP on the other hand has a very sophisticated security standard known as WS-Security or WSS. WSS defines how to protect SOAP messages that travel over a network. Prior to WSS, the only security available for SOAP was SSL and TSL which combine with HTTP to form HTTPS. Whereas HTTPS encrypts the entire message, WSS allows the encryption to be selectively applied and operates at the SOAP layer. WSS can travel with the message throughout the network and even persist when the message is queued or stored.

SCADA and DCS systems enabled with WSS inherit the following security capabilities:

- X.509 certificates
- Kerberos computer network authentication protocol
- SAML—security assertion markup language
- XML digital signatures

Gridlogix EnNET® is a complete SOA and web services framework for SCADA and DCS that fully supports REST, SOAP, HTTPS, and WSS.

- **Wireless Protocols**
  Identify any 802.11 (Wi-Fi) SCADA and DCS assets. Hackers can use a laptop with a wireless ethernet card along with freely downloaded software and learn the network's encryption key. For example, AirSnort is a wireless LAN (WLAN) tool which recovers encryption keys and operates by passively monitoring transmissions. The encryption key is derived after enough packets have been gathered. Once the hacker has the key, he can use a protocol scanner to capture an engineers login credentials. The hacker then listens in on the engineer's transmissions and captures command and control packets. Once the hacker has this info he can easily compromise the SCADA system.

There are emerging wireless technologies designed to be used at the sensor and actuator level inside SCADA and DCS systems. Hackers who possess wireless apparatus tuned to these new wireless standards can directly control actuators connected to critical equipment. Some of these new wireless standards use authentication and encryption but many do not.

ZigBee™ is one of the new wireless standards emerging for sensors and actuators and is based on the IEEE 802.15.4 standard. This standard supports 128-bit AES Encryption. ZigBee™ enabled sensors and actuators will become more prevalent over time due to the cost savings of being wireless. ZigBee™ and other competing wireless technologies will be found in buildings, power substations, industrial facilities, and security systems. Even though ZigBee™ is designed to be secure, as with 802.11 technologies, technology will surely emerge that can breach its boundaries.

Be aware of other sensor and actuator technology that uses wireless such as ultra wideband (UWB) and spread spectrum (SS). These technologies may not be encrypted. UWB and SS devices are emerging in SCADA and DCS because of the efficiencies they bring, but the security vulnerabilities may not have been addressed.

• **Business Partner Accessibility**

Business partner access is a concept that many SCA-DA engineers, vendors, and other business partners now desire. Business partner access to SCADA and DCS systems is typically made by a virtual private network (VPN) connection. Typically the granting organization provides the business partner its VPN client software.

Granting access to qualified 3rd party business partners has the following advantages:

— A SCADA vendor can remotely monitor the health of its customer's system and provide proactive technical support.
— A SCADA vendor can remotely apply software upgrades or patches—security or otherwise.
— An equipment vendor can remotely monitor their product being managed by the SCADA system and roll a truck before there is a critical failure.
— A regulatory agency can monitor critical aspects of an electrical grid or pipeline. For example, a utility can expose an interface to regulatory agencies so they can monitor real-time loads, voltages, currents, frequencies, and power factors at critical points along the electric grid.

Granting access to qualified 3rd business partners has the following disadvantages and considerations:

— Large organizations often require several members of IT management to agree on 3rd party access. It's often easier for IT management to say no to 3rd party access requests. IT management may perceive the rewards do not outweigh the risks. Business partners and in-house engineers need to be able to effectively convey the benefits and rewards, but IT ultimately has the responsibility of managing the risk.
— Because SCADA systems are often a mission critical component of a business' operation, don't assume that the business partner's employees are ethical. Some organizations require that the business partner's employees receive a background check.

— How does an organization know if the business partner employees have shared their access credentials with others inside or outside of their organization? For example, a business partner's employee with remote access might resign and share their access credentials with their new employer.
— Businesses must routinely change passwords of 3rd party access accounts, and automatically disable accounts that are rarely used.
— Restrict VPN access to only a few of business partner's IP addresses.

• **Modem or dial-up connections, and TCP/IP Gateways**

— Modem connectivity is still in wide-scale use because there are many older serial protocol based SCADA systems still in use that require remote access and can't use TCP/IP.
— It can be argued that modem based SCADA systems are relatively secure. They're physically separate and apart from the business network or the internet. Their main vulnerability is someone having a phone number and software application designed to communicate with the SCADA system.
— Engineers are replacing their modems with TCP/IP enabled gateways so as to better manage, improve performance, and extend the life of their SCADA assets. The serial protocol modem interface is replaced with a device that encapsulates the serial communications protocol within TCP/IP packets.
— It is not uncommon for IT management to approve connecting TCP/IP gateways to the business network because it's perceived that the gateway is an embedded industrial device that is immune to traditional business network vulnerabilities. However, gateways are typically compact PC's running commercial operating systems such as Windows or Linux. They're often packaged to resemble an industrial controller that would normally be installed in a locked control room cabinet. Gateways should be subjected to the same security policies as any PC or server.

SECURITY POLICIES AND COUNTER MEASURES

There are many ways to secure SCADA and DCS systems and the networks they reside on.

For example, here's a summarized list of security policies and counter measures an organization can take to secure its SCADA and DCS systems:

- Use strong passwords on the operator workstations, not only for the general operating system level authentication, but for the OWS SCADA software application as well.

- Routinely change passwords!

- Don't use the same user name and password on all operator workstations.

- Remove the vendor default user name and passwords from the OWS software.

- Place SCADA systems on their own subnet behind a firewall.

- Close all unnecessary ports on the firewall. Consider blocking all UDP communications.

- Configure SCADA firewalls to only accept connections from trusted IP addresses.

- Use a services oriented architecture (SOA) gateway between the SCADA firewall and the SCADA network. An SOA gateway such as Gridlogix EnNET can add even more security by encapsulating the SCADA protocol inside an encrypted SOAP envelope using WS-Security, HTTPS, or SSL. A SCADA network configured to use WS-Security can ensure all SCADA communications on the business network is encrypted.

- If allowing business partner access, know who your business partners provide your authentication credentials to. Do they give those credentials out to just anyone or any employee? Do they have former employees that still know your authentication credentials? Routinely change your business partner's passwords!

- Engineers should always inform their IT staff of the hardware, software, and operating system specifications of their TCP/IP and UDP enabled SCADA and DCS assets—DHCP or Static. Except for knowing a DHCP address was handed out, DHCP enabled devices are easily connected to networks and may go unnoticed on network management consoles because their data packet signatures are not recognized and therefore ignored.

- IT staff should regularly audit all existing and future TCP/IP and UDP enabled SCADA and DCS assets. Know which assets are using modern operating systems such as Windows. Keep an eye out for wireless assets—802.11a/b/g/n and emerging standards such as ZigBee™.

- Routinely do security vulnerability testing of the entire SCADA and DCS system.

SUMMARY

It is a good idea to allow SCADA and DCS systems to run on business networks. Engineers can remotely access field devices, centralize command and control, and provide high level dashboards to management. Equipment vendors can access runtime information on critical equipment that allows them to provide better support and prevent costly failures. Energy trading systems can be more efficient and profitable by accessing real-time data from SCADA systems. But in order for SCADA or DCS systems to be this open, an organization runs the risk of leaving backdoors to hackers, rogue employees, and viruses.

Organizations must guard against allowing hidden networks to evolve on their business network. So long as vendors, in-house engineers and IT staff can effectively communicate, then the likelihood of a hidden SCADA or DCS network evolving will be minimal. A hidden network of unsecured SCADA and DCS assets is not only a threat to life and property it's a back door to the business network.

All TCP/IP and UDP capable SCADA and DCS systems should be isolated from the business network and placed behind their own firewall. Tune the firewalls to accept only connections from trusted IP addresses. Use VPN tunnels within your intranet and especially the internet, or consider using an SOA gateway technology behind the firewall such as Gridlogix EnNET® that can encrypt communications between the client and SCADA system.

Keep in mind that as microprocessor power increases and packaging size decreases, and as wireless technologies improve, more and more SCADA and DCS components will become smarter—particularly the system's end points known as sensors and actuators. Controllers and RTUs are now running Windows which presents security vulnerabilities that the IT staff may not be aware of. So be aware that the sensors and actuators will eventually represent even more access points for hackers.

Finally, engineers and IT staff must work together and implement security policies specific to SCADA and DCS systems. Those policies should include a complete audit of all existing SCADA and DCS assets with ongoing audits and security vulnerability testing. Appendix A references security recommendations for SCADA and DCS systems as proposed by various organizations. Appendix B shows one such list of recommendations from the U.S. Department of Energy—"21 Steps to Improve Cyber Security of SCADA Networks."

References

Out of Control, CSO Online—www.csoonline.com, Todd Dantz, August 2004
The Rise of Smart Buildings, Part 2, Techworld—www.techworld.com, Robert L. Mitchell, March, 18th, 2005
SCADA vs. the Hackers, Mechanical Engineering Magazine Online—www.memagazine.org, Alan S. Brown, December 2002
SCADA Security and Critical Infrastructure, Infraguard Meeting, Eugene, Oregon, Joe St. Sauver, Ph.D., December 7th, 2004
DNPSec: Distributed Network Protocol Version 3 (DNP3) Security Framework, George Mason University and American University of Sharjah, Munir Majdalawieh, Francesco Parisi-Presicce, Duminda Wijesekera, December, 2005
NISTIR 7009—BACnet Wide Area Network Security Threat Assessment, U.S DEPARTMENT OF COMMERCE, National Institute of Standards and Technology, Building Environment Division, Building and Fire Research Laboratory, David G. Holmberg, July, 2003
Security Guidelines for the Electricity Sector—Version 1.0, North American Electric Reliability Council, June 14, 2002
CRITICAL INFRASTRUCTURE PROTECTION, Challenges and Efforts to Secure Control Systems, United States General Accounting Office, March, 2004
OPC CONSIDERATIONS FOR NETWORK SECURITY, The Online Industrial Ethernet Book, Donovan Tindill, Network Services Team Lead, Matrikon, Issue 23:35
Network Security for EIS and ECS Systems, Information Technology for Energy Managers, Barney Capehart, Joel Weber, November 2003
Connecting Energy Management Systems to Enterprise Business Systems Using SOAP and the XML Web Services Architecture, Web Based Energy Information and Control Systems, Barney Capehart, Lynn Capehart, Ron Brown, 2005

APPENDIX A

The following organizations have either published security standards or recommendations for SCADA and DCS systems:

- U.S. DEPARTMENT OF ENERGY
  — 21 Steps to Improve Cyber Security of SCADA Networks
  — Sandia Labs, Network Security Infrastructure Testing

- U.S DEPARTMENT OF COMMERCE
  — NISTIR 7009 -BACnet Wide Area Network Security Threat Assessment

- U.S. GENERAL ACCOUNTING OFFICE
  — Critical Infrastructure Protection, Challenges and Efforts to Secure Control Systems

- North American Energy Reliability Council (NERC)
  — CIP-002 to CIP-009

- American Gas Association (AGA) 12-1
  — AGA 12-1 Implementation

- The ZigBee™ Alliance
  — ZigBee Security Specification Overview

- The OPC® Foundation
  — The OPC Security Specification

APPENDIX B

"21 Steps to Improve Cyber Security of SCADA Networks" As published by the U.S. Department of Energy"

1.   Identify all connections to SCADA networks.
Conduct a thorough risk analysis to assess the risk and necessity of each connection to the SCADA network. Develop a comprehensive understanding of all connections to the SCADA network, and how well these connections are protected. Identify and evaluate the following types of connections:

- Internal local area and wide area networks, including business networks
- The internet
- Wireless network devices, including satellite uplinks
- Modem or dial-up connections
- Connections to business partners, vendors or regulatory agencies

2.   Disconnect unnecessary connections to the SCADA network.
To ensure the highest degree of security of SCADA systems, isolate the SCADA network from other network connections to as great a degree as possible. Any connection to another network introduces security risks, particularly if the connection creates a pathway from or to the internet. Although direct connections with other networks may allow important

information to be passed efficiently and conveniently, insecure connections are simply not worth the risk; isolation of the SCADA network must be a primary goal to provide needed protection. Strategies such as utilization of "demilitarized zones" (DMZs) and data warehousing can facilitate the secure transfer of data from the SCADA network to business networks. However, they must be designed and implemented properly to avoid introduction of additional risk through improper configuration.

3. Evaluate and strengthen the security of any remaining connections to the SCADA network.

Conduct penetration testing or vulnerability analysis of any remaining connections to the SCADA network to evaluate the protection posture associated with these pathways. Use this information in conjunction with risk management processes to develop a robust protection strategy for any pathways to the SCADA network. Since the SCADA network is only as secure as its weakest connecting point, it is essential to implement firewalls, intrusion detection systems (IDSs), and other appropriate security measures at each point of entry. Configure firewall rules to prohibit access from and to the SCADA network, and be as specific as possible when permitting approved connections. For example, an independent system operator (ISO) should not be granted "blanket" network access simply because there is a need for a connection to certain components of the SCADA system. Strategically place IDSs at each entry point to alert security personnel of potential breaches of network security. Organization management must understand and accept responsibility for risks associated with any connection to the SCADA network.

4. Harden SCADA networks by removing or disabling unnecessary services.

SCADA control servers built on commercial or open-source operating systems can be exposed to attack through default network services. To the greatest degree possible, remove or disable unused services and network daemons to reduce the risk of direct attack. This is particularly important when SCADA networks are interconnected with other networks. Do not permit a service or feature on a SCADA network unless a thorough risk assessment of the consequences of allowing the service/feature shows that the benefits of the service/feature far outweigh the potential for vulnerability exploitation. Examples of services to remove from SCADA networks include automated meter reading/remote billing systems, email services, and internet access. An example of a feature to disable is remote maintenance. Numerous secure configuration guidelines for both commercial and open source operating systems are in the public domain, such as the National Security Agency's series of security guides. Additionally, work closely with SCADA vendors to identify secure configurations and coordinate any

and all changes to operational systems to ensure that removing or disabling services does not cause downtime, interruption of service, or loss of support.

5. Do not rely on proprietary protocols to protect your system.

Some SCADA systems use unique, proprietary protocols for communications between field devices and servers. Often the security of SCADA systems is based solely on the secrecy of these protocols. Unfortunately, obscure protocols provide very little "real" security. Do not rely on proprietary protocols or factory default configuration settings to protect your system. Additionally, demand that vendors disclose any backdoors or vendor interfaces to your SCADA systems, and expect them to provide systems that are capable of being secured.

6. Implement the security features provided by device and system vendors.

Older SCADA systems (most systems in use) have no security features whatsoever. SCADA system owners must insist that their system vendor implement security features in the form of product patches or upgrades. Some newer SCADA devices are shipped with basic security features, but these are usually disabled to ensure ease of installation.

Analyze each SCADA device to determine whether security features are present. Additionally, factory default security settings (such as in computer network firewalls) are often set to provide maximum usability, but minimal security. Set all security features to provide the maximum level of security. Allow settings below maximum security only after a thorough risk assessment of the consequences of reducing the security level.

7. Establish strong controls over any medium that is used as a back door into the SCADA network.

Where backdoors or vendor connections do exist in SCADA systems, strong authentication must be implemented to ensure secure communications. Modems, wireless, and wired networks used for communications and maintenance represent a significant vulnerability to the SCADA network and remote sites. Successful "war dialing" or "war driving" attacks could allow an attacker to bypass all other controls and have direct access to the SCADA network or resources. To minimize the risk of such attacks, disable inbound access and replace it with some type of callback system.

8. Implement internal and external intrusion detection systems and establish 24-hour-a-day incident monitoring.

To be able to effectively respond to cyber attacks, establish an intrusion detection strategy that includes alerting network *administrators of malicious network activity originating from internal or external sources. Intrusion detection system monitoring is essential 24 hours a day; this capability can be easily set up through a pager. Additionally, incident response procedures must be in place to allow an effective response to any attack. To complement network monitoring, enable logging on all systems and audit system logs daily to detect suspicious activity as soon as possible.*

9.   Perform technical audits of SCADA devices and networks, and any other connected networks, to identify security concerns.
Technical audits of SCADA devices and networks are critical to ongoing security effectiveness. Many commercial and open-source security tools are available that allow system administrators to conduct audits of their systems/networks to identify active services, patch level, and common vulnerabilities. The use of these tools will not solve systemic problems, but will eliminate the "paths of least resistance" that an attacker could exploit. Analyze identified vulnerabilities to determine their significance, and take corrective actions as appropriate. Track corrective actions and analyze this information to identify trends. Additionally, retest systems after corrective actions have been taken to ensure that vulnerabilities were actually eliminated. Scan non-production environments actively to identify and address potential problems.

10.   Conduct physical security surveys and assess all remote sites connected to the SCADA network to evaluate their security.
Any location that has a connection to the SCADA network is a target, especially unmanned or unguarded remote sites. Conduct a physical security survey and inventory access points at each facility that has a connection to the SCADA system. Identify and assess any source of information including remote telephone/computer network/fiber optic cables that could be tapped; radio and microwave links that are exploitable; computer terminals that could be accessed; and wireless local area network access points. Identify and eliminate single points of failure. The security of the site must be adequate to detect or prevent unauthorized access. Do not allow "live" network access points at remote, unguarded sites simply for convenience.

11.   Establish SCADA "Red Teams" to identify and evaluate possible attack scenarios.
Establish a "Red Team" to identify potential attack scenarios and evaluate potential system vulnerabilities. Use a variety of people who can provide insight into weaknesses of the overall network, SCADA systems, physical systems, and security controls. People who work on the system every day have great insight into the vulnerabilities of your SCADA

network and should be consulted when identifying potential attack scenarios and possible consequences. Also, ensure that the risk from a malicious insider is fully evaluated, given that this represents one of the greatest threats to an organization. Feed information resulting from the "Red Team" evaluation into risk management processes to assess the information and establish appropriate protection strategies.

The following steps focus on management actions to establish an effective cyber security program:

12.   Clearly define cyber security roles, responsibilities, and authorities for managers, system administrators, and users.
Organization personnel need to understand the specific expectations associated with protecting information technology resources through the definition of clear and logical roles and responsibilities. In addition, key personnel need to be given sufficient authority to carry out their assigned responsibilities. Too often, good cyber security is left up to the initiative of the individual, which usually leads to inconsistent implementations and ineffective security. Establish a cyber security organizational structure that defines roles and responsibilities and clearly identifies how cyber security issues are escalated and who is notified in an emergency.

13.   Document network architecture and identify systems that serve critical functions or contain sensitive information that require additional levels of protection.
Develop and document a robust information security architecture as part of a process to establish an effective protection strategy. It is essential that organizations design their networks with security in mind and continue to have a strong understanding of their network architecture throughout its life cycle. Of particular importance, an in-depth understanding of the functions that the systems perform and the sensitivity of the stored information is required. Without this understanding, risk cannot be properly assessed and protection strategies may not be sufficient. Documenting the information security architecture and its components is critical to understanding the overall protection strategy, and identifying single points of failure.

14.   Establish a rigorous, ongoing risk management process.
A thorough understanding of the risks to network computing resources from denial-of-service attacks and the vulnerability of sensitive information to compromise is essential to an effective cyber security program. Risk assessments form the technical basis of this understanding and are critical to formulating

effective strategies to mitigate vulnerabilities and preserve the integrity of computing resources. Initially, perform a baseline risk analysis based on a current threat assessment to use for developing a network protection strategy. Due to rapidly changing technology and the emergence of new threats on a daily basis, an ongoing risk assessment process is also needed so that routine changes can be made to the protection strategy to ensure it remains effective. Fundamental to risk management is identification of residual risk with a network protection strategy in place and acceptance of that risk by management.

15. Establish a network protection strategy based on the principle of defense-in-depth.

A fundamental principle that must be part of any network protection strategy is defense-in-depth. Defense-in-depth must be considered early in the design phase of the development process, and must be an integral consideration in all technical decision-making associated with the network. Utilize technical and administrative controls to mitigate threats from identified risks to as great a degree as possible at all levels of the network. Single points of failure must be avoided, and cyber security defense must be layered to limit and contain the impact of any security incidents. Additionally, each layer must be protected against other systems at the same layer. For example, to protect against the insider threat, restrict users to access only those resources necessary to perform their job functions.

16. Clearly identify cyber security requirements.

Organizations and companies need structured security programs with mandated requirements to establish expectations and allow personnel to be held accountable. Formalized policies and procedures are typically used to establish and institutionalize a cyber security program. A formal program is essential for establishing a consistent, standards-based approach to cyber security throughout an organization and eliminates sole dependence on individual initiative. Policies and procedures also inform employees of their specific cyber security responsibilities and the consequences of failing to meet those responsibilities. They also provide guidance regarding actions to be taken during a cyber security incident and promote efficient and effective actions during a time of crisis. As part of identifying cyber security requirements, include user agreements and notification and warning banners. Establish requirements to minimize the threat from malicious insiders, including the need for conducting background checks and limiting network privileges to those absolutely necessary.

17. Establish effective configuration management processes.

A fundamental management process needed to maintain a secure network is configuration management. Configuration management needs to cover both hardware configurations and software configurations. Changes to hardware or software can easily introduce vulnerabilities that undermine network security. Processes are required to evaluate and control any change to ensure that the network remains secure. Configuration management begins with well-tested and documented security baselines for your various systems.

18. Conduct routine self-assessments.

Robust performance evaluation processes are needed to provide organizations with feedback on the effectiveness of cyber security policy and technical implementation. A sign of a mature organization is one that is able to self-identify issues, conduct root cause analyses, and implement effective corrective actions that address individual and systemic problems. Self-assessment processes that are normally part of an effective cyber security program include routine scanning for vulnerabilities, automated auditing of the network, and self-assessments of organizational and individual performance.

19. Establish system backups and disaster recovery plans.

Establish a disaster recovery plan that allows for rapid recovery from any emergency (including a cyber attack). System backups are an essential part of any plan and allow rapid reconstruction of the network. Routinely exercise disaster recovery plans to ensure that they work and that personnel are familiar with them. Make appropriate changes to disaster recovery plans based on lessons learned from exercises.

20. Senior organizational leadership should establish expectations for cyber security performance and hold individuals accountable for their performance.

Effective cyber security performance requires commitment and leadership from senior managers in the organization. It is essential that senior management establish an expectation for strong cyber security and communicate this to their subordinate managers throughout the organization. It is also essential that senior organizational leadership establish a structure for implementation of a cyber security program. This structure will promote consistent implementation and the ability to sustain a strong cyber security program. It is then important for individuals to be held accountable for their performance as it relates to cyber security. This includes managers, system administrators, technicians, and users/operators.

21. Establish policies and conduct training to minimize the likelihood that organizational personnel will inadvertently disclose sensitive information regarding SCADA system design, operations, or security controls.

Release data related to the SCADA network only on a strict, need-to-know basis, and only to persons explicitly authorized to receive such information. "Social engineering," the gathering of information about a computer or computer network via questions to naive users, is often the first step in a malicious attack on computer networks. The more information revealed about a computer or computer network, the more vulnerable the computer/network is. Never divulge data related to a SCADA network, including the names and contact information about the system operators/administrators, computer

operating systems, and/or physical and logical locations of computers and network systems over telephones or to personnel unless they are explicitly authorized to receive such information. Any requests for information by unknown persons need to be sent to a central network security location for verification and fulfillment. People can be a weak link in an otherwise secure network. Conduct training and information awareness campaigns to ensure that personnel remain diligent in guarding sensitive network information, particularly their passwords.

Chapter 36

Developing an Energy Information System: An Efficient Methodology

David C. Green, Green Management Services, Inc.
Fort Myers, Florida
dcgreen@dcgreen.com

ABSTRACT

Energy information systems (EIS) are the best way to understand the effects of energy consumption rates for any large campus of facilities. Developing a custom EIS is the most effective way to assure the ability to report meaningful, accurate and timely energy consumption information. This chapter explains a streamlined web-based software development methodology that works especially well for an EIS. The components of this development methodology are presented in the order in which they should be pursued by the developer. For a refresher on what a simple EIS for a school district may look like visit the following page: http://www.utilityreporting.com/udvm/[1]

INTRODUCTION

Developing an EIS is no easy task. Energy information systems collect huge amounts of data and relate them in such a way as to report on the patterns that exist over time. Many have asked the question, "where do I begin?." And, of course, "what path do I take?." This chapter breaks the EIS development process down into distinct components or "deliverables" and provides some goals to keep in mind throughout the development process. These goals relate directly to the efficient development of a web-based EIS application avoiding cost overruns and delays that are common in the software development world. Figure 36-1. shows the components of this EIS design methodology.

A good place to start is with a good mission statement defining the purpose of the EIS. This leads to an overall system design. From the system design, a prototype confirms that the system design meets the broad specifications of the project. The detail design component addresses all the individual specifications of

Components or Deliverables
1. Mission Statement
2. System Design
3. Prototype
4. Detail Design
5. Production Version

Figure 36-1.

the EIS which, by the way, are quite different from one organization to another. The final product is an EIS that takes energy consumption information from the meters to the conference room in a meaningful way, a way that everyone in the organization can understand. Using this methodology any qualified software engineer can efficiently develop an EIS to meet the specific needs of an organization. Let's begin with some goals to keep in mind throughout the development process.

GOALS

It's important to keep a few quality oriented goals in mind while developing an EIS. One of the main reasons for developing a *custom* EIS is to introduce new features as needed. The most efficient manner to develop software has come to be a technique known as "successive revision" or "iterative revision." The process depends a great deal on feedback from the users of the application provided during the development process. The two most desirable software development metrics, high cohesion and low coupling should be adhered to as much as possible. One main module using an iterative processing model lends itself to those desirable metrics well. The design should remain flexible since it is highly likely that new requirements will evolve the more the application is used. Using a centric data model with a consistent interface will help retain the flexibility to add new features without adding complexity.

This is why the successive revision technique is so appropriate for developing an EIS. The successive revision technique of software development simply means to introduce new features and test them one at a time until they pass the test. New features may be a specific report, calculation, menu item or any addition to the functionality of the application. Of course, one important test is that each new feature works the way the users expect it to. In order to pass that test a prototype must be available for the users to test the new feature. Early prototyping serves two purposes, one is to test new features and another is to help analyze the data as part of the system design component. So, the successive revision technique encourages parallel work on more than one of the components at a time and increases the likelihood that the developer can deliver each component in a timely manner. Not that "time" should be the controlling factor in which a component is considered complete. The software timetable should be driven by standards and not time. Those standards are defined not only by testing but also by well established software development metrics such as "cohesion" and "coupling." [2]

Cohesion is a term relating to the degree to which a software module performs one and only one task. Coupling relates to the degree to which a software module depends on other modules. Both together determine the overall complexity of the software code. [3] Obviously, the least complex the better code is. It makes the software more reliable and easier to maintain. For an EIS the iterative processing model lends itself to high cohesion and low coupling, the most desirable of conditions for limited complexity. The iterative processing model is used for extracting specific data from databases when it is not necessarily clear from the very beginning what conditions should be used to query the data. It's more commonly used in mathematical algorithms but works well for knowledgebase algorithms too. [4] A typical EIS user may produce a report, examine the report, refine the conditions then produce another report, and so on repeatedly until the desired report appears. So, one main module queries the database and returns a report based on a set of conditions specified by the user. As the user changes the conditions or supplements them the same module reports a different set of data. That module is highly cohesive since it performs only one action, producing a report from a set of conditions of which the module itself cannot change. As for coupling, that same main module has low coupling since it depends only on the set of conditions supplied to it directly as

input. That module can function on its own provided it has the set of conditions as input. In the case of a web-based EIS that set of conditions would be provided in a single web page address containing name-value pairs such as, "date=1/1/2006&sortdirection=ascending." That simple consistency of a single web page address producing each report helps to provide the flexibility needed for a good EIS design.

Flexibility means the ability to adapt to a changing situation. In the case of an EIS, the organization of the campus facilities may change or the reporting requirements may change as time goes on. Any report created from a simple web address allows a great deal of flexibility in how reports are distributed, perhaps by email or as a link in another web application. It also helps to create interactive reports with links that produce other reports or graphs.

A simple "centric" data model may help to accommodate changes in the organizational structure. See Figure 36-2. "Centric" means that the main tables in the database are all directly related to each other. In other words, any one main table can relate to any other main table directly without the need for intermediate table relationships. Other tables may stand alone or relate in singular to the circle of "most important" tables but no more than one level deep. This insures that changing the content of any of the main tables does not "trickle" down several levels into a maze of complex database structures.

Since most EIS applications are not required to report in real-time. A nightly pre-processing step may aid in the opportunity to accommodate some new features. Some calculations performed in this way speed up the response time of the reports in the web browser. Whether to pre-process the data or not depends on the overall goals of the project described in the mission statement.

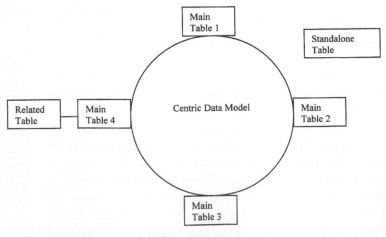

Figure 36-2.

MISSION STATEMENT

An EIS project mission statement is a clear and concise statement of purpose providing much needed focus throughout the development process. It must contain the purpose of the EIS or answer to the question "why is an EIS needed?" It also should contain a well defined action as the main idea of the statement. The mission statement will help keep any new requirements on track with the original purpose of the EIS. It is preceded by the creation of a statement of work, authored by the organization of which the EIS is benefiting and a requirements analysis conducted by a sampling of users. The requirements analysis results in a list of requirements to aid developers in the specific features required of the EIS.

So how can the mission statement help to provide focus? Each new requirement should be compared to the mission statement and the question asked, "does this requirement support our mission?." If the answer is "no" then the requirement should be omitted or modified. If the answer is "yes" then it is an acceptable requirement to add to the list of potential features. That is why a mission statement must contain a purpose, usually prefaced by a clause such as "in order to." A simple example might be "Our mission is to develop an EIS in order to save 5% per year on our energy costs…" Each requirement would then have to support the 5% per year savings goal to be acceptable. Of course, a real mission statement would not be that simple. The action clause "…develop an EIS…" should be much more defined either in the mission statement itself or by reference to some other document or accepted practice such as "…develop an EIS in accordance with our company software development practices…" In fact, the mission statement should provide answers to the "five W" questions of who, what, where, when and why. Here's an example of a well formed mission statement: "The mission of project EIS is for our information technology department to develop an energy information system as defined in the statement of work in order to aid our department managers to analyze and report on energy consumption and conservation measures over the next several years." Once the mission statement is complete and agreed upon developers can begin working on a system design that supports the mission.

SYSTEM DESIGN

System design is an overall outline of how the EIS will function and who will be involved with it. Developers consider who will take on certain roles and what their responsibilities will be. These roles must be well defined in order for the staff to stay focused on their tasks. Developers also conduct a thorough data analysis starting with the data received from the meters complete up through the organization infrastructure to the very top level of reporting. They derive the user interface design from the requirements list and user interviews with potential users. They have to make decisions about what database, operating system, network and development software to use that is consistent with web access throughout the organization, if it already exists. Someone has to decide where on the network to store all the data, programming code and documentation in such a way as to make them easy to revise and maintain. A good requirements analysis will already contain a timeline. [5] A development log coordinated with the items on the requirements list will provide traceability as to the progress of the project. Since this development methodology depends heavily on user feedback it is important to have well thought out plan for releasing each successive version of the prototype easily to the user for evaluation. Let's examine who might be performing all of these tasks.

A staff of three is well suited for EIS development. Their roles are well defined to prevent duplication of effort and allow them to stay focused on their tasks. See Figure 36-3. A "designer or programmer" can create the web site and programming code to query the database and produce the reports. This person designs the user interface, decides which software to use and organizes the file structure on the network. This person would also define the release plan deciding when and how to present the prototypes to the users. A "tester or data expert" would ensure the requirements are understood, monitor the timeline and help with the data analysis. This person is an expert at understanding the utility data and what needs to be done with them to produce the desired results. A "data manager" can assume the role of data analysis from the very beginning. This person would seek out the sources of the data, document their whereabouts and start organizing it in a way that is helpful to the programmer. This staff of three requires a specific skill set as shown below.

Data analysis for an EIS involves collecting data automatically and storing them in a convenient location. The data must belong to a well defined infrastructure that is hierarchically sound if possible. Any pertinent data that do not fit into a hierarchy can be stored in separate tables and related to the database as needed. To begin with, the data manager finds the source of the consumption data be it a utility company, meter data collection device or an energy management system.

Roles	Skill Requirements
Designer/ Programmer	1. Software engineering skills 2. Database skills 3. Organizational skills 4. Network administration skills
Data Expert/ Tester	1. Knowledge of utility industry 2. Understand numeric relationships well 3. Ability to demonstrate EIS to an audience
Data Manager	1. Basic computer skills 2. Organizational skills 3. Communication skills

Figure 36-3.

Those data are associated to a meter which in turn is associated to an account. Each account is associated to some location or building referred to by name or address and those locations are probably grouped together in any number of ways that make up the infrastructure of the EIS. There may be other information about the meters, accounts and locations that needs to be stored in some manner in order to appear in a report from time to time. These values typically go into reference or "lookup" tables. A good data analysis gives the programmer the ability to show the complete infrastructure from meter to the highest level of the organization. A sample infrastructure is shown in Figure 36-4 below.

This would be a good first step in developing the user interface. That is to show the users the complete infrastructure to see if anything is out of place and to see if they can navigate among the different levels as needed. Sitting down with the users and showing them the infrastructure is a good way to determine if they understand the features of a good interface. A good interface always keeps the options available to the user visible but has a minimum number of "buttons" to push to produce results. One test is to count the clicks required to produce a report. The fewer clicks required the better. The results should be consistent and not buried beneath several different pages. If this first test

of data analysis and user interface design is successful the programmer can make a final decision on which software to use to complete the EIS.

The database software is really the basic building block for an EIS. So it should be one that works well with the web. A database that allows the programmer to work with tables on their own and not tied up into a database container is better because it makes it easier to export and import data one table at a time. It might also help in debugging. Almost all of today's database engines support text indexing, that is using non-numeric values as the key field in a database table. This is a great feature that eliminates creating fields simply for the sake of acting as indexes to link tables together. The database engine should, of course, be very fast. Web data queries seem to take longer or maybe we expect them to be faster. In any case, a fast database engine is always desirable. The operating system and network software choices are most likely going to be the ones already in use at the organization. There is probably no advantage to using anything other than the systems already in place. The programming language should be a "higher level" language as opposed to any scripting language designed specifically for the web. Calculations and features of the EIS may very well require the versatility of a more robust programming language. It too must be compatible with the web, widely used and well supported. The choice of programming languages may have an effect on where files are stored and how they are organized.

File locations are a big part the organization of the EIS design. For the sake of speed and reliability the database files should be on the same machine as the web server even though that is not a strict requirement with database software these days. But, it does stand to reason that it will improve performance if all files are stored on the same server if possible. The programmer should organize the files by function and store them in folders named accordingly. All the configuration tables for instance should be stored together in one folder with some appropriate name. It helps to put all programming code in one file if possible. This goes against some outdated accepted practices of modularity, however for practical debugging purposes modularity is not neces-

Campus	Facility	Building	Account	Meter
ABC University	East Quadrant	Library	37465894	238756498
ABC University	West Quadrant	Cafeteria	37445698	548796321
ABC University	North Quadrant	Gymnasium	37433256	254896521
ABC University	South Quadrant	Science Building	37445871	935687412

Figure 36-4.

sarily a good thing. With the text search capability of the programming language interfaces we have now grouping all the procedures and functions into the same file is an advantage in finding and debugging specific code problem areas.

The timeline is an important part of the system design because it provides a means to stay focused on aspects of the project in the order in which the application needs them. It is a prioritized list of tasks that describe the project design. The accomplishment of those tasks can be recorded in a development log along with notes pertaining to any problems or changes that occur. See Figure 36-5. This gives traceability from the development course back to the project requirements ensuring that work did not drift off track. It also ensures that each requirement was addressed and that no shortcuts where taken. The development log can record the release of each successive revision and notes on its evaluation. Where applicable the log can track the hours spent on each task.

Date	Task	Hours	Revision
6/2/2006	Reviewed infrastructure requirements	6	1.01
6/3/2006	Designed data model	8	1.01
6/7/2006	Reviewed external sources of data	8	1.01

Figure 36-5.

The release plan is important in that it describes an easy way to move new features out to the data expert and users for evaluation. Each successive revision should be numbered and displayed so as to help in communicating differences between the revisions. Each revision should have a recorded release date in the development log. A good release plan will account for preserving any changes made to the configuration tables on the part of the users. Such changes as passwords, stored reports, etc. should not have to be recreated with each new release of the prototype.

PROTOTYPE

A prototype is a working version of the finished product with limited capabilities. Its purpose is to confirm that the system design meets the description set forth in the statement of work and the requirements list.

Many developers take a shortcut approach to prototyping by creating a slide show of pages to show how the application looks. However, this approach usually turns out to be wasted effort since the users cannot really experience the "feel" of the application without actually using it. An actual working prototype may evolve into the application itself as long as its initial assessment is positive and a change in development software is not necessary. This makes the time spent developing the prototype a valuable contribution to the overall effort. Users will be much more likely to communicate feedback after working with a functional prototype. It's also easier to demonstrate new features on an ongoing basis using a functional prototype. Over time it will act as a preliminary training tool to familiarize the users with the application. The initial prototype is the basis for further development at a more detailed level. Developers add specific features from the requirements list to the prototype as part of the detail design component of this methodology.

DETAIL DESIGN

All that remains for the detail design component is to integrate the desired report characteristics derived by the data expert and the users, install security features as required and organize the data tables. The report characteristics include what reportable values to display in the reports and how to filter or summarize the data across the different levels of the infrastructure. The security features depend on the organization. If the application is behind a firewall then security concerns may be minimal. Otherwise, a security certificate and encryption will be required. Data tables should be organized according to the desired reporting needs. But, since reported values will more likely span across time rather than locations it may be appropriate to divide the data tables up into focused data sets based on the infrastructure.

Start with the reportable values needed, for instance, electric consumption values such as kilowatt-hours. But there are many possible reportable values. There are demand values and charges, overall costs, transmission costs, billing charges, peak and off peak values. Then there are a number of values related to other utilities such as gas, water and oil. There could be as many as hundred possible reportable values or more. Along with those values comes the correct terminology for each, what units they are reported in and how many decimals to use. Then there are calculations to perform on these values to create other reportable values. So where does one put all of this information? Put the values information in one values table with a record for each reportable value. Only authorized users should modify the characteristics of the values in the values table.

Developers authorize users using an authorization table. Each user id is associated with what ever features are available and assigned a "yes" or "no" value for each one allowing or disallowing access to the feature. Password tables track the user's login password and associate it with their assigned user id. Once a user has logged in the features they have permission to see are available and the others are not. Also, the developers can control access to certain data sets in this way as well.

Developers should not program compound queries that merge several data tables together into a *single data set* every time a user clicks on a link to view a report. They are much too inefficient for web-based applications. Summarizing operations performed in a pre-processing step are much more efficient. Data can be stored in *focused data sets* based on the infrastructure levels such as campus, facility, building, account and meter. For instance, to show the total kilowatt-hours per year for all the meters at particular campus over several years it would be easy to summarize the total kilowatt-hours for all meters and store the result in a table for the campus as a whole. The campus table would have one record for each month with the total kilowatt-hours. Then the only

summarization needed at the time the user requests the report would be to summarize the months into totals for each year. Figure 36-6 shows the difference between a *single data set* and *focused data sets*. With this functionality the application is getting very close to becoming the finished product and ready to be released as a production version rather than a prototype.

PRODUCTION VERSION

The difference between the prototypes and the production version other than the fact that the users have accepted the usefulness of the application is with configuration features and a user guide. It is best to hold off on designing forms to access configuration items until the end of the development cycle since the number and type of configuration items is likely to change a lot during development. The same goes for the user guide since many features are added during development.

Administrative users can modify the configuration items held in a configuration table once developers release the application for production use. This is simply a lookup table of name/value pairs that control the behavior of the application. These particular configuration items usually do not apply to any one user but rather

Single Data Set

Year	Month	Campus	Facility	Building	Account	Meter	kWh
2006	1	ABC University	East Quadrant	Library	37465894	238756498	12,365
2006	1	ABC University	West Quadrant	Cafeteria	37445698	548796321	15,235
2006	1	ABC University	West Quadrant	Storage Room	37444596	857496521	9,215
2006	2	ABC University	East Quadrant	Library	37465894	238756498	11,456
2006	2	ABC University	West Quadrant	Cafeteria	37445698	548796321	14,254
2006	2	ABC University	West Quadrant	Storage Room	37444596	857496521	8,114

Focused Data Sets

Year	Month	Campus	kWh
2006	1	ABC University	36,815
2006	2	ABC University	33,824

Year	Month	Campus	Facility	kWh
2006	1	ABC University	East Quadrant	12,365
2006	1	ABC University	West Quadrant	24,450
2006	2	ABC University	East Quadrant	11,456
2006	2	ABC University	West Quadrant	22,368

Figure 36-6.

control the application as a whole based on the current conditions. For instance, this is where the version number is stored and the paths to file locations.

The users guide should be a living document that includes basic instructions, context sensitive help and a knowledgebase of frequently asked questions. The basic instructions should be readily available to the novice users. The context sensitive help feature can be a help button that displays information about the page the user is currently using or a toggle link that displays help information on each page or not depending on which state the toggle is in. It is important to update the frequently asked questions knowledgebase regularly. Developers could tie it to a form used for asking the questions themselves. A user could enter a question into a web form and submit it. When the developers answer the question the application submits the question and the answer automatically to the knowledgebase for review by others.

CONCLUSION

Developing a custom EIS is the most effective way to assure the ability to report meaningful, accurate and timely energy consumption information. It is not easy. But, any qualified software engineer can do it provided there is a data expert to help. It's important to keep in mind the quality goals; use the successive revision technique to refine the design, use good software development metrics such as high cohesion and low coupling and make the design as flexible as possible using a simple data model. The mission statement should be the easily accessible guide to deciding which new features are appropriate for the system. Without a doubt someone will want to add a feature that really does not fit and

will take the development team on a wayward course wasting time and money. In those cases, let the mission statement demonstrate that the request is not reasonable avoiding lengthy debates. The overall system design should capture the essence of the requirements for the EIS. It should describe the foundation for the application and the mechanism the users take part in to access the EIS and produce results. The initial prototype demonstrates this essence or overall system design. Also, prototyping is the best way to facilitate the successive revision process and it adds so much more to the development effort. It gets potential users involved early and is a great communication tool to discuss features, analyze data and conduct preliminary training. The detail design component is simply the process of repeatedly adding new features and testing them. Most of these features should have been listed in the requirements list but do not have to be ideas that were there in the beginning. The finished product is a revised application accepted by the users that contains access to all the configuration items necessary to maintain the EIS as well as a good user guide.

References

[1] Capehart, Barney; "Using Virtual Metering to Enhance an Energy Information System"; Information Technology for Energy Managers, Vol II; Fairmont Press, Inc., Lilburn, Ga. 2005.

[2] Tallett, Internet page, http://codeguru.earthweb.com/Cpp/misc/misc/applicationcontrol/article.php/c10347/; Jupitermedia Corporation, September 23, 2005

[3] Internet page, http://www.c2.com/cgi/wiki?CouplingAndCohesion; Cunningham & Cunningham, Inc., June 5, 2006

[4] Kurland, Lee, Domshlak; "Better than the real thing? Iterative pseudo-query processing using cluster-based language models"; Internet page, http://www.cs.cornell.edu/~kurland/pf.pdf; Cornell University, Ithaca NY, accessed June 9, 2006

[5] Capehart, Barney; "Developing an Energy Information System: Rapid Requirements Analysis"; Information Technology for Energy Managers, Vol II; Fairmont Press, Inc., Lilburn, Ga. 2005.

Chapter 37

Developing an Energy Information System: Working with the Budget

John B. Broughton, CEM, Cape Canaveral, FL
David C. Green, Green Management Services, Inc., Fort Myers, FL

ABSTRACT

THE PURPOSE OF THIS CHAPTER is to help energy managers (EM) to become more creative in finding dollars for implementing an effective energy information system (EIS) within the constraints of their existing budgetary structure. Generally due to the cost and scope of an EIS project it never makes the budget.

However, some groups within an organization may have larger budgets and be willing to share in the expense of an EIS project. An EM may be able to partner with these groups on projects that have similar methods or goals to an EIS.

INTRODUCTION

Typically each year companies allocate funds for the various functions required to conduct business, in what is known as a fiscal year budget. The budget consists of an itemized list of required expenses that the company projects that will be needed. Each of these budgeted items is reviewed during long arduous meetings by the key decision makers of the company to determine what their associated costs will be and the expected rates of return. The creation of a fiscal budget provides for the efficient operations of a company as well as the necessary checks and balances to ensure corporate growth and shareholder profits.

Line items such as marketing, capital improvements, maintenance, and utilities typically consume the majority of the budget. The remaining balance of the budget is available to the many groups within an organization for new and improved projects. To be approved for this money the groups must vie for attention and present justifiable projects to the financial planners in the hopes of theirs being chosen for funding. It is left up to the expertise of the energy department or energy manager (EM) to prove that an (EIS) is a feasible project for implementation.

By far the largest obstacle an EM must overcome is the company's budget. From the purchasing department to the boardroom businesses are constantly aware of the increasing costs associated with energy [1]. If a business's management has not already made the commitment to energy management then it becomes the crusade of the EM to convince them of its necessity and worth. By preparing documentation, educating management, and providing estimates to financial planners the EM is constantly struggling to acquire the sufficient funding for implementing an EIS system.

In an ideal world corporations would see the relevance of being able to track their utilities consumption so that they could efficiently project future trends and determine areas for cost savings initiatives. The ideal world would follow the recommendations of the Energy Star's Guidelines for Energy Management diagram shown in Figure 37-1.

In Figure 37-1, Energy Star's Guideline for Energy Management starts with a commitment from the organization at the upper levels and therefore implies some sort of funding is allocated to an organizations budget for implementing this process. An integral part of this process is the implementation of an EIS system to provide for proper oversight and controls.

However, in the real world most companies have not yet made the first step toward successful energy management, making the commitment. The commitment is made when funding for its implementation is part of the budget. Whether it has been due to inadequate cost justification to the financial planners, or misconceptions by the facilities department about an EIS, energy management seldom makes the cut.

Energy managers understand that their projects are only a small part of a larger picture within their organization. An effective EM works closely with the companies financial decision makers to determine other sources of possible funding. By co-partnering with other groups

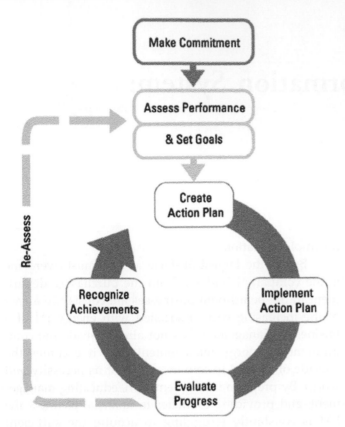

Figure 37-1. Energy Star's Guidelines for Energy Management Diagram

within the organization the EM can accomplish energy management projects that will help to build an EIS.

Figure 37-2, lists some common organizational groups that usually have larger budgets and may be open to co-partnering energy saving projects.

Typical Organizational Budgets by Group

Repairs/Facilities Maintenance
Training/Education
Security
Capital Improvement/New Construction
Utilities/Telecommunications
Emergency Repairs
Rebates/Found Money

Figure 37-2

REPAIRS/FACILITIES MAINTENANCE

Each year a major portion of a corporation's budget is allocated to the maintenance and upkeep of equipment and property. Numerous energy savings opportunities exist in the expense of those funds. This money is spent via the facilities department and sometimes without any regard to energy efficiency or energy goals. The facilities department is held accountable for maintaining assets in a cost effective and expeditious manner within their assigned budget.

A creative EM will frequently attend facilities meetings to gain a better understanding of the daily operations of a business and to stay informed of any upcoming facilities projects. The constant interaction with the facilities group and maintenance personnel will expose areas where the maintenance or repair of existing assets may offer potential energy improvements. Providing support to the facilities group that helps them to meet or exceed their group's goals may reveal possible future energy projects. Figure 37-3 is a list of priority items that facilities groups face over the course of the year. These activities usually require the most time and effort for the facilities department.

Since many manufacturers now offer a variety of equipment with many real time monitoring capabilities the EM should strive to recommend and procure those whenever possible as long as they fit their proposed EIS system. These devices will provide for the useful tracking of the equipment for maintenance purposes and the data for an EIS system. As long as the incremental cost of the equipment is not unreal then the facilities group will be amiable to purchasing. The installation of equipment that can someday be linked into an EIS system at a later date will only reduce the cost of that overall system. Therefore the installation using facilities maintenance dollars can go a long way toward reaching the goal.

The crucial concept for the EM is to have an established EIS plan which involves an open ended protocol for a wide variety of equipment. The EM should be fully aware of the information technology (IT) industry and be specific with the requirements as how they will apply to the overall EIS system. The EM should also be redundant in his EIS schedule to allow for various computer languages or platforms that can ultimately be interconnected to the desired EIS.

TRAINING/EDUCATION

Often times each subgroup within an organization is allocated money to train and educate their personnel on a yearly basis. This budgeted line item provides an excellent morale booster within a company and the necessary education to keep the employees abreast of technology advances in their respective professions. Since energy management encompasses many aspects of computers, construction, equipment, maintenance, man-

Typical Repairs/Facilities Maintenance Items	Energy Conservation Opportunities
Lighting	EIS compatible equipment, motion sensors, lighting upgrades
HVAC	EIS compatible equipment, energy efficient equipment
Electrical Motors	EIS compatible equipment, energy efficient equipment
Utilities Distribution	EIS compatible equipment

Figure 37-3

agement and more, it offers a portal for possible cross training within the various groups.

Since energy conservation is a priority with the government there are plenty of free courses that apply to the various groups within the company. Typically these various groups are not aware of these events and should be notified by the energy community well in advance in the hopes of them attending. Knowledge is Power, by educating the masses the energy manager finds out that projects that they where requesting are now being requested by the other groups who have better budgets then the energy department. Some of the larger training courses can be found by visiting the Department of Energy's (DOE), Energy Efficiency and Renewable Energy (EERE) website, and the DOE Federal Energy Management Program (FEMP) website. These courses provide for low cost no cost training that hopefully will insight action by their attendees in reducing energy consumption. With regard to EIS systems, by making personnel aware of their scope and benefits it will hopefully create a reactive environment where they too are requesting the EIS system.

SECURITY

In recent years budgets have been increased for security purposes. This increase in funding has set security groups scrambling for potential projects that will improve the safety of the workplace. Typically security personnel like to see or record activities in real time so that they can monitor and determine threats as quickly as possible. Most of these monitoring systems require power and data connection back to a central location. They require the manpower and equipment for a security system that is analogous to the EIS system proposed by EMs. This similarity provides opportunities for an EM to co-partner with the security group to support their projects and specify equipment.

CAPITAL IMPROVEMENTS/NEW CONSTRUCTION

Companies are constantly looking at increasing their assets which in turn will maximize their net worth on the balance sheet. New construction is a big part of a corporate strategy. New construction projects are a high priority with energy managers because they allow them the flexibility in specifying the proper equipment for installation. In recent years the United States Green Building Council, Leadership in Energy and Environmental Design, has offered an excellent guideline to new and existing construction projects to ensure minimal environmental impact on the environment in the construction of new and existing projects. Some of the larger corporations, school boards and government agencies have already requested that new building be designed to certain levels of certification by LEED. Therefore the EM has viable budget dollars to implement energy projects. [2]

Once capital improvement projects have been identified, the EM should become an integral member of the design team to make recommendations that will not only meet the requirements of LEED but also to help them to achieve their energy goals.

UTILITIES/TELECOMUNICATIONS

Almost all companies have large expenditures for telecommunications. This budgetary line item requires yearly funding and is understood to be a necessity for operating a business by the financial community. With little to no scrutiny each year a large portion of a company's operating budget gets allocated for these devices.

With recent advances in cell phone technologies an EM can attempt to upgrade their energy data collection processes by recommending the procurement and use of certain cell phones. Today's cell phones are more than just communication devices. Some of these high tech gadgets provide internet access, e-mail, simplified office functions like excel spreadsheets and word documents, and as of recently radio frequency identification (RFID) capabilities.

This latest technology, RFID provides a system for meter reading via the cell phone. By using something called near field communication (NFC). An RFID tag is placed on equipment such as a utility meter. When the RFID enabled phone is brought within close range of the RFID tag, the phone automatically boots up a pre-programmed routine. In this case a meter reading screen would appear for the user to input data in the field. Once the data are keyed into the phone and verified for transmittal they are sent via an internet service provider (ISP) to a central server. The data can then be accessed by the EM for review and future use. At present Nokia has an RFID kit that attaches to the Nokia 5140 phone. More information regarding this device can be obtained from Nokia or their representatives. Figure 37-4 is a conceptual diagram of the wireless capabilities of RFID and cell phones. [3]

Figure 37-4. Conceptual Diagram of a RFID Phone Meter Reader System

Since cell phones are already considered a necessary budget item, an EM can propose the procurement of these devices in an attempt to increase the company's energy data collecting capabilities.

This is only a small step to bringing together a complex network of sub-metering throughout a company, but once this process of recording data is established and data are being collected then the next phase of presenting and proving the value of these kinds of data can

be achieved. Typically all energy information systems start small, allowing the energy manager to begin a unique data collection process that can lead to obtaining additional funding.

Another useful aspect to the dual use of cell phones for meter readers is that in recent years the Internal Revenue Service has become increasingly aware of the fact that many of these devices are being used for non business activities. Some industry experts believe that the IRS may be looking at greater accountability standards for these devices. Therefore the dual function of using a phone as a meter reader will only add to the validity of these expenditures.

EMERGENCY REPAIRS

One of the most obvious opportunities for an EM is when repairs are required in the aftermath of a catastrophic event. Disasters such as hurricanes, tornadoes, and floods all seem to be occurring with more frequency in recent years. Whether you're a proponent of global warming or not, the fact is that with their increased activity companies have begun to increase budget money for them. The repairs that take place in the wake of these types of events are typically paid for by insurance claims or other capital asset reserves. Typically these repairs occur quickly and are "short tracked" through the accounting department to ensure the business is operational and back to a profitable venture as soon as possible.

During these emergency situations the EM must become pushy and involve themselves in all aspects of the rebuild. Since actual replacement of equipment is typically not available due to obsolescence or lead time an alternate is sought. The EM can provide support and assistance to the groups that are most active in the emergency repairs. By helping them to complete the work as quick as possible the EM may find quick solutions to upgrades or repairs they have been seeking over many years. An EM can recommend the best fit equipment for motors, heating-ventilation-air conditioning systems, or other utility grade components that will later easily interconnect with their desired EIS.

REBATES & FOUND MONEY

Many times the money for an EIS or equipment leading up to its design and implementation can be found by conducting an in-depth and thorough audit of a company's paid utility bills. Utility companies are not infallible and have been known to make errors in billing. By checking paid bill against rate schedules and verifying meter readings, tax charges, and franchise fees an EM may reveal a situation where too much was paid to the utility. If this is true then the EM should notify the utility of this error and request a refund. Typically utility companies will provide a credit to the account if they are in agreement with the error. That credit leaves extra money in the utilities budget that the EM should request for energy projects.

Keep in mind errors are just that and that utility companies are not out to "get us" so to speak. They gladly work with EMs to correct errors as well as point out areas where they think energy can be conserved. A special note here is to make sure that you fully understand the billing process because sometimes when an error in your favor is pointed out another error in their favor may result. Hence you owe them more!

This so called "found money" which companies usually hold in a reserve account or put into a general fund for use by various disciplines should be set aside for energy projects. The "found money" at the time of receipt could be transferred to an account for EIS or sub-metering with an aggressive stance by the energy manager's input to the company's financial management.

Other opportunities are available with utility rebate incentives. Utilities offer rebates for the purchase and installation of energy efficient equipment. This money again is refunded to the customer through a credit to the utility bill. This leaves extra money in the budget since it was not necessary for bill payment. The EM should request that the financial department set that money aside for energy projects.

CONCLUSION

The goal of an energy manager should be to assist commercial and government property owners and managers in gaining control over their largest operating expense, their utilities. With energy costs continually increasing, these services can prove to be very effective. Other benefits such as enhanced property values, improved tenant satisfaction, and superior corporate image may also occur. Energy information systems tend to be an important step in the right direction toward that goal. All energy engineers should have a basic plan on what their final energy information system model will look like, how it will function and what they expect it to do upon completion of its installation. An EIS requirement statement should be an easily accessible guide to deciding what new features are appropriate for a system.

Once the system requirements of the EIS have been established it is almost guaranteed that there will be no room in the fiscal budget to fund such an enormous project at one time. Therefore, it is left up to the EM to come up with ingenious ways to incrementally gather funds to piece their system together over the years. The facilities and maintenance department may be willing to install some real-time monitoring equipment as part of a repair project or new construction. Training opportunities may bring new awareness of the benefits of an EIS. Security and telecommunications projects could provide network access or innovative data collection devices. And then there is always the "found money" created by examining the utility bills with a "fine tooth comb." Researching some of these ideas could help in finding money to implement a comprehensive EIS.

References

[1] Bennett, Charles, J.; *"Navigating Energy Management: A Roadmap for Businesses"*; Internet page, http://www.energystar.gov/ia/business/guidelines/Navigating_Energy_Management.pdf; Department of Energy, (2005).
[2] "LEED: Leadership in Energy and Environmental Design"; Internet page, http://www.usgbc.org/DisplayPage.aspx?CategoryID=19; United States Green Building Council, (2006).
[3] Romen, Gerhard; "NOKIA Unveils RFID Phone Reader"; Internet Page, http://www.rfidjournal.com/article/articleview/834/1/1; RFID Journal, (2004).

Section VII

Future Opportunities for Web Based Enterprise Energy and Building Automation Systems; and Conclusion

Chapter 38

My Virtual Value Visions for Building Automation Beyond 2006

Ken Sinclair, Editor/Owner AutomatedBuildings.com

MY WIFE JANE AND I were very pleased to be honored with last year's Vision award at BuilConn 2005, in Dallas. This award chosen by more than 400 of our peers was given to us as the folks who best support the vision of networked building systems. As BuilConn Europe in Amsterdam and BuilConn 2006 in Palm Springs, CA, in May 2006 have now passed, we feel compelled to dig out our future vision glasses and peer into our IT-centric future to try and restore the faith of those who called us visionaries.

All my visions beyond 2006 include a new ingredient I like to call virtual value. I think we all know what the word value means, but adding the word virtual requires some clarification. There are a lot of definitions for the word virtual but the one I like is "existing only in software." It is a simple definition and lets me drive home what I want to talk about—virtual value and my vision of it.

What do BuilConn Visionaries know about virtual value? Jane and I have been making a living for the last seven years selling virtual holes for advertising in the virtual media website we call AutomatedBuildings.com. Ten years ago, I would not have thought this possible, but now it seems to work and does demonstrate the concept of added virtual value. It has value, yet it only exists in software.

The process of assembling AutomatedBuildings.com along with the reading and editing of the large amount of monthly insightful content, as well as absorbing the information that evolves out of BuilConn and the myriad industry events we attend, propels me to extend my thinking forward to provide you my vision of the importance of virtual value to our industry in the next few years. The fact that I am assembling these visions in another hemisphere, half a world away, in the day before that time will occur in my North American home is a testament to the anywhere-ness value of virtual mediums. My visions for building automation beyond 2006 follow, with a brief description of each.

True Convergence in Our IT Centric World Will Occur with or without Us.

Our success is dependent on our ability to step out of the comfort zone of our industry and enter into the convergence industry. The new convergence industry is built on virtual values that exist only in software but bring great value to our clients. I love this quote from Henry Ford, the automotive pioneer: "If I had asked people what they wanted, they would have said faster horses." I believe there are those in the traditional building automation industry who are designing faster horses, rather than trying to get their minds around the fact we are moving into a new era when all the rules will change and everything will be done differently.

The convergence industry will lead us to new solutions and new ways, whether we are part of it or not. Trust me, we all do want to be part of this exciting new era. In my December 2005 Engineered Systems Building Automation Column "Guess Who's Coming To Dinner" and my January 2006 column "The New Kids on the Block are Giants" I talk about Cisco and Hewlett-Packard (HP) entering our arena. This is a very positive step for our industry but it is the harbinger of a new era. It is no longer about horsepower, it is all about change and successful convergence. Do not fear change, embrace it. These giants will need all the partners they can get to help define our virtual value future.

Presentation and Interaction of Our Industry Information Will Occur Only in Accepted IT Formats, such as Web-based Browsers, Cell Phones, VOIP, Video Broadcasting, Digital Signage, etc.

The amount of man hours of virtual value added when we become part of the IT-centric world has made it impossible and undesirable to keep up the development of custom and unique interfaces for building automation. New IT power and new mediums provide unheard of reach and power to the building automation industry. We must now conform to evolving IT standards.

Network Knitting Now Greatly Amplifies the Virtual Value Model.

Open protocol supplied equipment is making connection to the virtual world easier, but hurdles still remain to provide a truly integrated project. A good example of this is that over the years, building automation has become pointless. Years ago, DDC points were added to a system to improve energy performance and make the facility easier to operate. The tide has since turned, and now almost all devices are equipped with DDC points as an integral part of the device. If the manufacturers and designers of the HVAC and other building-related devices have done a good job of "future proofing," they will have adhered to one or all of the open communication protocols.

These integral points are accessible as part of an open network of points which results in a radical drop to the building infrastructure cost. A large part of today's large building automation task is connecting the networks of the points supplied with each device or subsystem into a useful integration that becomes a valued part of our client's enterprise.

The Web Will Continue to Lead the Way with Strong Examples and Models to Follow for Adding Virtual Value to Everything in Our Industry.

Most of the methods of presenting information now exist on the web; on-demand video, sound, radio, news feeds, narrowcasting, PowerPoint and Flash presentations, etc. All point clearly to the way the intelligent building communities can be created with displays of static and dynamic information on a network of interactive I/O such as digital signage.

To add the desired virtual value to our clients projects we simple have to understand the information flow requirements and determine how best to achieve them. We have all the tools and examples we need, and in the future, new methods are being developed, not by or for our industry, but to improve the overall process of IT communication.

The advertisers and the content provided in Engineered Systems and AutomatedBuildings.com supply vivid examples of the race to build virtual value into products and services. Providing links back to legacy systems and leaping forward to powerful, but cost-effective, web services are great examples of added virtual value.

The focus for efforts beyond 2006 will be on providing the greatest virtual value. We now have the capability to add virtual value to almost anything and web-based IT-centric solutions allow us to leap frog costs effectively.

The Puzzle Piece Players Will Provide Powerful Software and Web Services that Will Arm Our Virtual Value Tool Kits.

One of the most visible pieces will be flat screen networked interactive digital signage with multi-function displays customized to our intended function. Narrowcasting will allow targeted, on-demand broadcasts to be delivered anywhere in our virtual value environment. This service will carry all of the functions and interfaces of today's systems combining web and video, but will also include powerful VOIP phone integration and whatever else the IT giants bring to the party.

> *I love this quote from Henry Ford, the automotive pioneer: "If I had asked people what they wanted, they would have said faster horses." I believe there are those in the traditional building automation industry who are designing faster horses, rather than trying to get their minds around the fact we are moving into a new era.*

LARGE INTERNATIONAL COMPANIES BUY COMPANIES THAT CAN ADD VIRTUAL VALUE

News releases on the Automated Building Web site show movement by the major companies buying IT solutions to add virtual value. My vision beyond 2006 is that this trend will continue. I have captured a few past examples;

- Nov. 30, 2005—Honeywell announced today that it has acquired Tridium, Inc., a privately held software-development company based in Richmond, Va. with subsidiaries in London and Singapore.

- May 19, 2004—Schneider Electric announces today the acquisition of US-based Andover Controls Corporation, a Balfour Beatty (LSE: BBY) company

- Schneider Electric announced that it will acquire Sweden's TAC, a major player in the global building automation and control market.

- Schneider Electric has agreed to acquire Canadian-based Power Measurement, Inc. (PMI),

- April 23, 2004—Carrier Corporation, a subsidiary of United Technologies Corporation has reached a definitive agreement to acquire Automated Logic Corporation, a technology leader in building automation systems.

- Novar plc of the United Kingdom is to acquire the trade and assets of Alerton Technologies, Inc. of Redmond, WA.

Conclusion

IT innovation for companies that provide virtual value leads to acquisition.

VIRTUAL VALUE ARCHITECTS

To provide the necessary multiple-industry cross-over necessary to bring true virtual value to the enterprise, new players will be required. This virtual architect is not one who works with brick and mortar, but a new entity who uses IT systems and the correct underlying protocols to bring virtual value to the enterprise and the affected community. As we now depend on the traditional architects to turn art into function while preserving community, the new architect will find valuable functions in multi-media and make this more valuable to the online community and his client's enterprise.

This is actually starting to happen; from the December issue of AutomatedBuildings.com, an article titled "Flexible Master Program for Intelligent Buildings" I have extracted the following;

http://www.automatedbuildings.com/news/dec05/articles/reading/reading.htm

Intelligent buildings are about building management, space management and business management. One recent definition is:

An intelligent building is a dynamic and responsive architecture that provides every occupant with productive, cost effective and environmentally approved conditions through a continuous interaction amongst its four basic elements: places (fabric, structure, facilities; processes (automation, control systems;. people (services, users) and management (maintenance, performance in use) and the inter-relation between them.

There are other definitions but they all emphasize integration, responsiveness, flexibility, process and management in business, places and people. Reading University offers a master's level program in intelligent buildings with an emphasis on these issues.

The architect and the owner, with the chief information officer, will control what information is presented, but the people will feedback what information they actually require with the help of their virtual value mentors.

PERSONAL VIRTUAL VALUE MENTORS

These mentors will be necessary because we are entering into a time warp where we do not know what we want from virtual value because we do not know what it can provide. There is a strong place for virtual value mentors to show us the available virtual systems and information that is available to us so we can understand how to best use these new ways to provide our own optimized and personalized best virtual value.

A "digital dashboard" simply means aggregating a wide variety of information from a variety of applications and systems to give executives the collective "overview" needed to make educated business decisions. These virtual value presentations are a great example of what I am talking about.

I have extracted the following virtual values from Paul Ehrlich's (2004's BuilConn Vision winner) article, Part 2: Intelligent Building Construction and Operation." This article appeared in *Engineered Systems'* November 2005 issue in a special supplement entitled What Is An Intelligent Building? and on AutomatedBuildings.com November issue as part of the same supplement

System Dashboards

Like the dash of your car, a system dashboard provides a summary of critical building alarms, energy information and key maintenance items at a glance. The dashboard is responsible to summarize all of the critical building information and present it in the proper format for different members of the facility management team. For example, the operating engineer requires detailed information about specific mechanical systems, while the property manager needs a summary of energy and operating expenses over the last 4 weeks. The difference between a system dashboard and a typical user interface for an integrated building automation system is in focus. Dashboards are more focused on sorting and filtering data to provide the information needed to perform specific roles. Building automation systems tend to be much more generalized and designed for the operating engineer

Virtual Concierge

It is even possible to centralize "high touch" functions such as visitor management. Services are available that utilize video and voice over internet communications to allow for real time interaction with visitors, allowing them to be greeted, present identification and receive building badges all from a centralized, often remote, location!

Tenant Portals

Tenant portals are another method of providing virtual value: One critical element is providing a method for the building occupants (tenants, employees, associates, students, patients, etc.) to interact with the building and building management. In the past, this has been done with phone calls, face to face meetings, and faxes. Today, it is most effectively done with an internal website called a portal. Tenant portals provide information about the facility, contact information, directories, energy efficiency, emergency preparedness and a central place to enter issues. Information from the portal can then be used to drive maintenance and operation requests. Since the portal is a two way communications channel, it can also be used to collect critical feedback on occupant satisfaction and comfort levels.

COMMUNITY VIRTUAL VALUE

Community virtual value simply grows out of the collection of a community of well-executed virtual values. The online community learns from each well-executed lesson and also will learn quickly what does not work and what is not accepted in the community. The presentation of public information such as weather and news is being fashioned into virtual value web services that can be added easily to personal dashboards or home gateways.

Some hardware changes are aimed at making virtual value easier to achieve, such as power over ethernet (PoE). The prospect of PoE will change the balance of this equation dramatically, since the removal of the need to run power to building automation devices will bring significant benefits overcoming most, if not all of the negatives of ethernet. IPv6 will also further drive IP-based architectures for smart devices in buildings.

The large number of existing and readily available non-IP devices, as well as the massive installed non-IP legacy systems base, will dictate a need for gateways and other protocol translation devices. The only logical architecture of such interfaces is to ensure that data and connectivity on the ethernet and IP side is structured using XML and web services. Anyway you look at it, web services will play a crucial role in both the implementation of native IP solutions and the adoption of IP infrastructure as the integration backbone for building automation systems.

Wireless will become batteryless using new energy-harvesting techniques. Stand-alone sensors connected to virtual networks will wake up on demand and send us information only when required. As the number of wireless devices available for building applications such as HVAC, lighting, and security control grows, so does the challenge of how to power them efficiently and independently.

New energy-harvesting techniques, which scavenge minuscule amounts of ambient energy present in the environment, are quickly being developed to power a variety of wireless networked devices including sensors, switches, and the radio electronics necessary to transmit their signals.

Conclusion

Virtual value and our command of it will determine our success for 2006 and beyond.

Chapter 39

Cisco Connected Real Estate

David Clute and Wolfgang Wagener, Cisco Systems

ABSTRACT

THIS CHAPTER WAS DEVELOPED in response to a number of business drivers that are taking place in the industry. Cisco's own experience in managing a global portfolio of approximately 20 million square feet and almost 400 buildings has provided opportunities to streamline its own web-based facility management and energy management systems. Cisco's Workplace Resources Organization (WPR) is responsible for complete life-cycle management of the Cisco global real estate portfolio including strategic planning, real estate transactions and lease administration, design and construction, building operations and maintenance and for the safety and security of Cisco employees worldwide.

The Cisco Connected Real Estate (CCRE) program was initiated in the WPR organization and with the support of many cross-functional stakeholders in various business units, particularly Cisco's IT organization, this program has gained tremendous visibility across the company and now throughout the real estate industry. The successful deployment of the technologies discussed in this chapter depend heavily on the interaction of people, process and tools across a complex "eco-system" of employees, partners, vendors and suppliers. The methodologies presented here are changing the way that real estate is developed, used and managed, and is shifting the basis of the real estate business model from one based solely on space to the provision of service.

INTRODUCTION

Responsiveness. Innovation. Agility. Adaptability. All qualities that organizations must possess in order to thrive in today's highly competitive global economy. Until recently, however, these qualities have not been readily associated with the real estate that organizations use. That is changing. The real estate sector is in a state of transformation, driven largely by customers from a broad range of industries, demanding more from their assets.

These demands and needs are converging to create a shift in the way that buildings are both conceived and used. The changing real estate business climate is being driven by:

- Customers searching for ways to achieve visibility, transparency, and control over their entire real estate portfolio.

- A drive for innovation and sustainable capabilities to reduce capital and operational expenditure.

- Key stakeholders searching for opportunities to optimize value in the real estate life cycle.

- Industry searching for means to improve competitiveness and differentiation of its offering.

- Saving on energy consumption and achieving environmental sustainability.

- Technology adoption accelerating transformation

- Cisco Connected Real Estate drives value by transforming the way real estate stakeholders—for example developers, landlords, tenants, and others—design, build, operate, and use real estate.

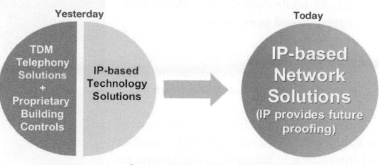

Telephony, IT technology, and Building Controls Converge

Figure 39-1. Building and Technology Solutions are Merging

- In the workplace, Connected Real Estate simplifies the business of providing real estate and allows landlords and owners to deliver effective work environments that drive workforce productivity. This relates directly to the ability of IP communications and innovative technology to drive higher productivity and greater cost savings.

- In a building, Connected Real Estate enables the delivery of powerful services or revenue-generating opportunities that drive business growth by combining real estate and IT. This is possible due to IP being installed as the fourth utility over which these services can be delivered, in-house or as managed services.

- In most parts of the world, energy is expensive and excessive consumption is becoming socially unacceptable; Connected Real Estate enables energy efficiency through planning, visibility, monitoring, and control.

- Last but not least, Connected Real Estate improves safety and security by transforming how building operators and owners can protect their people and assets.

Cisco Connected Real Estate does all of this by harnessing and integrating the power of IP networks. Connected Real Estate is predicated on the three fundamental principles relating to real estate and network interactions:

- Creating a "building information network" or flexible and scalable network foundation as the facility's fourth utility

- The convergence and integration of ICT and building systems onto a common IP network, reducing OpEx and CapEx while optimizing building management and operations.

- Transformation of the physical environment; delivering space differently, and introducing new ways of working.

These three principles form the basis of the Connected Real Estate solution. Together, and interacting in different combinations according to the various needs of developers, owners and occupiers, the principles are

driving the next wave of transformation in the construction and use of real estate.

This chapter explores those principles in detail and demonstrates how Connected Real Estate is delivering huge financial and operational advantages not only to the construction, real estate and property services industries but also to their customers such as hotel operators, multiplexed retail outlets, and corporate tenants in sectors as diverse as hospitality, healthcare, education and retail finance. It is changing the way that real estate is developed, used and managed, and is shifting the basis of the real estate business model from one based solely on offering space to service provision.

This new approach looks along the entire building life cycle, from concept, design and construction through to maintenance and operation. The network and the transformational capabilities it delivers is at the heart of this new approach. And to use it successfully means doing things differently, literally from the drawing board.

DELIVERING BUILDING INFORMATION NETWORKS: THE NETWORK AS THE FOURTH UTILITY

"After four years of market research it was time to renovate our business model. Our strategy was to differentiate this building and all our assets in an already saturated market. One of the methods to achieve this goal was the creation of a unique communications network that would connect Adgar tower with all our buildings in Canada and Europe. The Cisco Connected Real Estate initiative matched our aims."

Roy Gadish, CEO
Adgar Investment and Developments
Tel Aviv, Israel.

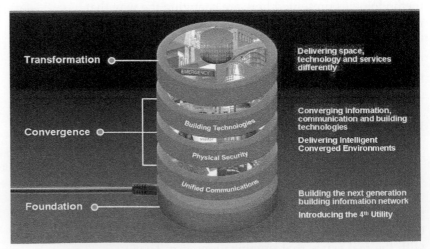

Figure 39-2. Developing Building Infrastructure

Power, water and heating are taken for granted in the construction of a building. To date however, the provision of communications and information networks has been left to tenants. This means that in multi-occupancy buildings, a number of parallel networks are likely to be installed on a piecemeal basis, with each tenant responsible for meeting its own requirements.

Today, however, the demand for connectivity creates a new business model for landlords and developers. The network becomes part of the fabric of the building, supplied to tenants just as water, light and heating are today. By providing the network infrastructure as part of the building's platform, developers provide a point of competitive differentiation to attract prospective tenants. This applies to all forms of real estate whether commercial office space, retail developments, hotels or even residential developments. Connected Real Estate provides landlords and owners with the ability to provide services that respond to the needs of their users, shifting the business model from space to service.

Traditional commercial buildings offer landlords limited opportunities for generating additional revenue from their tenants. And generally the only way that they can compete for prospective tenants is through location and lease rates. In short, landlords sell space. The integration of the network in the fabric of the building, becoming the building information network as the facility's fourth utility, removes those barriers and provides numerous new revenue generation opportunities. Rather than simply selling space, the network allows landlords to move into a business model based on service provision.

As an example, tenants in One America Plaza in San Diego have instant, secure access to communications and data networks providing them with the connectivity that they would otherwise have to acquire themselves.

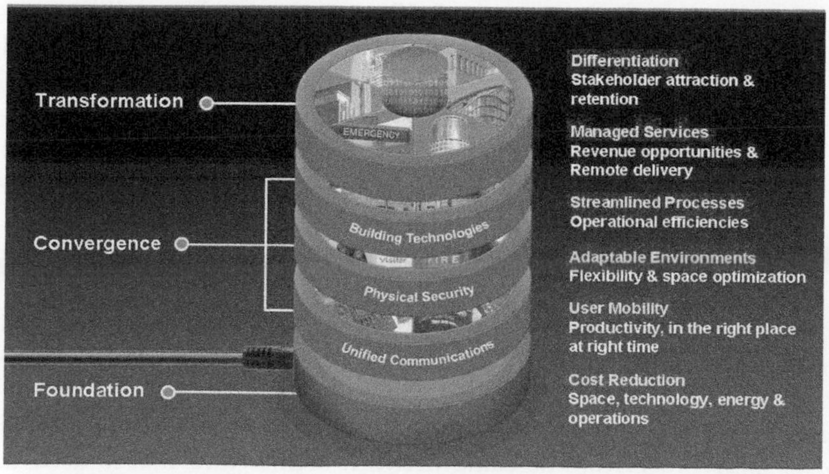

For the building owners, the provision of the network as a utility means that they have a real source of competitive differentiation in the property market, attested to by the occupancy rates of 95 percent compared to an 88 percent average in the San Diego region. One America Plaza tenants are able to connect their operations to the network almost instantaneously (compared to the average 30-day turnaround when ordering from an ISP) and at far lower cost than sourcing provision independently. One America Plaza has a wireless network, meaning that literally the whole building is connected, allowing tenants the mobility and flexibility of working styles that characterize business today.

It's not simply commercial office space that benefits from the integrated offering of IP networks in the fabric of the building. One of the largest real estate developments in the Arabian Gulf, the $1.3 billion Greenfield, mixed-use development, Amwaj Island, will provide residential, commercial, and hospitality tenants immediate access to a range of communications and data services through the creation of a single IP infrastructure backbone that will provide connectivity to every home, business and hotel room on the island. This means that Amwaj Island will be able to provide added security and extra services to tenants such as video on demand for apartments and hotels and digital signage in retail and hospitality facilities over a single network, enhancing the revenue potential for landlords and providing enhanced services at lower costs for tenants.

Having an IP network at the heart of a building does more than deliver service to tenants. It enables significant reduction of move-in and retrofit costs and increases the speed at which they can set up their businesses. Such a network:

- Reduces tenant operating costs
    - Creates more productive work environments
    - Improves flexibility and start-up
    - Enhances user/landlord responsiveness
    - Enables 24-hour availability using Contact Center
    - Provides the ability to optimize building and tenant management (such as track work-order status)

A network that is integrated in the fabric of the building also provides landlords and users with greater control and security of the building's operations. For all building owners integrating communications, security, and building systems into one IP network

Figure 39-3. The Connected Real Estate Value Foundation

creates significant financial and operational advantages.

Boston Properties is a real estate investment company (REIT) that owns and operates 120 properties in major centers in the US, covering more then 40 million square feet. Boston Properties uses an integrated network approach to property management that connects all of its buildings systems to a single converged network infrastructure. This approach provides Boston Properties with the ability to monitor systems including energy management, security, ventilation, and access control around the clock from a single control center. The tenants of the buildings in Boston Properties' portfolio also benefit from this integrated approach. The system allows them to make requests for service over the web, greatly enhancing the efficiency with which their requests are handled and allowing Boston Properties to manage its workflow far more effectively. More than 60 percent of all requests from tenants are now received through the dedicated website. The benefits of centralized building management for Boston Properties became even more apparent when they added a new building to their existing portfolio. They achieved a payback period for "connecting" this new property into the portfolio of less than five months by adding the building's operation to the centralized management infrastructure.

As well as transforming the way that buildings are managed, the network also allows landlords to create services that respond to tenants' needs. For example, in a retail development digital signage can be provided that carries promotional material for a particular store, and content can be tailored to fit the precise requirements of each retailer. Landlords are able to provide instant access to a range of communications services—such as connectivity, internet access, and IP telephony. The landlord, in effect, becomes a service provider responding to the needs of its customers as they develop over time and opening new sources of revenue in addition to rent.

CONVERGENCE AND INTEGRATION: IT AND BUILDING SYSTEMS ON A COMMON IP NETWORK

"The move towards increasing enterprise integration enhances the need for advanced BAS solutions. Companies across all vertical building markets are striving to increase integration across the entire enterprise to improve information management and optimize the strategic decision-making process. As BAS increasingly adopt IT standards, they are increasingly converging with traditional IT infrastructures. Adoption of IT standards in the BAS industry, and the inherent cost savings regarding BAS integration, is causing many building owners to rethink the value proposition of integrated BAS"

(Building Automation Systems Worldwide Outlook, Market Analysis and Forecast Through 2009)

A key element of the business case for the Cisco Connected Real Estate framework is based upon the convergence of information technology and communication systems, security, and building systems onto a single IP network. This next wave of convergence creates opportunities for key stakeholders in the building value chain.

Most buildings and campuses today are constructed with multiple proprietary networks to run systems such as heating, ventilation and air conditioning (HVAC), security and access controls, lighting, and fire and safety as well as separate voice, video, and data telecommunications networks. As a result we see buildings that are complex to operate, with high installation, integration and ongoing maintenance costs, and sub-optimal automation functionality. Typically these generate constraints and inefficiencies such as:

- High CapEx for design, engineering, and installation
- Building performance not optimized, limited functionality
- Expensive maintenance (OpEx)
- High integration cost when linking different devices
- Reduced management capabilities—limited reporting and monitoring options (isolated views)
- Less flexibility with closed systems, proprietary networks, custom processes, vendor dependency

The Cisco Connected Real Estate solution unites the disparate—and often proprietary—networks and systems over a single IP network that allows all communication, security, and building systems to be monitored and managed centrally.

This so-called "building and IT convergence" creates new opportunities to reduce a building's total cost of ownership (TCO), enhance the building's performance, and deliver new building services to tenants.

The Connected Real Estate approach applies not simply to the network within one building. Cisco Connected Real Estate allows identical levels of oversight and control to be exercised across a geographically dispersed portfolio of facilities, or unrelated remote properties when offering building management as a managed service. The same network architecture that is used to allow an organization to communicate and share in-

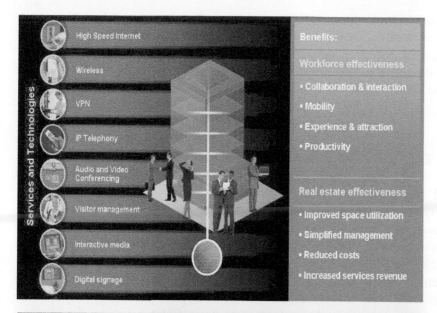

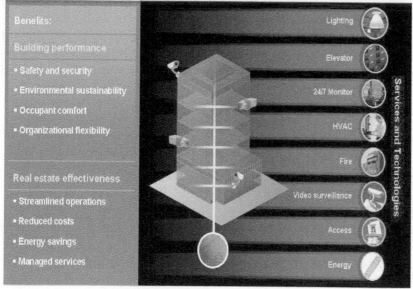

Figure 39-4. Moving from Multiple to a Single IP Network for Communications and Building Systems

formation is also used to distribute information about activity within a building over the internet to any point on the network where it is needed. Security, for example, can be monitored and controlled across a broad campus or group of buildings from a single, central point. Furthermore, security personnel in a firm do not need to be worried about laying in their infrastructure since their cameras will be using IP. It is also easier and more cost-effective to add, replace and move IP cameras between locations.

A study commissioned by the Converged Buildings Technology Group (CBTG), a consortium of building system manufacturers including Tour Andover Controls, Molex Premises Networks and ADT Fire &

Security, set out to measure the advantages of converged systems against those built on the traditional, separate model. By creating a model of an eight-storey office building suitable for 1,500 people, they were able to measure the capital and operating expenditure levels that both approaches would generate. The study found that the converged approach generated CapEx savings of 24 percent in the construction phase and reduced operating expenses (OpEx) by 30 percent over the economic life of the building.

Creating centralized capabilities for building management has a direct impact on the ongoing costs of building operation and maintenance. Centralization introduces economies of scale by which fewer staff can monitor and control far more properties in one or more real estate portfolios. For specific services, such as energy management, an intelligent network can provide constant visibility and monitoring of temperature and energy consumption and ensure that the system is adjusted to suit the demand. Because the system is constantly monitored, reaction times to unexpected developments such as energy surges and loss of power are significantly improved. In addition, centralized building management for energy issues eases the increasing regulatory pressure to comply with stringent environmental standards. An intelligent networked approach ensures that the risks of non-compliance can be managed effectively with a preventative rather than reactive approach.

Real estate assets spread across a wide geographical area benefit from this integrated approach, and in much the same way, existing campuses of many buildings in one physical location can also derive considerable operational and financial benefits from adopting this novel approach to building control.

Pharmaceutical company Pfizer operates 4 million square feet of R&D and manufacturing facilities on its site in the UK. More than 70 buildings are in active use and were operating with multiple building systems that created inconsistent and unwieldy control. Pfizer decided to link its building control network to its corporate IT, and the business now has consistent control from over 70,000 data points over one IP network and the internet. Pfizer has achieved considerable savings on annual energy costs (5 percent or euro 8.6 million) as well as a 15 percent reduction in annual maintenance costs. Business continuity in mission critical facilities has been enhanced with facilities subject to far less downtime, and overall control of all real estate has been considerably enhanced.

CONVERGING SAFETY, SECURITY AND IT

The convergence of IT networks and buildings systems has a significant impact on the ability to create safer environments for building owners, operators, and users. Interoperability of devices and networks ensures that critical real-time data can be acted upon. Video surveillance, access control and asset management over the IP network can be used to drive more sophisticated and comprehensive physical security strategies.

Cisco Connected Real Estate provides a secure platform for consistent, real-time communication of emergency status and instructions through data, voice and video formats to multiple devices including PCs, IP telephones, and even information display and public address systems. This allows rapid communication of emergency information to tenants, visitors, and employees. For example, IP telephony applications allow security personnel rapidly to inform building occupants of security breaches that may require building evacuation.

Cisco Connected Real Estate enables state-of-the-art access control to buildings and car parks using a variety of recognition technologies. This not only allows close control of people and physical assets (such as workforce, car parking, equipment) but also lets building owners regulate access to buildings or areas, for example excluding personnel who are listed as on holiday or on sick leave (and may have been victims of identity theft).

IP-enabled closed circuit television (CCTV) has been proven to reduce vandalism and other forms of lawlessness. One example comes from the UK, where the local education authority in Newport, Wales installed IP-enabled CCTV to monitor school premises that were suffering from a high incidence of vandalism. By using CCTV installed over an IP network, the education authority has been able drastically to reduce the incidence of vandalism and reports considerable savings of both cash and teachers' time. The end result is a safer, happier and more productive school environment in which teachers can focus on teaching, rather than dealing with the

aftermath of vandalism. The flexibility of an IP network enables clear images to be monitored at any distance and instant action to be initiated. Digital storage of IP CCTV enables archived images to be instantly recalled without laboriously searching through videotapes.

Cisco Systems has implemented a worldwide security system that uses its IP network to provide video surveillance, security monitoring and access control in all of its 388 sites around the world. Each site is centrally monitored from one location in either the UK, United States, or Australia. Not only does CCTV over IP create an instantaneous reaction time so that facilities can be monitored in real time, but also other elements of the building operations (such at temperature control and leak detection) are visible on the network so that a potentially damaging change in the physical environment of a sensitive area can be spotted and controlled before long-term damage is done. Cisco saves on the physical presence required to control security without diminishing the level of security available to all of its sites around the world. The return on investment that Cisco generates from this integrated approach is $10 million (euro 8 million) each year.

TRANSFORMING HOW WE USE OUR ENVIRONMENT: DELIVERING SPACE DIFFERENTLY, INTEGRATING SPACE, TECHNOLOGY, AND SERVICES

"The great agent of change which makes new ways of working inevitable is, of course, information technology, the power, reliability, and robustness of which are already evident in their impact not only on work processes within the office but on every train, in every airport lounge, at every street corner, in every classroom, library and café. Office work, no

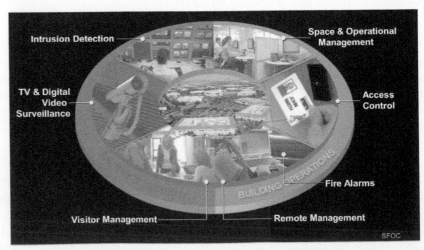

Figure 39-5. Convergence of Safety, Security and IT

longer confined to office buildings, is everywhere."
Frank Duffy, Reinventing the Workplace

The nature of work is changing. Knowledge workers are mobile. They collaborate. They are no longer desk-bound by the processes they execute but are brought together by the projects they develop. This fundamental shift has not been widely reflected in the way that the space these workers occupy is designed and arranged. For many organizations today, much of the space they maintain is largely underutilized. An increasingly mobile workforce may mean that many of a building's intended occupants are absent for much of the time. Dedicated offices and cubicle spaces are empty, and spaces for meetings or other forms of interaction lie dormant. Yet, all these assets have to be maintained as if they were being occupied to maximum capacity. Improving the efficiency of the design and use of physical spaces will limit the need for space and result in cost reduction (rent and operations), while enhancing the productivity of the workforce.

Transforming the Workplace

Cisco's converged network for voice, video and data is enabling transformation of the workplace by helping organizations create flexible real estate portfolios, and supporting new workplace designs—at the same time improving organization-wide productivity, collaboration and mobility. A converged network:

- Improves employee mobility and remote working
- Delivers flexible and efficient workspace
- Enables new working practices and better collaboration
- Improves space optimization and use
- Reduces real estate costs

Robust, scalable and secure networks for voice, video and data improve employee productivity. Unified communications, wireless access and VPNs create flexible work environments, employee mobility and remote working initiatives. A converged network allows secure synchronous and asynchronous collaboration, email, voice mail, conference calls, video conferencing, knowledge management initiatives, intranets and instant messaging. This enables new working practices while reducing overall real estate requirements through, for example, hot-desking and VPN-based remote working.

IT and real estate executives can take advantage of a fully converged intelligent information network to create virtual workspaces that provide more flexible and efficient work environments. This converged intelligent information network provides wireless work areas that connect people to important corporate assets and building amenities to provide greater mobility, productivity and communication capabilities. This allows owners to achieve greater operational gains from the much more flexible and efficient use of their existing space and to use the network to derive more from their assets.

Cisco Connected Workplace in Action

Cisco Systems has put these principles into action with the redevelopment of office space at its headquarters in San Jose and around the world (Bangkok, Taipei, Charlotte, New York, and growing). Cisco's employees, like many in other organizations work differently than they did even as recently as a few years ago. An increasingly global work-force and customer base makes it more likely that employees need to work at nontraditional hours, leaving their offices vacant at other times. More complex business and technology issues increase the need for collaboration with team members in the same building or at various sites worldwide. Employees are often away from their desks, in meetings or workgroup discussions. Even Cisco employees who work on site are likely to be mobile within the building.

The Cisco Connected Workplace approach is built on the use of space rather than its allocation in accordance with head count. The design of the offices reflects the way that people work, i.e. collaborative and mobile both within and beyond the building and different types of space have been created that reflect the work modes that an individual employee may cycle through in the course of a working day. These range from quiet spaces for working alone, to areas designed for collaborative and more social ways of working. And critically, employees are supported at all times with access to data and communications through an IP network that provides various forms of connectivity. In Cisco's Building 14 Proof of Concept at headquarters, the result is that a space which under traditional configuration would have accommodated 88 workers, is now used by 200. Every aspect of space usage has been investigated and the approach modified to take account of needs as they develop. As employees have responded to the flexibility, so the workplace has evolved. For example, wireless access points have been installed in greater numbers in areas of the building where people tend to congregate, ensuring that no single access point is overloaded and thereby impairing employees' productivity.

The success of the Connected Workplace solution comes in two forms: reduced costs and improved employee satisfaction and productivity across a range of indicators. The costs of cabling and IT infrastructure have both been halved compared to a traditional office, as has spending on furniture. The savings generated by having more people productively using space more effectively means that rent and construction costs are also cut considerably (by 37 percent and 42 percent respectively). A more cost-effective solution is accompanied by higher levels of employee satisfaction with the new arrangements. Nearly 80 percent of employees say that they prefer the new environment, citing factors such as the greatly increased ease of finding a meeting space. Nearly two-thirds say that they enjoy coming to work more. In short, the connected workplace offers better and more productive use resources at lower cost; a proposition that few organizations can afford to ignore.

One example of an organization that has put this approach to building usage to work is Hillingdon Borough Council in the UK. There, the housing department identified that some 70 percent of its staff could work remotely provided that they had the necessary levels of connectivity eliminating the need to maintain the same level of office space. Staff now have access to the information, services and applications they need whenever, and wherever they are—whether working from home or on the road. The council has saved more than euro 4.3 million in annual office costs.

Another example is the UK's largest telecommunications service providers, British Telecom (BT). It has recognized the significant savings and productivity improvements available from rethinking the space it uses. Today, about 9,500 BT staff are contractually employed to work primarily from home and more than 63,000 others are able to work independently from any location at any time. BT now saves some euro 6.5 million annually on property costs, absenteeism has fallen by 63 percent and staff retention has increased dramatically

CISCO CONNECTED REAL ESTATE AND THE NEED TO DO THINGS DIFFERENTLY

New thinking about the use of the network in the deployment of real estate assets creates a wide range of exciting possibilities for the development of both the business models for the property value chain and the way that owners, operators and the building users will be able to use the buildings they inhabit.

But for these new models to become reality the design of the IT network will require a new place in the elaborate property development process. Though the design and consequences of water, electricity and gas infrastructures (the first three utilities of a building) are included in the very early stages of a property's con-

ception, network and communications needs are rarely given the same early attention. The network needs to become an up-front consideration from the planning stages and drawing board onwards.

Understanding the Building Life Cycle

A building life cycle comprises four phases: conceptualize, design, construct, maintain and operate.

- Conceptualize: The phase in which the building is scoped and financed, conceptualization, consumes about 2 percent of the total costs of the building life cycle and marks the beginning and end of each building life cycle.

- Design: During the design phase, architects and engineers plan the detailed layout, structure, and execution of the building.

- Construct: In the construct phase the building is erected to its design specifications. Together, the design and construct phases account for some 23 percent of the total costs of the building life cycle.

- Maintain and Operate: The maintenance and operation phase represents the time during which the building is used, typically 25 to 30 years in today's fast-moving environment—marked by its economic of functional life. It accounts for 75 percent or more of the total costs of the building life cycle.

With more than three-quarters of the total expense of a building arising during the maintain and operate period, rather than as initial capital expenditure, decisions taken in the design and construct phases can have far-reaching financial and operational effects. Therefore during those phases key stakeholders should carefully consider a building's network; especially because the Connected Real Estate framework will positively affect the functional use and design of the building and thus support the transformation of space and businesses. Decisions made during the early stages can effectively create the infrastructure that reduces ongoing operations costs over the life cycle of the building, and improves the opportunity to create revenue streams in the appropriate markets.

The inclusion of an IP network in the building design process, and its installation as early as possible in the construction process, provides immediate gains for building owners. The single IP network reduces capital costs during the construction process, because infrastructure can be laid more easily (rather than being retrofitted with consequent cost and disruption) and the single open standards cabling infrastructure reduces the requirement for multiple closed proprietary networks and the associated costs of installing them. Secondly, by installing networks early, building owners can extract value from the network over a longer period of time, increasing overall return on investment.

Cisco Connected Real Estate also helps lower operating expenses over the building's life-cycle. An open standards based building infrastructure encourages a centralized (and/or remote) approach to monitoring, maintenance and control of the building environment.

Higher levels of connectivity between building systems provides an array of benefits through access to and sharing of real time data including:

- Optimized remote control, monitoring and reporting of building systems including centralized management of a distributed property portfolio.

- Intelligent heating and lighting and cooling systems that reduce costs through increased energy efficiency.

- Improved staff productivity (maintenance, facilities and security personnel) and enhanced health and safety compliance.

- Improved asset management and tracking together with automated work scheduling, billing and help

desks linked to existing enterprise resource planning (ERP) systems.

The Cisco Connected Real Estate IP framework features embedded technologies that guarantee quality of service and high levels of security and resilience further reducing maintenance and repair costs. Furthermore, all components of the network are built entirely on open standards. Hardware, software, and services are designed using roadmaps that anticipate and support constantly changing business requirements.

Acknowledgements

There are many people that have been involved in the development and deployment of the CCRE program at Cisco, too many to mention here. There are several people however, that must be mentioned as key contributors to the success of this program.

Mark Golan—VP, WW Real Estate and Workplace Resources
Wolfgang Wagener—Manager, Workplace Resources
Andrew Thomson—Business Development Manager
Rick Huijbregts—WPR Program Manager
Oscar Thomas—Marketing Manager
Cori Caldwell—Marketing Manager
Agnieszka Jank—Integrated Marketing Communications Manager
Ray Rapuano—Strategic Account Manager

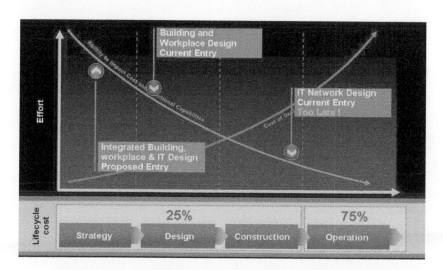

Figure 39-6. Understanding Building Lifecycle Costs

Chapter 40

Final Thoughts and Conclusion

Barney L. Capehart, Anto Budiardjo and Ken Sinclair

FROM BUILDINGS TO THE
ENTERPRISE VIA IT—ANTO BUDIARDJO

THE HUMBLE BUILDING has stood on its own for many decades. Though most often owned and operated by large organizations, buildings have tended to be disconnected from the corporate nerve center, this has left the down-playing of buildings as a distant cost that can only be managed by groups that often seem to be beyond the close control of organizations. All that could often be expected is that the cost is kept low, the occupants of them be happy then simply feed the historical cost of the building as a line item in the corporation's P&L sheets.

One of the fascinating benefits of the convergence of building systems and IT is that the very mechanism that is bringing much needed control of many facets of organizations of today is now being more and more connected to building systems. We are talking of course of IT, all of the technology involved with connecting information generating, storage, analysis as well as reporting entities around the corporation.

The collection of all of these systems, connected by IT, is what most people are referring to as "the enterprise." By their near real-time connectivity and ample storage and analysis power, the information from all corners of the organization is brought together for the benefit of the organization.

IT's focus on human resources, supply chain, customer relationship and finance has brought about in recent decades, an increase in the efficiency of the organization unheard of in the middle of the 20th century. Supply chain is now so efficient that most of the world's stock do not sit in warehouses; they sit in container ships, ports, rail and trucks. They are designed to arrive at their destination just when they are needed. Customer relationship systems tell organizations when their customer's are likely to need products, little time is spent chasing worthless leads, producing sales organization efficiencies previously unheard of.

It is time to plug the building into this nerve center, time to discover the efficiencies that can be extracted out

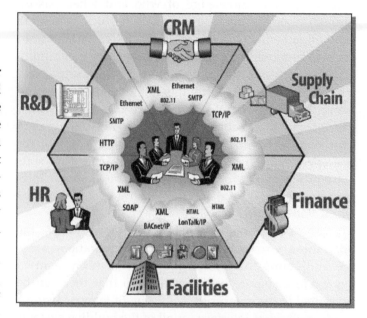

of connecting these buildings to the organization's functions that depend on facilities for their success.

As attendees and speakers regularly comment at BuilConn conferences; connectivity is no longer the challenge, neither is the technology. The challenge is to search for how the information that comes from buildings can be useful to the enterprise.

The building industry needs to focus on that word "information" because it is the information that can be extracted out of buildings that will provide the value to the enterprise. Just look for a moment at the "information" that brings value to supply chain and customer relationship; it's information like the location of a shipment, the time it will take to get to destination and the quality of the environment of the container. It's the time since a customer purchased a product, the customer's demographics, and the customer's satisfaction of a product (yours or your competitor). All of this information is somehow useful to the enterprise. Before connectivity and IT, all of these information pieces were there, just not connected and not available to the enterprise, and thus not available for the enterprise to make decisions based on them.

Now let's go back to buildings, think about that

humble temperature sensor on the wall or the card reader by the door or that humble light switch. In the past they were only there to switch on and off some HVAC equipment, to allow or deny access through the door and to allow humans to illuminate the space.

In the connected building, those devices that we are all familiar with have greater value yet to be determined. If a well programmed and structured enterprise based system can have access to the temperature in all of the buildings, have knowledge of who and when passed through door and have knowledge of which spaces are being illuminated (consuming energy), then surely it can use such information to some good for the organization?

For decades, the buildings systems industry have strived to integrate, with the function of integrating being the end-game of fancy closed and open systems based technologies able to do magic in transporting data from one device and system to another. The author for one spent many years developing and marketing software products just for that purpose. The value for integration for its own sake is marginal at best, the author learnt that the hard way!

The effort was noble, the technology works, the only thing wrong is the objective. Integration is not the end-game. Integration is purely a mechanism to aggregate disparate pieces of data so that they can be presented to the enterprise in a coherent manner, using a single method of interface for all of the building systems that are found installed in today's facilities.

Without integration, without some way to gather all of the information together, it would be impossible or at least very hard for enterprise systems to make use of the disparate pieces of data found in buildings. So the decades of work on integration and open systems was not in vain after all, it is for a purpose, one could argue for an even greater purpose than anyone would have foresaw.

The term IT gives us strong clues, the term was created when most computing industry knew itself as data processing. The move to the IT term is the realization in the late 80's that it was the information that was (and still) valuable and that technology needs to focus on the collection, management, transmission, storage, analysis and presentation of information. IT (information technology) is what the building industry needs to be as we start to focus on the information itself in addition to the core functions of managing and controlling buildings.

So, now that we know we can collect the information, we know that we can transport the information to enterprise systems, we also know that organizations large and small are eager to squeeze as much efficiency out of all of their resources—why cannot information from buildings be fed into this information hungry enterprise machine to figure out how it can be used for the benefit for the building owner.

IT is the enabling connection between buildings and the enterprise. IT is the same enabling connection to all the other corners of the organization's functions to the enterprise. The opportunity is now to figure out just how this information can benefit building owners. Once found, it will change how we look at that sensor on the wall or card reader forever.

FINAL THOUGHTS—KEN SINCLAIR

"Oh What a Wonderful Web We Weave"

To bring conclusion to this book we need to understand that the web and its present connections to the enterprise will change forever enterprise energy and facility management systems. If you look back over all the chapters you have read, you will see that they all revolve around the web and applying present day information technology to building management. The web and web ways implemented rapidly with web services, and IT open procedures to connect to the enterprise, are revolutionizing the industry.

In my forward for this book I talk about the real renaissance that is reshaping the building automation industry, and now that you have completed reading this book I think you can better understand this statement. Can we see the end of all this change? I do not think so. Powerful forces unleashed by web services built globally on open IT standards will reset the hinges of our thoughts about energy and facility management. These web services will cross over traditional building boundaries into blended cities and even blending nation wide control concepts. Open standards that we believed belong to our industry such as oBIX, BACnet, and LonMark will become global IT standards that will allow the world to develop web services applications completely outside of the box. These powerful forces will rapidly evolve over the next few years. The interest of the large IT industry players in all of this will fuel this reality. If you think about all the chapters you will see a common message that states:

"The web is the glue and fully converged connections to enterprise our goal."

It will take all of our combined efforts to organize, educate, and communicate a new level of integration. Building information networks will be created where the network infrastructure (including building control networks) and interactive layers provide a secure and robust platform for all types of applications from lighting, HVAC to instant messaging, IP phones as well as traditional data and information applications.

If there is a next book in this series, it will talk about server side agents that are providing great value to building automation for a very low monthly subscription cost. Blended services built on new financial models will be provided for extremely low costs to improve building energy management and the complete facility management. It will talk about and provide examples of new devices that authenticate a user's identity with their fingerprint and enables secure access to buildings and other physical facilities, local and remote computers and networks, and online or on-site financial transac-

tions. The boundaries of control in a blended city will be explored and demonstrated.

Professor Joroff defined a blended city as commercial real estate that brings together a physical and digital environment to enable human capital to contribute to the global economy. He described communities where intelligent buildings are helping the tenants in them to live and work better as they impact a universal marketplace.

The connectivity of wireless or Wi-tivity specifically related to controls and monitoring of device systems, covering everything from wireless mesh, sensor networking such as ZigBee as well as WiFi, WiMax, GPRS and G3 types of wireless technologies would all be discussed in a future book.

CONCLUSION—BARNEY CAPEHART

With the three books in this area called "Information Technology for Energy Managers," we have proceeded from the basic ideas of metering, automated meter reading, data collection, and data processing and storage, to the use and value of web-based energy information and control systems, building automation systems, enterprise energy and facility management systems, and finally enterprise resource management systems. The wave of new IT, TCP/IP and web-based equipment technology is being followed by an even stronger new wave of technology implementation directed to the tasks of enterprise energy and facility management, and even enterprise resource management.

The goal of these three books has been clear and consistent throughout the project:

To help prepare energy and facility managers to understand some of the basic principles of IT, so that they can successfully:

- purchase or develop
- install
- operate
- improve, and
- capture the facility operational cost savings and improvements from

Web based energy information and control systems, including BAS systems and enterprise energy management systems.

Good luck with your own projects using and benefiting from this new technology. The more success you achieve with this new technology, the more we—as the editors—feel that we have accomplished something of significant benefit to your company and your organizations, and to our economy, and to our world.

About the Authors

Bill Allemon is an Energy Program Manager at Ford Land, a wholly owned subsidiary of the Ford Motor Company. His responsibilities include managing capital projects, administering energy performance contracts, and driving energy awareness activities for North American Vehicle Operations. Bill represents Ford on the ENERGY STAR Motor Vehicle Manufacturing Focus Group. Bill's team submitted a winning application for a 2006 ENERGY STAR Partner of the Year Award in Energy Efficiency. During his 16 years with Ford, Bill has held various positions in the design, construction, operation and maintenance of industrial, institutional, and commercial facilities. Bill holds a Bachelors degree in Electrical Engineering from Lawrence Technological University and a Master of Science in Administration from Central Michigan University.

Paul J. Allen is the Chief Energy Management Engineer at Reedy Creek Energy Services (a division of the Walt Disney World Co.) and is responsible for the development and implementation of energy conservation projects throughout the Walt Disney World Resort. Paul is a graduate of the University of Miami (BS degrees in Physics and Civil Engineering) and the University of Florida (MS degrees in Civil Engineering and Industrial Engineering). Paul is also a registered Professional Engineer in the State of Florida. The Association of Energy Engineers (AEE) selected Paul as the 2001 Energy Manager of the Year. (paul.allen@disney.com)

Rick Avery is the Director of Marketing at Bay, LLC. In this role, Rick is responsible for bringing to market Bay's control and management products, while promoting the energy savings message to industry. Prior to joining Bay in 2004, Rick held sales and marketing management positions in the consumer products sector. Rick holds a Bachelors degree from the University of Rochester and a MBA from the Weatherhead School of Management at Case Western Reserve University.

John Avina, Director of Abraxas Energy Consulting, has worked in energy analysis and utility bill tracking for over a decade. Mr. Avina performed M&V for Performance Contracting at Johnson Controls. In later positions with SRC Systems, Silicon Energy and Abraxas Energy Consulting, he has taught well over 200 software classes, handled technical support for nearly a decade, assisted with product development, and written manuals for Metrix Utility Accounting System™ and Market-Manager™, a building modeling program. Mr. Avina managed the development of new analytical software that employed the weather regression algorithms found in Metrix™ to automatically calibrate building models. In October 2001, Mr. Avina, and others from the defunct SRC Systems founded Abraxas Energy Consulting. Mr. Avina has a MS in Mechanical Engineering from the University of Wisconsin-Madison, where he was a research assistant at the Solar Energy Lab. He is a Member of the American Society of Heating Refrigeration and Air-Conditioning Engineers (ASHRAE), the Association of Energy Engineers (AEE, and a Certified Energy Manager (CEM).

Sanjyot Bhusari, LEED AP is a systems integration engineer for the Technology Group of the Gainesville, FL, office of Affiliated Engineers, Inc. (AEI), a consulting, engineering and technology firm. Sanjyot has lead design efforts on several system integration projects. His experience also includes commissioning, mechanical systems design, writing guide specifications and developing control standards for various clients. Sanjyot has expertise in open system protocols like BACnet and LonWorks as well as application of Information Technology standards like Web Services in Building Automation. He has organized "plug-fests" intended to demonstrate interoperability between different building automation systems. Sanjyot has published several papers in the areas of system integration and energy conservation. He has also presented in several conferences such as BACnet, BuilConn USA, BuilConn Europe and International Congress of Refrigeration. He holds a Master of Science degree from the University of Florida.

Michael Bobker, M.Sc., CEM, is a Senior Fellow at the City University of New York Institute for Urban Systems, and development director of CUNY's new building performance lab. He has nearly 30 years of experience with varied energy efficiency technologies and service delivery mechanisms, including 10 years as a principal and director of operations for a NYC-based energy services company.

Michael R. Brambley, Ph.D., manages the building systems program at Pacific Northwest National Laboratory (PNNL), where his work focuses on developing and deploying technology to increase the energy efficiency of buildings and other energy using systems. His primary research thrusts in recent years have been in development and application of automated fault detection and diagnostics and wireless sensing and control. He has been with PNNL for nearly 17 years before which he was an assistant professor in the Engineering School at Washington University in St. Louis.

Michael is the author of more than 60 peer-reviewed technical publications and numerous research project reports. He holds M.S. (1978) and Ph.D. (1981) degrees from the University of California, San Diego, and the B.S. (1976) from the University of Pennsylvania. He is an active member of the American Society of Heating, Refrigerating, and Air-Conditioning Engineers (ASHRAE) for which he has served on technical committees for computer applications and smart building systems. He has been the organizer of numerous seminars and symposia at ASHRAE's semi-annual meetings and is a member of ASHRAE's Program Committee. In addition to several other professional organizations, Michael is a member of the Instrumentation, Systems, and Automation Society (ISA) and Sigma Xi, The Scientific Research Society.

David Brooks, P.E., is a project manager for the Technology Group of the Gainesville, FL, office of Affiliated Engineers, Inc. (AEI), a consulting, engineering and technology firm. Serving as a leader within the AEI-Gainesville Technology Group, **Mr. Brooks** works directly with AEI systems engineers, contractors, and client staff to ensure the latest technology is being incorporated into the design process. He is intrinsically involved in the overall design approach, process, and implementation of all Technology projects. Mr. Brooks oversees the development of all system integration, control system, telecommunication, security, fire alarm and other intelligent building system designs. He has commissioned numerous intelligent building systems including complex building automation, laboratory control, chilled water, steam plant, HVAC, fire alarm, lighting, and security systems throughout his career. With over 17 years of experience, Mr. Brooks is a recognized expert in the area of intelligent building systems and the methods used to seamlessly bring these systems together into a single integrated package. His specialized technical expertise, project management skills, creative approach to system problem resolution, and ability to communicate effectively with both owners and facility staff is what he attributes his success to.

Ron Brown is the CTO and co-founder of Gridlogix, Inc. Mr. Brown is the primary architect behind Gridlogix technology strategies and solutions that include EnNET® an XML web services remote device management and integration framework. Prior to co-founding Gridlogix, Mr. Brown co-founded Automated Energy, Inc. (AEI) where he served as president and chief information officer. During his tenure, he recruited and led the team that designed and engineered AEI's industry leading Enterprise Energy Management (EEM) System. He has several patents pending.

Mr. Brown's 20 plus years of experience have been in the management, consulting, software development, implementation and integration of GIS, energy management, building management, distribution automation, and MIS projects for various corporations and utilities.

Mr. Brown holds a B.S. Degree in engineering physics from the University of Central Oklahoma, and an A.A.S. Degree in general engineering from Oklahoma State University. He is a senior member of the Association of Energy Engineers (AEE), member of the Institute of Electrical and Electronics Engineers (IEEE), and a member of the Geospatial Information & Technology Association (GITA).

John B. Broughton has over 20 years experience in construction, energy management and renewable energy. John has a Master Degree in Business, a Bachelors of Science in Mechanical Engineering. He is a Certified Energy Manager and Building Professional Engineer with the Association of Energy Engineers. He has been a certified energy rater in Florida, served on the ASTM E-44 committee for Renewable Energy, and active in the construction of Florida's Energy Code over the years. John has inspected over a million square feet of government, commercial and residential construction projects. (jb@nrgmanager.com)

Anto Budiardjo is president & CEO of Clasma Events Inc., and is a seasoned marketing and product development professional specializing in the HVAC, security, and IT disciplines. Mr. Budiardjo has more than two decades' experience within these industries and has fashioned his expertise into an energetic, visionary, and dynamic approach to business management.

Mr. Budiardjo has held executive-level marketing and product development positions with various controls companies where he was responsible for product management and marketing communications. His rare combination of marketing and technology practices has enabled him to fine tune and soften the often daunting task of transitioning the product development process from an engineering-centric focus to a market-centric focus.

Mr. Budiardjo's entrepreneurial and creative spirit has afforded him international business opportunities throughout Europe, Americas, and Asia/Pacific further enabling him a global perspective on our market.

Mr. Budiardjo is a frequent speaker at industry events and is a contributing editor of AutomatedBuildings.com. He is also the recipient of the Frost & Sullivan 2005 Building Technologies CEO of the Year award.

Barney L. Capehart, Ph.D., CEM, is a Professor Emeritus of Industrial and Systems Engineering at the University of Florida in Gainesville, FL. He has broad experience in the commercial/industrial sector having served as the founding director of the University of Florida Energy Analysis and Diagnostic Center/Industrial Assessment Center from 1990 to 1999. He personally conducted over 100 audits of industrial and manufacturing facilities, and has helped students conduct audits of hundreds of office buildings, small businesses, government facilities, and apartment complexes. He regularly taught a University of Florida course on energy management, and currently teaches energy management seminars around the country for the Association of Energy Engineers (AEE). He is a Fellow of IEEE, IIE and AAAS, and a member of the Hall of Fame of AEE. He is the editor of Information Technology for Energy Managers: Understanding Web-Based Energy Information and Control Systems, Fairmont Press, 2004. He is the co-author of Guide to Energy Management, 4th Edition, author of the chapter on Energy Management for the Handbook of Industrial Engineering, and is co-author of the chapter on Energy Auditing for the Energy Management Handbook, 5th Edition. He can be reached at Capehart@ise.ufl.edu

Lynne C. Capehart, BS, JD, is a consultant in energy policy and energy efficiency, and resides in Gainesville, FL. She received a B.S. with High Honors in mathematics from the University of Oklahoma, and a JD with Honors from the University of Florida College of Law. She is co-author of Florida's Electric Future: Building Plentiful Supplies on Conservation; the co-author of numerous papers on PURPA and cogeneration policies; and the co-author of numerous papers on commercial and industrial energy efficiency. She was project coordinator for the University of Florida Industrial Assessment Center from 1992 to 1999. She is a member of Phi Beta Kappa, Alpha Pi Mu, and Sigma Pi Sigma. She is president of the Quilters of Alachua County Day Guild, and has two beautiful grandchildren. Her email address is Lynneinfla@aol.com

Noshad A. Chaudhry is a Project Manager at Tridium Inc. managing multi million dollar projects with Tridium's large customers such as Honeywell, Johnson Controls etc. He has been involved in project development, software development and technical support for Tridium's Niagara AX, Niagara Portal and Vykon Energy Suite products. Mr. Chaudhry has provided in-depth understanding in evaluating solar and distributed generation, energy monitoring and cost saving analysis at major Niagara deployed sites in the US and Canada. Mr. Chaudhry has a B.S in Management Information System from Southern Polytechnic State University, Marietta, GA and a Professional Accounting Diploma from Conestoga College, Canada. He is a member of the Institute of Management Accountants

David Clute joined Cisco's Advisory Services group in June 2005. Mr. Clute has served in several capacities during his tenure at Cisco Systems including Manager, eSolutions and Manager, WPR Global Operations. In his current role, he provides consulting expertise for Cisco-Connected Real Estate and "Next Generation" building design for converged real estate and information technology solutions. In addition to his primary role within Advisory Services, Mr. Clute also serves on the Executive Board for OSCRE Americas, the Open Standards Consortium for Real Estate, promoting data exchange standards for the real estate industry. Mr. Clute has over 25 years of experience in architecture, engineering, systems development and implementation of applications for the infrastructure management and corporate real estate industry. He is recognized in the industry as a leading authority involving the integration of Computer Aided Design (CAD) Computer-Integrated Facilities Management (CIFM) and Geographic Information Systems (GIS) for large-scale corporate, government and military clients. Clute received his B.S. Architectural Engineering from the University of Colorado—Boulder

Gregory Cmar, CTO and cofounder, Interval Data Systems, Inc., is an authority in demand side management technologies and utilities, complemented by expertise in database, application, and automation technologies. With 30 years of experience, in facilities operations, energy conservation, energy analytics, energy auditing, monitoring and control systems, and utility billing, he is one of the foremost experts in world on the application of interval data. At IDS, Greg leads the product definition and development efforts as well as the energy management services team.

Prior to IDS, Greg was a cofounder and director of engineering at ForPower, an energy conservation consulting firm serving utilities, property managers, and engineering companies. As engineering manager at Coneco, an energy services company and subsidiary of Boston Edison, he established the energy audit department and later developed Boston Edison's real-time pricing application. Greg served as vice president of Enertech Systems, an energy monitoring and control systems contractor, where he did some of his earliest work with interval data and databases in the early 1980s. Earlier in his career, Greg held various roles at Johnson Controls, the Massachusetts Energy Office, and Honeywell.

Greg holds patent #5,566,084 for the process for identifying patterns of electric energy, effects of proposed changes, and implementing such changes in the facility to conserve energy.

Bruce K. Colburn, Ph.D., P.E., CEM, is Executive Vice President and Chief Operating Officer of EPS Capital Corp., which is an international consulting firm involved in energy studies, development of energy efficiency and productivity improvement programs, and financing of performance projects via different financing and contract structures. He has been active in energy projects in 24 countries. He is a former Associate Professor of EE at Texas A&M University, and was a Visiting Professor at Baylor University. He has won numerous professional awards, been an editor and reviewer for many professional journals, and is a widely published author. He has been active in energy efficiency for over 30 years, and is on the Board of the Distributed Generation Institute of AEE, as well as being inducted into the AEE Hall of Fame. Dr. Colburn may be contacted at bcolburn@epscapital.com; the firm web site is www.epscapital.com.

Michael Cozzi, Executive Vice President As cofounder, executive vice president and partner of Cirro Energy Services, Mr. Cozzi is responsible for leading client energy consulting engagements in regulated and deregulated natural gas and electricity markets. Mr. Cozzi is considered a market expert on demand-response programs and has assisted a number of large corporations, government, and universities in developing and implementing electric procurement strategies that have saved millions of dollars in energy costs. Prior to the opening of the deregulated Texas electric market, Mr. Cozzi served the Senior Energy Marketing Manager at Reliant Energy, and most recently Director of Major Accounts at Direct Energy Business Service, a division of Centrica North America. Prior to working for Reliant Energy, Mr. Cozzi worked for The Dow

Chemical Company in various sales and marketing roles, including a position as the North American Business Analyst for natural gas and electricity procurement for Dow's manufacturing facilities. With experience leading market development strategies and large commercial and industrial sales teams, Mr. Cozzi has been recognized as one of the industry's top sales executives. An engineering graduate of Purdue University with an MBA from the University of Detroit, Mr. Cozzi spends his free time with his wife and two children in addition to working with children in the YMCA's Big Brother Program, basketball coach, and church youth group leader. He is a member of Success North Dallas and the Plano Chamber of Commerce.

Paul Ehrlich, PE, a well-known industry stakeholder and advocate of integrated and intelligent buildings. In 2004 he formed the Building Intelligence Group LLC, an independent consultancy, whose primary purpose is to help system suppliers as well as building owners and managers, maneuver their operations through the vast changes prompted by open systems, convergence and enterprise building management. The main focus is in the areas of facility and IT integration, convergence and intelligent buildings. Clients include developers, property management firms and major manufacturers of Building Systems and associated technologies.

Previously Paul was with Trane where he served as Business Development Leader guiding the global direction and development of Trane's line of building control products. Prior to working for Trane, Paul worked for Johnson Controls providing solutions to building owners.

Throughout his career, Paul has been actively involved with various industry groups involved in the creation of new automation standards and technologies. Previous roles include chairing the ASHRAE Guideline 13 committee on how to specify DDC controls, chairing the BACnet sub-committee on interoperability, and acting as the inaugural chair for the oBIX committee to establish XML standards for building controls.

At BuilConn 2004, Paul was honored with a Buildy Vision Award for his perseverance in promoting whole building integration and interoperability through advocacy, promotion, educational and training endeavors.

Paul has a Bachelors degree in mechanical engineering from the University of Wisconsin and a Masters of Business Administration from the University of St. Thomas. He is a licensed engineer in the State of Wisconsin and lives with his family in White Bear Lake Minnesota.

Khaled A. Elfarra, Ph.D., CEM, DGCP is currently General Manager of Engineering/Projects Dept. at the National Energy Corporation—Egypt (NECE). His role is the technical studies and designs of energy efficiency technologies. The focus of the current studies is on distributed generation, cogeneration and the gas fired technologies. He has extensive experience in the field of Energy and environmental technologies. Over the last two years, Mr. Elfarra has conducted over 20 audits in technology innovation and business strategies for many industrial facilities in Egypt and the Middle East region. He has a PhD in management and decision sciences-engineering management. As a former technical manager assistant of the Energy Conservation and Environment project (ECEP), USAID funded, he was involved in techno-economic feasibility studies, capacity building, technologies implementation, and projects management. Mr. Elfarra has conducted more than 80 audits, 20 feasibility studies, and has supervised projects construction for more than 14 million US$. He gave training in energy auditing, environmental auditing, energy technologies, energy management systems and pollution control for many industrial facilities staff. He also received training on energy and environmental issues in the US in multi disciplines. Mr. Elfarra has participated partially in the Egyptian environmental policy and energy reforming policy. He has conducted many market study surveys regarding the gas sales strategy, potential for energy and environmental market in Egypt, and solid waste management programs. He worked as a short-term consultant for DANIDA JAICA, USAID and UNDP in many of the Egyptian Projects. Mr. Elfarra is a certified energy manager by US Association of Energy Engineers as well as a distributive generation certified professional. He was also certified as a second party auditor of Environmental management systems (ISO 14000) by British excel partnership Inc. Mr. Elfarra is certified by AEE as a local instructor for the Certified Energy Managers Course in Egypt. He is chair of the Egypt CEM Board.

Dr. Clifford Federspiel is the president of Federspiel Controls, a consulting firm that provides energy services and energy management control products to the commercial buildings industry. Previously he held an academic staff appointment at UC Berkeley, where he was affiliated with the Center for the Built Environment (CBE) and the Center for Information Technology Research in the Interest of Society (CITRIS). At Berkeley, Dr. Federspiel managed several projects on the application of wireless sensor networks (motes) to building automation. Prior to his appointment at UC Berkeley,

Dr. Federspiel was a senior member of the technical staff at Johnson Controls. Dr. Federspiel received his Ph.D. and SMME from the Massachusetts Institute of Technology, and his BSME from Cal Poly, San Luis Obispo. cf@federspielcontrols.com

Kevin Fuller is responsible for marketing and product development for IDS. He brings over 20 years of technical and marketing experience in database, data warehouse, OLAP, and enterprise applications to his role as executive vice president. Kevin has a strong appreciation of how businesses use data to their advantage, and focuses on how to apply technology to solve real business problems. He can be reached at kevin@intdatsys.com.

Keith E. Gipson has been a technologist for almost two decades. Starting out as a Technician with Honeywell Inc. in 1987, graduating to an Engineer at Johnson Controls in the mid-90's and at Pacific Gas and Electric in 1997. A successful entrepreneur and business professional, Mr. Gipson co-founded in 1997 the world's first; Internet based Enterprise Energy Management company, Silicon Energy Corp (*www.siliconenergy.com*). The privately held company grew from three persons to a 200 plus employee, multi-million dollar company. Itron Corp. acquired Silicon Energy in March 2003 for $71M.

Mr. Gipson was Awarded United States Patent number 6,178,362, Jan 23, 2001 as Co-inventor of: an Energy Management System and Method utilizing the Internet to perform Facility and Energy Management of large corporate enterprises. This was the first EEM or "Enterprise Energy Management" system. In February 2006, as a result of his contribution to the field of technology, and specifically the electric industry, Mr. Gipson was recognized by Southern California Edison as a "modern day African-American" inventor. Edison utilizes EEM software co-invented by Mr. Gipson as the basis for the software it offers to its large commercial and industrial customers. Presently, Mr. Gipson is the CTO and Co-Founder of Impact Facility Solutions, Inc. (*www.myfacility.com*).

Bill Gnerre is the CEO and cofounder of IDS. With over 20 years of information technology entrepreneurial experience, he has an exemplary record of bringing enterprise software applications to market and dealing with user adoption of new technology. In addition to facilities operations and enterprise energy management, his background includes experience with CAD/CAM, engineering document management, PDA data collection, and other customized enterprise applications. Bill provides leadership and strategy for IDS and works closely with clients to ensure their success. He can be reached at bill@intdatsys.com.

David C. Green has combined experience in Intranet/Internet technology and database queries and has developed programming for Energy Information Systems. David has been the president of his own consulting company, Green Management Services, Inc., since 1994. He has a Bachelor of Science degree in Chemistry and a Master of Arts degree in Computer Science. David is also a Lieutenant Colonel in the Illinois Army National Guard and has 18 years of military service. David has successfully completed major projects for The ABB Group, Cummins Engine Company, ECI Telematics, M.A.R.C. of the Professionals, Walt Disney World and The Illinois Army National Guard. (dcgreen@dcgreen.com)

Chris Greenwell is Director of Business Development at Tridium Inc and has over 20 years experience in the energy industry. His experience includes serving in a variety of capacities for Central and Southwest and American Electric Power including overseeing, testifying and designing multi electric companies' rates, developing and evaluating demand-side management programs and helping build C3 Communications and Datapult, their un-regulated energy information service companies. More recently he has implemented new market development strategies for Green Mountain Energy, a competitive retail renewable energy provider. He is an engineering graduate from Texas Tech University, holds an MBA from Sul Ross State University and is a registered professional engineer in Texas.

Randolph L. Haines is the Energy Manager for Thomas Jefferson University in Philadelphia. Randy has twenty-nine (29) years experience in project, maintenance and facilities management. The first seventeen (17) years were in the metal working industry before moving to Thomas Jefferson University, an urban teaching hospital in center city Philadelphia. In 1997, Randy accepted the newly created position of Energy Manager for the Jefferson Health System, an alliance of hospitals and research buildings in the Philadelphia region covering more than 10 million square feet with 22,000 employees. This position was established to address the complexities created from the deregulation of energy in Pennsylvania and to save utility consumption and costs.

In January 2001, Jefferson awarded a $10 million contract to Alliant Energy to install energy efficient lighting, variable frequency drives, additions to their building automation system and the installation of an advanced metering system throughout the health system, for which Randy is the primary overseer.

A University of Pittsburgh Industrial Engineering graduate, Randy has worked at Eaton Corporation, Mitchell Industries, Delaval Condenser Division and Thomas Jefferson University. He is a past president of the Greater Philadelphia Chapter of the Association of Energy Engineers; past President of the Trenton—Princeton Chapter of the Association for Facilities Engineering; and a member of ASHRAE and the Hospital Engineers. In 2004, Randy received the "International Energy Manager of the Year" from the Association of Energy Engineers. Randy can be contacted by: Phone: (215) 503-6099

Rusty T. Hodapp, P.E., CEM, CEP, GBE, LEED™ AP, has over 23 years of experience in energy, facility and infrastructure asset engineering and management with two Fortune 100 companies and one of the world's premier commercial airports. He is the Vice President of Energy & Transportation Management at the Dallas/Fort Worth International Airport where he is responsible for the operation, maintenance, repair and renewal of the airport's energy, utility and transit systems.

Under his leadership, DFW's energy efficiency and air quality initiatives have been widely acclaimed winning a U.S. Department of Energy Clean Cities Excellence award in 2004 and the prestigious Star of Energy Efficiency award from the Alliance to Save Energy in 2005.

Hodapp holds a Bachelor of Science in Chemical Engineering from Colorado State University and a Master of Business Administration from the University of Texas at Arlington. He is a Registered Professional Engineer in the State of Texas, holds professional certifications in Energy Management, Energy Procurement, Green Building Engineering and is a LEED Accredited Professional.

In 2003, the Association of Energy Engineers named him "International Corporate Energy Manager of the Year." For more information, contact Rusty Hodapp at: rhodapp@dfwairport.com

Michael Ivanovich has been editor-in-chief, of HPAC Engineering Magazine since 1996, when he made a career change from research to publishing. Under his direction, the magazine has been revitalized, a website established, and two regular supplements initiated (Boiler Systems Engineering and Networked Controls). Prior to joining HPAC Engineering, Mr. Ivanovich was a senior research scientist at Pacific Northwest National

Laboratory, working on projects involving internet-technology development, green buildings and residential building codes. His background also includes working on the ozone hole project for NOAA Aeronomy Laboratory, developing a network of solar-powered weather stations for Colorado and development of a protocol for investigating IAQ problems in Minnesota homes. He has a graduate degree in energy engineering and undergraduate degrees in computer science.

Srinivas Katipamula, Ph.D., got his M.S. and Ph.D. in mechanical engineering in 1985 and 1989, respectively, from Texas A&M University. He has been working as a senior research scientist at Pacific Northwest National Laboratory, in Richland, WA, since January 2002. He managed the analytics group at the Enron Energy Services for 2 years (2000 through 2001). Before joining EES, he worked at PNNL for 6 years and prior to that he worked for the energy systems lab at the Texas A&M University from 1989 to 1994.

He has authored or co-authored over 60 technical publications, over 25 research reports, and made several presentations at national and international conferences. He has recently written a chapter, "Building Systems Diagnostics and Predictive Maintenance," for CRC Handbook on HVAC. He is an active member of both ASHRAE and the American Society of Mechanical Engineers (ASME).

Sila Kiliccote is a scientific engineering associate at Lawrence Berkeley National Laboratory in Building Technologies Department with the Lighting Group. She has an electrical engineering degree fro University of New Hampshire, with a minor in illumination engineering and a master's in building science degree from Carnegie Mellon University.

Ward Komorowski, P.E., is director, facilities and building services, with Johnson Controls, Inc. During 23 years with Johnson Controls, Komorowski has applied innovative strategies to enable high building performance at the company's headquarters campus in Milwaukee, Wis. He was the winner of the 2005 Facility Executive of the Year award, presented by Today's Facility Manager magazine.
Komorowski is recognized as a Certified Plant Engineer by the Association of Facility Engineers and is a past president and board member of that organization. He is also a lecturer in the Architectural Engineering and Building Construction Department at the Milwaukee School of Engineering. He can be reached at ward.p.komorowski@jci.com.

Selly Kruthoffer, Director of Marketing, joined WebGen Systems in 2001 as Account Manager for the beta site project. In 2003 she was appointed Director of Marketing and is responsible for planning and executing the organizations marketing objectives.

Prior to joining WebGen, Selly worked with Pepsi-Cola International in their Marketing department, leading research projects in the Caribbean markets. Before that, she served as Director of Marketing and Customer Service for a business to business leasing organization.

Selly has a Masters in Business Administration from the University of Miami.

James M. Lee—Chief Executive Officer, is the founder of Cimetrics and has acted as its CEO since its formation. Mr. Lee has been a leader in the embedded control networking and building automation community for over 15 years. As founder and former President of the BACnet Manufacturers Association, the leading open systems networking consortium in the building automation industry, Mr. Lee's aggressive promotion of the BACnet open protocol standard has helped make Cimetrics a high-profile player in the arena. Mr. Lee has a B.A. in Physics from Cornell University.

Mr. Jim Lewis is the CEO and co-founder of Obvius, LLC, in Portland, OR. He was the founder and president of Veris Industries, a supplier of current and power sensing products to BAS manufacturers and building owners. Prior to founding Veris, Mr. Lewis held several positions at Honeywell including Branch Manager. He has extensive experience in knowing the needs of building owners, integrating existing metering and sensing technologies and developing innovative products for dynamic markets.

For more information or a demonstration, contact Obvius Corporation at (503) 601-2099, (866) 204-8134 (toll free), or visit the website at: http://www.obvius.com

Eric Linkugel works for Pacific Gas & Electric Company as a Business Customer Specialist for Demand Response Programs. He graduated from California State Polytechnic University—San Luis Obispo, with a B.S. in Industrial Technology and an M.S. in Industrial and Technical Studies.

Dr. Dirk Mahling, Chief Technology Officer, joined WebGen in 2000 as chief technology and chief information officer. He is a co-creator of WebGen's IUE® System. Mahling currently has a faculty appointment at the University of Pittsburgh, School of Information Sciences.

Mahling previously served as chief knowledge officer and vice president of e-solutions for Primix Solutions, as a knowledge management consultant for Ernst & Young, LLP, and as a chief architect for A.T. Kearney.

He holds a bachelor's of science degree in computer science and a master's degree in cognitive engineering and psychology from Carolo-Wilhemina University in Brunswick, Germany, and a master's degree in computer and information science and a doctorate degree in computer and information science from the University of Massachusetts at Amherst.

Aimee McKane is the Deputy Group Leader at the Lawrence Berkeley National Laboratory's (LBNL) Washington, DC office. Ms. McKane works with Best-Practices, a voluntary industrial partnership coordinated by the Department of Energy's (DOE) Office of Energy Efficiency and Renewable Energy's Industrial Technologies Program to encourage increased energy efficiency of industrial motor-driven, steam, and process heating systems. She is responsible for building industry partnerships, creating program strategies, developing the technical portfolio, and data analysis. On behalf of DOE, she provided leadership in the formation of the Compressed Air Challenge, an extensive industry partnership that seeks to improve compressed air system performance through training and education. She served as Chief Technical Advisor for a five-year motor systems program pilot in China administered by the United Nations Industrial Development Organization (UNIDO) in cooperation with the Chinese government. Aimee has also managed LBNL work in support of the Federal Procurement Challenge through the Federal Energy Management Program.

In 2003, Ms. McKane received the "Champion of Energy Efficiency" award from the American Council for an Energy Efficient Economy in recognition of her contributions to the field. Ms. McKane has a BA with a major in architecture from Washington University and an MA in Business and Policy from the State University of New York.

Jim McNally, P.E., Manager, Utility Information Services, manages the Utility Information Services at Siemens Building Technologies, Inc. This service has customers throughout North America. Jim is a registered professional engineer with more than thirty years experience in the design and analysis of mechanical and electrical systems in buildings. He has been managing Siemens Utility Information Services and directing its development for the past eight years. Prior to assuming his position with Siemens, he was in private practice as a consulting engineer. He writes articles and conducts seminars on building energy use and metering.

He is an applied researcher whose technical innovations include: 1) development of an on-line meter reporting system, 2) the pre-packaged wattmeter [DEM], and 3) the multi-variant, non-linear [MVNL] load forecasting technology. He consults with various universities, hospitals, and manufacturers in the development of their metering systems. Key metering projects include: Brown University, HJ Heinz, the University of Nevada at Reno, Carle Hospital, and Chicago Public Schools.

Education: MS, Bldg. Sc. [University of Sheffield, England], B'Arch. [Univ. of Illinois].

Jim McNally can be reached at *jim.mcnally@siemens.com*.

Naoya Motegi is a Senior Research Associate at the Commercial Buildings Systems Group in the Building Technologies Department, Lawrence Berkeley National Laboratory. His research emphasis is on building commissioning, energy information systems, and demand response. He has a Master of Science in Architecture from UC Berkeley, and a Master of Science in Architecture and Environmental Engineering from Waseda University in Tokyo, Japan.

William O'Connor, Deployment Manager, is a graduate of Northeastern University with a Bachelors degree in Mechanical Engineering and holds his certificate as a Certified Energy Manager (CEM). Before joining WebGen Systems, he had worked for large and small Automatic Temperature Control Contractors in the Boston area for more then 20 years.

In working for both large and small firms, William has developed a unique combination of experiences in every aspect of the Control Industry from Control Technician to Operations manager. Whether for clean rooms, laboratories, office buildings or schools, William has been successful for many years in designing and installing control systems.

Richard R. Paradis, CEM, Senior Energy Analysts-WebGen Systems. is a Certified Energy Manager, and has been in the Energy Efficiency industry since 1978. Rick has worked for utilities, design/consultant firms, non-profit management consultant firms and energy services companies working primarily in the commercial and industrial market sectors. This work included writing technical assistance audit reports, developing design alternatives for HVAC, lighting, thermal storage, and alternative energy projects, providing construction observation and review services as well as monitoring and verification protocols.

Rick has also managed and supervised technical potential studies and various technical assessments of end use equipment for the natural gas utilities in Massachusetts and New Jersey for developing utility Demand Side Management (DSM) programs. For his work in this area and setting the technical evaluation standard for DSM programs in Massachusetts in the early 1990s, Rick received the nomination and induction into the Marquis Who's Who in Science and Engineering 1994/1995 edition as well as the Millennium edition.

Jim Peedin is a lead strategist for integrated energy management services at Honeywell Building Solutions. With 35 years of utility, independent power production and utility consulting experience, he is involved in utility restructuring throughout the country and helps major energy users develop long-range management strategies.

Peedin has worked with the Fort Bragg Directorate of Public Works to develop and implement an energy management modernization program since 1999. He holds a bachelor's degree in nuclear engineering from North Carolina State University. Jim can be reached at 919-550-2113 or jim.peedin@honeywell.com.

Mary Ann Piette is the research director of the California Energy Commission's PIER Demand Response Research Center and the deputy group leader of the Commercial Building Systems Group. She has been at Berkeley Laboratory for more than 20 years, with research interests covering commercial building energy analysis, commissioning, diagnostics, controls, and energy information system. Her recent work has shifted toward developing and evaluating techniques and methods to improve demand responsiveness in buildings and industry. She has a Masters in mechanical engineering from UC Berkeley, and a Licentiate in building services engineering at the Chalmers University of Technology in Sweden.

Robert Proie has over 20 years of private and public sector facilities and operations management experience. Robert earned a BS Mechanical Engineer/PE at the University of Pittsburgh, and an MBA at Ohio State University with focus in operations management. He is responsible for Operations & Maintenance of Orange County Public Schools, the 12th largest school district in the US. Their current energy costs are projected to top $35 million in FY 2007. Using the FSP, along with other key innovative initiatives such as digital controls for the energy management of the school district's 4,500

portable classrooms he has reduced consumption per square foot while the district continues to grow by over 5% per year.

Sam Prud'homme is a freelance technical writer and computer programmer. He has been affiliated with Bay, LLC, since 1993, where he produced operation manuals and product literature for the company's line of air compressor controls. While at Bay, he also created software utilities used to estimate compressed air system energy expenses and the potential savings from updated system controls. In 1994, he authored the EPRI Compressed Air Handbook, a joint project between the Electric Power Research Institute and Bay. Sam has a computer science degree from Yale University.

Thomas S. Riley, P.E., has 20 years experience in the energy industry. Prior to forming Cogent Energy (*www.cogentenergy.com*), Mr. Riley was Vice President for a large energy services company where he held a variety of management positions overseeing such functions as energy engineering, project development and account management.

Mr. Riley is a seasoned energy professional who previously started and ran a successful energy services business for an engineering and project management firm. Mr. Riley was responsible for all aspects of the business including business development, engineering, project management and operations. Mr. Riley is past President of the Association of Energy Engineer's Bay Area Chapter and served two years on the California State University Board of Governors. He holds a B.S. in Mechanical Engineering from the University of California, Berkeley, and is a registered professional engineer in the state of California.

Rich Rogan is an energy system specialist for Honeywell Building Solutions. In this role, he is responsible for developing solutions that allow customers — in the commercial, federal and municipal sectors — to actively manage their entire energy supply chain via integrated energy information systems.

Rogan has been with Honeywell for 20 years. He is a licensed professional engineer in Pennsylvania, as well as a certified energy manger. Rogan received his bachelor's in mechanical engineering technology from Spring Garden College. Rich can be reached at 570-443-9961 or richard.m.rogan@honeywell.com.

Joseph H. Rosenberger, P.E., CEM-President/CEO-SophNet Inc.

As founder of SophNet Inc in 1996, Joseph Rosenberger brings high end solutions to energy issues using Dashboard Technology and machine-machine interfaces for Industrial clients. SophNet is a Tridium developer presently deploying large-scale "SCADA based hardware/software in Demand Response for ERCOTT ISO and CAISO.

Joe has 18 years of experience in all aspects of the power and energy field. He is a licensed Professional Electrical Engineer in New York State, North Carolina He graduated from Union College in Engineering and completed graduate work at New York University in MSEE. Mr. Rosenberger brings his 15 years of experience with NY-SEG Corporation as a Power System Consultant, Supervising Engineer, and Eastern Region Account Manager to bear on his work with industry. Joe's respect throughout the field and his broad energy background has placed him on the cutting edge of promoting changes in today's energy marketplace. Mr. Rosenberger brings a background to the design and building of automated electric-metered and power services. As President/CEO, Mr. Rosenberger will oversee the installation and design of energy saving services. His knowledge and experience in the industry will better position SophNet to respond quickly to changing market demands. Joe spends his time between NY and NC with his wife and 2 daughters who are college graduates and working in the Southeast.

Chris Sandberg is Principal Engineer, Energy Management Systems at Reedy Creek Energy Services (a division of the Walt Disney World Co.) and is responsible for the design, installation and commissioning of building automation systems for Disney theme park projects throughout the world. Chris is a graduate of Purdue University (BS in Construction Engineering and Management) and the University of Florida (Master of Building Construction). (chris.d.sandberg@disney.com)

Ken Sinclair has been in the building automation industry for over 35 years as a service manager, building owner's representative, energy analyst, sub-consultant and consultant. Ken has been directly involved in more than 100 conversions to computerized control. Ken is a founding member and a past president of both the local chapter of AEE and the Vancouver Island chapter of ASHRAE. The last five years his focus has been on AutomatedBuildings.com, his online magazine. Ken also writes a monthly building automation column for Engineered Systems and has authored three industry automation supplements: Web-Based Facilities Operations Guide, Controlling Convergence and Marketing Convergence.

Travis Short is the director of technology for Lynxspring, Inc., and has a very diverse background centered on the design, programming, installation, and commissioning web enabled energy management systems. He has been involved with the project management and engineering of the installation of advanced building management systems for Mission Critical Facilities. In addition he has been the lead commissioning agent for several building management systems for several high profile Mission Critical Facilities. While at Lynxspring, Inc. he has been in charge of managing the development of web enabled OEM solutions for Lyxnspring's JENEsys™ product line. Development duties include the use of Java and XML as software platforms for JENEsys™ to create manufacturer specific applications. Current OEM solution strategies are appliance based and are powered by the Niagara AX Framework developed by Tridium, Inc. Lynxspring provides solutions by combining state-of-the-art hardware with design, programming, and deployment of interoperable protocol control solutions to fit the most simple to the most complex criteria that a building owner or manufacturer expects in today's market.

Rahul Tongia, Ph.D., is a faculty member in the School of Computer Science (ISRI) and the Department of Engineering and Public Policy at Carnegie Mellon University, Pittsburgh. His research explores interdisciplinary issues of infrastructure (especially telecom and power), spanning technology, policy, regulation, and security. He has years of global experience in technology and policy analysis, and has dozens of publications and conference presentations. Dr. Tongia is active in global advisory and analysis, having been the Vice-Chair of the United Nations ICT Task Force Working Group on Enabling Environment (formerly, Low-Cost Connectivity Access), and he is on the Technology Advisory Board of a major US utility advanced metering project. Dr. Tongia can be reached at *tongia@cmu.edu*.

Wolfgang Wagener, Ph.D., Architect AIA, RIBA
Wolfgang Wagener is Head of Real Estate and Construction solutions within Cisco's global Real Estate and Workplace Resources organization. Wolfgang's primary area of expertise is in working with occupiers, developers, and property owners to deliver innovative real estate, design and technology solutions that enhance business performance. An architect by profession, he lectures regularly across Europe, North America, and Asia.

Prior to joining Cisco Systems, Wolfgang was a practicing architect, urban planner, educator and author. He worked with Murphy/Jahn in Chicago, Richard Rogers Partnership in London, and he had a private practice in Los Angeles, where he was also a Visiting Professor at the University of California in Los Angeles (UCLA) and the University of Southern California. He managed research, planning, design and construction of mixed use urban and residential developments, transportation buildings, and corporate headquarters throughout Europe, Asia Pacific and North America. His research and education areas are 19th and 20th century architecture and urban development, workplace design, environmental sustainability and the impact of technology innovations in the real estate and construction industry.

Wolfgang Wagener received a Ph.D. in Architecture from the RWTH Aachen, one of Europe's leading technology institutions, and an Advanced Management Degree in Real Estate Development from Harvard University.

Rahul Walawalkar, CEM, is a Research Analyst with Customized Energy Solutions, Philadelphia and a Ph. D. candidate in Department of Engineering and Public Policy at the Carnegie Mellon University in Pittsburgh. He obtained a Masters in Energy Management & Advanced Certification in Energy Technology at New York Institute of Technology. He worked for Tata Infotech for 5 years, in various roles, including coordinator for company's initiatives in energy management and program manager for new product development initiative called iGems. He is recipient of 'Demonstration of Energy Efficiency Development—DEED' Scholarship from American Public Power Association during 2002-06. He is also a recipient of the Computer Society of India's Young IT Professional Award for 2 years and Golden Web Award by International Association of Webmasters & Designer for 3 years. He has written over 30 papers, and is a member of AEE. Mr. Walawalkar can be reached at *rahul@walawalkar.com*. Website: *www.walawalkar.com*.

David Watson has over 15 years experience designing, programming, and managing the installation of control and communications systems for commercial buildings, industrial processes and remote connectivity solutions. At LBNL, he is working with innovative building technologies such as demand response systems, energy information systems and wireless control networks. Prior to joining LBNL, David held engineering, project management and product development positions at Coactive Networks, Echelon, York International and Honeywell. He designed and managed the installation of hundreds of projects including: internet based control and monitoring of thousands of homes and businesses, communication systems for micro turbine based distributed power generation systems and industrial process controls for NASA wind tunnels and biotech manufacturing. Mr. Watson graduated from California Polytechnic University, San Luis Obispo with a degree in mechanical engineering.

Tom Webster, P.E., is a Research Specialist at the Center for Environmental Design Research at the University of California, Berkeley. He has been engaged in building energy, controls, and communications for almost thirty years, focusing on building energy analysis and simulation, commercial and residential HVAC systems engineering, distributed control systems design, and digital control product development. Tom's current research focus is on underfloor air distribution systems (UFAD) including topics such as room air stratification, energy simulation, field and case studies of underfloor systems, and cost analysis. Tom has participated in the California Energy Commission's (CEC) High Performance Building Project where he developed a large fan diagnostics tool, and the Federal Energy Management Program/New Technology Demonstration Program where he performed technology assessment studies for energy management systems. Mr. Webster is an active member of ASHRAE (American Society of Heating, Refrigeration and Air-Conditioning Engineers). You can contact Tom at twebster@berkeley.edu or visit our UFAD website at http://www.cbe.berkeley.edu/underfloorair/.

Index

Printed and bound by CPI Group (UK) Ltd, Croydon, CR0 4YY

17/10/2024

01775658-0004